AF722657

The ecology and management of biological invasions in southern Africa

The ecology and management of biological invasions in southern Africa

Proceedings of the
National Synthesis Symposium on the
ecology of biological invasions

Edited by
I A W Macdonald
F J Kruger
A A Ferrar

1986
OXFORD UNIVERSITY PRESS
CAPE TOWN

Oxford University Press
OXFORD NEW YORK TORONTO
MELBOURNE AUCKLAND NAIROBI
DAR ES SALAAM CAPE TOWN PETALING JAYA
SINGAPORE HONG KONG TOKYO DELHI
BOMBAY CALCUTTA MADRAS KARACHI

AND ASSOCIATES IN
BEIRUT BERLIN IBADAN NICOSIA

QH
195
.A323
N38
1985

Copyright © 1986 Oxford University Press

OXFORD is a trademark of Oxford University Press

ISBN 0 19 570417 7

All rights reserved. No part of this publication may be reproduced, stored in a retrieval system, or transmitted in any form, or by any means, electronic, mechanical, photocopying, recording, or otherwise, without the prior written permission of Oxford University Press, Southern Africa

Printed and bound by Citadel Press
Published by Oxford University Press, Harrington House,
Barrack Street, Cape Town 8001, South Africa

B157/-OS

Preface

This volume constitutes the first attempt to synthesize available information on the invasion of southern Africa's natural ecosystems by introduced organisms. It represents the culmination of the South African contribution to the international SCOPE project on 'The Ecology of Biological Invasions'.

SCOPE, the Scientific Committee on Problems of the Environment, is a committee of the International Council of Scientific Unions. SCOPE's objectives are the advancement of knowledge on the influence of humans on their environment and the synthesis of this knowledge into forms useful to the international and national bodies responsible for the resolution of environmental problems. In particular, SCOPE seeks to identify problems of global concern which are amenable to interdisciplinary synthesis and which show some likelihood of improved resolution in the short and medium term. The invasive spread of plants, animals, and micro-organisms introduced by man into areas remote from their centres of origin was identified as being such a problem at the SCOPE General Assembly in Ottawa in June 1982.

The project has entailed the convening of symposia in several countries as well as other activities intended to develop topics of joint concern, such as international working groups on modelling invasions and on the invasion of nature reserves. The specific questions being addressed by the project are:

(1) What are the factors which determine whether a species will be an invader or not?
(2) What are the site properties which determine whether an ecological system will be relatively prone to, or resistant to, invasion?
(3) How should management systems be developed using the knowledge gained from answering these questions?

The idea for the project was initially conceived at the Third International Conference on Mediterranean Ecosystems held in South Africa in 1980. South African scientists were instrumental in preparing the project proposal that was put to SCOPE for approval in 1982. The Invasive Biota Working Group of the National Programme for Ecosystem Research has directed the South African contribution to the project in liaison with the South African National Committee for SCOPE. A programme for the South African project was published in 1983 (Ferrar and Kruger 1983).

Perhaps because the problems of invasive biota have been more acutely felt here, the South African programme followed a course different to those of other countries. Firstly, it produced regional bibliographies on invasive alien organisms, including vascular plants in general (Moran and Moran 1982) as well as those invading the fynbos biome (Macdonald and Jarman 1984), aquatic animals (Bruton and Merron 1985) and birds (Brooke in preparation). Secondly, four workshops have been convened to collate unpublished information on the subject. Two of these have centred on the fynbos biome, one to review invasions over all taxonomic groups (Macdonald and Jarman 1984) and the other to analyse the management of alien vascular plants (Macdonald et al 1985). A third dealt with plant invasions in Natal (Macdonald and Jarman 1985) and the fourth reviewed all taxonomic groups in South West Africa/Namibia (Brown et al 1985).

The third component of the programme has been the identification and funding of short-term research projects in the field of biological invasions. In general these projects have still to yield their published results but the interim findings of several of these have been drawn upon in compiling this synthesis, which is the final national contribution to the SCOPE project.

The chapters of this volume comprise a structured set of invited papers presented at the National Synthesis Symposium on the Ecology of Biological Invasions in Stellenbosch, 27 to 29 November 1985. Papers were solicited to provide a synthesis of current knowledge and management expertise on the topic for South Africa and its neighbouring territories. Following the symposium all the chapters were reviewed at an editorial workshop held from 30 November to 3 December 1985 at Houw Hoek in the western Cape, with the principal purpose of identifying overlaps and omissions. Revised chapters were then refereed by internationally recognized authorities in each specific field. As a result of the different levels to which various facets of the study of alien organisms have advanced in southern Africa, the chapters vary from summaries of qualitative observations, to theoretical analyses of the processes involved in these invasions and their control. The wide range in the depth of treatment of the various topics usefully mirrors the variety of scientific attention that has been paid to the problems resulting from alien invasions in southern Africa.

In editing and collating these proceedings we have attempted to follow a standard terminology and a uniform style and format, while retaining the individual character of each contribution. The indulgence of the authors in accommodating these editorial pressures is appreciated, although complete uniformity was of course not achieved. For the convenience of the reader some authors have chosen to refer to certain products by brand names, this implies no endorsement of such products by either the authors or the editors.

The following are thanked for making possible the production of this volume: the National Programme for Ecosystem Research which has funded all the activities of the Working Group for Invasive Biota; the members of this Working Group, F J Kruger (Convener), Dr C C Appleton, Dr P J Ashton, Professor M N Bruton, A A Ferrar, S K B Godschalk, I A W Macdonald, Professor V C Moran, Dr S Neser and M J Wells; the organizing committee for the national synthesis symposium which comprised together with the Editors, F Bierman and D M Richardson; the participants at Houw Hoek Inn, who were, in addition to most of the senior authors, E R Ashton, M L Jarman, F M Pressinger, D M Richardson, P Holmes and M W Fraser; all the international referees whose names are listed at the back of this book; the staff of the Foundation for Research Development who wordprocessed, collated and edited to a quite unreasonable deadline, L Horn, M L Jarman, D Koekemoer, E Mantle, M Orton and particularly L van Niekerk who edited the copy; P C Fraser, G Baker, R C R Edgar and K R Kelly for final text preparation and R Seagrief and the staff of Oxford University Press for the indexing, and their general support.

August 1986

Ian Macdonald
Fred Kruger
Tony Ferrar

Brooke R K (in preparation). Bibliography of alien birds in southern and south-central Africa. South African National Scientific Programmes series. CSIR, Pretoria.

Brown C J, Macdonald I A W and Brown S E (eds) 1985. Invasive alien organisms in South West Africa/Namibia. South African National Scientific Programmes Report No 119. CSIR, Pretoria.

Bruton M N and Merron S V 1985. Alien and translocated aquatic animals in southern Africa: a general introduction, checklist and bibliography. South African National Scientific Programmes Report No 113. CSIR, Pretoria.

Ferrar A A and Kruger F J 1983. South African Programme for the SCOPE project on the ecology of biological invasions. South African National Scientific Programmes Report No 72. CSIR, Pretoria.

Macdonald I A W and Jarman M L (eds) 1984. Invasive alien organisms in the terrestrial ecosystems of the fynbos biome, South Africa. South African National Scientific Programmes Report No 85. CSIR, Pretoria.

Macdonald I A W and Jarman M L (eds) 1985. Invasive alien plants in the terrestrial ecosystems of Natal, South Africa. South African National Scientific Programmes Report No 118. CSIR, Pretoria.

Macdonald I A W, Jarman M L and Beeston P (eds) 1985. Management of invasive alien plants in the fynbos biome. South African National Scientific Programmes Report No 111. CSIR, Pretoria.

Moran V C and Moran P M 1982. Alien invasive vascular plants in South African natural and semi-natural environments: bibliography from 1830. South African National Scientific Programmes Report No 65. CSIR, Pretoria.

Table of Contents

Preface v

Introduction xiii

Part I. Historical perspectives

Chapter 1
Human settlement in South Africa and archaeological evidence for alien plants and animals .. 3
J Deacon

Introduction
The fossil and palaeoclimatic record
The spread of Stone Age populations in South Africa
The impact of pastoralism in the west
The influx of Iron Age farmers in the east
European settlement patterns
Patterns from the past

Chapter 2
The history of introduction of invasive alien plants to southern Africa 21
M J Wells, R J Poynton, A A Balsinhas, K J Musil, H Joffe, E van Hoepen and S K Abbott

Introduction
Possible introductions before 1652
Species introduced after 1652
Conclusions

Chapter 3
A case study of some woody plant introductions to the Cape Town area 37
G L Shaughnessy

History of introductions
Objectives or motivation for introductions
Success in achieving objectives
Analysis of invasive success

Part II. Faunal invasions and their control

Chapter 4
Faunal invasions of aquatic ecosystems in southern Africa, with suggestions for their management 47
M N Bruton and J G van As

Introduction
Extent of invasions
Impact of invasive aquatic animals in southern Africa
Management of faunal introductions
Conclusions

Chapter 5
Alien and translocated terrestrial vertebrates in South Africa 63
R K Brooke, P H Lloyd and A L de Villiers

Introduction
Amphibians and reptiles
Birds
Mammals
Discussion

Part III. The invasion of southern Africa's biomes

Chapter 6
Alien species in terrestrial ecosystems of the fynbos biome 77
I A W Macdonald and D M Richardson

Introduction
Patterns of land use and disturbance
History of introduced species
Patterns of invasions by some successful alien species
The nature of the invasions
Ecological impacts
History of control
Conclusions

Chapter 7
Invasive alien organisms in the Namib Desert, Upper Karoo and the arid and semi-and savannas of western southern Africa 93
C J Brown and A A Gubb

Introduction
Methods
Invasive alien plants
Invasive alien animals
Conclusions
Appendices listing alien taxa

Chapter 8
Alien plant invasions in the grassland and savanna biomes 109
L Henderson and M J Wells

Introduction
Alien species and their history
The nature of the invasions
Ecological impacts
The history of management and control

Chapter 9
Alien invasions in indigenous evergreen forest 119
C J Geldenhuys, P J le Roux and K H Cooper

Introduction
History of forest disturbance by man and of alien introductions
Distribution of invasive species in forest habitats
Processes of invasion and ecological impacts
Management and control
Synthesis

Chapter 10
Alien plants and animals on South African continental and oceanic islands: species richness, ecological impacts and management 133
J Cooper and R K Brooke

Introduction
Description of the islands
Species richness
Ecological impact of aliens
Management implications and priorities

Part IV. Towards understanding the ecology of biological invasions

Chapter 11
Processes of invasion by alien plants 145
F J Kruger, D M Richardson and B W van Wilgen

Introduction
Entry of plant species to South Africa
Invasion of new habitat by the established alien
Demographic factors involved in the process of invasion by plants
Conclusions

Chapter 12
Seed biology of invasive alien plants in South Africa and Namibia 157
S J Dean, P M Holmes and P W Weiss

Introduction
Canopy-stored seed
Soil-stored seed
Short-lived seed
Seedless invaders
Hybridization
Unsuccessful aliens
Conclusions

Chapter 13
A comparative analysis of fleshy fruit displays in alien and indigenous plants 171
R S Knight

Introduction
Methods
Results and discussion
Conclusions
Appendices listing potentially vertebrate-dispersed taxa

Chapter 14
Photosynthesis and carbon allocation in invasive plants of South African biomes 179
B H Whiting, G C Bate and D J Erasmus

Introduction
Photosynthesis and carbon allocation
Photosynthetic pathways of alien and indigenous vegetation occurring in South African biomes
Photosynthesis in specific alien invaders
Carbon allocation in specific alien invaders
General
Conclusions and needs for further research

Chapter 15
Standing crops, growth rates and resource use efficiency in alien plant invaded ecosystems 189
M C Rutherford, F M Pressinger and C F Musil

Introduction
Comparative growth rates
Concluding remarks

Chapter 16
The life history styles of invasive fishes in southern Africa 201
M N Bruton

Introduction
Life history characteristics of fishes
Life history styles of invasive fishes
Conclusions

Chapter 17
The differential invasion of southern Africa's biomes and ecosystems by alien plants and animals 209
I A W Macdonald, F J Powrie and W R Siegfried

Introduction
Background to the present study
Methods
Patterns of incidence of alien rodents in owl pellets
Results of the nature reserve analyses
Conclusions

Part V. Impacts of invasions

Chapter 18
Impacts of alien organisms on terrestrial communities with emphasis on communities of the south-western Cape 229
G J Breytenbach

Introduction
Impact on biological diversity
Impact on community structure and function
Impact on processes
Impacts due to control measures
Discussion and conclusions

Chapter 19
Impact of woody aliens on ecosystem properties 239
D B Versfeld and B W van Wilgen

Introduction
Vegetation structure
Litterfall and decomposition rates
Fire hazard
Nutrient cycling
Energy balance
Hydrology
Conclusions

Chapter 20
Ecological impacts and economic consequences of alien invasive organisms in southern African aquatic ecosystems 247
P J Ashton, C C Appleton and P B N Jackson

Introduction
Plants
Invertebrates
Vertebrates
Conclusions

Part VI. Managing invasive alien plants

Chapter 21
The potential of insect herbivores for the biological control of invasive plants in South Africa 261
V C Moran, S Neser and J H Hoffmann

Introduction
Alien or native target weeds
Abundance of the weed species in its native land
Annual and biennial plants
Taxonomic isolation of the target weed
The aggressive attributes of the weed
The more insect herbivores the better
Adaptations, long associations and coevolution
Conclusions

Chapter 22
Insect herbivores as determinants of the present distribution and abundance of invasive cacti in South Africa 269
H G Zimmermann, V C Moran and J H Hoffmann

Introduction
Opuntia vulgaris
Opuntia ficus-indica
Opuntia aurantiaca
Insect predators
Conclusions

Chapter 23
Costs and benefits of the Australian blackwood *Acacia melanoxylon* in South African forestry 275
C J Geldenhuys

Introduction
The conflicting interests
Resolving the conflicts
Control methods
Research

Chapter 24
The importance of seed-attacking agents in the biological control of invasive alien plants 285
S Neser and R L Kluge

Introduction
Unchecked, prolific seed production as a factor in invasiveness
Feasibility of the use of seed-attackers in biological control
Present projects involving seed-attackers in South Africa
Seed reduction as a useful compromise in biological control
Conclusions

Chapter 25
Integrated control for invasive alien weeds 295
R L Kluge, H G Zimmermann, C J Cilliers and G B Harding

Introduction
Understanding the target plant
Optimizing the use of control options
Determining the necessary level of control
The future of integrated control of invasive alien plants
Summary

List of authors and referees 305

Biological index 307

Subject index ....

Introduction

The intention of this volume is to review the phenomenon of invasion by introduced organisms of natural or nearly natural ecosystems in southern Africa. Authors have focused on southern African data and examples, only using information from further afield to emphasize or contrast the southern African experience. We chose to define southern Africa as that portion of the continent south of the Cunene and Zambezi Rivers and the northern border of South West Africa/Namibia (Figure 1). Most of the authors are South Africans and most of the information available on the subject of biological invasions is derived from research carried out in this country. The emphasis of the book is thus mainly on South Africa. The other countries of southern Africa were colonized by Europeans more recently than South Africa and their invasion by introduced species has in general been less extensive.

We have addressed the subject of biological invasions in terms of the biome classification proposed for South Africa by Huntley (1984). This was extrapolated into South West Africa/Namibia using the units defined by Brown et al (1985) based on the vegetation classification of Giess (1971). For the re-

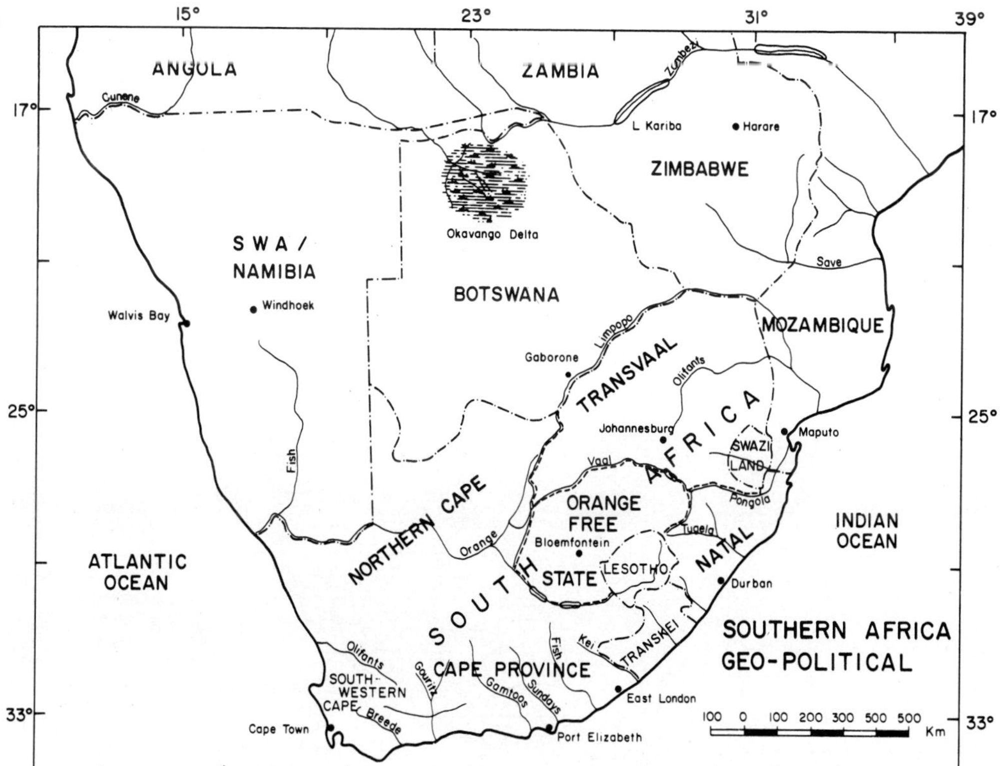

Figure 1. The area defined here as southern Africa, showing the countries, provinces, rivers and major towns.

maining countries of southern Africa the vegetation types on the map prepared for the Flora Zambesiaca area (Wild and Barbosa 1967) were reclassified according to the biomes of Huntley (1984). The biomes were mapped by quarter degree areas, classifying each according to the predominant biome (Figure 2).

The treatment of alien invasions as a topic is restricted in the papers that follow, to exclude the large number of introduced species that thrive only in plantations, agricultural fields and gardens. The problems of these weed species in 'transformed ecosystems' (sensu Poore 1978) are seen to be sufficiently different ecologically and economically, to merit separate consideration.

The study of biological invasions is complicated by a diversity of terms. In many cases the words used in this field have other meanings in popular usage and have been used with several different meanings within the field of biological invasions. The dialogue generated in compiling this synthesis achieved a fair measure of consensus in the definition and use of terms, but the debate is far from over. The diversity of opinion persists in certain inconsistencies of language that will be found in this multi-authored text. In the interests of clarity we include a brief explanation of the terms that are used by our authors.

An introduced species is one that has been brought intentionally or unintentionally by people to areas outside of its natural range of distribution. This is synonymous with the terms alien, exotic and in some instances translocated. An introduced species may either succeed or fail to establish self-sustaining populations in areas of natural or near-natural vegetation or habitat, ie in original or modified ecosystems (sensu Poore 1978). Introduced species that are successful are sometimes referred to as naturalized or established aliens. Such species differ widely in their tendency to spread. Although the establishment of a self-sustaining population is the minimum criterion for an alien species to be termed invasive, the term more generally denotes those species most evidently successful in spreading unaided into untransformed ecosystems.

It is tempting to classify species by their degree of

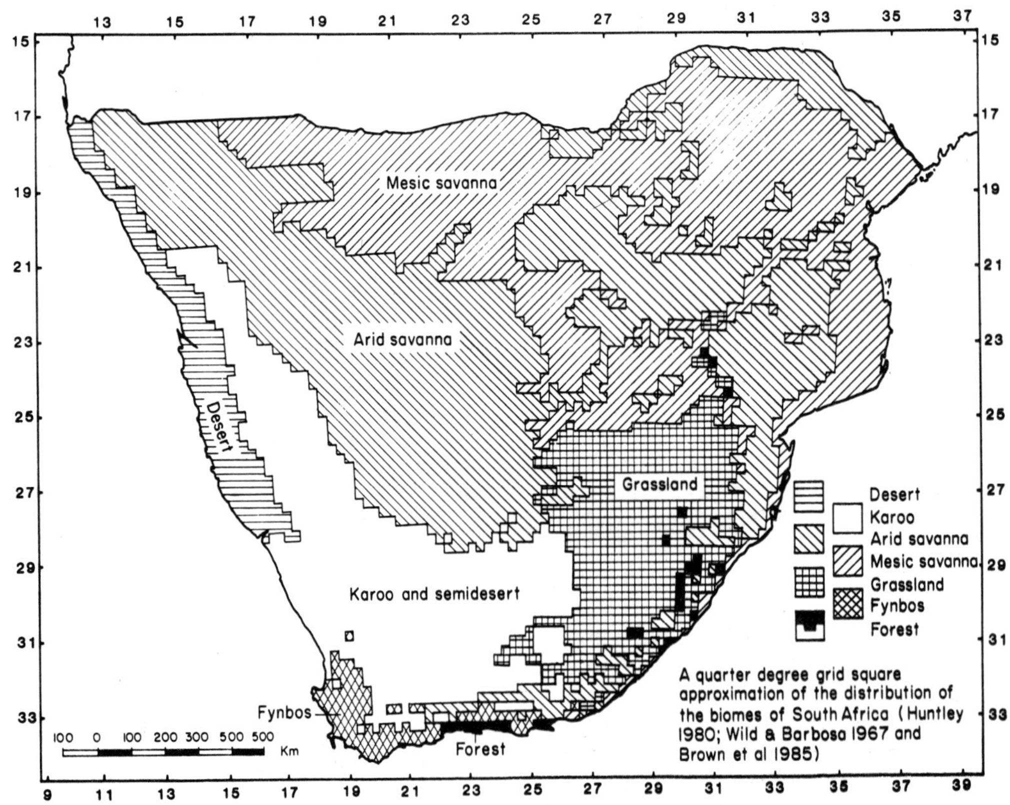

Figure 2. The seven major biomes of southern Africa mapped according to the quarter degree grid system.

invasiveness, sometimes in anthropomorphic terms such as passive, active or aggressive invaders. However, no such classification is really valid without a comparative measure of their rates or extents of spread. It is premature and not particularly useful to categorize these species according to some intuitive evaluation of their status in terms of this complex and poorly understood ecological process.

Although most of the introduced species that have become invasive in southern Africa come from other continents, there are some species which have been introduced from elsewhere in Africa. These are also termed aliens or exotics. In a few cases, species that were naturally limited to one portion of southern Africa have been introduced to other areas within the region, and have become invasive there. These too are termed alien or, sometimes, translocated species. In the case of aquatic species the translocation can have been effected over relatively short distances, as between adjacent catchments. This type of introduction is now widespread, greatly facilitated by the construction of intercatchment water transfer schemes which are a prominent feature of water resource development in South Africa. Species that have been re-established in areas from which they have become locally extinct (often incorrectly termed 're-introduced') are also referred to as translocated species but are specifically excluded from this synthesis.

Similarly, the volume does not cover native, i e indigenous, species whose distribution or density may have increased in response to man-induced changes in the environment. These "indigenous invasions" are fundamentally different in their origins, and are referred to only where they have occurred in response to environmental alterations brought about by aliens or, alternatively, where they shed light on the processes involved in analogous alien invasions.

The ecosystems and peoples of southern Africa have suffered major invasions by introduced pathogens. For example, the ecological systems of our tropical and subtropical regions still bear the imprints of the rinderpest outbreaks of 1896 and 1904. Regrettably we were not able to enlist contributors who could deal competently with this topic.

This southern African synthesis is likely to differ substantially from those carried out elsewhere as part of the SCOPE project, such as the other Southern Hemisphere synthesis, that of Australia (Groves and Burdon 1986). Southern Africa has been occupied, and to some extent modified, by people for approximately 1,5 million years as against the mere 40 thousand years of Australia. Many vegetation types in southern Africa have few or no trees, in contrast with Australia and other continents and possibly because of its long history of human occupation. As a result European colonists had a real need to introduce trees from the start. There were, by contrast, far fewer intentional releases of alien terrestrial vertebrates in southern Africa than in Australia. Southern Africa has a much larger biota than countries in the Northern Hemisphere and its communities are proportionately richer in native species. Another important difference is that, apart from modifications through the use of fire, major transformations of southern African ecosystems have only taken place in the fairly recent past, (within the last 1600 years) relative to the Northern Hemisphere and in particular, Europe.

The southern African situation is also possibly unique, in that we have both well developed and less developed economies and land-use systems intermingled within the same region. This juxtaposition has resulted in a situation where a wide range of alien species are invading rangelands which do not contribute much to the cash economy of the region. The cost-effectiveness of control programmes in these areas is difficult to evaluate. Where the yield of commodities from the land is low in relation to the needs of the people, the value of an invasive alien is often seen to be positive or at least neutral, and certainly not sufficiently negative to warrant expensive control activities. Another example of the uniqueness of the southern African situation is that control programmes are currently being implemented, which integrate the latest advances in biological control with labour-intensive manual control operations, the extent of which would be inconceivable in a first-world region. It is hoped that in many ways the southern African experience will bridge the gap between countries rich in expertise but with limited human resources, and the needs of those huge areas in the less developed world where expertise in biological invasions is virtually absent, but where alien invasions on the scale found in southern Africa have just begun.

Brown C J, Macdonald I A W and Brown S E (eds) 1985. Invasive alien organisms in South West Africa/Namibia. South African National Scientific Programmes Report No 119. CSIR, Pretoria.

Giess W 1971. A preliminary vegetation map of South West Africa. Dinteria 4, 1-144.

Groves R H and Burdon J J (eds) 1986. Ecology of biological invasions: an Australian perspective. Australian Academy of Science, Canberra.

Huntley B J 1984. Characteristics of South African Biomes. In: Booysen P de V and Tainton N M (eds) Ecological effects of fire in South African Ecosystems. Springer-Verlag, Berlin. pp 1-17.

Poore D 1978. Ecosystem conservation. In: Sourcebook for a world conservation strategy. IUCN General Assembly Paper 78/10 add 4. IUCN, Morges.

Wild H and Barbosa L A G 1967. Vegetation map of the Flora Zambesiaca area. Collins, Salisbury.

Historical perspectives

This section includes three papers which review the role of people in introducing new species to South Africa and aiding their spread through the region. **Deacon** also gives a broad picture of how prehistoric people may have modified the native ecosystems so as to favour the spread of introduced species, and what the trends of such influences may have been.

It is clear from **Deacon's** account that the hand of man has lain lightly or heavily on the landscapes of South Africa for a great span of time, perhaps as much as 1,5 million years. Human influence during this period has been masked by major changes in climate, and is likely to have been small until the advent of the controlled use of fire. The evidence suggests that this may have occurred about 150 000 years ago, although the extent and intensity of consequent disturbance to the natural ecosystems of the region is debatable. There is no archaeological evidence for the introduction of currently naturalized species until the arrival of domesticated animals with the pastoral Khoi peoples about 2000 years ago and with the Bantu-speaking peoples of the subsequent Early Iron Age. The Bantu speakers also brought with them domesticated and semi-domesticated crop plants and, perhaps, weeds.

So the picture for South Africa is much the same as it is for the Americas and Australia, with the principal introductions of plants and animals occurring rapidly after settlement by European people. South Africa was also affected by the tradition of British Victorian naturalists and gardeners, which resulted in a series of rapid introductions in the nineteenth and early twentieth centuries (see also Kruger et al this volume). These introductions were often motivated by the desire to make an unfamiliar landscape more hospitable or to provide the sports of home, as much as by utilitarian needs arising from the dearth of natural resources such as timber, as **Shaughnessy** and **Wells and co-authors** point out. This free traffic in introductions, involving any species which could reasonably be expected to succeed in the new habitats, persisted until well into the middle of the twentieth century. But the first warning voices were raised around the turn of the century, when naturalists and agriculturalists began to publicize the consequences of the spread of *Opuntia* species (see Moran and Neser and Zimmermann et al this volume).

Shaughnessy's account shows clearly how the present distributions of some of the introduced plant species depend principally on their history of propagation, rather than on ecological factors, although the opposite holds in other cases. Ecologists and resource managers will need to take more careful account of historical factors in dealing with invasives.

Chapter 1

Human settlement in South Africa and archaeological evidence for alien plants and animals

J Deacon
University of Stellenbosch.

Introduction

The aim of this book is to examine the ecology of invasive plants, animals and microorganisms, emphasizing those that have disrupted natural ecosystem processes in southern Africa. This chapter summarizes when and how people, who were responsible for introducing these alien plants and animals, colonized South Africa over the past 1,5 myr (million years) or more (Table 1). The survey ends in 1911 AD, a year after Union when the present-day boundaries were established. During this time people spread throughout the subcontinent and exploited the plant and animal food resources in much the same way as other diversivores. Prior to 2 000 BP (before present), animal extinctions and changes in plant and animal communities resulted more from climatic changes than from the impact of human populations, whereas within the last 2 000 years, people actively introduced alien plants and animals. A summary of the major stages of human settlement in southern Africa helps to place present-day distributions in perspective.

The fossil and palaeoclimatic record

South Africa has an impressively long fossil record with some of the earliest forms of life, but it was not until 14-2,5 myr that the modern mammalian bovid taxa evolved in the African savanna. It has been suggested (Brain 1981; Vrba 1985) that bursts of speciation and extinction during this period were triggered by changing climatic conditions resulting from continental drift and the fluctuating growth of the Antarctic ice cap. From about 2,5 myr, global cycles of cooler and warmer temperatures, referred to generally as the Pleistocene Ice Age, have had a periodicity of about 100 000 years. The climate in South Africa for about 70% of this time span was 1-6°C cooler than at present. For the remaining 30% of the time, temperatures were similar to those of the present. These changes have had profound effects upon biota throughout southern Africa.

Address for correspondence: J Deacon, Department of Archaeology, University of Stellenbosch, Stellenbosch, 7600.

Changes in the fossil record with the appearance of man

Comparison of the 5 myr mammalian fauna from the Langebaanweg fossil site about 100 km north of Cape Town with that of historic times gives some idea of the scale of change that has taken place. The larger mammals in the older assemblages include a number of extinct genera such as *Agriotherium africanum*, a carnivorous bear, *Sivatherium*, a primitive antlered giraffe, and a primitive gomphothere elephant, all of which suggest that the vegetation along the river included more trees and open grassland than it does today (Hendey 1981, 1982). In later assemblages dating within the last 3,5 myr, the numbers of genera and species of large mammals that have become extinct (28) is significantly greater than the number of new ones (three genera, namely *Panthera*, *Equus* and *Homo*) (Klein 1984) (Table 1). As Klein (1984) points out, this phenomenon is not exclusive to South Africa and seems to be the result of the pre-Pleistocene fauna adjusting to the cyclic changes of climate and vegetation of the last 2 myr.

Whereas carnivores are well represented, primates are rare in south-western Cape fossil assemblages dating to about 5 myr, although ancestors of man were present in eastern Africa at this time. The fossil faunas at Sterkfontein, Swartkrans, Kromdraai and Makapansgat in the Transvaal that date from 1-3 myr, however, contain abundant remains of australopithecines (the immediate ancestors of the genus *Homo* or man the toolmaker) and other primates, together with more carnivore taxa than are present in modern assemblages. It has been suggested that the subsequent reduction in large carnivore taxa reflects the entry and expansion of hominids into a meat-eating niche (Hendey 1978; Klein 1984).

Table 1. Summary of major trends in economy, distribution, extinctions and alien introductions in South Africa during the period of human occupation.

Time (BP)	Technological stage	Economy	Distribution	Extinctions	Alien introductions
1,5 Myr to 150 000	Earlier Stone Age	Hunter-gatherer-scavenger	Throughout South Africa	About 20 genera	None known
150 000 to ca 60 000	Middle Stone Age	Hunter-gatherer-fisher	Throughout South Africa	About 8 genera	None known
60 000 to ca 35 000	Middle Stone Age	Hunter-gatherer-? fisher	Sites very rare	None known	None known
35 000 to 2 000	Later Stone Age	Hunter-gatherer-fisher	Rare sites pre-14 000 but thereafter widespread	*Equus capensis,* *Megalotragus priscus,* *Pelorovis antiquus* *Antidorcas bondi*	? *Ricinus communis,* castor oil plant
2 000 to present	Khoi pastoralists	As above plus domesticated stock	West of 400 mm isohyet		*Ovis aries,* sheep *Bos taurus,* cattle *Capra hircus,* goat *Canis familiaris,* dog *Medicago polymorpha,* burweed
300 AD to 1000 AD	Early Iron Age	Mixed farming	Natal, Eastern Transvaal, Central Transvaal		*Sorghum* sp *Eleusine* sp, millet *Pennisetum* sp, bullrush millet ? *Gallus gallus,* domesticated chicken *Rattus rattus,* house rat
1000 AD to present	Late Iron Age	Mixed farming	East of 400 mm isohyet		? *Gossypium* sp, cotton ? *Nicotiana tabacum,* tobacco ? *Zea mays,* maize
1652 AD to 1806	Early European settlement	Cash crops plus subsistence farming	Cape south of Orange River, Natal	*Hippotragus leucophaeus*	Vines, European cereals, other crops and associated weeds; *Pinus pinaster,* cluster pine *Opuntia ficus-indica,* prickly pear *Mus musculus,* house mouse *Oryctolagus cuniculus,* domestic rabbit *Felis catus,* domestic cat *Equus caballus,* horse *E asinus,* donkey

(continued)

Table 1. Summary of major trends in economy, distribution, extinctions and alien introductions in South Africa during the period of human occupation (continued).

Time (BP)	Technological stage	Economy	Distribution	Extinctions	Alien introductions
1806 AD to 1910	British occupation	As above plus mining	Throughout South Africa	*Equus quagga*	Australian acacias, *Hakea* spp, hakea, *Nerium oleander*, oleander *Opuntia aurantiaca*, jointed cactus *Prosopis* spp, mesquite *Lantana camara*, lantana *Nicotiana glauca*, wild tobacco *Eichhornia crassipes*, water hyacinth, and many weeds and ornamental plants *Iridomyrmex humilis*, Argentine ant *Hemitragus jemlahicus*, Himalayan tahr *Sciurus carolinensis*, grey squirrel

The effect of changing climates on human populations

There is convincing evidence worldwide, both from ocean and continental sedimentary sequences, for cyclic changes in climate over the last 2 myr or more (Shackleton and Opdyke 1976; Imbrie and Palmer 1979). The most recent glacial-interglacial cycle (from 125 000 BP to the present) is the best known and the best dated and allows generalizations to be made about the effect that alternating cool and warm climates had on human populations. It is no coincidence that the periods that are better known (because there are more occupation sites) are interglacials or interstadials when temperatures were warmer worldwide (Figure 1b). The number of sites dating to the cooler period of the last glacial between about 70 000 and about 16 000 BP is much lower (Figure 1c). Deacon and Thackeray (1984) have hypothesized that foraging strategies in which there is a high reliance on underground plant foods and territorial game, were more appropriate in interglacial times when environments were more heterogeneous. On the other hand, mobile mixed strategies, in which more search-time is invested in obtaining larger migratory herd animals, were used under conditions of relatively low habitat diversity such as appear to have developed over large parts of southern Africa during glacial times (Deacon et al 1984). The consequence of this latter strategy is lower archaeological visibility of sites dating to cooler episodes. This hypothesis introduces the concept of human population fluxes in the past 125 000 years that must have operated during other glacial cycles too.

Late Pleistocene faunal assemblages in the southern Cape (dating from about 100 000 to 10 000 BP) are dominated by large grazing antelope. In this region the area of grassland during glacial times was extended by the exposure of the continental shelf as sea level was lowered worldwide by at least 100 m placing the coastline south of Cape Agulhas as much as 200 km south of its present position. By contrast, faunal assemblages that date within the Holocene (the last 10 000 years) are dominated by small, browsing, nongregarious antelope, implying that the vegetation in the last 10 000 years was bushy rather than grassy. Such shifts in vegetation and hunting strategies would have occurred many times over the past 2 myr as interglacials alternated with glacials.

Inland there is not as great a contrast evident in the faunal assemblages that pre- and post-date 10 000 BP. This is partly because there are few well-dated, long-sequence sites with bone preserved that date to the last glacial, between 70 000 and 16 000 BP. However, where faunal assemblages at sites with deposits older and younger than 10 000 BP are compared, as at Redcliff in Zimbabwe (Cruz-Uribe 1983) and Border Cave in KwaZulu (Klein 1977), only the northward extension of the modern range of the blesbok, springbok and mountain reedbuck and the southward extension of the lechwe has been documented (Klein 1984). These range extensions are indicative of grassland expansion at the expense of woodland.

During the Holocene, temperatures gradually rose. The period between about 8 000 and 5 000 BP was slightly warmer than at present with lower effective precipitation that is reflected in a lower human population density in the dry interior of South Africa (Figures 1d, 1e and 1f). The composition of plant and animal communities in historic times is thus a response to climatic conditions of the last 5 000 years when temperatures have remained relatively stable around the present-day mean.

The spread of Stone Age populations in South Africa

Location and distribution of Earlier Stone Age sites

The Earlier Stone Age (ESA) is that time period from the beginning of stone toolmaking (which in South Africa is poorly dated but is likely to be between 1,5 and 1,0 myr), and a significant shift in the technique of stone tool manufacture which took place about 150 000 BP. The ESA is characterized by an astonishing lack of technological change over a period of more than 1 myr. Technological innovation virtually stagnated once the handaxe had been invented and perfected.

The distribution of ESA sites through time shows a gradual increase in the number of sites and in the range of habitats exploited. Whereas there are very few (less than five) sites which are clearly from the earliest stages of the ESA, there are hundreds which can be assigned to the middle and later stages. The trend is towards an expansion outwards from the savanna and bushveld areas into the drier environments such as the Namib desert, into the cooler regions of the southern Cape coast and into the higher rainfall regions of Natal and the Transvaal Lowveld. This territorial expansion must signal not only the ability of people to exploit a wider range of habitats, but also a gradual increase in population numbers and greater sociality (Turner 1982). While a number of genera and species (about 20) became extinct during the ESA, including the three-toed horse *Hipparion* and the genus *Elephas*, there is no evidence to suggest that this was the result of human predation.

That people were able to adapt to such a variety of habitats and to spread so widely in the subcontinent highlights the fact that they were habitat generalists rather than specialists. Diamond (1984) notes the factors common to Pleistocene extinctions and concludes that carnivores and large or 'giant' mammals are more prone to extinction than herbivores, habitat specialists are at greater risk than generalists, social species need larger population numbers to survive than solitary species and population size is the overwhelming factor that shapes extinctions. The lower the population, the greater the risk of extinction. The fact that the genus *Homo* not only survived drastic habitat changes, but also extended its habitat range and apparently eased out several competitors, suggests that people became opportunists and habitat generalists, and that their population numbers during the ESA must have kept well above the critical level.

The earliest evidence for fire

The vast majority of ESA sites are in the open, preventing preservation of evidence for the control of fire. The evidence for fire in the small number of cave sites with ESA occupation in South Africa (fewer than ten) is far from conclusive (Deacon 1975; Volman 1984). It is only from the Middle Stone Age (MSA), dating from about 150 000 BP, that we have repeated and undoubted evidence for the regular use of fire in cave sites where substantial deposits of ash have built up and burnt bone is found.

The Middle Stone Age and regional differentiation

The Middle Stone Age (MSA) represents the next major stage in human physical and technological evolution. It is during this time that anatomically modern people, *Homo sapiens*, first appear. The MSA is characterized by a change in the range of stone tools, the methods by which they were made, the introduction of hafting, the range of sites occupied, the range of foods exploited and in the degree of regional differentiation. The first evidence that people exploited sea foods (fish, shellfish, marine birds and marine mammals) together with terrestrial resources is found at MSA coastal sites in the southern Cape, although it is possible that ESA people collected marine foods and the evidence has been obliterated by changes in sea level. The longest MSA sequence in the world in which this has been recorded is that at Klasies River Mouth on the southern

Cape coast (Singer and Wymer 1982) where shellfish remains are found with MSA artefacts at the base of the sequence (estimated to date to about 115 000 BP).

The Later Stone Age

Between 30 000 and 20 000 years ago a change took place in the design of stone tools that was accompanied by the gradual introduction of a range of other tools for which we have direct analogues in the toolkits of the present-day Khoisan peoples of southern Africa. These 'modern' Later Stone Age (LSA) artefacts include parts of bows and arrows, bored stones, ostrich eggshell water containers with incised decorations, shell beads, and painted and engraved stones.

Large shell middens are a feature of LSA sites along the coast and Buchanan et al (1978) have measured the impact of systematic collecting on shellfish resources. Where particular species were exploited over long periods, the mean size of the shellfish collected decreased significantly. Fish, shellfish, crustaceans, marine mammals and marine birds are all represented in LSA coastal middens and inland sites in the eastern Cape, Lesotho and Natal (mostly dating within the last 4 000 years) also contain fish bones and shells of freshwater mussels (Deacon 1984a).

Three large grazers, the Cape horse, the giant buffalo and the giant hartebeest, became extinct by 10 000 BP and the hunters focused instead on small non-gregarious browsers (Deacon et al 1984). Amongst the floral remains, charcoals have been analysed at LSA sites in the southern Cape and show that *Acacia karroo,* which was absent in deposits dating to 10 000 BP, was present by 6400 ± 75 BP, and by 1700 ± 55 BP accounted for 36% of all firewood (Deacon et al 1983). This expansion of its range pre-dates the appearance in the area of domesticated stock but may have been accelerated in the last 2 000 years by deliberate burning of the veld to promote grazing.

With better conditions for preservation, LSA

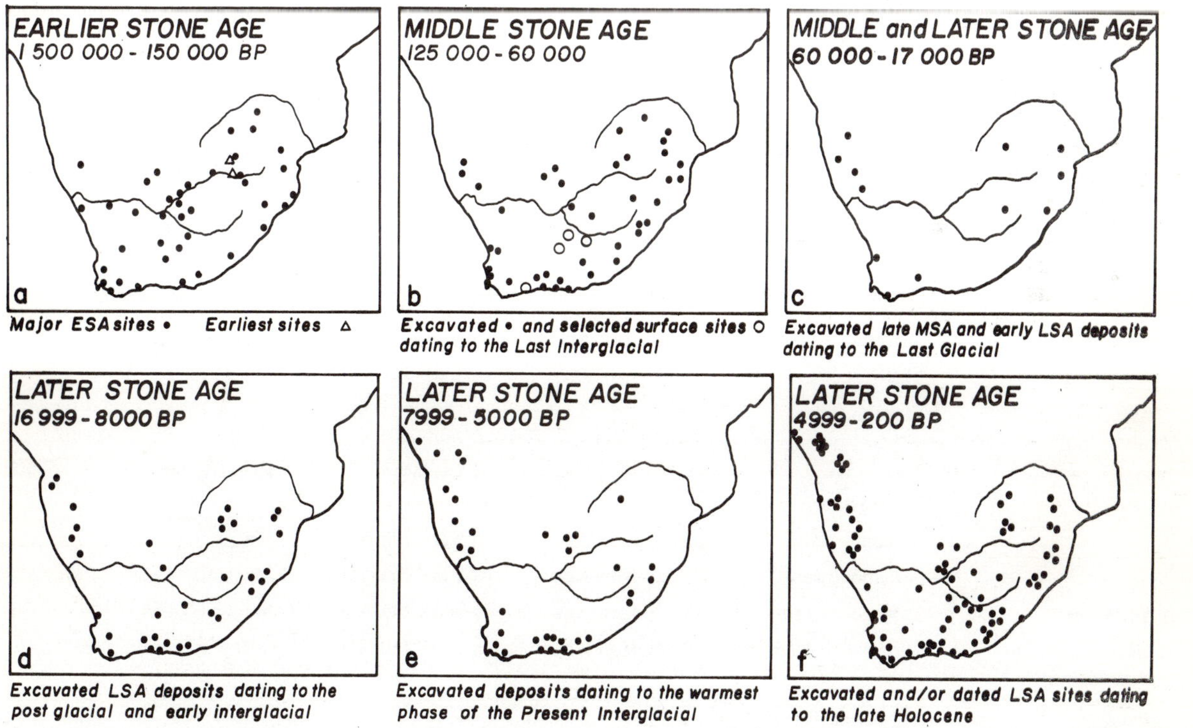

Figure 1. The distribution of Stone Age sites in South Africa. Data from Clark (1967); Deacon (1984a,b); Deacon and Thackeray (1984).

deposits also contain macroscopic plant remains. In pre-10 000 BP deposits there is one instance of well-preserved plant remains at Border Cave in KwaZulu where leaves, twigs, grass and seeds were found in deposits dated on associated charcoal to about 38 000 BP. However, *Diospyros dichrophylla* seeds from the same level gave dates ranging from 1963 AD ± 1 to 1971 AD ± 1 indicating some disturbance of the deposits. Beaumont (1978) has separated the samples he considers could have been disturbed from those which he considers are in situ, but this may be risky because the in situ sample includes two seeds and one shell identified as *Ricinus communis* (Beaumont 1978), the castor oil plant, which has long been held to be a recent immigrant to South Africa. Before rejecting this evidence for the pre-European occurrence of *Ricinus communis*, we should note that it has also been identified in a Later Stone Age context at Big Elephant Shelter in Namibia dated to 3130 ± 40 BP and in Early Iron Age levels at Shongweni South Cave in Natal where associated materials are dated to between the second and eighth centuries AD (O Davies unpublished). It has also been found in deposits dating between the ninth and seventeenth centuries AD at Renbaan, De Hangen and Elands Bay in the western Cape (Parkington 1977) but it is

Table 2. Earliest dated occurrences of alien plants known in archaeological contexts.

Species	Site	Associations	Date	Reference
Ricinus communis Castor oil	Border Cave, KwaZulu	Later Stone Age	? 37 500 BP	Beaumont 1978
	Big Elephant Shelter, SWA	Later Stone Age	3130 ± 40 BP (1180 BC)	Wadley 1979
	Renbaan Cave, W Cape	Later Stone Age	1150 ± 50 BP (800 AD)	C Liengme personal communication
	De Hangen, W Cape	Later Stone Age	390 ± 45 BP (1560 AD)	Parkington 1977
	Elands Bay Cave, W Cape	Later Stone Age	315 ± 45 BP (1635 AD)	Parkington 1977; C Liengme personal communication
	Shongweni Cave, Natal	Early Iron Age	ca 180 – 780 AD	O Davies unpublished
Medicago polymorpha Burweed	Scott's Cave	Later Stone Age	1190 ± 100 BP (760 AD)	Wells 1965; Deacon 1967
Sorghum sp	Schroda, N Transvaal	Early Iron Age	ca 800 AD	Hanisch 1981
	Magogo, Natal	Early Iron Age	ca 500 – 700 AD	Maggs 1984
Eleusine coracana Millet	Magogo, Natal	Early Iron Age	ca 500 – 700 AD	Maggs 1984
Pennisetum glaucum	Silver Leaves, E Transvaal	Early Iron Age	ca 300 AD	Klapwijk 1974
	Magogo, Natal	Early Iron Age	ca 500 – 700 AD	Maggs 1984
	Ndondondwana	Early Iron Age	ca 500 – 700 AD	Maggs 1984
Citrullus sp	Magogo, Natal	Early Iron Age	ca 500 – 700 AD	Maggs 1984
Vigna unguiculata var *sinensis*	Shongweni, Natal	Early Iron Age	ca 180 – 780 AD	O Davies unpublished
	Lanlory, Zimbabwe	Early Iron Age	ca 800 – 900 AD	Huffman 1972, 1979b
Voandzeia subterranea	Shongweni, Natal	Early Iron Age	ca 200 – 800 AD	O Davies unpublished
Cajanus cajan	Shongweni, Natal	Early Iron Age	ca 180 – 780 AD	O Davies unpublished

hard to determine either here or at the other sites whether the seeds work their way down into these deposits or whether they are truly associated with the dated charcoal. This problem could be solved by the relatively new accelerator dating technique that can determine the age of the seeds themselves.

Another assumed alien, *Medicago polymorpha*, known from historical evidence to have been here at least 120 years ago (Phillips 1938), has been identified at the base of a Later Stone Age deposit at Scott's Cave in the Gamtoos Valley where it dates to 1190 ± 100 BP (760 AD) and is associated with domesticated sheep (Wells 1965; Deacon 1967). Because the burs become entangled in sheep wool, this plant may well have been introduced together with domesticated sheep by the ancestors of the Khoi. Unlike the *Ricinus communis* seeds, there is a much stronger case for the *Medicago polymorpha* burs from Scott's Cave being truly associated with the radiocarbon dated material because the deposit is sealed by a pre-European roof fall.

The impact of pastoralism in the west

The bringing of domesticated animals to South Africa by Stone Age people about 2 000 years ago marks the first known deliberate introduction of alien fauna to the subcontinent. No caprines or cattle are indigenous to South Africa and the breeds kept by the Khoi at the time of European contact were derived from those bred by the earliest pastoralists in north-eastern Africa and the Near East over a period of 7 000 years or more. Sheep *Ovis aries* seem to have been introduced first and the earliest remains in South Africa that are undoubtedly associated with charcoal dated from the same stratigraphic units are from Die Kelders, Hawston, Kasteelberg, Boomplaas and Nelson Bay Cave along the south-western and southern Cape coast (Figure 2a; Table 3). The radiocarbon dates place their occurrence between 1960 ± 85 and 1700 ± 55 BP (ie 10 BC to 250 AD). Earlier dates for sheep, ranging between 3060 ± 60 and 2645 ± 95 BP, have been published, but are not generally accepted because of doubts expressed by the excavators about the precise association between dates and sheep bones (Deacon 1984a,b). The sheep kept by the Khoi were fat-tailed and hairy and were used for their fat and meat rather than for their wool (Epstein 1971).

Cattle *Bos taurus* remains are rare in all Stone Age assemblages probably because they were slaughtered less often than sheep. This makes it difficult to assess whether the apparently later appearance of cattle in the archaeological record (after 100 AD) reflects their later introduction or not. Also, without key body parts their bones are difficult to distinguish from those of buffalo (Table 3). Similarly, goats *Capra hircus* have not been identified in Later Stone Age contexts earlier than 1100 AD, although they are present in Early Iron Age assemblages in the Transvaal and Natal (Table 3). The domestic dog *Canis familiaris* is recorded in a LSA shell midden associated with pottery dating to 1150 ± 40 BP (800 AD) at Cape St Francis (Voigt 1983) and is possibly also present at Die Kelders and Nelson Bay Cave in the south-western and southern Cape at about the same time or earlier (Voigt unpublished). The archaeological evidence available at present is that dogs were absent from prepottery sites suggesting that they may have been introduced at the same time as sheep and/or cattle because they were used by the Khoi to help round up flocks and ward off predators.

The route taken by early pastoralists into South Africa is not well established. There is some controversy as to whether they came due south from Botswana and Zimbabwe to the coast and then moved westwards to the south-western Cape, or whether they came southwards along the west coast from Namibia and subsequently moved eastwards (Elphick 1977; Deacon et al 1978; Robertshaw 1978). Although there are no dated occurrences earlier than about 1100 AD in Namaqualand on the north-west coast (Webley 1984), sheep may have been present in the Namib at Falls Rock Shelter by 600 BC (Kinahan 1984) and were certainly at Mirabib shelter in the fifth century AD (Sandelowsky 1977).

The impact of pastoralism on ecosystems in South Africa is difficult to assess with accuracy. There is no evidence for the extent of good grazing in prepastoralist times, but by 1652 AD the Dutch settlers noticed that the Khoi in the south-western Cape moved their sheep and cattle seasonally, wintering in the area around St Helena and spending the summer around the Cape Peninsula to obtain full benefit from the local pastures (Avery 1976; Smith 1983). The historic records indicate that the majority of the Khoi lived in the west in Namaqualand and in the south-western and southern Cape between the Cape Fold Belt and the coast, in other words in the winter and all-year rainfall regions west of the 400 mm isohyet (Figure 2a).

The vast majority of the sheep and cattle sold to the Dutch colonists at the Cape by Khoi herders from 1662-1713 AD came from the coastal plains between Cape Town and the Gouritz River (Elphick 1977). Sales from the region amounted to about 10 500

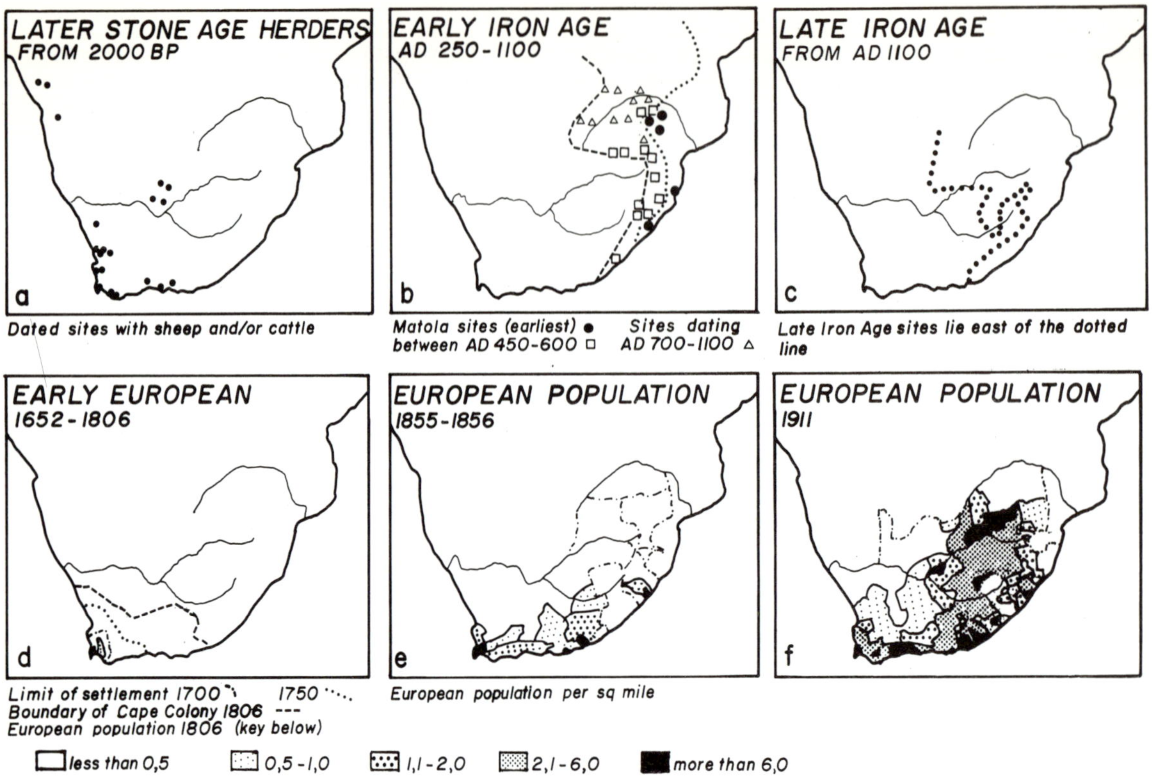

Figure 2. The distribution of Stone Age herder and Iron Age sites and of the European population between 1652 and 1911. Data from Christopher (1982); Deacon (1984a,b); Maggs (1984).

sheep and about 6 500 cattle during this 50-year period. For the entire western and southern Cape, over 32 000 sheep and over 14 000 cattle were sold in the same period. This would have been only a fraction of the Khoi cattle holdings, however, because they were notoriously reluctant to sell their cattle to the Dutch.

One of the changes in the archaeological record that apparently came about with the spread of Khoi pastoralists was the extinction of the blue antelope *Hippotragus leucophaeus* or bloubok which took place within the last two hundred years. Klein (1974) has suggested that the blue antelope preferred much the same grazing as sheep and became a target for both Khoi and European hunters as a result, although the numbers of blue antelope represented in pastoralist sites are not large. The extinction of the quagga *Equus quagga* however, may have had more to do with low initial populations and overexploitation by hunters than with competition for the same resources as domesticated animals.

The influx of Iron Age farmers in the east

Dating of the earliest Iron Age settlements

Farmers have been present in South Africa at least since the third century AD. Their arrival from eastern and central Africa, with exotic crops such as *Pennisetum* (millet) and *Sorghum* species (Table 2), meant not only a change in the method of food production, but also profound changes in settlement patterns, technology and social and political organization. Early Iron Age (EIA) villages from the Transvaal to Transkei (circa 250-1100 AD) are associated with iron slag and iron and copper artefacts, grindstones, occasional domesticated plant remains, sherds from large pottery vessels used both for storage and cooking, and middens that include bones of domesticated cattle, sheep and goats and a significant proportion of wild game. The dog, house rat *Rattus rattus* and chicken *Gallus gallus* are so far known from only one or two sites.

Distribution of Early and Late Iron Age communities

The earliest EIA sites in South Africa date to the third and fourth centuries AD and are located in the eastern and southern Transvaal and in Natal (Maggs 1984) (Figure 2b). From their distribution it has been inferred that the tradition of pottery-making associated with these sites (known as Matola) derived from an 'eastern stream' of agricultural immigrants who introduced farming along the coast from Mozambique and eastern Zimbabwe. Similar sites along 3 200 km of coastline from Kenya to southern Natal date within 150 years of each other (Maggs 1984) suggesting a remarkably rapid spread. Many of these early village sites, such as Enkwazini and Mzonjani in Natal, are usually about two hectares in extent. They are situated just inland of the coastal dune cordon in Natal and Mozambique (Maggs 1980a; Hall 1981) on soils that, when cleared initially of forest cover by 'slash and burn', were high in humic content. It is assumed that the sandy soils were rapidly leached making it necessary for the farmers to move on to new plots after a few years.

There is some evidence for domesticated crops in the form of casts of bullrush millet seeds, *Pennisetum glaucum*, pressed into potsherds at Silver Leaves in the eastern Transvaal lowveld (Klapwijk 1974), but few Matola sites have as yet yielded faunal remains in any quantity. EIA farmers moved gradually westwards into a new range of environments across the Great Escarpment and inland into Natal and the Transvaal plateau, where all sites known at present occur below the 1 000 m contour (Maggs 1980b). The earliest sites with well preserved and well dated faunal remains are Broederstroom in the southern Transvaal (Mason et al 1983) and Magogo in Natal where sheep, goats, cattle and the domestic dog have been identified in deposits dating to the sixth century AD (Maggs 1984; Voigt 1984) (Tables 2 and 3). At Broederstroom south-west of Johannesburg, Mason (1981) has excavated a series of villages, built between about 350 AD and 600 AD. Slag heaps show that iron was worked there and the faunal remains show that wild fauna provided most of the meat, but also included sheep and goats as the most common domestic animals with smaller numbers of cattle. Cattle, sheep and goats have been identified also at Happy Rest in the northern Transvaal between the fourth and fifth centuries AD (Voigt and Plug 1984). Domesticated dogs are recorded from Magogo and are dated to the eighth century AD at Ndondondwane in Natal where sheep, goats and cattle as well

Table 3. Earliest dated occurrences of domesticated and other alien animals known in archaeological contexts.

Species	Site	Associations	Date	Reference
Ovis aries, sheep	Die Kelders, SW Cape	Later Stone Age	1960 ± 85 BP 10 BC	Schweitzer 1974, 1979
	Kasteelberg, SW Cape	Later Stone Age	ca 100 AD	Klein unpublished
	Hawston, SW Cape	Later Stone Age	1860 ± 60 BP (90 AD)	Avery 1975
	Boomplaas, S Cape	Later Stone Age	1700 ± 50 (250 AD)	Deacon et al 1978
	Nelson Bay, S Cape	Later Stone Age	ca 250 AD	Inskeep and Vogel 1985
	Mirabib, SWA	Later Stone Age	1550 ± 50 BP (400 AD)	Sandelowsky 1977
	Broederstroom, S Transvaal	Early Iron Age	ca 350 – 600 AD	Mason et al 1983
	Happy Rest, N Transvaal	Early Iron Age	ca 350 – 470 AD	Voigt and Plug 1984
	Magogo, Natal	Early Iron Age	1360 ± 50 BP (1590 AD)	Voigt 1984
	Ndondondwane	Early Iron Age	ca 700 – 800 AD	Voigt and von den Driesch 1984

(continued)

Table 3. Earliest dated occurrences of domesticated and other alien animals known in archaeological contexts (continued).

Species	Site	Associations	Date	Reference
Capra hircus, goat	Broederstroom, S Transvaal	Early Iron Age	ca 350 – 600 AD	Mason et al 1983
	Happy Rest, N Transvaal	Early Iron Age	ca 350 – 470 AD	Voigt and Plug 1984
	Magogo, Natal	Early Iron Age	1360 ± 50 BP (590 AD)	Voigt 1984
Bos taurus, cattle	Kasteelberg, SW Cape	Later Stone Age	ca 100 AD	Klein unpublished
	Die Kelders, SW Cape	Later Stone	ca 400 AD	Klein and Scott 1974
	Broederstroom S Transvaal	Early Iron Age	ca 350 – 600 AD	Mason et al 1983
	Happy Rest, N Transvaal	Early Iron Age	ca 350 – 470 AD	Voigt and Plug 1984
	Magogo, Natal	Early Iron Age	1360 ± 50 BP (590 AD)	Voigt 1984
	Ndondondwane, Natal	Early Iron Age	ca 700 – 800 AD	Voigt and von den Driesch 1984
? *Bos*	Scott's Cave, S Cape	Later Stone Age	1190 ± 100 BP (760 AD)	Klein and Scott 1974
? *Bos*	Blinkklipkop, N Cape	Later Stone Age	1160 ± 50 BP (790 AD)	Thackeray et al 1983
Rattus rattus, house rat	Ndondondwane, Natal	Early Iron Age	ca 700 – 800 AD	Voigt and von den Driesch 1984
	Pont Drift, N Transvaal	Early Iron Age	ca 900 – 1000 AD	Plug et al 1979
? *Gallus gallus* chicken	Ndondondwane, Natal	Early Iron Age	ca 700 – 800 AD	Voigt and von den Driesch 1984
Canis familiaris, dog	Magogo, Natal	Early Iron Age	1360 ± 50 BP (590 AD)	Voigt 1984
	Cape St Francis, S Cape	Later Stone Age	1150 ± 40 BP (800 AD)	Voigt 1983
? dog	Nelson Bay, S Cape	Later Stone Age	1500 ± 55 BP (450 AD)	Voigt unpublished
? dog	Die Kelders, SW Cape	Later Stone Age	1590 ± 80 BP (360 AD)	Voigt unpublished

as the house rat and, possibly, the domestic chicken have also been identified (Voigt and Von den Driesch 1984). The last two aliens imply contact with Arabs along the coast which could have wide implications for the reconstruction of EIA trade networks.

No plant remains have been found to indicate conclusively that crops were grown at Broederstroom, although the size and length of occupation of the site would suggest that cultivated fields surrounded the villages. Cultigens are known from other sites of similar age, however. Sorghum has been identified at Schroda on the Limpopo River (Hanisch 1981) and at Magogo in Natal where millet, *Eleusine coracana* and cf *Pennisetum glaucum*, and the tsama melon *Citrullus* species are dated to the sixth or seventh century AD (Maggs 1984). A larger range of cultivated plants exotic to South Africa has been identified at Shongweni South Cave and includes both millets mentioned above, ground beans *Voandzeia subterranea*, cow peas *Vigna unguiculata* var *sinensis*, *Ricinus communis* and *Cajanus cajan* in deposits dated to between the third and eighth centuries AD (O Davies unpublished). All these cultigens are found today to the north of Natal in eastern Africa, reinforcing the hypothesis that it is from this direction that the immigrant farmers came (O Davies unpublished).

Early Iron Age settlement did not extend into the region of all-year rainfall west of East London at 28°E, nor into the cooler highlands or the winter rainfall region (Figure 2b), because adequate summer rainfall was essential for growing the staple tropical crops such as sorghum and millet (Maggs 1980a). The settlements ranged from about eight hectares to as much as 20 ha (Maggs 1984) which can give some idea of the scale of forest and bush clearance that accompanied this new lifestyle. Some wood was used in the construction of pole and mud huts, but far larger quantities were collected to make charcoal for iron smelting furnaces. A group of seven furnaces dating within a few hundred years of each other at a smelting complex south-east of Phalaborwa in the eastern Transvaal was surrounded by an estimated 180 tons of iron slag from which about 48 tons of iron had been produced. Over 800 tons of charcoal from about 7 000 trees had been used in the smelting process (Van der Merwe and Killick 1979). There is abundant evidence for iron working at EIA sites, but so far no relatively complete furnace sites are known, perhaps because the furnaces were dismantled after each firing (Maggs 1984). Copper and soapstone were also mined.

By 800 AD there were farming villages through most of the savanna areas of the Transvaal, the Natal Midlands (Maggs 1984) and the coastal region and inland valleys below the 1 000 m contour in Transkei (Granger and Feely 1985). There appears to have been a strong preference for river valley sites with permanent water, deep colluvial soils for crops and sweet grassveld for stock.

After 800 AD there is a marked increase in the number of settlements with a strong pastoral emphasis along the Limpopo in the northern Transvaal and westwards into Botswana (Figure 2b). The sites are easy to identify from growth of the grass *Cenchrus ciliaris* on cattle kraals (Denbow 1979). The increased importance of cattle is interpreted as indicative of an increasing inequality of wealth among Iron Age communities that led ultimately to more complex societies. Another manifestation of this increasing wealth is the placement of village sites on hilltops, often with defensive stone walling and stone-walled settlements. The first signs of Arab trade in this area are seen in imported glass beads at sites such as Schroda on the Limpopo dated to the ninth century AD (Hanisch 1981). The beads were exchanged for ivory and skins, gold being important a few centuries later in Zimbabwe. These characteristics heralded the beginning of a change that in archaeological terms is called the Late Iron Age (LIA).

The LIA has greater regional diversity and therefore less unity than the EIA and different hypotheses have been proposed to explain the abrupt change between the two traditions at some sites. Phillipson (1977) suggests that the LIA origins lie in Zaire to the north and that the change indicates the southward migration of a new wave of Bantu-speaking immigrants. Huffman (1978, 1979a), on the other hand, interprets the sequences of pottery styles as evidence for a migration of people from Natal northwards to the Limpopo Valley and beyond. The evidence for both hypotheses is slender and is based on inferences from pottery styles. The implications for the study of alien invasions that may be linked to these hypothetical population movements are obvious. Phillipson's explanation could mean new strains of stock and crops and their related pests, parasites and weeds, whereas Huffman's explanation would not.

The impact of mixed farming on the ecosystem

Whatever the origins of the LIA people, from the eleventh century AD onwards the eastern half of South Africa east of the 400 mm isohyet was gradually occupied by Iron Age farmers (Figure 2c). They preferred sites on hillslopes and hill tops rather than in valleys and the location of sites depended also on the specialization of the particular community: pastoralists preferred the drier western areas, agriculturalists needed reliable rainfall and metal-workers needed both ore sources and wood for their furnaces. However, the lack of these resources did not always prevent people from carrying on their activities: iron ore was carried up to 20 km from its source at some sites (Van der Merwe and Killick 1979).

Higher population densities in the LIA are reported for the grassland margins where they interfaced with savanna (Maggs 1984). Where this led to erosion and overexploitation, there is some evidence for terracing and the construction of walled tracks for livestock (Marker and Evers 1976; Collett 1979). However, as Maggs (1984) has argued, archaeological distributions may be anomalous because despite the higher density of LIA sites in upland areas in Natal, the Zulu wars of the 18th and 19th centuries were concerned with a struggle for control of lowland areas which clearly were important to the protagonists. The apparently denser distribution of sites in hilly areas may thus simply reflect higher archaeological visibility, lowland sites having been destroyed by subsequent European farming activities.

Evidence for new alien plants and animals after the initial introduction of domesticated grains and stock is not available. However, there are some indications that Arab traders along the east coast may have been instrumental in bringing new ones. The complex of sites in the Limpopo Valley around Mapungubwe Hill was an important centre for trade in the tenth and eleventh centuries AD. As early as the ninth century glass beads and cowrie shells were being traded nearby at Schroda (Hanisch 1981) and at K2 beads were imported and ivory was exported via Arab traders from the tenth to the twelfth century. Finds of cotton cloth at five sites in Zimbabwe and spindle whorls in Zimbabwe and at Mapungubwe have been taken as evidence for the introduction of weaving by Arab traders at least as early as the thirteenth century AD (Huffman 1971). However, it has not been possible to determine whether the cloth was made from the indigenous cotton plants *Gossypium herbaceum* var *africanum* and *G arboreum*, or from imported cotton. At Pont Drift on the Limpopo, there is further evidence for Arab influence in the presence of *Rattus rattus* (Plug et al 1979).

The question whether the Late Iron Age people were cultivating *Zea mays* before European farmers came to South Africa has not been answered clearly. The earliest date for a maize cob is 1780 AD ± 20 from Border Cave in KwaZulu (Beaumont 1978) and they have also been dated to the mid-nineteenth century in Natal (Maggs 1984). There is no archaeological evidence for tobacco *Nicotiana tabacum* having been grown before the arrival of Europeans.

While the presence of Iron Age people in particular areas is useful in assessing the potential impact they may have had on the environment, their absence from areas that were previously occupied may be equally revealing. From the end of the twelfth century, Mapungubwe and the other sites in the Limpopo Valley were abandoned and were not reoccupied and there appears to have been a general depopulation of the region. Changes in the trading network with the Arabs undoubtedly played their part for after this time the archaeological evidence shows a shift in the trade focus with sites like Great Zimbabwe becoming centres for the gold trade through to the fourteenth and fifteenth centuries and little direct evidence for links between the Arabs and people to the south of the Limpopo. This shift in trade centres placed South Africa out of the mainstream of contact with the Arabs and may well have retarded the spread of Arabian goods and technology. By the end of the fifteenth century the importance of Great Zimbabwe had waned and in the centuries that followed the Portuguese overtook the Arab trade initiative.

The pattern of establishment and subsequent abandonment of Iron Age villages and more substantial settlements is a graphic illustration of the length of time it takes for pastoralists and agriculturalists to exhaust the natural resources of an area. Cattle-based communities were able to stay in one area for two or three centuries where trade boosted the economy. Agriculturalists in Natal, on the other hand, seem to have had to abandon their settlements after a few decades, but could return to them at a later date. Both economies would have had a marked impact on woodland and a considerable amount of clearance would have been necessary in forested areas. However, there is no evidence that alien plants were replacing indigenous ones at this time and the impact on the environment seems to have been in terms of clearing and redistribution rather than replacement. Clearing of fields for agriculture helped open up the woodland areas for cattle in particular, but the domesticated sheep, goats and cattle never became feral. Woodland clearance took its greatest toll in Botswana and neighbouring territories to the west where cattle keeping became far more important than agriculture. There is no evidence that pests or parasites associated with domesticated stock affected game animals, but stock may well have played a part in distributing such organisms outside their natural habitat ranges, particularly after about 1400 AD when livestock farmers colonized even the higher-lying areas of Lesotho and the Drakensberg. By the time Jan van Riebeeck landed at the Cape in 1652, the Iron Age communities east of the 400 mm isohyet had already been established as mixed farmers for over a thousand years.

European settlement patterns

European settlement, particularly within the last 150 years, is unique in the number and range of deliberate and accidental alien introductions. The introduction of alien plants and animals into South Africa in historic times is documented in Chapters 2-5. Here, a broad outline of the speed and direction of spread of European people into the landscape is given up to an arbitrary date of 1911 AD, the year after Union, when the present-day boundaries were established and the first Union census was held. Because the emphasis in this section is on the impact of European settlers, the figures quoted do not include land and stock owned by indigenous people who continued for the most part with a mixed agricultural economy in the rural areas of the Transvaal, Orange Free State, Natal, Transkei and Ciskei.

The cultivation frontier

The purpose of the first Dutch settlement at the Cape in 1652 was primarily to grow fresh vegetables for passing ships; livestock was purchased from the Khoi. In 1699, the European population totalled 1 265 and was spread over some 4 000 km^2 from the Cape Peninsula to the Hottentots Holland.

There were 5 000 ha under cultivation (Christopher 1976) but as early as 1720 complaints were received about soil exhaustion. After the smallpox epidemic of 1713 when large numbers of local Khoi labourers died, it became increasingly necessary to import labour to work on the farms and by 1721 there were more slaves than Europeans in the colony. This ratio was maintained throughout the Dutch period of administration (Christopher 1976). Many of these slaves came from Malagasy and the Dutch East Indies and brought with them a new range of plant foods and spices.

The most successful crops were wheat and vines, and vegetables close to Cape Town. As farmers moved further afield, the problem of transporting produce to markets at the Cape became greater and livestock consequently increased in importance. By 1855, wheat remained the major crop covering 37,3% of the cultivated area, followed by oats (27,4%), barley and rye (9,7%), maize (8,1%), and vines (6%) (Christopher 1976).

Whereas under Dutch rule the Cape Peninsula remained the only large centre, the British began a concerted effort to expand the settlement. In the first fifty years of British rule, the European population increased fivefold to reach about 142 000 by the mid-1850s (Christopher 1982) (Figure 2c,d). The 1820 settlers in the eastern Cape established a second growth point around Port Elizabeth and Grahamstown. By 1850 Natal had become a third growth point and after the discovery of diamonds at Kimberley in 1870 and of gold on the Witwatersrand in 1886, the interior opened up for development on a scale that had not been seen before in South Africa (Figure 2e,f). After the construction of a railway network between 1890 and 1911, it became economic to grow crops on a commercial basis in the interior. With the railway and the increase in population related to mining in the interior, the area cultivated by European farmers increased from 20 000 ha in 1806 to 2 812 000 ha in 1911 (Christopher 1982). Figure 3 illustrates graphically this sudden increase in European expansion after the discovery of diamonds and gold. In 1875/1880, districts in which more than one per cent of the land was under cultivation were confined to the south-western Cape and the eastern Cape. By 1911, all districts of the Transvaal and Natal and the coastal plain from Cape Town to the eastern Cape had more than one per cent of land under cultivation. Maize as a cash crop developed from the 1870's and 1880's in the north-eastern Orange Free State, but became important only after the Anglo-Boer War. Cultivation of sugar and timber in Natal also expanded markedly only after 1902, although sugar had been introduced in the 1850's and commercial wattle *Acacia mearnsii* plantations began in 1886 (Christopher 1976, 1982). Attempts were made to start indigo, cotton, arrowroot and rubber plantations in Natal from the 1850's, but these failed (Christopher 1976).

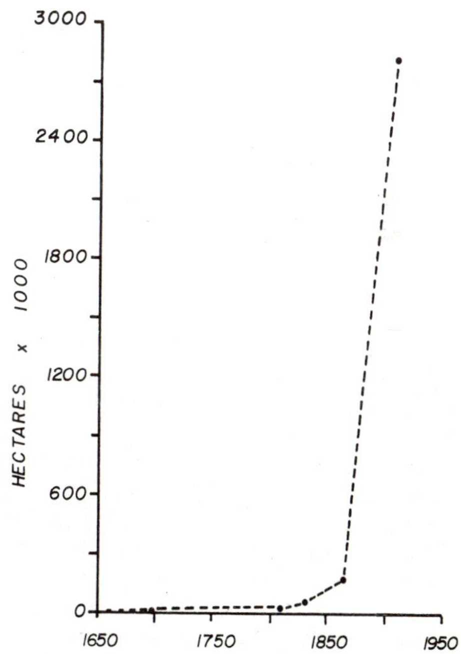

Figure 3. European expansion in South Africa from 1652 to 1911, in terms of hectares under cultivation (data from Christopher 1976, 1982).

The impact of exotic domesticated crops and of the weeds and pests that came with them must have been highly restricted before the 1870's. Many weeds, for example, were introduced into the Transvaal with fodder imported from Argentina during the Anglo-Boer War (Wells et al this volume) and therefore have a history of less than 100 years in South Africa. The slow start made by agriculture in South Africa was the result of the relatively poor physical environment and undeveloped transport system; its rapid growth after 1870 was the result of a dramatic increase in the European population and of the vastly increased efficiency of the transport system. Technology, irrigation schemes and scientific agricultural methods, on the other hand, were largely responsible for the growth of agriculture after the First World War.

The livestock frontier

Christopher (1982) has pointed out that prior to the First World War, farming in South Africa was essentially livestock orientated. The expansion of colonial frontiers was, until the discovery of gold and diamonds, at the behest of livestock farmers seeking more grazing lands. Initially, however, farmers merely added livestock to their agricultural farms in the immediate vicinity of Cape Town. Whereas in 1716 only nine per cent of all agricultural producers were stock farmers, in 1770 this had risen to 66%. By this time, European stock farmers had already established themselves in the eastern Cape and had pressed northwards into Namaqualand and north-eastwards into the Great Karoo. The first major clash with the Xhosa took place in 1779 along the Great Fish River (Christopher 1976), and the frequency of these clashes increased once the British government sought to increase control and, subsequently, to stabilize the frontier with settlers between 1811 and 1853.

As livestock farmers expanded the frontier, particularly in the northward movement of the Great Trek in the 1830's and 1840's, woolled sheep increased dramatically as the merino was introduced and the export trade expanded in the 1840's and 1850's (Table 4).

Whereas there were over one million sheep in the Orange Free State in 1856, they were concentrated in the south, in the north cattle, although more important than sheep, were not numerous. No comparable figures are available for the Transvaal. It was a time of no fences and large herds of wild game with the major income derived from ivory, skins and hides. Money was used to buy guns and ammunition, coffee and sugar. There is no record of systematic bush-felling or cultivation and as late as 1880 only half the farms in the Transvaal were thought to be occupied (Christopher 1976), yet most of the desirable land had been appropriated. This pattern changed dramatically with the discovery of gold in 1886: the estimated population of 43 260 Europeans in the Transvaal in 1880 had increased tenfold by 1911. In the Orange Free State, most of the land had been appropriated by 1856 and the European population in 1880 of 61 022 escalated to 175 189 in 1911.

In 1860 the Crown Lands Act of the Cape of Good

Table 4. Indices of European agricultural and pastoral development in South Africa from 1806 – 1911 (after Christopher 1976 and 1982).

Date	Wooled sheep (000)	Other sheep (000)	Cattle (000)	Area cropped (000 ha)	Land alienated (000 000 ha)
1806	14	1 240	208	20	1,8
1815	15	1 577	258	–	–
1824	15	2 192	352	–	–
1830	–	1 906	312	53	3,2
1839	–	2 339	307	65	–
1849	2 283	2 115	525	78	–
1855	4 828	1 626	451	78	15,2
1865	8 370	–	–	182	–
1875	9 986	–	580	–	–
1890	13 631	–	–	–	29,6
1911	11 052	–	2 062	2 800	36,2

Hope recognized that sheep farmers needed larger farms in arid areas and after 1879 the government issued temporary grazing licences that enabled farmers to utilize grazing in good years without tying them permanently to areas where rainfall was erratic. These factors, together with the invention and distribution of the windmill pump and barbed wire, had much to do with the opening up of the north-western Cape, but even so the rate of progress was slow: by 1911, only one borehole with a windmill pump was recorded in Namaqualand (Christopher 1976). By the early 1880's major advances had been made into the desert margins after the annexation of Griqualand West in 1871. Despite its relatively higher rainfall, Natal remained underdeveloped until after 1911, partly as a result of high prices for land and partly because of the inaccessibility of most of the remaining Crown Lands. Wool was the most important export in pre-Union Natal with sheep farms concentrated in the north adjacent to the Orange Free State and in the west adjacent to the Eastern Cape.

This very brief overview helps to highlight several general points. First, the marked and widespread impact of European settlers in South Africa took place only within the last 100 to 150 years. Prior to that time, settlement was confined to restricted areas around Cape Town, present-day Port Elizabeth and, to a lesser extent, around present-day Durban. Second, the drier western half of South Africa was settled later and by stock rather than crop farmers and so, it may be predicted, has not been invaded by alien plants to the same extent as the areas with more than 400 mm of rain per annum. Third, as Macdonald and Jarman (1984) and Brooke et al (this volume) have pointed out, there is a general lack of naturalized animal invaders in South Africa and those that are successful are confined to built-up areas where the system has been severely disturbed (for example, the squirrel *Sciurus carolinensis*, Himalayan tahr *Hemitragus jemlahicus* and various exotic birds). The only representatives of the animal kingdom that appear to have spread more widely are the house rat and the Argentine ant *Iridomyrmex humilis*. Finally, there must surely be a close correlation between the density of human population in an area (that is ultimately related to the trade and transportation network) and the degree of invasion by aliens.

Patterns from the past

The first known deliberate alien introduction to southern Africa is that of domesticated sheep which came ultimately from northern and eastern Africa about 2 000 years ago, followed by cattle, goats and dogs that were introduced by the pastoralist Khoi in the west and by the Iron Age farmers in the east. Cereals and other food crops indigenous to central and northern Africa were also introduced by the Iron Age people who occupied the eastern half of South Africa after about 250 AD. The European invasion, nearly 1 400 years later, began in the south-west, spread eastwards to other coastal points and then, only latterly, into the interior.

The generalizations that can be drawn from the gross changes in human population distribution over the past million or more years may seem to be obvious but could be of practical use in identifying areas where alien plants and animals are likely or unlikely to become widespread. The driest regions of South Africa were never growth points or regions of intensive occupation. During Stone Age times they were depopulated when temperatures were at their coldest and also at their warmest, and later were the last places to be colonized by European farmers. For Stone Age people dependent on natural resources, highest population densities built up under conditions such as those which prevailed in the early Holocene (12 000 to 9 000 years ago) and in the late Holocene after about 4 500 BP and in regions where rainfall was reliable rather than high as, for example, in the southern Cape and eastern Transvaal. Pastoralists became successful in the winter rainfall regions of the southern and western Cape coast, whereas agriculturalists with tropical crops were confined to the summer rainfall regions. It was only once European farmers entered the scene with well developed trade objectives and, later, sophisticated transport and water distribution technology, that inhibiting factors such as low rainfall and distance from markets were overcome.

The Dutch settlers imported food as well as non-food plants, but even then the South African vegetation in regions other than the south-western Cape remained well-nigh pristine for another 150 years or more. It was the Victorians, the nineteenth century British colonists of the Cape and Natal who, in the space of less than a century, brought seeds and seedlings from Europe, the Americas and Australia to change irrevocably the composition of many plant communities in South Africa.

Acknowledgements

I should like to thank C. Boucher, A. J.

Christopher, H. J. Deacon, Martin Hall, Christine Liengme, Andrew B. Smith, Elizabeth Voigt, and M. J. Wells for assistance in gathering the information for this paper and for comments on drafts of the manuscript.

Avery G 1975. Discussion on the age and use of tidal fish-traps (visvywers). South African Archaeological Bulletin 30, 105-113.

Avery G 1976. A systematic investigation of open station shell midden sites along the south-western Cape coast. M A Thesis, University of Cape Town, Cape Town.

Beaumont P B 1978. Border Cave. M A Thesis, University of Cape Town, Cape Town.

Brain C K 1981. The evolution of man in Africa: was it a consequence of Cainozoic cooling? Geological Society of Southern Africa, Alex du Toit Memorial Lecture 17.

Buchanan W F, Hall S L, Henderson J, Olivier A, Pettigrew J M, Parkington J E and Robertshaw P T 1978. Coastal shell middens in the Paternoster area, south-western Cape. South African Archaeological Bulletin 33, 89-93.

Christopher A J 1976. Southern Africa. Dawson, Folkestone.

Christopher A J 1982. Towards a definition of the nineteenth century South African frontier. South African Geographical Journal 64, 97-113.

Clark J D (ed) 1967. Atlas of African prehistory. Chicago University Press, Chicago.

Collett D P 1979. The archaeology of the stone-walled settlements in the eastern Transvaal, South Africa. MSc Thesis, University of the Witwatersrand, Johannesburg.

Cruz-Uribe K 1983. The mammalian fauna from Redcliff cave, Zimbabwe. South African Archaeological Bulletin 38, 7-16.

Deacon H J 1967. Two radiocarbon dates from Scott's cave, Gamtoos valley. South African Archaeological Bulletin 22, 51-52.

Deacon H J 1975. Demography, subsistence and culture during the Acheulian in southern Africa. In: Butzer K W and Isaac G Ll (eds) After the australopithecines. Mouton, The Hague. pp 543-569.

Deacon H J, Deacon J, Brooker M and Wilson M L 1978. The evidence for herding at Boomplaas cave in the southern Cape, South Africa. South African Archaeological Bulletin 33, 39-65.

Deacon H J, Deacon J, Scholtz A, Thackeray J F, Brink J S and Vogel J C 1984. Correlation of palaeoenvironmental data from the Late Pleistocene and Holocene deposits at Boomplaas cave, southern Cape. In: Vogel J C (ed) Late Cainozoic palaeoclimates of the southern hemisphere. Balkema, Rotterdam. pp 339-351.

Deacon H J, Scholtz A and Daitz L D 1983. Fossil charcoals as a source of palaeoecological information in the fynbos region. In: Deacon H J, Hendey Q B and Lambrechts J J N (eds) Fynbos palaeoecology: a preliminary synthesis. South African National Scientific Programmes Report No 75. CSIR, Pretoria. pp 174-182.

Deacon H J and Thackeray J F 1984. Late Pleistogene environmental changes and implications for the archaeological record in southern Africa. In: Vogel J C (ed) Late Cainozoic palaeoclimates of the southern hemisphere. Balkema, Rotterdam. pp 375-390.

Deacon J 1984a. Later Stone Age people and their descendants in southern Africa. In: Klein R G (ed) Southern African prehistory and palaeoenvironments. Balkema, Rotterdam. pp 221-328.

Deacon J 1984b. The Later Stone Age of southernmost Africa. British Archaeological Reports, Oxford.

Denbow J R 1979. *Cenchrus ciliaris*: an ecological indicator of Iron Age middens using aerial photography in eastern Botswana. South African Journal of Science 75, 405-408.

Diamond J M 1984. Historic extinctions: a Rosetta Stone for understanding prehistoric extinctions. In: Martin P S and Klein R G (eds) Quaternary extinctions: a prehistoric revolution. University of Arizona Press, Tucson. pp 824-862.

Elphick R 1977. Kraal and castle. Khoikhoi and the founding of white South Africa. Yale University Press, New Haven.

Epstein H 1971. The origin of the domestic animals of Africa. Africana Publishing Corporation, New York.

Granger J E and Feely J M 1985. Progress with research on the distribution of the Iron Age in Transkei. Paper presented at the Biennial Conference of the Southern African Association of Archaeologists, Grahamstown.

Hall M 1981. Settlement patterns in the Iron Age of Zululand. British Archaeological Reports, Oxford.

Hanisch E O M 1981. Schroda: a Zhizo site in the northern Transvaal. In: Voigt E A (ed) Guide to archaeological sites in the northern and eastern Transvaal. Transvaal Museum, Pretoria. pp 37-53.

Hendey Q B 1978. Late Tertiary Hyaenidae from Langebaanweg, South Africa, and their relevance to the phylogeny of the family. Annals of the South African Museum 76, 265-297.

Hendey Q B 1981. Palaeoecology of the late Tertiary fossil occurrences in 'E' Quarry, Langebaanweg, South Africa, and a reinterpretation of their geological context. Annals of the South African Museum 84, 1-104.

Hendey Q B 1982. Langebaanweg. A record of past life. South African Museum, Cape Town.

Huffman T N 1971. Cloth from the Iron Age in Rhodesia. Arnoldia (Rhodesia) 5(14), 1-19.

Huffman T N 1972. The Early Iron Age in the Urungwe District, Rhodesia. Rhodesian Prehistory 8, 12-16.

Huffman T N 1978. The origins of Leopard's Kopje: an 11th century Difaquane. Arnoldia (Rhodesia) 8(23), 1-23.

Huffman T N 1979a. African origins. South African Journal of Science 75, 233-237.

Huffman T N 1979b. Test excavations at NaBa and Lanlory northern Mashonaland. South African Archaeological Society Goodwin Series 3, 14-46.

Imbrie J and Palmer K I 1979. Ice ages: solving the mystery. Enslow, Short Hills, New Jersey.

Inskeep R R and Vogel J C 1985. Radiocarbon dates from the Holocene levels at Nelson Bay Cave and an interim report on their associations. South African Archaeological Bulletin 40.

Kinahan J 1984. On the relative homogeneity of a short Holocene sequence of stone tool assemblages from the central Namib desert. South African Journal of Science 80, 273-276.

Klapwijk M 1974. A preliminary report on pottery from the north-eastern Transvaal, South Africa. South African Archaeological Bulletin 29, 19-23.

Klein R G 1974. On the taxonomic status, distribution and ecology of the blue antelope, *Hippotragus leucophaeus* (Pallas 1766). Annals of the South African Museum 65, 99-143.

Klein R G 1977. The mammalian fauna from the Middle and Later Stone Age (later Pleistocene) levels of Border Cave, Natal Province, South Africa. South African Archaeological Bulletin 32, 14-27.

Klein R G 1984. The large mammals of southern Africa: Late Pliocene to Recent. In: Klein R G (ed) Southern African prehistory and palaeoenvironments. Balkema, Rotterdam. pp 107-146.

Klein R G and Scott K 1974. The fauna of Scott's cave, Gamtoos valley, south-eastern Cape Province. South African Journal of Science 70, 186-187.

Macdonald I A W and Jarman M L (eds) 1984. Invasive alien organisms in the terrestrial ecosystems of the fynbos biome, South Africa. South African National Scientific Programmes Report No 85. CSIR, Pretoria.

Maggs T M O'C 1980a. Mzonjani and the beginning of the Iron Age in Natal. Annals of the Natal Museum 24, 71-96.

Maggs T M O'C 1980b. The Iron Age sequence south of the Vaal and Pongola rivers: some historical implications. Journal of African History 21, 1-15.

Maggs T 1984. The Iron Age south of the Zambezi. In: Klein R G (ed) Southern African prehistory and palaeoenvironments. Balkema, Rotterdam. pp 329-360.

Marker M E and Evers T M 1976. Iron Age settlement and soil erosion in the eastern Transvaal, South Africa. South African Archaeological Bulletin 31, 153-165.

Mason R J 1981. Early Iron Age settlement at Broederstroom 24/73, Transvaal, South Africa. South African Journal of Science 77, 401-415.

Mason R J, Brown A J V, Fatti L P and Beardall G M 1983. Cluster and correspondence analysis of Iron Age faunal assemblages from southern, western and eastern Transvaal. South African Journal of Science 79, 189-203.

Parkington J E 1977. Follow the San. PhD Thesis, University of Cambridge, Cambridge.

Phillips E P 1938. The weeds of South Africa. Union of South Africa Division of Botanical Services, Pretoria.

Phillipson D W 1977. The later prehistory of eastern and southern Africa. Heinemann, London.

Plug I, Dippenaar N J and Hanisch E O M 1979. *Rattus rattus* (house rat) from Pont Drift, an Iron Age site in the northern Transvaal. South African Journal of Science 75, 82.

Robertshaw P T 1978. The origin of pastoralism in the Cape. South African Historical Journal 10, 117-133.

Sandelowsky B H 1977. Mirabib - an archaeological study in the Namib. Madoqua 10, 221-283.

Schweitzer F R 1974. Archaeological evidence for sheep at the Cape. South African Archaeological Bulletin 29, 75-82.

Schweitzer F R 1979. Excavations at Die Kelder's, Cape Province, South Africa: the Holocene deposits. Annals of the South African Museum 78, 101-233.

Shackleton N J and Opdyke N D 1976. Oxygen-isotope palaeomagnetic stratigraphy of Pacific core V28-239, late Pliocene to latest Pleistocene. Geological Society of America Memoir 145, 449-464.

Singer R and Wymer J 1982. The Middle Stone Age at Klasies River Mouth in South Africa. University of Chicago Press, Chicago.

Smith A B 1983. Prehistoric pastoralism in the southwestern Cape, South Africa. World Archaeology 15, 79-89.

Thackeray A I, Thackeray J F and Beaumont P B 1983. Excavations at the Blinkklipkop specularite mine near Postmasburg, northern Cape. South African Archaeological Bulletin 38, 17-25.

Turner A 1982. Hominids and fellow travellers. South African Journal of Science 78, 231-237.

Van der Merwe N J and Killick D J 1979. Square: an iron smelting site near Phalaborwa. South African Archaeological Society Goodwin Series 3, 86-93.

Voigt E A 1983. Mapungubwe: an archaeozoological interpretation of an Iron Age community. Transvaal Museum Monograph 1, Pretoria.

Voigt E A 1984. The faunal remains from Magogo and Mhlopeni: small stock herding in the Early Iron Age of Natal. Annals of the Natal Museum 26, 141-163.

Voigt E A and Plug I 1984. Happy Rest: the earliest Iron Age fauna from the Soutpansberg. South African Journal of Science 80, 221-227.

Voigt E A and Von den Driesch A 1984. Preliminary report on the faunal assemblage from Ndondondwane, Natal. Annals of the Natal Museum 26, 95-104.

Volman T P 1984. Early prehistory of southern Africa. In: Klein R G (ed) Southern African prehistory and palaeoenvironments. Balkema, Rotterdam. pp 169-220.

Vrba E S 1985. African Bovidae: evolutionary events since the Miocene. South African Journal of Science 81, 263-266.

Wadley L 1979. Big Elephant shelter and its role in the Holocene prehistory of central South West Africa. Cimbebasia B 3(1), 1-75.

Webley L E 1984. Archaeology and ethnoarchaeology in the Leliefontein Reserve and surrounds, Namaqualand. M A Thesis, University of Stellenbosch, Stellenbosch. 216 pp.

Wells M J 1965. An analysis of plant remains from Scott's cave in the Gamtoos valley. South African Archaeological Bulletin 20, 79-84.

Chapter 2

The history of introduction of invasive alien plants to southern Africa

M J Wells*, R J Poynton†, A A Balsinhas*, K J Musil*, H Joffe*, E van Hoepen* and S K Abbott†

*Botanical Research Institute and †South African Forestry Research Institute.

Introduction

In this chapter it is our aim to review the history of how man's actions have, or may have, resulted in the introduction of alien plants that have invaded natural vegetation in the subcontinent. Of necessity, our geographic and taxonomic terms of reference are somewhat wider than for other chapters in this volume.

Central to any history of plant introductions is the need to know what has been introduced. It is convenient if a researcher can start with a list of undoubted alien species and then investigate related historical matters such as their origin, time and place of arrival and what factors influenced their introduction. This review is substantially based on such a preliminary list (Wells et al 1986), which identifies nearly 800 species introduced to southern Africa.

Methods

A preliminary herbarium and literature survey revealed that there were at least 1 000 candidate species to be considered and several thousand publications likely to be directly relevant, excluding many unpublished documents such as private letters and horticultural society minutes. Much of the latter material is not easily accessible and some of it is in the archives of the same foreign countries from which our alien species have originated. The older material is in languages as diverse as Greek, Arabic, Chinese and Portuguese.

The magnitude of the task of reviewing so broad and diverse a field in a period of a few months determined both the level of the study and the methods to be used. Information was gathered for as many species as possible but the information about individual species or factors influencing their introduction could not possibly be complete.

Address for correspondence: M J Wells, Botanical Research Institute, Private Bag X101, Pretoria, 0001.

The method followed was: (a) to evaluate our knowledge of which alien species have been introduced to southern Africa; (b) to draw on existing information in order to characterize them in terms of eg families, life forms, regions of origin, and likely reasons for introduction; (c) to supplement this with an herbarium and literature search for early records of introduction and occurrence; (d) to search readily accessible historical literature for information that could be linked to plant introduction; and (e) to attempt to correlate introduction with causative or influencing historical factors. During this study a great deal of interesting information was collected on the introduction, establishment and spread of particular alien plants. All except a few examples have had to be omitted, for want of space and in order to obtain an even treatment. They now form part of a data bank for a much more detailed future study.

The problem plants catalogue

The most comprehensive listing of alien plants in southern Africa is included under the heading 'naturalized exotics', by Wells et al (1986) in their, 'Catalogue of problem plants of southern Africa'. Inclusion in this list is based on the taxon having been recorded in literature or herbaria as being 'of foreign origin, but established and reproducing itself as though a native' (Jackson 1905; see also Onions 1980). On this basis the catalogue lists 789 alien plant taxa of species or lower rank, henceforth referred to as 'species'. It also lists 91 species whose naturalization is uncertain and 104 species whose alien origins are in doubt. The former may be innocuous escapes from cultivation or important invasive aliens of the future - their status cannot be determined without further surveys and collecting.

The existence of so large a group of species whose origins are in doubt creates concern about the extent of our knowledge of our alien flora, especially as

Wells et al (1986) acknowledge that their coverage of tropical areas of the subcontinent is incomplete. This concern increased when during the literature survey for this chapter it became clear that there are a number of species that botanists regarded as alien to southern Africa 50-100 years ago, which are now often accepted as being indigenous, eg *Crotalaria laburnifolia*, *Eleusine indica* and *Rhynchelytrum repens* (McGibbon 1858; Burtt Davy 1904; Smith 1930). It was therefore necessary to investigate the possibility that a number of plant species usually regarded as indigenous, could be invasive aliens introduced before the existence of adequate botanical records and perhaps prior to colonization of the Cape by the Dutch in AD 1652.

Possible introductions before 1652

For the period before 1652, we gave particular attention to the extent to which early contacts with peoples from other continents and other parts of Africa provided opportunities for plant (and livestock) introduction, as well as the extent to which crop and other plant introductions were made. An attempt was made to determine the extent to which human colonization and particularly the spread of agriculture resulted in the introduction of alien plants.

The extent of disturbance of the natural vegetation was evaluated, and the degree to which this was likely to have favoured alien invaders. Species of doubtful origin were also assessed, with present distributions that include possible donor countries, and with characteristics and uses linking them to human activities.

In this volume, Deacon has summarized archaeological evidence of contacts between peoples of the Near East and North Africa circa 5000 BC, and has documented the southward migrations of pastoralists and agriculturalists and of the first signs of Arab trade in the area around 800 AD. To this must be added historical, linguistic and other evidence that Mediterranean and Asiatic peoples had been trading along the east coast of Africa since 1000 BC, that they had settled it as far south as Sofala by 1000 AD, were active in trading (and probably also in mining) in the interior, cultivated crops at least along the coast (and possibly in the interior), and had almost certainly rounded the southern point of Africa by about 1400 AD (Bulpin undated; Coetzee 1954; Davidson 1961; Raven Hart 1967; Hromnik 1981). According to Herodotus, Phoenicians circumnavigated Africa in about 600 BC, pausing to grow crops along the way (Rawlinson 1952). Dart (1925) reviews evidence, which includes that from bushman paintings, concluding that 'not only medieval Arabian, but also Indian, Chinese, pre-Koranic Arabian, Palestinian, Phrygian and even Babylonian influences have played a part in moulding the destinies of the primitive peoples of southern Africa'.

Crop introductions

The exact sequence and importance of outside cultural influences on Africa are often disputed but there is an ever increasing body of evidence regarding the origin and introduction of crops to Africa (eg Goodwin 1939; Jeffreys 1950; Murdock 1959; Fordham 1965; McMartin 1969; Phillipson 1977; Barraclough 1978; Hromnik 1981). In summary: barley *Hordeum vulgare* and wheat *Triticum aestivum* as well as various other cereals, legumes, root crops and vegetables were introduced to North Africa from Europe and the Near East between 4500-300 BC. Asian and Indonesian crops including bhang *Cannabis sativa*, coconut *Cocos nucifera*, Asian yams *Colocasia antiquorum* and *C esculenta*, coco yams *Dioscorea* species, plantain *Musa paradisiaca*, banana *M sapienta*, rice *Oryza sativa* and sugar cane *Saccharum officianarum* were probably introduced to the east coast of Africa between 1000 BC and 1500 AD. It is usually accepted that plants of American origin eg peanut *Arachis hypogaea*, sweet potato *Ipomoea batatas*, physic nut *Jatropha curcas*, cassava *Manihot* species, tobacco *Nicotiana tabacum*, guava *Psidium guajava* and maize *Zea mays*, were introduced after 1500 AD to the west coast (of Africa) by Europeans engaged in the slave trade. But there is some evidence that *Zea mays* and *Ipomoea batatas* were cultivated in Africa in pre-European times. This evidence of earlier introductions to Africa from America is strengthened by a report recorded by Padre Gomes in the 1640's of maize having been introduced to Portugal (via Africa) by the Moors during their occupation, ie at some time prior to 1147 AD when they were driven out of that part of Portugal (E Axelson personal communication).

These crops became widely dispersed in sub-Saharan Africa and it has been argued (Murdock 1959) that the Asian crops particularly contributed to agriculture and to human nutrition, thereby stimulating the increase in and migration of human populations. Most of these crops were certainly being used by Bantu peoples in the eastern parts of southern Africa when the first European records were made (MacKeurtan 1930; Herman 1936; Raven Hart 1967).

The role of 'pre-historic' human colonization

Observations from many parts of the world eg Moore (1967); Good (1969); Crosby (1972) and Numata (1975) suggest that human colonization of any but the most inhospitable deserts or icecaps is followed inevitably by alien plant invasions although the kind of plants involved and the extent of the invasion varies greatly. In this case several of the early crop plants are themselves invasive in southern Africa eg *Cannabis sativa*, *Cocos nucifera*, *Jatropha curcas* and *Psidium guajava*. Dart (1925) points out that early European observers 'called attention to the great number of plants, fruits and trees of Indian habitat to be found together on the (Zimbabwean) gold mines area', and to similar occurrences on the terraced hills of Inyanga. Marloth (1939) also hypothesized that 'the rough lemon (*Citrus limonia*) thickets growing wild (in Zimbabwe) and the scattered presence of seedling citrus trees in eastern central Africa might well be the outcome of introductions over 500 years ago'. Wild (1968) links the introduction of American crops along the west coast of Africa, to the spread of invaders, such as *Acanthospermum* and *Bidens* species, via central to eastern Africa and perhaps as far south as the Transvaal, prior to the arrival of European colonists in South Africa, ie perhaps by 1600 AD. Other aliens from America whose establishment in continental Africa antedates the beginning of this century are *Alternanthera pungens*, *Apium leptophyllum*, *Argemone mexicana*, *Duranta repens*, *Euphorbia chamaesyce*, *Lantana camara*, *Mirabilis jalapa* and *Physalis angulata* (Wild 1968).

Disturbance to the native habitats

Sagar and Harper (1961) have demonstrated the importance of disturbance as a factor affecting the penetration of natural communities by alien species. It is clear from the data provided by Deacon (this volume) that iron age pastoralists and agriculturalists could not have practised mixed farming (and particularly slash and burn techniques associated with shifting agriculture) on the eastern side of the subcontinent for over 1 000 years, without substantial disturbance of natural communities. This is confirmed by Feely (1980) who observed widespread secondary affects in vegetation previously assumed to be in a relatively undisturbed natural state. It may be surmised that any alien invader plants, especially those from elsewhere in Africa and Asia, that were well established in these communities at the time of arrival of the first Europeans, would have been regarded as indigenous.

Possible early introduced species

There are several hundred widespread or tropical African species which have uses or weedy characteristics related to human activities, and which may have been introduced to southern Africa. Among these are medicinal, religious, food, fibre or barrier plants including *Borassus aethiopicum*, *Echinochloa stagnina*, *Eleusine corocana*, *Euphorbia tirucalli*, *Gossypium herbaceum*, *Hibiscus sabdariffa*, *Lablab purpureus*, *Oxytenanthera abyssinica*, *Ricinus communis*, *Sesamum indicum* and *Tamarindus indica*. Crop weeds which may also have been introduced include *Brachiaria eruciformis*, *Commelina benghalensis*, *Cyperus rotundus*, *Eleusine indica*, *Sorghum verticilliflorum* and *Striga asiatica*. Burweeds with tropical distributions that could have been spread southwards by domestic stock include *Achyranthes aspera*, *Aristida junciformis* var *junciformis*, *Cenchrus ciliaris*, *Setaria verticillata*, *Tragus berteronianus* and *Triumfetta pilosa* var *tomentosa*. Other widely distributed species, which could also have been early introductions, include *Catharanthus roseus*, *Centella asiatica*, *Juncus bufonius* and *Rhipsalis baccifera*.

Of these, a few species such as *Euphorbia tirucalli* seem almost certain to be alien (Wells 1959; Leach 1960), but most would require genetic, morphological and ecological studies (Pritchard 1960; Baker and Stebbins 1965), as carried out by Pierce (1982) on *Spartina*, to elucidate their origins. Wide distribution ranges need not be due to introduction by man, but could stem from efficient natural dispersal, or even from Gondwana origins. Nevertheless the evidence reviewed points strongly to the introduction to southern Africa of alien plants as having started one to two thousand years ago and having possibly involved many species before the arrival of European colonists in 1652.

Species introduced after 1652

Of the 789 invasive alien plant species listed by Wells et al (1986) 530 have been catalogued with regard to family, region of origin, life form, life cycle, purposes for which planted, habitats favoured, kind of weed and undesirable characteristics. These species were those most frequently referred to in the literature. These alien invaders have not been rated according to importance.

However, 47 of the species were identified as weeds of concern to conservationists, that transform habitats or landscapes (see Appendix), ie change the character, condition, form or nature of a natural ecosystem over a substantial area. Aspects transformed would be those such as species richness, structure or appearance of the plant community, substrate quality, climate, or the faunal component of the ecosystem. The classification of characteristics in the catalogue (Wells et al 1986) is not complete and neither the alien nor the transformer species dealt with here should be regarded as the only ones in southern Africa.

Principal families

An earlier analysis (Wells et al 1983) based on a smaller sample, showed that approximately 50% of the plants introduced to southern Africa belong to four families: Poaceae, Fabaceae, Asteraceae and Solanaceae (in that order) with Brassicaceae supplying many agrestal species that are marginally invasive. Analysis of the larger sample now available gives substantially the same result, but the transformer species according to family show that approximately 50% belong to just three families: Fabaceae, Myrtaceae and Pinaceae (in that order), with most of the remainder belonging to Cactaceae, Poaceae, Proteaceae, Salicaceae and Solanaceae.

Regions of origin

The results of an analysis of the regions of origin of the 530 alien plants catalogued are shown in Table 1. In many ways it has been most unsatisfactory to have to link the temperate, largely agrestal, weed flora of Europe with the tropical weed flora of eg India, in one region of origin, 'Europe and Asia'. On the other hand there were too many shared species for it to be feasible to allocate them to separate regions without an in-depth study.

The two Americas, despite their geographic near-separation, share many species whose exact origins are difficult to pin-point. This forced us to employ the following classification procedure:

(1) if a species came from South and Central America, we chose 'South';

(2) if it came from North and Central America we chose 'North'; and

(3) if it came from all three we placed it in 'Americas'.

This resulted in an under-representation of 'Central America' as a region of origin of our plant invaders. It is also probable that island origins have been under-represented, as a result of their taxa now occurring on and being credited to nearby mainlands.

Table 1. Regions of origin of alien plant species.

Region of origin	Number of species	% of species
Europe and Asia	243	45,8
South America	139	26,2
North America	46	8,7
Australia	38	7,2
Elsewhere in Africa	20	3,8
Pantropical	16	3,0
Central America	12	2,3
Americas (region indeterminate)	9	1,7
Cosmopolitan	4	0,7
Atlantic Ocean Islands	1	0,2
Indian Ocean Islands	1	0,2
Pacific Ocean Islands	1	0,2
Total	530	100

It is noteworthy that almost as many alien species originated from Europe and Asia as from the rest of the regions put together, particularly since most of the species come from Europe, which does not have a very rich flora. South America has supplied us with three times more species than has North America, and four times as many species as we have received from Australia. The apparently small contribution of invaders from elsewhere in the continent of Africa is probably misleading, since some species brought in by tribal people and their stock in the prehistoric period are likely to have been regarded as indigenous.

A similar survey carried out by Burtt Davy (1904) of the alien flora of the Transvaal yielded a slightly higher figure of 48% for imports from Europe and Asia, but a very much lower figure of only 17% for aliens of American origin. This supports the conclusion, reached later in this chapter, that many of the American species are relatively recent introductions to the area south of the Limpopo River. Wild (1968), in a survey of the origin of weeds, rather than invaders, in 'South Central Africa' ie Botswana, Zambia, Zimbabwe, Malawi and Mozambique, with some references to Angola, found that of the extra-African species, only 17% were from Europe and Asia, 44% from the Americas and 21% pantropical or cosmopolitan. This apparent difference in the

Table 2. Transformer species: regions of origin, frequency amongst all alien species, relative to frequency of trees and shrubs amongst all alien species.

Region of origin	Number of species	Transformers as a percentage of of all alien species from the region of origin	Trees and shrubs as a percentage of all alien species from the region of origin
Australia	17	44,7	73,6
South America	17	12,2	27,3
Europe and Asia	8	3,3	13,6
North America	4	8,7	21,7
Elsewhere in Africa	1	5,0	20,0
Other regions	0	0,0	18,2
Total	47	8,9	22,8

representation of American as compared with European and Asian introductions, may well reflect a habitat-linked discontinuity in invasive alien plant floras somewhere near the tropic, but could also be partially due to historical factors.

An analysis of the regions of origin of transformer species provides a very different picture from that provided by the alien plants as a whole (Table 2). The success of Australian and South American species, which between them supply 72% of the transformer species is probably a consequence of the comparable environments provided by the three southern continents, but could also be the result of more deliberately selective introductions.

A further measure of success is obtained by comparing the percentage of alien plants from the various regions of origin that have achieved transformer status (Table 2). The great disparity in the ability of alien species to become transformers, and especially the high percentage from Australia, requires explanation. Most of these species are trees and shrubs (Table 3). It is therefore axiomatic that the areas of the world that have supplied us with most trees and shrubs have also supplied us with most transformer species.

Trees and shrubs are far more prevalent amongst the aliens that we have received from Australia, than they are amongst aliens from any other region. The percentages of trees and shrubs and of transformer taxa received from the various regions of origin correlate well (Table 2). The success of Australian trees and shrubs in southern Africa has been ascribed by Macdonald (1985) to that country having provided 'a unique evolutionary environment for fire-adapted trees and shrubs suited to temperate climatic conditions and soils of low nutrient availability'. Perhaps it could equally be linked to the lack of evolutionary stimuli sufficient to develop competitive, cold-hardy savanna trees and shrubs in southern Africa (Brain 1985, 1986), or to the scarcity of competitive indigenous trees in the fynbos (Oliver 1977; Hall et al 1980).

Table 3. Life forms of transformer species.

Life form	Number of species	% of species
Tree/shrub	41	87,2
Erect herbs	2	4,3
Free floating	2	4,3
Floating attached	1	2,1
Shrub/climbers	1	2,1
Others (including dwarf shrubs, climbers, epiphytes, parasites)	0	–
Total	47	100

Why and how introductions took place

Where species are cultivated, they are likely to have been introduced intentionally. Conversely, where species are not cultivated for any purpose, their introduction is likely to have been accidental. These assumptions will not always be correct in respect of the introduction of individual species. The cornflower *Centaurea cyanus* may have arrived as garden plants or as impurities in corn seed. However, linking cultivation to probable reason and mode of entry should provide an overall picture of why and how alien plants are likely to have entered the subcontinent.

The ratio of cultivated to uncultivated alien plant species from each region of origin is shown in Table 4. For most regions of origin, the percentage of cultivated taxa amongst the alien species is within 10% of the average of 58,1%. For Australia, the high percentage cultivation (81,6%) may be linked to the high percentage of tree species imported. Fewer impurities are likely to have been imported with tree seed than with agricultural crop seed. Also the individual planting of tree seeds does not favour the germination of contaminants as much as do the broadcast sowing methods used for agricultural crops. It is also possible that the establishment in southern Africa of herbaceous Australian introductions may have been prevented by the effects of defoliation and trampling by ungulate herbivores (Macdonald 1985).

Table 4. Cultivated and uncultivated species relative to all alien plant species from various regions of origin.

Region of origin	Number of species: Uncultivated	Cultivated	% cultivated
Australia	7	31	81,6
Elsewhere in Africa	4	16	80,0
South America	54	85	61,2
Europe and Asia	112	131	53,9
North America	22	24	52,2
Central America	6	6	50,0
Pantropical	10	6	37,5
Others	7	9	56,3
Total over all regions	222	308	58,1

For 'elsewhere in Africa', the figure (80%) is probably deceptive, since crop plants are the only ones usually recognized as having been introduced from the tropics. By contrast, the low figure (37,5%) for species of pantropical origin probably reflects the fact that the origins of most crop plants are better known than are those of ubiquitous ruderals and agrestals.

Out of a total of 47 transformer species 44 (94%) are cultivated and in most cases their introduction is known to have been deliberate. Of the three exceptions, all from South America, two, *Stipa trichotoma* and *S tenuissima*, were almost certainly accidental imports in fodder. The third, *Solanum mauritianum*, was recorded as long ago as 1862 in Natal and its manner of introduction is not known.

Many introductions are planted for more than one purpose (Table 5). The variety of purposes for which introduced species are cultivated is one factor that

Table 5. Classification of alien plant species from various regions of origin by cultivation purposes.

Region of origin	Number of species cultivated for the different purposes: Fodder crop	Agricultural crop	Silvicultural crop	Ornamental	Barrier	Cover/binder	Species cultivated	Average number of cultivation purposes per species
Europe and Asia	26	43	5	70	19	26	131	1,44
South America	9	14	2	71	21	10	85	1,49
Australia	0	0	11	21	24	16	31	2,32
North America	0	3	4	20	8	6	24	1,95
Elsewhere in Africa	8	5	0	9	2	8	16	2,00
Pantropical	1	1	0	5	1	1	6	1,50
Central America	0	2	0	6	1	1	6	1,66
Others	2	0	0	6	1	2	9	1,22
Total, all species	46	68	22	208	77	70	308	1,61
% of all purposes	9,4	13,8	4,5	42,4	15,7	14,2		
Total, transformer spp % of all purposes	Nil	5,4	11,7	34,1	27,9	20,7	47	–

could affect the variety of habitats and land-use systems to which they are exposed, which in turn could affect the plant's opportunities for successful establishment and spread. The high number of purposes for which Australian plants are cultivated may have had a direct and positive effect on their success rate.

By far the most common purpose for planting alien species is for ornamental purposes, followed by barriers, cover/binders, fodder crops, silvicultural crops and other agricultural and horticultural crops. In many cases, especially that of trees used for afforestation, as barriers, and as ornamentals, it is difficult to determine the real motivation for importing the species in the first place. The more successful and rewarding species were often reintroduced a number of times with various uses in mind.

Place of introduction

The search for early herbarium records which might indicate the place and time of introduction of alien plants is not complete, but has covered the National Herbarium, the Bolus Herbarium, the Herbarium of the Forestry Research Institute and part of the Natal Herbarium, ie over two thirds of the relevant specimens in South Africa. The literature search has covered the main 'Floras', taxonomic revisions

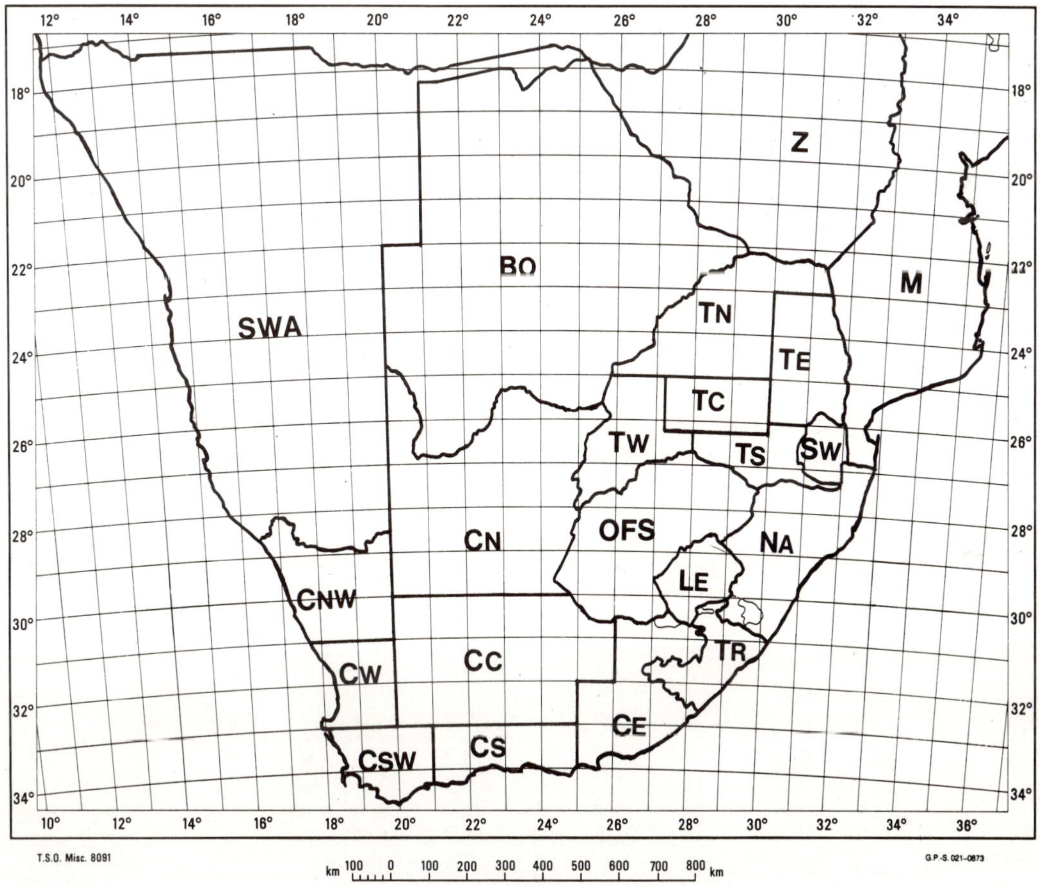

Figure 1. The subdivisions of southern Africa used for describing the history of plant introductions to the subcontinent. (Bo = Botswana, Cc = central Cape, Ce = eastern Cape, Cn = northern Cape, Cs = southern Cape, Csw = south-western Cape, Cw = western Cape, Cnw = north-western Cape, Le = Lesotho, M = Mocambique, Na = Natal, OFS = Orange Free State, Sw = Swaziland, SWA = South West Africa/Namibia, Tc = central Transvaal, Te = eastern Transvaal, Tn = northern Transvaal, Tr = Transkei, Ts = southern Transvaal, Tw = western Transvaal, Z = Zimbabwe).

and accounts by pioneer botanists and travellers, but is also far from complete. The localities of the first records of all 530 alien plant species have been classified according to the 21 introduction areas depicted in Figure 1.

These data show that the first records of plant introductions are most frequent for the south-western Cape. This was not only when Cape Town was the main centre for European colonization of southern Africa, but also continued up until 1909, long after the opening of other ports in the eastern Cape, Natal and Mozambique. Since 1910 most first introduction records have come from the central Transvaal. This change was probably influenced by centralization of the government Departments of Agriculture and Forestry in Pretoria, where new plants were brought for trial; the human population growth around the Johannesburg goldmining area (Deacon this volume), and the growth of air traffic through Johannesburg since 1945.

Records indicate that virtually all first introductions of alien plants were in the south-western Cape up until 1819. First introductions were made to progressively more areas from that date, until, in the period 1910-1985, when first introductions were recorded in 17 out of the 21 introduction areas. The only areas not involved were the north-western Cape, the northern Transvaal, Mozambique and Zimbabwe.

From 1820 onwards the second- and third-most first plant introductions have usually been recorded from either the eastern Cape or Natal. However, it is noteworthy that the importance of the central Transvaal as an introduction area grew steadily from 1890 when it occupied third position to 1910 when it came first. The eastern Transvaal has been an important introduction area since 1870. This could be connected to the fact that gold was discovered there before it was discovered in the central Transvaal and also that the eastern Transvaal lay on the main transport rider's route between Maputo (then Lourenco Marques) and the central Transvaal.

Time of introduction

In interpreting herbarium and literature records of time of introduction, as presented here, it must be kept in mind that there is often an unpredictable lag between the time of introduction and the first collection or literature record. It is nevertheless the best information available at present.

The four regions of origin that have provided us with most alien species, namely Europe and Asia, South America, North America and Australia, have been selected for comparison. The earliest records of

Table 6. First southern African records of alien plant species from four regions of origin, noted in periods between 1650 and 1985, corresponding to the five major periods referred to in the text.

	Region of origin							
	Europe & Asia		South America		North America		Australia	
Period	Species	%	Species	%	Species	%	Species	%
1650 – 1699	39	16,0	5	3,5	2	4,3	–	–
1700 – 1749	7	2,9	1	0,7	–	–	–	–
1750 – 1799	23	9,5	4	2,8	–	–	–	–
1800 – 1819	4	1,6	2	1,4	–	–	–	–
1820 – 1839	17	7,0	9	6,3	2	4,3	3	7,9
1840 – 1859	20	8,2	18	12,7	8	17,4	14	36,8
1860 – 1869	8	3,3	1	0,7	1	2,2	2	5,3
1870 – 1879	10	4,1	2	1,4	–	–	2	5,3
1880 – 1889	24	9,9	11	7,7	5	10,9	1	2,6
1890 – 1899	24	9,9	16	11,3	3	6,5	6	15,8
1900 – 1909	19	7,8	16	11,3	5	10,9	3	7,9
1910 – 1985	48	19,8	57	40,2	20	43,5	7	18,4
Total	243	100	142	100	46	100	38	100

alien plant species from these regions have been tabulated against a time scale and expressed as a percentage of the total number of records from that region recorded between 1650 and 1985 (Table 6). The longer periods at the start of the time scale reflect the paucity of records, which came mainly from a few collectors or writers such as Oldenland and Hartog (1727), Mentzel (1785) and Thunberg (1823).

The initial period of European settlement

Plant introductions from Europe and Asia during the period 1650-1799 reflected the concern of the Dutch colonists to cultivate as wide as possible a variety of agricultural and horticultural crops, mainly from their homeland in Europe or their possessions in the East. Smith (1930) records that the first alien plant invader in South Africa, with a well documented history, is probably sporry *Spergula arvensis*. Jan van Riebeeck, who founded the first European settlement at the Cape in 1652, imported it from Holland as a potential fodder plant. In 1659 he sent Fiscal Gabbema to Robben Island to report on how the plants being established there, including sporry, were flourishing. It was subsequently tried at the company's 'Buyten posten' where its horses were kept. Kolben (1731) described it in 1705-1713 as 't ergste onkruid is', and Mentzel (1785) recorded during his stay (1733-1740) that it was a 'noxious weed' spread in cattle manure (in Smith 1930).

Within the first few years of settlement at the Cape, the Dutch had introduced 50 or more crop plants, including a number of present-day invader species such as *Portulacca oleracea*, *Foeniculum vulgare*, and *Rorippa nasturtium-aquaticum*.

Common names give a clue to some other early introductions. Smith (1930) concluded from Dutch common names in the South Africa literature that 'stinkkruid' *Datura stramonium*, the 'karmedik' *Cnicus benedictus*, and 'kiesieblaar' *Malva parvifolia* were early introductions as medicinal plants and that 'muskuskruid' *Erodium moschatum*, 'roerkruid' *Pseudognaphalium luteo-album*, 'warkruid' *Cuscuta* species, 'hondekruid' *Euphorbia peplus*, 'muggekruid' *Mollugo nudicaulis*, and 'skaapsuring' *Rumex angiocarpus*, all accompanied early Dutch colonists. He also speculates that 'turksche vijgen' *Opuntia ficus-indica* may have entered South Africa in 1656, ie in the same year and by the same route that the Dutch imported 'turksche koring' *Zea mays* from Morocco.

Accidental introductions are seldom well documented. The first weed to earn official wrath was 'dravick' *Lolium temulentum*. In 1659 it was the subject of a 'plakkaart', ie a notice or regulation posted in a public place, issued by van Riebeeck's council.

Meanwhile a few useful or ornamental plants from South America and elsewhere were introduced by keen gardeners such as Governor W A van der Stel. It is not always possible to be sure of the identities of the species involved but *Canna indica*, *Manihot* species, *Psidium guajava*, *Verbena bonariensis* and *Prosopis* species, (ie '*Acacia americana*') seem to have been introduced in this way. *Xanthium spinosum* was also introduced, presumably by accident (Kolben 1731; Smith 1930).

Period of further settlement in the nineteenth century

Many settler groups came to the country during the period 1800-1869, mainly to the eastern Cape and Natal (Figure 1). They included British settlers in 1820 and 1848-1852, Mauritian sugar planters and German cotton planters in 1847, Javanese indigo workers in 1854, German settlers in 1857, Indian sugar workers in 1860-1866 and were followed by thousands of immigrants of many nationalities who flocked to the diamond fields between 1867-1870 (Russell 1911; Bulpin 1950; Gordon 1981;). Most of the economic plants that they imported were reintroductions of species such as *Hordeum vulgare* and *Triticum aestivum* already introduced by the Dutch to the south-western Cape. But the reintroductions were often different strains from different parts of Europe or Asia and no doubt brought different weed seeds with them. Military activity probably resulted in accidental imports of weeds with animals and bedding. The importance of the dispersal of weed seed as impurities in crop seeds and fodder is well known and has been treated at some length by Salisbury (1961). Prior to 1850 and the introduction of modern winnowing and screening machinery, we can assume that seed imported to Africa was more or less contaminated by weed seeds (Wild 1968). Some of these weeds that became invasive include *Conyza canadensis*, *Gnaphalium pennsylvanicum* and *Heliotropium curassavicum* from North America, *Bidens pilosa*, *Coronopus didymus*, *Nicandra physaloides* and *Physalis peruviana* from South America, and *Agrimonia odorata*, *Bidens biternata*, *Polygonum aviculare* and *Taraxacum officinale* from Europe and Asia.

Some of the first records from this period are from the interior of the country, eg *Polygonum aviculare* was first found in the central Cape (Figure 1). In the south-western Cape there was active importation by private plant collectors and for the first botanical gardens, nursery and afforestation enterprises (Shaughnessy this volume). One collector, Baron von Ludwig is credited with having imported over a thousand species (Bradlow 1965), including representatives of genera such as Australian *Acacia*, *Eucalyptus* and *Hakea*, and South American *Cereus*, *Chromolaena* and *Opuntia*, which have provided several important invasive and transformer species.

Other intentionally introduced plants that became invaders include *Albizia lophantha*, *Leptospermum laevigatum* and *Pittosporum undulatum* from Australia, *Gleditsia triacanthos*, *Pinus radiata* and *Robinia pseudoacacia* from North America, *Cassia tomentosa*, *Lantana camara*, *Nicotiana glauca* and *Parkinsonia aculeata* from South America, and *Ailanthus altissima*, *Rosa eglanteria*, *Spartium junceum* and *Ulex europaeus* from Europe and Asia.

Most first records were still of plants from Europe and Asia, but the emphasis was less strong, with vigorous South American and Australian plants attracting a lot of attention. A higher percentage of the first records were weeds, possibly a legacy of the earlier crop imports. There were many reintroductions by settler groups. Missionaries, who were often interested in botany and in bringing new crops and novelties to their parishioners, spread introduced plants such as *Opuntia aurantiaca* to remote mission stations (Wells 1971).

Newly introduced plants were valuable commodities. The poorly paid curator of the Cape Town Botanical Garden propagated and sold plants to increase his income (Wells 1971). In 1853 Bishop Breutel exchanged seeds of blue gum trees (*Eucalyptus* species) for victuals, when travelling from Genadendal to the eastern Cape (Kruger 1966) and passers-by begged seed of *Acacia mearnsii* from John van der Plank in Natal (Bulpin 1950). *Xanthium spinosum* became in 1860 the first 'proclaimed weed' in southern Africa, interpreted by Smith (1930) as a sign of deterioration of the natural grasslands.

Development of the interior

The population boomed as immigrants came in great numbers to South Africa during 1870-1899 following the discovery of gold and diamonds (Deacon this volume).

They almost certainly introduced plants such as *Eragrostis tef*, which Smith (1930) records fell from bales and was soon spread alongside the newly constructed railway lines from the coast to the interior. Isolated observations may be misleading, but it is certainly true that in this period many weedy species like *Bidens bipinnata*, *Bidens biternata*, *Linaria vulgaris*, *Rumex angiocarpus* and *Sonchus oleraceus*, as well as cultivated species such as *Opuntia ficus-indica*, *Prunus persica* and *Salix babylonica* were recorded in areas far inland. *Aira caryophyllea* was first recorded in the western Cape, *Lithospermum arvense* and *Malva pusilla* in the north-western Cape, *Aeschynomene indica* in the northern Transvaal, and *Pityrogramma calomelanos*, *Richardia brasiliensis*, *Sorghum halepense* and *Trichodesma zeylonicum* in the eastern Transvaal, *Alternanthera pungens* and *Boerhavia diffusa* in SWA/Namibia, and there were many records from the central Transvaal (Figure 1).

The charter of the British South Africa Company to trade and mine as far north as Tanzania greatly increased contacts with tropical Africa and resulted in a number of intentional importations. A vigorous tropical form of *Chloris gayana* was brought to the Cape. A number of South American species such as *Conyza bonariensis* and *Richardia brasiliensis* which first appeared in the Transvaal could have been introduced in this way, ie from the north, or as a result of fodder imports, ie through southern ports. Plants that probably owe their introduction to military activity include *Achyranthes sicula*, *Amaranthus deflexus*, *Oenothera tetraptera* and *Parthenium hysterophorus*. One Australian species *Physalis minima*, first recorded in 1895, could also have been introduced in fodder.

The same series of events that took place in the south-western Cape in the first half of the 19th century (Shaughnessy this volume) was repeated 20-40 years later in the eastern Cape and Natal (Anonymous undated). Alien invader plants introduced during this period include *Cassia hirsuta*, *Cestrum aurantiacum* and *Leucana leucocephala*. First records of naturalized aliens from Europe and Asia still predominated in this period, with South and North America well represented. There was an apparently reduced interest in new Australian plants. However, *Blackiella inflata*, many *Eucalyptus* species and *Myoporum serratum* were recorded for the first time.

The Anglo-Boer War and afterwards

The plant introduction record for the period 1900-1909 was dominated by accidental imports in fodder during the second Anglo-Boer War (1899-1902), and in the period after the war by the enthusiastic efforts of agriculturalists and foresters to import fodder and timber species with which to rebuild the economy. Amery (1909) records: 'During ing the war (the British obtained) from the Argentine ... 138 000 tons of hay and 2 500 tons of oats; and from Australia and New Zealand 193 000 tons of hay and 120 000 tons of oats'. Canada is not mentioned but substantial quantities came from there too (Smith 1930). Orders of fodder continued to arrive and had to be accepted long after hostilities ceased. 'At the end of the war, forage on hand totalling nearly 500 000 tons was held at 150 depots in different parts of South Africa. This was liable to deterioration and was sold to local contractors' (Amery 1909).

There are records of aliens such as *Tagetes minuta* and *Bidens formosa* spreading in the wake of the military (Burtt Davy 1904; Smith 1930; L E W Codd personal communication; C L Forssman personal communication). Other American and European aliens, probably introduced in fodder include: *Acanthospermum australe*, *A hispidum*, *A glabratum*, *Ambrosia artemisiifolia*, *Flaveria bidentis*, *Inula graveolens*, *Oenothera parodiana*, *Physalis viscosa*, *Solanum sisymbrifolium*, *Stipa neesiana*, *S tenuissima* and *S trichotoma*. The *Stipa* species were only collected many years later and we have to rely on accounts that they were well established in the 1920's, and their distribution was associated with military fodder stores (Wells unpublished).

During the period 1904-1908 the Transvaal Botanist imported several thousand seed samples annually from all over the world. In most years about 13 tons of farm seeds were distributed to about 700 cooperating farmers in various parts of the Transvaal (Burtt Davy 1904-1908). Smith (1930) records that many European grasses were tried out as sources of winter fodder in Australia. When glowing reports reached South Africa seed was requested. As a result a spectrum of European grasses go under the name 'Australiese gras' in South Africa, ie *Bromus diandrus*, *Dactylis glomerata*, *Phalaris caerulescens* and *P tuberosa*. Some of the grasses imported from Australia at the turn of the century, were actually African grasses doing a round trip eg *Panicum maximum* and *Rhynchelytrum repens*. The Australian plants *Atriplex semibaccata* and *Chenopodium carinatum* imported at this time have both proved to be invasive. During the same period the Conservator of Forests for the Transvaal was distributing about half a million tree seedlings annually, including species that have subsequently become invasive such as *Acacia dealbata*, *Acacia melanoxylon*, *Hakea suaveolens*, *Caesalpinia decapetala*, and *Eucalyptus* and *Pinus* species (Legat 1904-1908).

In this period the series of events previously described for the south-western Cape, eastern Cape and Natal was repeated in the Transvaal. Burtt Davy's reports (1904-1908) record every year that invasive burweeds and other aliens such as *Cuscuta campestris* had reached the Transvaal. In the country as a whole there were nearly as many new records of naturalized exotics from South America as from Europe and Asia, with only three from Australia. Among the alien invader plants most important to conservation managers introduced at this time were *Pinus patula* and *Schinus terebinthifolius*.

The modern phase

During the 75 years from 1910 large industrial developments have been accompanied by a human population explosion, by increased pressure on natural resources, and by a greatly increased demand for ornamental plants to satisfy an affluent gardening public. The improvement of sea and air transport has greatly facilitated plant imports, whilst it is uncertain to what extent this has been countered by the imposition of restrictive import permits.

A large number of aliens, first recorded early in this period, could have resulted from Anglo-Boer War fodder and grain imports ie taken a decade to be discovered. These include *Anchusa azurea*, *Anthoxanthum odoratum*, *Centaurea repens*, *Cynosurus echinatus* and *Echium vulgare* from Europe and Asia, *Ibicella lutea*, *Lepidium bonariense*, *Stipa clandestina* and *Stipa neesiana* from South America, and *Cenchrus incertus* and *Oenothera biennis* from Central and North America. Other aliens first recorded early in this period could well have stemmed from earlier pasture plant introductions eg *Agrostis avenacea* and *Atriplex muelleri* from Australia and *Brassica juncea* and *Chenopodium pumilio* from Europe and Asia. A very substantial list of seed contaminants compiled by the South African Division of Seed Control and incorporated in Wells et al (1986) shows how easily plants can be introduced this way even after the application of seed screening technology. Introductions as impurities in crop seed probably included *Alhagi camelorum*, *Amaranthus viridis*, *Avena byzantina*, *Cirsium arvense*, *Hypericum perforatum* and *Orobanche minor* from Europe and Asia, and several *Amaranthus* species and *Gaura sinuata* from the Americas.

The direct effects of industrial development on plant introductions are difficult to assess, but several species have been first recorded in the immediate vicinity of factories and railway yards. Most recently, in 1981, a small but dense, isolated stand of a highly unpalatable Australian grass *Stipa variabilis* was found near a new tractor engine factory on the west coast (National Herbarium).

The demand for ornamental plants is clearly reflected in first records of numerous invasives among garden species, eg *Aleurites fordii*, *Duchesnea indica*, *Lilium formosanum* and *Plectranthus barbatus* from Europe and Asia, *Araujia sericifera*, *Aristolochia elegans*, *Cassia bicapsularis*, *C corymbosa*, *Cortaderia selloana*, *Eupatorium macrocephalum*, *Harrisia martinii*, *Opuntia vulgaris*, *Sesbania punicea* and *Solanum seaforthianum* from South America, *Montanoa hibiscifolia*, *Opuntia dillenii*, *O imbricata*, *Tithonia diversifolia* and *T rotundifolia* from Central and North America, and *Acacia podalyriifolia* and *Homalanthus populifolius* from Australia. Water garden and fish tank enthusiasts were probably responsible for the introduction of invasive water plants, eg *Azolla filiculoides*, *Eichhornia crassipes*, *Myriophyllum aquaticum* and *Salvinia molesta*, all from South or Central America.

Agricultural crop plants that have become invasive include several *Rubus* species from both Europe and North America (notably *R cuneifolius*) which hybridize with indigenous species. These *Rubus* species were introduced to relatively cool mountain areas such as those in Swaziland. Pasture, crop and lawn grasses from tropical Africa that have been introduced since 1910 and become naturalized include *Cynodon aethiopicus*, *C nlemfuensis*, *C plectostachyus*, *Digitaria didactyla*, *Pennisetum clandestinum*, *P purpureum*, *P setaceum* and *P villosum*. Accidental introductions via tourism or the moving of agricultural stock are not well documented. However, two alien tropical grasses, *Eleusine tristachya* and *Eragrostis tenuifolia*, have appeared in the northern Transvaal in recent years at caravan parks and stock pens.

The volume of plant imports to the subcontinent can be gauged from the fact that in 1983-1984 approximately 2 300 phytosanitary investigations were made in respect of 2,2 million tons of imported plant products in South Africa alone (Anonymous 1985). Precautions such as these might help control some unwanted plant imports but can manifestly not exclude all. It is still possible to legally import potentially dangerous species. For example despite the record of aggressive invasiveness shown by many Australian *Acacia* species, *A mangium*, which has the reputation of being more vigorous than *A melanoxylon*, was recently imported for plantation trials (S Neser personal communication). Plant smugglers do not advertise their methods but it is obvious that illegal imports by gardeners and nurserymen have been greatly facilitated by the ease and volume of air transport. It is also obvious that many vegetatively propagated, decorative cultivars belonging to the Berberaceae, Cactaceae and Cupressaceae, whose import is prohibited, have nevertheless been introduced to the subcontinent in the last 25 years. Fortunately none of them are aggressive invaders.

The speed of air travel also increases the likelihood of seeds surviving as accidental imports. A J Healy (personal communication) in New Zealand, filled a glasshouse with the crop from one pair of muddy boots, taken from an airways passenger.

Roughly 20% of all first introductions of plants from Australia, Europe and Asia was recorded between 1910 and 1985. The percentage figure for South American and for North American introductions was over 40%, and for the first time, new records of plant introductions from South America outnumber those from Europe and Asia. There is no proof that the South American introductions reflect the legacy of the war years (1899-1902) but this seems the most likely explanation, in view of the observations recorded by Burtt Davy (1904), Smith (1930) and others.

Conclusion

Although there are inadequacies in the historical record and in the collecting coverage, a considerable amount of data is now available on the introduction of alien plants to southern Africa. This points to plants from elsewhere in Africa, from Europe and Asia, and perhaps also from the Americas, having invaded our flora prior to Dutch colonization of the Cape in 1652. Some of these plants may well have been established here long before the Portuguese reached the east coast of Africa in 1498.

Southern Africa has been subject to a two-directional invasion by alien plants (sometimes involving the same species); via tropical Africa (Wild 1968) and via the southern coastal ports (which supplied most of the human colonists after 1652). Between 1652 and 1819 almost all the southern introductions entered via the south-western Cape (Cape Town), and this entry point remained the most important until 1910 when the central Transvaal recorded more first introductions.

Over 50% of the introduced species belong to four families. Europe and Asia have supplied over 50% of the species introduced up until 1910, but since then South America has supplied more. By far the highest proportion of transformer species (those most important to conservation managers) have come from Australia and South America. Most of the invasive alien plants were at some stage cultivated and their introduction probably intentional. The strongest general motivation for introduction was probably ornamental gardening, however the most successful invaders tend to be those used for a variety of planting purposes.

Ports such as Cape Town, Port Elizabeth, Durban and Maputo were foci of colonization and therefore of plant introductions. The same species were often introduced to these ports thousands of kilometres (and sometimes centuries) apart. Introductions to each area tend to follow the same sequence, with economic (food) plants being followed by ornamental (garden) plants and then timber (forestry) plants. This is certainly true for the south-western Cape, and for Durban (D P McCracken personal communication). A relatively small percentage of all species introduced have become truly invasive, including all transformer species. These are the most important species from a conservation viewpoint, and are mostly trees and shrubs, many of them used for forestry, barrier and ornamental purposes.

Climatic and other environmental differences between the various introduction areas are great, and it is not surprising that introductions should fail in one area and succeed in another. An example is that of *Chromolaena odorata* which was first recorded in the south-western Cape in 1858 and not recorded again until 1947 in Natal, where it is now very widespread. Each area merits a separate history of plant introductions, but this study is restricted to first records for the subcontinent, with very little attention given to reintroductions.

In conclusion, it seems that there is hardly an event controlled by man, be it the building of a railway or the launching of a missionary crusade, that has not favoured disturbance and the establishment of alien plants. There is also a certain inevitability about human nature, as witness Jan van Riebeek's call to his masters in Holland in 1652, 'Send us anything that will grow' (Fairbridge 1931). Witness also the thoughts of Nicholas Downton, who visited the Cape 40 years before van Riebeeck. As he wandered on the slopes of Table Mountain he pondered: 'It much repenteth me that I came unprovided of all sorts of garden seeds which might be helpful or necessary for relief of any Christians, which hereafter might come hither . . . also acorns which in time may do good to posterity'. These sentiments, if not these exact words, have been echoed down the years by botanists, gardeners, foresters and farmers. Their urge to alter, tame or improve their new home was certainly encouraged by colonial authorities. At times powerful individuals such as Sir Joseph Hooker at Kew, were extremely active in distributing 'new and interesting' plants to enthusiasts in southern Africa and in effecting an exchange of plants between the colonies. As a result, most of our alien invader plants were intentional introductions and it is extremely likely that many of the rest were transported in association with them. Since these imports were made to establish industries and gardens, the problems associated with today's plant invaders can be seen as a tax being paid by the community for the prosperity and amenities that have been and are still being enjoyed.

Amery L S (ed) 1900-9. History of the war in South Africa 1899-1902. The Times, London.

Anonymous 1985. Annual report of the Department of Agriculture and Water Supply. Government Printer, Pretoria.

Anonymous undated. (Correspondence and reports relative to the state of Botanical Enterprise in Natal). Unpublished reports in Killie Campbell Library, Durban.

Baker H G and Stebbins G L (eds) 1965. The genetics of colonizing species. Academic Press, New Nork.

Barraclough G 1978. The Times Atlas of World History. Times Books, London.

Bradlow F R 1965. Baron von Ludwig and the Ludwigsburg Garden. Balkema, Cape Town.

Brain C K 1985. Temperature induced environmental changes in Africa as evolutionary stimuli. In: Vrba E S (ed) Species and Speciation. Transvaal Museum Monographs, Pretoria. pp 45-52.

Brain C K 1986. Prospects for a southern African continental record of cainozoic temperature changes. South African Journal of Science 82, 94-96.

Bulpin T V 1950. To the shores of Natal. Howard Timmins, Cape Town.

Bulpin T V undated. Islands in a forgotten sea. Howard Timmins, Cape Town.

Burtt-Davy J 1904. Alien plants spontaneous in the Transvaal. Report of the South African Association for the Advancement of Science 2, 252-299.

Burtt-Davy J 1904. Report(s) of the Division of Botany and Agrostology. Report(s) of the Transvaal Department of Agriculture, Pretoria.

Coetzee C G 1954. Die stryd om Delagoabaai en die suidooskus, 1600-1800. DSc Thesis, University of Stellenbosch, Stellenbosch.

Crosby A W 1972. The Columbian exchange. Biological and cultural consequences of 1492. Greenwood, Wesport, USA.

Dart R A 1925. The historical succession of cultural impacts upon South Africa. Nature 115, 425-429.

Davidson B 1961. Old Africa rediscovered. Gollanz, London.

Fairbridge D 1931. Historic farms of South Africa. Oxford University Press, London.

Feely J M 1980. Did iron age man have a role in the history of Zululand's wilderness landscapes. South African Journal of Science 76, 150-153.

Fordham P 1965. The geography of African affairs. Pelican. Cited by McMartin (1969).
Good R 1969. The geography of flowering plants. Longmans Green, London.
Goodwin A J H 1939. The origins of certain African foodplants. South African Journal of Science 36, 445-463. Cited by McMartin (1969).
Gordon R 1981. The place of the Elephant. Shuter and Shooter, Pietermaritzburg.
Hall A V, De Winter M, De Winter B and Van Oosterhout S A M 1980. Threatened plants of southern Africa. South African National Scientific Programmes Report No 45. CSIR, Pretoria.
Herman L 1936. Nathanial Isaacs: Travels and adventures in Eastern Africa I, II. Van Riebeeck Society, Cape Town.
Hromnik C A 1981. Indo-Africa. Juta, Cape Town.
Jackson D J 1905. A glossary of botanic terms with their derivation and accent. Duckworth, London.
Jeffreys M D W 1950. Negro agricultural origins. South African Journal of Science 45. Cited by McMartin (1969).
Kolben P 1731. The present state of the Cape of Good Hope. Vol II. W Innys, London. pp 261-288. (also referred to as "Kolb" and "Kolbe").
Kruger B 1966. The pear tree blossoms: a history of the Moravian Mission Stations in South Africa 1737-1869. Genadendal, South Africa.
Leach L C 1960. *Euphorbia tirucalli* L.: its typification, synonymy and relationships, with notes on "Almeidina" and "Cassoneira". Kirkia 9, I, 69-86.
Legat C E 1904-1908. Report(s) of the conservator of Forests. Report(s) of the Transvaal Department of Agriculture, Pretoria.
Macdonald I A W 1985. The Australian contribution to southern Africa's invasive alien flora: an ecological analysis. Proceedings of the Ecological Society of Australia 14, 225-236.
MacKeurtan G 1930. The cradle days of Natal. Shuter and Shooter, Pietermaritzburg.
Marloth R H 1939. South Africa's oldest living citrus trees. South African Journal of Science 36, 189-192.
McGibbon J 1858. Catalogue of plants in the Botanic Garden, Cape Town, Cape of Good Hope. Solomon, Cape Town.
McMartin A 1969. The introduction of sugarcane to Africa and its early dispersal. The South African Sugar Year Book 1969-70, 10-17.
Mentzel O F 1785. Vollständige und zuverlaszige geographische und topographische Beschreibung des berühmten und in aller Betrachtung merkwürdigen Afrikanischen Vorgebirges der Guten Hoffnung, vol 1. Gunther, Glogau.
Moore R M 1967. The naturalization of alien plants in Australia. IUCN Publication. New Series 9, 82-97. Morges, Switzerland.
Murdoch G P 1959. Africa: its peoples and their culture. History. McGraw- Hill, New York. Cited by McMartin (1969).
Numata M 1975. Kikashokubutsu (Introduced Plants Jap.). Dainihontosho- Publication, Tokyo.
Oldenland H B and Hartog J 1727. Naaukeurige en uitvoerige beschryving van de Kaap de Goede Hoop I, Appendix to Flora Zeylanica by J Burmann. Jan Waesberg, Amsterdam.
Oliver E G H 1977. An analysis of the Cape Flora. Proceedings of the 2nd National Weeds Conference of South Africa. Balkema, Cape Town. 1-18.
Onions C T 1980. The Shorter Oxford English Dictionary. Clarendon, Oxford.
Phillipson D W 1977. The later prehistory of eastern and southern Africa. Heinemann, London.
Pierçe S M 1982. What is *Spartina* doing in our estuaries. South African Journal of Science 78, 229-230.
Pritchard T 1960. Race formation in weedy species with special reference to *Euphorbia cyparissias* L. and *Hypericum perforatum* L. In: Harper J L (ed) The Biology of Weeds. Blackwell, Oxford.
Raven Hart R 1967. Before van Riebeeck. Rustica Press, Cape Town.
Rawlinson G 1952. History of Herodotus. Great books of the Western World. Hutchins R M (ed) 6. Encyclopaedia Britanica.
Russell R 1911. Natal, the land and its story. P Davis, Pietermaritzburg.
Sagar G R and Harper J L 1961. Controlled interference with natural populations of *Plantago lanceolata*, *P major* and *P media*. Weed Research 1, 163-176.
Salisbury E 1961. Weeds and Aliens. Collins, London.
Smith C A 1930. 'History of alien plants and their folk names, in South Africa'. Unpublished manuscript. Botanical Research Institute, Pretoria.
Thunberg C P 1823. Flora Capensis (complete edition). List of alien species at the Cape. Cotta, Stuttgart.
Wells M J 1959. An account of the plant ecology of the Nagle Dam area of Natal. MSc Thesis, University of Natal, Pietermaritzburg.
Wells M J 1971. The introduction of *Opuntia aurantiaca* to South Africa. Unpublished report. Botanical Research Institute, Pretoria.
Wells M J, Engelbrecht V M, Balsinhas A A and Stirton C H 1983. Weed flora of South Africa 2: Power shifts in the veld. Bothalia 14 (3, 4), 961-965.
Wells M J, Balsinhas A A, Joffe H, Engelbrecht V M, Harding G and Stirton C H 1986. Catalogue of problem plants of southern Africa. Memoirs of the Botanical Survey of South Africa, 53. Government Printer, Pretoria.
Wild H 1968. Weeds and aliens in Africa: the American immigrant. University College of Rhodesia, Salisbury.

Appendix

The 47 invasive alien species that transform habitats and landscapes, arranged according to country of origin.

Australia, 17 species:
Acacia baileyana, A cyclops, A dealbata, A decurrens, A longifolia, A mearnsii, A melanoxylon, A podalyriifolia, A pycnantha, A saligna, Eucalyptus cladocalyx, E grandis, E lehmannii, Hakea gibbosa, H sericea, H sauveolens, Leptospermum laevigatum

South America, 17 species:
Azolla filiculoides, Cereus peruvianus, Cestrum laevigatum, Chromolaena odorata, Eichhornia

crassipes, Jacaranda mimosifolia, Lantana camara, Myriophyllum aquaticum, Nicotiana glauca, Opuntia aurantiaca, O ficus-indica, Prosopis glandulosa, Psidium guajava, Sesbania punicea, Solanum mauritianum, Stipa tenuissima, S trichotoma

Europe and Asia, 8 species:
Arundo donax, Melia azedarach, Pinus pinaster, P pinea, Populus alba, P canescens, Pyracantha angustifolia, Salix babylonica

North America, 4 species:
Pinus patula, P radiata, Prosopis velutina, Rubus cuneifolius

Elsewhere in Africa, 1 species:
Euphorbia tirucalli

Chapter 3

A case study of some woody plant introductions to the Cape Town area

G Shaughnessy
Australian National Parks and Wildlife Service.

History of introductions

Because Cape Town was established as a victualling station for ships en route between Holland and the East Indies, it was inevitable that exotic plants for food and other purposes would be introduced from the beginning of European settlement in 1652. To describe all plant introductions since that time would be a major task which is not attempted here. This review is based on a study by Shaughnessy (1980) which concentrated on thirteen species of alien woody plants which are today classed as pest plants (Hall 1979) in the Cape Town region.

The thirteen species to which attention is directed are listed in Table 1. They include eleven Australian species in four genera: *Acacia* (six species), *Albizia* (one species), *Hakea* (three species) and *Leptospermum* (one species) and two species of *Pinus* from the Mediterranean region of Europe.

Pines

From an examination of early records, beginning with the diary of Jan van Riebeeck, first commander of the Cape (Thom 1954), and of secondary sources, including the survey by Appel (1966) of tree-planting at the Cape between 1652 and 1795, Shaughnessy (1980) concluded that *Pinus pinaster* was almost certainly introduced to Cape Town at some time in the period 1685 to 1693, under the name *P sylvestris*. Another conifer, which was referred to at the time as *Abies,* but which the evidence suggests was probably *P pinea*, was introduced in that period also. These conclusions are based primarily on an account by Kolb (1719) of his sojourn at the Cape from 1705 to 1713. He described three specimens (one of the former species and two of the latter) growing in the Company's Garden. Other contemporary evidence is provided by Valentijn (1726) and in Nienaber and Raven-Hart (1970). The Company's Garden was the site of Van Riebeeck's original produce garden; it later became the Cape Town Botanic Garden and is now the Cape Town Public Gardens.

In 1733 there were reported to be four coniferous trees in the Company's Garden (Mentzel 1785). From the description, it is likely that they were *P pinea*. By 1772, reports indicate that *P pinaster* was quite widespread in the Cape Town region and by 1810 both *P pinaster* and *P pinea* were conspicuous features of the Cape Town landscape (Shaughnessy 1980).

The earliest known record of *Pinus halepensis* comes from a report in 1857 that at Elsjes Kloof farm on the Riviersonderend, near Caledon, there was a row of trees including *P pinaster*, *P pinea* and *P halepensis* (Smith 1857d). They were judged to be 30 years old from their size. If correct, this means that *P halepensis* must have been introduced some years before 1830.

Australian plants

The earliest record of Australian plants growing at the Cape appears to be Burchell's inclusion of *Casuarina stricta* in a list of plants that he saw growing there during his visit in 1810 (Burchell 1822), but as this species occurs widely on islands of the Pacific and Indian oceans, the plants seen by Burchell did not necessarily originate from Australia. Other early references include that of Latrobe (1821) who described the garden of the country seat of the Deputy Governor in 1816 as containing "some curious plants, from Botany Bay and China".He referred also to the garden of the minister of the Lutheran church, which contained "a great variety of singular plants, trees, and shrubs, the produce of this and other southern regions".

The oldest known account of an actual introduction of Australian plants is by James Bowie, who settled in Cape Town in 1827. Bowie had previously been employed by William Hooker as a plant collec-

Address for correspondence: G Shaughnessy, Australian National Parks and Wildlife Service, GPO Box 636, Canberra ACT 2602, Australia.

tor for Kew Gardens, in which occupation he had travelled in both South Africa and South America. Bowie (1832) described his efforts to grow "seeds of various species of the Australian *Acaciae*" which he had brought with him from England and sown immediately on arriving in Cape Town. Most of the seeds had failed but after three years several of them germinated. Bowie reported also that he had sown seeds of *Acacia longifolia*, collected from a specimen that he had brought from England.

It is not surprising that the first introduction of Australian acacias to the Cape was made from England rather than directly from Australia, for many Australian plants were known as horticultural specimens in Britain by that time and all but one (*Acacia pycnantha*) of the Australian plants being considered in this report are recorded as having been introduced to Britain before 1825 (Hereman 1868).

All of the other Australian species under consideration here were first recorded as present in the southwestern Cape during the period 1835 to 1865 (Table 1).

Objectives or motivation for introductions

Lack of indigenous trees

The motivation for the introduction of the first *Pinus* species can be attributed to a desire to establish trees in a landscape where indigenous trees were not abundant (Sonnerat 1782; Shaughnessy 1980). Trees were needed for fuel, shelter, construction materials and aesthetic purposes. By the first half of the 19th century avenues of pines and oaks were common along roads and around farmhouses in the Cape Town region (Latrobe 1821; Burchell 1822; Backhouse 1844; Bunbury 1848).

Botanical gardens

The Australian species brought to the Cape were part of an influx of exotic plants consciously introduced in the 19th century as botanical curiosities or for their potential usefulness for a variety of horticultural endeavours.

In 1830 Baron C F H von Ludwig, a Cape Town merchant, established a botanical garden on private land situated between Cape Town and Table Mountain (Bradlow 1965). Over the next 17 years and with the assistance of the same Bowie who had earlier introduced *Acacia longifolia*, he introduced some 1 660 species from all over the world to establish the botanical showcase of Cape Town. Although no complete list of species established in the garden is available, various partial listings (Bradlow 1965) in-

Table 1. The 13 woody alien plant species relevant to this study.

Family	Species	Vernacular name in Cape Town	Earliest known occurrence in SW Cape (Shaughnessy 1980)
Original Provenance: Australia			
Mimosaceae	*Acacia cyclops*	rooikans	1857
	A longifolia	long-leaved wattle	1827
	A mearnsii	black wattle	1858
	A melanoxylon	blackwood	1848
	A pycnantha	golden wattle	1865
	A saligna	Port Jackson willow	1848
	Albizia lophantha	stinkbean	1835
Proteaceae	*Hakea gibbosa*	rock hakea	1835
	H sericea	silky hakea	1858
	H suaveolens	sweet hakea	1850
Myrtaceae	*Leptospermum laevigatum*	Australian myrtle	1850
Original Provenance: Mediterranean Region of Europe			
Pinaceae	*Pinus halepensis*	alleppo pine	ca 1830
	P pinaster	cluster pine	1685 – 1693

clude eight Australian genera. Of the pest plants under consideration here, only *A melanoxylon* is clearly mentioned, but circumstantial evidence suggests that *Albizia lophantha* and *Hakea gibbosa* were introduced during this period also (Bowie 1835; Evans et al 1969; Shaughnessy 1980).

By the middle of the 19th century, under the influence of the Hookers of Kew, there was great enthusiasm in the British colonies for botanic gardens. The Cape Governor, Sir Harry Smith, shared that enthusiasm and succeeded in 1848 in establishing the Cape Town Botanic Garden on the site of the former Company's Garden (Bradlow 1965). At the time of its establishment it was intended that plants from von Ludwig's garden should be transferred there (Montagu 1848), but it is not clear whether this was actually done.

Experimentation

The new Botanic Garden became the focus for experimentation with potentially useful plants from many sources, but there was a particular emphasis on Australian plants in the 1850's and 1860's. This resulted from the direct interest of George Grey, Governor of the Cape between 1854 and 1861, who had previously served as Governor of the colony of South Australia, and also from the activities of Baron F J H von Mueller, Government Botanist in the Australian colony of Victoria, who is well known for his widespread proselytisation of Australian plants during the 19th century (Cape Town Botanic Garden 1856, 1858, 1861; Smith 1857a,b; Rees 1892; Willis 1949). A catalogue of the plants growing in the Garden in 1858 included 25 Australian species of *Acacia* and six of *Hakea* (McGibbon 1858).

Dissemination of plants to the public was an important role of the Botanic Garden. By 1855 a good stock of "young plants of timber trees, shrubs, climbers and ornamental plants of all kinds" was available for distribution (Cape Town Botanic Garden 1856). The success of the Garden in this enterprise encouraged the establishment of private plant nurseries which raised and sold "such easily produced things as gum-trees, hakeas, etc" that the Garden's own business began to suffer (Cape Town Botanic Garden 1860).

Sand stabilization

The Botanic Garden was an important source of plants for a scheme initiated in the mid-nineteenth century to stabilize drift sand on the Cape Flats, the low-lying sandy plain east of Cape Town. The need to control the sand became acute during the construction of the first solid road across the Flats and the first planting was carried out in 1845 (Montagu 1845). Acacias were used in the scheme in 1847 (O'Farrell 1847), on the site of the present-day Athlone power station (Shaughnessy 1980), but their specific identity is not known.

By 1850 this major area of drift sand, which was known as the White Sands at that time and eventually became the suburb of Pinelands, had been stabilized by the planting of the indigenous Hottentot fig (*Carpobrotus edulis*), rye and "rush" (Central Board of Commissioners of Public Roads 1849, 1850a). An effort was then made to establish woody species. Application was made to the Director of the Botanic Garden for "as many forest tree plants as can be spared ... especially of the South Australian kind" (Central Board of Commissioners of Public Roads 1850b). Species known to have been established during this phase include: *Acacia longifolia*, *A saligna*, *Hakea suaveolens*, *Leptospermum laevigatum*, *Pinus pinea* and *Callitris* species. Others, for which there is weaker evidence, include *Acacia cyclops*, *A melanoxylon* and *Hakea gibbosa* (Newman 1855; Hemming 1857; Smith 1857c; Shaughnessy 1980).

Economic gain

In 1857 the same site became an experimental forestry area, the Roads Board having decided to seek a financial return for its efforts there (Central Board of Commissioners of Public Roads 1857). A new superintendent, who had previously been gardener at the farm near Caledon where *P halepensis* and other *Pinus* species had been planted before 1830 (Smith 1857d), began experimenting with exotic woody species obtained from both the Botanic Garden and private farms. He was particularly impressed by the rate of growth of *Acacia saligna* but, as his purpose was to establish trees, he concentrated on eucalypts, pines and various deciduous tree species. He concluded that *Pinus pinea* and *P pinaster* were the most promising, while *P radiata* and *P halepensis* should also do well (Smith 1858). From surplus stock in the nursery at the White Sands, seeds and seedlings of *A saligna*, *Hakea suaveolens* and *Leptospermum laevigatum:*, as well as "Blue gums, conifers and tree-acacias" were supplied to the public (Smith 1857d, 1861).

The operations at the White Sands site were pivotal for the spread of alien woody plants in the Cape Town region (Shaughnessy 1980). In this harsh environment many species were tried. Those that succeeded became the source of seeds used to establish government plantations in other parts of the Cape

Flats for drift sand control and for the production of tan bark, fuel, timber and railway sleepers, as well as general afforestation. They were used also to supply the public with plants for establishment in gardens and on farms as hedges, wind breaks, shelter belts and for tan bark and firewood. At the White Sands the government forestry service had its beginnings and the success of *P pinaster* led to its subsequent widespread planting on Table Mountain as a forest tree.

The appointment in 1858 of a Colonial Botanist, who was charged inter alia with investigating the suitability of the Cape for useful exotic trees and other plants, maintained the public's interest in Australian plants. The second incumbent of the post, J C Brown (1863-1866), encouraged the planting of anything that would grow and stressed the usefulness of shrubs and trees such as *Acacia melanoxylon*, *A mearnsii*, *A saligna*, *Albizia lophantha* and *Hakea* species (Colonial Botanist 1864, 1865, 1866).

Climatic amelioration

Brown considered the most important justification for planting trees to be the improved hydrological regime that he believed would result. While he advocated planting indigenous species where possible, he felt obliged to recommend the planting of the rapid-growing Australian trees during the severe drought of the 1860's because of the desirable effects he believed would follow (Colonial Botanist 1865).

Twenty to thirty years later, foresters justified the establishment of trees, particularly *Pinus pinaster*, on Table Mountain by stating that they would increase the water supply (Superintendent of Woods and Forests 1883; Conservators of Forests 1897).

Success in achieving objectives

Widespread planting

By 1865 all of the 13 species of particular concern to this report, together with many other species of woody plants from Australia and elsewhere, had been introduced to the Cape Town region. Experiments had shown the most suitable species for local conditions, particularly on the sandy soils of the Cape Flats. The stage was now set for widespread planting of the most appropriate species.

Shaughnessy (1980) conducted an analysis of available historical sources, including archival manuscript records, maps and photographs, parliamentary and municipal papers, periodicals and newsapers, and published primary and secondary sources, and documented the records of planting of these 13 species in the Cape Town region from 1845 to 1975. The major findings of that study for the period after 1865 are presented below.

Forestry authority

The study showed that the government forestry authority (under various titles) played a major role in attempting to establish many of these plants both on the Cape Flats and on Table Mountain and the adjoining mountains. In the area now known as Pinelands and Langa, the Uitvlugt Plantation (formerly the White Sands) was extensively planted to *Pinus pinaster* with *Acacia saligna* interplanted to provide shelter for the young pines. A general need for forest trees was the primary motivation in the 1880's, but subsequently the justification was the potential production of railway sleepers. Neither intended use was actually realized and the plantations were gradually taken over for other purposes.

When it was realized that the bark of *Acacia mearnsii*, *A pycnantha* and *A saligna* was rich in tannin, used in the production of leather, extensive plantations of these species were commenced. *Acacia saligna* was the most successful on the Cape Flats. However, the overseas market for the bark collapsed in 1893 because of competition from the bark of *A mearnsii* exported from Natal and many stands of *A saligna* were abandoned or managed as firewood plantations instead.

Extensive areas of the Flats were seeded with *A cyclops* and *A saligna* by the forestry authority, firstly in the cause of drift sand control, in conjunction with the disposal of Cape Town's refuse, but later simply in the name of "afforestation".Such planting on the Flats continued until as late as 1947 on the Eerste River plantation, where 3 529 ha of *A cyclops* and *A saligna* were planted.

Meanwhile, the forestry authority had extended its operations to the mountains adjacent to Cape Town in the 1880's. Most of the lower slopes of Table Mountain were privately owned by that time and in many cases were already afforested with *P pinaster* as a result of encouragement by forestry officials. The first government plantations were therefore on the high plateaux of Table Mountain, above the buttresses. In the case of Devil's Peak, adjacent to Table Mountain, the forestry authority had access to virtually the whole mountain and systematically proceeded to afforest it. At Tokai, south of Table Mountain, afforestation began on the lower slopes of Constantiaberg and then moved to the summit plateaux and to the western slopes of the mountain above Hout Bay. When water supply reservoirs were built on the plateaux of Table Mountain in the 1890's the forestry authority abandoned its plantations there and commenced operations on the eastern buttresses and a small plateau near the summit.

Besides the need to produce timber, forestry officials justified the afforestation of Cape Town's mountains in terms of improved water supply, aesthetics, and fire control. According to the forester in charge in the 1880's, J S Lister, the "bleak and naked appearance" of Table Mountain in its natural state was "a subject of daily comment' (Superintendent of Woods and Forests 1883). Likewise, D E Hutchins saw the opportunity in the 1890's to afforest Devil's Peak as the means to cover the "bare and stony slopes" which were "a reproach and eyesore to Cape Town" (Conservators of Forests 1893). Even as late as 1936, the forestry authority claimed that "a large section of the public do not see in a background of bare slopes and Gibraltar-like rock the proper aesthetic setting for Cape Town and Table Bay" (Division of Forestry 1936).

The establishment of the Devil's Peak forestry plantation, was justified also on the grounds that a dense cover of forest would be more fire resistant than the natural vegetation (Secretary for Agriculture 1892).

The method of afforestation on the mountains caused a great deal of disturbance to the natural vegetation, involving ploughing, digging of pits and ridging of the soil. While transplants were used initially, the faster method of direct broadcasting of seed was adopted in the 1890s. Extraordinary densities of seeds were sown, up to 50 kg ha^{-1} being quite usual (as against the 6 kg ha^{-1} that would be considered sufficient today).

In the first two decades of the 20th century the plantations on the upper mountainslopes were gradually abandoned when it became obvious that they would not be very successful. This abandonment left stands of *P pinaster*, with *P halepensis*, *A mearnsii* and *A melanoxylon*, as remnants of a past land use available to act as a seed source for future generations of these species.

The significance of the forestry authority's role to this study lies in its widespread planting activity followed by gradual abandonment of plantations. Of the 15 reserves owned by the authority in the Cape Town region during the period 1845-1975, only four remained by 1975. In those 15 reserves 114 km^2 were affected by afforestation and other planting activities. Of those 114 km^2 only 11,6 km^2 (10,2%) still belonged to the forestry authority in 1975 (Shaughnessy 1980). Thus, present-day patterns of land ownership give little hint of the major role played by the authority in the past in altering the natural environment of the Cape Town region by the planting of aliens. The major clue to this role is provided by the widespread distribution of certain exotic species, the legacies of former plantations.

Municipality of Cape Town

The Municipality of Cape Town was involved in tree-planting on Cape Town's mountains also, mainly in the 1890's and the first decade of the 20th century, again using *P pinaster* as the major species and employing the direct sowing technique. Later, when areas of the Cape Flats came under its jurisdiction, planting with *A cyclops* and *A saligna* was carried out in some areas. As with the forestry authority, but on a smaller scale, these former land-use activities have left their mark on the modern landscape.

Private individuals

Private individuals, encouraged by officialdom, continued to grow aliens in large numbers. D E Hutchins, who was in charge of government forestry in Cape Town through most of the 1890's, published in 1893 a catalogue of suitable trees (Hutchins 1893). He advised that it was useless to attempt to plant trees until the indigenous vegetation was eradicated by burning, ploughing and cross-ploughing. He suggested densities of 100 kg ha^{-1} for *P pinaster* on sand drifts 1 and 5 kg ha^{-1} "for good wattle seed on good soil". All but two of the 13 pest plants were listed in his catalogue.

In the 1880's and 1890's farmers on the Cape Flats were encouraged to establish trees on their properties by a system of prizes awarded for the best plantations. The major species used was *Acacia saligna*. Initially, the major justification was the need to establish a stable vegetation cover to minimize the threat of drifting sand, but later the aim was to produce tan bark. By 1891 private growers were producing more bark than the government forestry plantations. In many cases, to ensure the successful establishment of the tanning species, the natural vegetation, which was considered "useless bush" was cleared away first. The forestry authorities supplied massive quantities of seeds and transplants to the public. For example, in the period 1882-1893, the following estimated numbers of seeds were distributed: 14 million *A cyclops*, three million *A mearnsii*, 1,5 million *A pycnantha*, 300 million *A saligna*, 16 million *Leptospermum laevigatum*, one million *P halepensis* and 34 million *P pinaster* (Shaughnessy 1980).

It can be inferred that numerous planting activities were conducted by individuals, although few systematic records are available. The first director of the Kirstenbosch National Botanical Gardens, which were established on the eastern slopes of Table Mountain in 1913, stated that large areas of the land set aside for the Gardens were occupied already by hakea, which he claimed resulted from regular deliberate scattering of seeds by a person "actuated no doubt by the best of benevolent intentions" (Pearson 1915). Similarly, it has been stated that mountaineers scattered pine seeds on the eastern buttresses of Table Mountain (Cowling 1976; Shaughnessy 1980).

Analysis of invasive success

The role of human agency

Shaughnessy (1980) concluded that deliberate establishment by human agency had been an important process in the spread of certain alien woody plants in the Cape Town region, particularly in the period 1884 to 1917, but in certain areas until as recently as 1947. Such establishment frequently included major disturbance or even removal of the original vegetation cover. The subsequent abandonment of many plantations has tended to disguise the fact that many stands of alien species are actually relicts of a past man-made landscape which was consciously created to replace the natural landscape.

In considering individual species, Shaughnessy (1980) concluded that human agency had played a major role in the spread of four of the 13 species under consideration: *Acacia cyclops*, *A saligna*, *Hakea suaveolens*, and *Pinus pinaster*. For three other species little evidence for deliberate spread by humans was found: *Acacia longifolia*, *Albizia lophantha* and *Hakea sericea*. The other six species were considered to belong to an intermediate category: *Acacia melanoxylon*, *A mearnsii*, *A pycnantha*, *Hakea gibbosa*, *Leptospermum laevigatum* and *Pinus halepensis*.

Ecological implications

The present-day distribution of those species for which human dispersal has been an important factor must largely reflect that dispersal. Therefore, one cannot say that they are invasive simply because they are widespread today. Evidence for invasive tendencies must be sought by repeat mapping studies or other means, such as controlled planting. Thus, an ecological study of a stand of any of the ten pest plant species placed above in the category where human influence was major and in the intermediate category, should consider the possibility that the plants may have their origins in an earlier plantation at that site. Furthermore, the possibility that considerable disturbance of the original vegetation may have occurred before the plantation was established, should be taken into account.

Conversely, the other three pest plant species, for which this study showed little evidence of active dispersal by human agency (*Acacia longifolia*, *Albizia lophantha* and *Hakea sericea*) must, if found to be widespread, owe their distribution pattern largely to their own dispersal capacity and therefore could be considered to be aggressive invaders.

Appel A 1966. Die geskiedenis van houtvoorsiening aan die Kaap, 1652-1795. M A Thesis, University of Stellenbosch, Stellenbosch.

Backhouse J 1844. A narrative of a visit to the Mauritius and South Africa. Hamilton Adams, London.

Bowie J 1832. Hints on raising the leguminous plants of Australia and the Cape of Good Hope from seeds, on acclimatizing them in Europe and on their native habits. Gardener's Magazine 8(36), 5-9.

Bowie J 1835. (No title). Cape of Good Hope Literary Gazette 5, 54.

Bradlow F R 1965. Baron von Ludwig and the Ludwigsburg Garden. Balkema, Cape Town.

Bunbury C J F 1848. Journal of a residence at the Cape of Good Hope, with excursions into the interior, and notes on the natural history and the native tribes. Murray, London.

Burchell W J 1822. Travels in the interior of southern Africa, Vol 1. Longman, London.

Cape Town Botanic Garden 1856. Report of the Committee of Management of the Botanic Garden, Cape Town, for the year 1855. Cape of Good Hope: Annexures to the Votes and Proceedings of the House of Assembly: G13-1856.

Cape Town Botanic Garden 1858. Report on the Botanic Garden, Cape Town, for the year 1857. Cape of Good Hope: Annexures to the Votes and Proceedings of the House of Assembly: G26-1858.

Cape Town Botanic Garden 1860. Report on the Botanical Garden, Cape Town, for the year 1859. Cape of Good Hope: Annexures to the Votes and Proceedings of the House of Assembly: G40-1856.

Cape Town Botanic Garden 1861. Report of the Botanical Garden, Cape Town, for the year 1859. Cape of Good Hope: Annexures to the Votes and Proceedings of the House of Assembly: G31-1861.

Central Board of Commissioners of Public Roads 1849. Report for the year 1848. Cape of Good Hope Government Gazette, 3 May 1849.

Central Board of Commissioners of Public Roads 1850a. Report for the year 1849. Cape of Good Hope Government Gazette, 11 April 1850.

Central Board of Commissioners of Public Roads 1850b. (Letter to Director of the Botanical Garden, 14 June 1850). Cape Archives Depot, Cape Town: 1/27 Central Board of Commissioners of Public Roads (1843-1859): CRB 90 Letters Despatched: p 446.

Central Board of Commissioners of Public Roads 1857. Report for the year 1856. Cape of Good Hope: Annexures to the Votes and Proceedings of the House of Assembly: G22-1857.

Colonial Botanist 1864. Report for the year 1863. Cape of Good Hope: Annexures to the Votes and Proceedings of the House of Assembly: G23-1864.

Colonial Botanist 1865. Report for the year 1864. Cape of Good Hope: Annexures to the Votes and Proceedings of the House of Assembly: G24-1865.

Colonial Botanist 1866. Report for the year 1865. Cape of Good Hope: Annexures to the Votes and Proceedings of the House of Assembly: G1-1866.

Conservators of Forests 1893. Report for the year 1892. Cape of Good Hope: Annexures to the Votes and Proceedings of the House of Assembly: G27-1893.

Conservators of Forests 1897. Report for the year 1896. Cape of Good Hope: Annexures to the Votes and Proceedings of the House of Assembly: G17-1897.

Cowling R 1976. The ecological status of the understorey communities of pine forests on Table Mountain. In: Moll E J and Campbell B M (eds) The ecological status of Table Mountain, Appendix. Department of Botany, University of Cape Town, Cape Town.

Division of Forestry 1936. Report for the year April 1935 to March 1936. Union of South Africa: Annexures to the Votes and Proceedings of the House of Assembly: UG53-1936.

Evans D S, Deeming T J, Evans B H and Goldfarb S (eds) 1969. Herschel at the Cape; diaries and correspondence of Sir John Herschel, 1834-1838. Balkema, Cape Town.

Hall A V 1979. Invasive weeds. In: Day J, Siegfried W R, Louw G N and Jarman M L (eds) Fynbos ecology: a preliminary synthesis. South African National Scientific Programmes Report No 40. CSIR, Pretoria. pp 133-147.

Hemming J S 1857. The Cape Flats and how they may be improved. Cape Monthly Magazine 1, 263-271; 353-363.

Hereman S 1868. Paxton's botanical dictionary. Revised edition. Bradbury, Evans, London.

Hutchins D E 1893. Tree-planting, 1893; a descriptive catalogue of the best trees to plant in Cape Colony, with brief instructions for planting. Richards, Cape Town.

Kolb P 1719. Caput Bonae Spei Hodiernum. Monath, Nuremburg. 2 v.

Latrobe C I 1821. A journal of a visit to South Africa. Second edition. Seeley, London.

McGibbon J 1858. Catalogue of plants in the Botanic Garden, Cape Town, Cape of Good Hope. Solomon, Cape Town.

Mentzel 0 F 1785. Vollstandige und zuverlaszige geographische und topographische beschreibung des beruhmten und in aller betrachtung merkwurdigen Afrikanischen vorgebiges der Guten Hofnung, Vol 1. Gunther, Glogau.

Montagu J 1845. Government notice. Cape of Good Hope Government Gazette, 25 July 1845.

Montagu J 1848. (Letter to the Commissioners of the Botanical Garden, 4 May 1848). Cape Archives Depot, Cape Town: 1/16 Colonial Office (1806-1912): CO 5309: Letter Book, Miscellaneous, Vol 8, 1847-1848.

Newman W A 1855. Biographical memoir of John Montagu. Harrison, London.

Nienaber G S and Raven-Hart R 1970. Account of the Cape; brief description of Natal; journal extracts on East Indies by J D Buttner. Balkema, Cape Town.

O'Farrell B 1847. (Letter to Central Roads Board, 15 June 1847). Cape Archives Depot, Cape Town: 1/17 Central Board of Commissioners of Public Roads (1843-1859): CRB 38 Letters Received: NO 496.

Pearson H H W 1915. The National Botanic Gardens. Mountain Club of South Africa Journal 18, 18-23.

Rees W L 1892. The life and times of Sir George Grey, KCB. Second edition. Hutchinson, London. 2 v.

Secretary for Agriculture 1892. Correspondence. Cape Archives Depot, Cape Town: 9/2/1-9/2/3 Secretary for Agriculture (1892-1911): BRON-AGR 132 Correspondence Files, Folio No 553 Vol 5: Peak Plantation.

Shaughnessy G L 1980. Historical ecology of alien woody plants in the vicinity of Cape Town, South Africa. PhD Thesis, University of Cape Town, Cape Town. 421 pp.

Smith R 1857a. (Letter to Central Roads Board, 29 May 1857). Cape Archives Depot, Cape Town: 1/27 Central Board of Commissioners of Public Roads (1843-1859): CRB 71 Letters Received.

Smith R 1857b. (Letter to Central Roads Board, 26 October 1857). Cape Archives Depot, Cape Town: 1/27 Central Board of Commissioners of Public Roads (1843-1859): CRB 71 Letters Received.

Smith R 1857c. (Letter to Central Roads Board, 30 December 1857). Cape Archives Depot, Cape Town: 1/27 Central Board of Commissioners of Public Roads (1843-1859): CRB 71 Letters Received.

Smith R 1857d. Notes on planting. Cape Monthly Magazine 1, 327-33.

Smith R 1858. The Cape Flats. Cape Monthly Magazine 4, 303-308.

Smith R 1861. (Letter to Chief Commissioner of Roads, 26 May 1861). Cape Archives Depot, Cape Town: Public Works Department: PWD 246 Letters Received from Superintendent of Plantations.

Sonnerat P 1782. Voyage aux Indes Orientales et à la Chine, fait par ordre du roi, depuis 1774 jusq'en 1781. Paris. 2v.

Superintendent of Woods and Forests 1883. Report for the year 1882. Cape of Good Hope: Annexures to the Votes and Proceedings of the House of Assembly: G105-1883.

Thom H B (ed) 1954. Journal of Jan van Riebeeck. Balkema, Cape Town. 3 v.

Valentijn F 1726. Beschryvinge van die Kaap der Goede Hoop. Van Braam, Dordrecht.

Willis M 1949. By their fruits: a life of Ferdinand von Mueller, botanist and explorer. Angus and Robertson, Sydney.

Faunal invasions and their control

The invasion of untransformed ecosystems by introduced animals has received scant attention from southern African zoologists. This is thought by **Brooke and co-authors** to accurately reflect the limited extent of the problem posed by these species in the region. If this is the case, it introduces a host of ecologically interesting speculations. Why for instance, have the same mammalian vertebrates been so much more successful as invaders in Australia? Why do certain species invade successfully on the offshore islands but fail to establish on the mainland? Is there any significance in the preliminary observation that intra-continental translocations seem to have given rise to more successful 'invasions' than introductions from other continents?

The account of **Bruton and Van As** indicates that it is in southern Africa's aquatic ecosystems where animal invasions have been significant. This may simply be an artefact of the increased number of introductions that have been made to these ecosystems, or it may be that the aquatic systems of the region are more susceptible to invasions than the terrestrial ones. If they are, can this be attributed to the 'insular' nature of aquatic systems in this predominantly arid region, or to the high levels of artificial disturbance to which these systems have been subjected in recent years?

One area where the information on animal invasions is undoubtedly deficient is that of invertebrate species in terrestrial ecosystems. A few of the better known examples, such as the snail *Theba pisana* and the ant *Iridomyrmex humilis*, are mentioned in other chapters (Macdonald and Richardson; Breytenbach). A chapter devoted entirely to this important group should have formed an integral part of this volume but the knowledge on invertebrate invasions was considered too incomplete to justify such a review. Rectifying this omission is one of the major challenges facing zoologists in the field of alien invasions of southern Africa.

Chapter 4

Faunal invasions of aquatic ecosystems in southern Africa, with suggestions for their management

M N Bruton* and J van As†

*J L B Smith Institute of Ichthyology and †Rand Afrikaans University.

Introduction

Whereas the problem of terrestrial invasives in South Africa has long been realized, that of aquatic invasives has largely been ignored. Because some aquatic invasives, here and elsewhere, have had deleterious impacts, there has been a change of attitude, but new problems are likely to arise. Renewed interest in aquaculture, associated with the need to feed increasing human populations, has resulted in further introductions and indigenous translocations. The rapid development of the aquarium trade and improved methods of transporting live aquatic animals have increased the risk of invasion from this source. Furthermore, the plethora of intercatchment connections via tunnels, pipes and canals has facilitated many translocations. The problem of aquatic invasives is thus increasing in scope and urgent steps will have to be taken to contain it.

Freshwater ecosystems of southern Africa

Southern Africa is an arid subcontinent with a mean annual precipitation (MAP) of about 500 mm (Alexander 1985), which is low compared to the world average of 860 mm (Noble and Hemens 1978). Most rain falls near the eastern and southern coasts but the western half of the region receives less than 300 mm yr^{-1} (Davies and Day 1986). Furthermore, stable, dry high-pressure atmospheric systems persist for long periods over the major part of the subcontinent, producing little rain but causing high evaporation rates. The precipitation which does fall is highly seasonal and has to satisfy high soil moisture deficits before runoff takes place. On average over the region, the proportion of MAP which emerges from catchments as streamflow is very low (8,6%) compared to that in high latitude countries (25-75%) (Alexander 1985). High ratios of streamflow to rainfall occur only on the uplands of the eastern seaboard and southern and western coastal mountains, where the ratio may exceed 50%. Mean annual streamflow in South Africa increases as a power function of annual rainfall in the catchments (Midgley and Pitman 1969). The extensive central plateau, which represents more than half the surface area of the country, receives only 27% of the MAP, most of which falls in the extensive catchment of the Orange/Vaal system (650 000 km^2) (Davies and Day 1986).

Several coastal lakes are found along the south-west, south and east coasts of the subcontinent, and floodplains, vleis, marshes and bogs are mainly scattered in the moist eastern regions. All large rivers and many small rivers have been impounded, with the five largest man-made lakes each having an area of over 100 km^2 (Davies and Day 1986). Except for spillage from dams, which is becoming rare as more water is removed for agricultural, domestic and industrial usage, most of the flow in these rivers is regulated. In order to convey water from the moist east to the arid west, or from the Orange-Vaal system into adjacent catchments, numerous intercatchment water transfers have been made.

Southern Africa can be divided into three broad ichthyofaunal provinces: (a) the subtropical subtraction zone at the southern end of the African tropics; (b) a temperate transitional zone; and (c) a Cape ichthyofaunal region. The latter is the only province which lies wholly in southern Africa, and consequently contains many endemics (Skelton 1980). Several of these endemics are rare or endangered (Skelton 1977). The subtropical subtraction and temperate transitional zones also contain some endemics or fishes which are rare or endangered in southern Africa, but the incidence is lower (Skelton 1980). In addition, there is an archipelago of montane localities which extends from the south-west Cape through east Africa which has a relatively high level of endemicity of fishes, other animals and plants (Skelton 1980).

The protection of endemics and rare and endangered species should obviously be a priority when

Address for correspondence: M N Bruton, JLB Smith Institute of Ichthyology, Private Bag 1015, Grahamstown, 6140.

considering the management of invasive animals. A real threat exists in the Cape ichthyofaunal region due to the widespread introduction of trout and bass in the upper and lower reaches of rivers, and the increasing colonization of sharptooth catfish *Clarias gariepinus* in the middle reaches. This problem requires urgent attention. The limited but highly varied aquatic ecosystems of southern Africa are thus subject to increasing pressures of human use and many are subject to progressive transformation.

The geographical area considered here as 'southern Africa' is that part of the African continent south of and including the Cunene-Zambezi catchments (but excluding the Shire River), which is widely regarded as a more or less distinct zoogeographical area by aquatic biologists (eg Roberts 1975; Bowmaker et al 1978). The movement of an aquatic animal within this area is regarded as a translocation whereas one from outside is regarded as an alien introduction. An invasive species is defined here as an

Table 1. Summary of the number of alien and translocated indigenous aquatic animals in southern Africa (from Bruton and Merron 1985).

Taxonomic group		Alien species	Translocated indigenous species
Ciliophora	ciliates (1)	2	0
Cnidaria	coelenterates (2)	1	0
Platyhelminthes	flatworms (3)	1	0
Arthropoda	arthropods (4)	6	0
Mollusca	molluscs (5)	7	0
Chordata			
Osteichthyes	fishes (6)	20	44
Amphibia	frogs (7)	0	2
Reptilia	reptiles (8)	1	1
Aves	birds (9)	5	2
	Total	43	49

Alien species:

(1) *Ichthyophthirius multifiliis, Trichodina acuta,*

(2) *Craspedacusta sowerbyi,*

(3) *Bothriocephalus acheilognathi,*

(4) *Argulus japonicus, Carcinus maenas* (M), *Paulinia acuminata, Cyrtobagous salviniae, C singularis, Neochetina eichhorniae,*

(5) *Thais haemostoma* (M), *Bedeva paivae* (M), *Physa acuta, Biomphalaria* sp, *Helisoma duryi, Lymnaea columella, Mytilus galloprovincialis* (M),

(6) *Salmo gairdneri, S salar, S trutta, Salvelinus fontinalis, Carassius auratus, C carassius, Cyprinus carpio, Tinca tinca, Gambusia affinis, Poecilia reticulata, Xiphophorus helleri, Lepomis cyanellus, L macrochirus, Micropterus dolomieu, M punctulatus, M salmoides, Perca fluviatilis, Serranochromis* cf *steindachneri, Serranochromis robustus robustus, Tilapia rendalli rendalli,*

(8) *Chrysemys scripta,*

(9) *Anas platyrhynchos, A rubripes, Cairina moschata, Cygnus olor, Tadorna tadorna.*

M = marine

Translocated indigenous species:

(6) *Protopterus annectens brieni, Marcusenius macrolepidotus, Petrocephalus catostoma, Kneria auriculata, Alestes lateralis, Hydrocynus vittatus, Rhabdalestes maunensis, Hemigrammocharax machadoi, Barbus aeneus, B anoplus, B barnardi, B bifrenatus, B eutaenia, B gurneyi, B mattozi, B natalensis, B tangandensis, Coptostomabarbus wittei, Labeo capensis, L cylindricus, L umbratus, Mesobola brevianalis, Austroglanis sclateri, Clarias gariepinus, Eutropius depressirostris, Synodontis leopardinus, S macrostigma, S woosnami, Nothobranchius orthonotus, Oreochromis andersonii, O macrochir, O mossambicus, Pharyngochromis darlingi, Pseudocrenilabrus philander, Serranochromis codringtoni, S giardi, S macrocephalus, S robustus jallae, S thumbergi, Tilapia rendalli swierstrae, T sparrmanii, Mugil cephalus, Myxus capensis, Sandelia capensis,*

(7) *Xenopus laevis, Phrynomerus bifasciatus,*

(8) *Crocodylus niloticus,*

(9) *Aythya fuligula, Cygnus atratus.*

alien or translocated species which has established self-sustaining populations in natural or seminatural communities. Invasives may be alien species introduced from remote sources, or translocated indigenous species which are found naturally within southern Africa but were translocated by man into catchments in which they were not naturally distributed. Introductions may be intentional or unintentional. Intentional introductions are those in which a species is introduced into the subcontinent for a particular purpose eg to stock a seminatural or natural water body for sport fishing or biocontrol. Unintentional introductions include those in which the species is accidentally brought into the subcontinent (eg parasites on fishes) or the species was purposefully introduced into a captive situation in the subcontinent and then escaped into the wild (eg aquarium or aquaculture species).

Extent of the invasions

According to current knowledge, 92 invasive animals have established populations in marine and freshwater aquatic ecosystems in southern Africa, of which 43 are alien and 49 translocated (Table 1). The coypu *Myocastor coypus*, which was reported by Bruton and Merron (1985) to have escaped from fur farms, has apparently been eradicated in the wild and is no longer regarded as an invasive species. The majority of the invasives (69,6%) are fishes. Four of the alien invasives are parasites. Another 21 species are suspected to be successful invaders and a further 19 alien aquatic species which occur in captivity in the subcontinent are regarded as a potential threat to indigenous species or ecosystems if they escape (Bruton and Merron 1985).

South Africa and southern Africa have a higher proportion of alien species than some other continental countries (USA, Canada, Mexico) but a lower proportion than islands (New Zealand, Hawaii, Puerto Rico, all of which have a depauperate indigenous ichthyofauna) (Table 2). The proportion of alien species in Australia is higher than in southern Africa but lower than in South Africa.

Alien fishes were first introduced into southern Africa over 260 years ago and are now widely distributed. The most widespread species is the common carp *Cyprinus carpio* which occurs in 10 of the 13 major drainage basins, with largemouth bass (9 basins), bluegill (8), mosquitofish and smallmouth bass (7), rainbow trout (6), and brown trout and goldfish (5) also very common (Table 3). The number of alien fish species is higher in the more southerly

Table 2. The numbers of alien and translocated indigenous fish species in various countries or continents, and these numbers expressed as a percentage of the total number of indigenous species of freshwater fishes ie those which spend their entire life cycle in fresh water (data from Scott and Crossman 1973; Courtenay and Stauffer 1984; W R Courtenay personal communication; P R Miller personal communication and M N Bruton unpublished). Here, southern Africa includes South Africa, Transkei, Ciskei, Venda, Bophuthatswana, Lesotho, Swaziland, Botswana and South West Africa/Namibia but not Zimbabwe or Mozambique.

	Indigenous species I	Alien species	% of I	Translocated indigenous species	% of I
Southern Africa	176	20	11,4	44	25,0
South Africa	107	20	18,7	20	18,7
Canada	177	9	5,1	37	20,9
Continental USA	665	42	6,3		
Mexico	275	26	9,5	29	
Australia	145	22	15,2		
New Zealand	27	18	66,7		
Hawaii	10(a)	31	310		
Oceania	?	24			
Puerto Rico	3	23	766,7		

(a) there are no indigenous fishes in Hawaii that are restricted to freshwater, but about 10 species spend much (but not all) of their lives in freshwater (W R Courtenay personal communication).

Table 3. Date of first introduction and distribution of alien fishes in southern Africa (from Crass 1964; Jubb 1965, 1967; Bell-Cross 1972, 1976; Bowmaker et al 1978; Gaigher et al 1980; Kleynhans 1984; Schrader 1985; Bruton and de Moor in press; personal communications from J Alletson, O Bourquin, M M Coke, I de Moor, I G Gaigher, C J Kleynhans, J Minshull and T Pike, and collection records of the JLB Smith Institute of Ichthyology, Albany Museum and National Museum, Bulawayo, Zimbabwe.

Species	Common name	Date of first introduction into southern Africa	Distribution by drainage basin													
			South coastal*	Orange/Vaal	Natal; Umtamvuna to Mkuzi	Pongolo & Incomati	Limpopo	Sabi & Lundi	Buzi & Pungwe	Lower Zambezi**	Middle Zambezi**	Upper Zambezi**	Okavango	Cunene	Dams in SWA/Namibia	Total
Salmo gairdneri	rainbow trout	1897	X	X	X	X	X		X							6
Salmo salar	salmon	1896	X													1
Salmo trutta	brown trout	1890	X	X	X	X			X							5
Salvelinus fontinalis	brook trout	1890	X		X		X		X							4
Carassius auratus	goldfish	before 1726	X	X	X		X				X					5
Carassius carassius	Crucian carp	?	X													1
Ctenopharyngodon idellus	grass carp	1967		X	X	X										3
Cyprinus carpio	common carp	before 1726	X	X	X	X	X	X		X	X		X		X	10
Hypophthalmichthys molitrix	silver carp	1977			X											1
Tinca tinca	tench	1910	X		X					X						3
Gambusia affinis	mosquitofish	1936	X	X	X		X	X		X	X					7
Poecilia reticulata	guppy	1980s			X											1
Xiphophorus helleri	swordtail	before 1974			X	X										2
Lepomis cyanellus	green sunfish	?	X													1
Lepomis macrochirus	bluegill	1938	X	X	X	X	X	X		X	X					8
Micropterus dolomieu	smallmouth bass	1937	X	X	X	X	X			X	X					7
Micropterus punctulatus	spotted bass	1939			X		X									2
Micropterus salmoides	largemouth bass	1928	X	X	X	X	X	X		X	X					9
Perca fluviatilis	perch	1915	X	X	X											3
Oreochromis niloticus × *O aureus*	hybrid	1978			X											1
Serranochromis sp	Cunene happy	1970s												X		1
S robustus jallae	nembwe	1962			X			X								2
S robustus robustus	sungwa	1960s				X										1
Tilapia r rendalli	redbreast tilapia	1970s												X		1
		Total	14	10	18	8	10	5	3	6	6	0	1	2	2	

*Olifants River in western Cape to Umtamvuna River in Natal

**Lower Zambezi: mouth to Cahora Bassa. Middle Zambezi: Cahora Bassa to Victoria Falls. Upper Zambezi: Victoria Falls upstream drainage basins and decreases towards the north, which is in inverse relationship to the diversity of the freshwater fish fauna as a whole (Bowmaker et al 1978). The higher numbers of alien fish species in the south coastal and Natal drainages may be a result of the early arrival of British settlers there and their tendency to 'improve' depauperate fish faunas by introducing species from the Northern Hemisphere. Alternatively, the more complex fish communities to the north could have resisted alien invasions more effectively. Our knowledge of fish distribution in the more southerly drainages is also better.

The Eerste River in the western Cape, which is adjacent to a long established nature conservation station, is inhabited by six species of alien fishes. The Olifants (five species) and Berg Rivers (five) in the western Cape are also heavily invaded. These river systems have the highest number of endangered indigenous fish species in the subcontinent, many of which are now at risk of population reduction or extinction because of the invasions (Skelton 1977). Ten of the 26 fish species (38%) in the Great Fish River in the eastern Cape are aliens (Laurenson and Hocutt 1986).

The larger river systems to the north have relatively few alien fishes (Table 3). No alien fishes have been recorded from the Upper Zambezi River. The Okavango system has no invasive fishes except for translocated *Oreochromis mossambicus* (Skelton and Merron 1984) and common carp *Cyprinus carpio* (Schrader 1985), both of which occur in the Omatako Omuramba, a 500-km ephemeral southern tributary of the Okavango River. No alien fishes occur in the coastal lakes or rivers of Maputaland (Bruton and Kok 1980) although an unsuccessful introduction of *Micropterus salmoides* was made into Lake Sibaya in 1935 (Harrison 1936) and *Barbus gurneyi* were translocated from central Natal to a small lake near Kosi (Bruton and Kok 1980).

Alien fishes frequently colonize turbid impoundments in southern Africa and may form an important component of the fish fauna. The National Institute for Water Research (1985) estimated that the standing stock and biomass of adult common carp in Hartbeespoort Dam (Transvaal) in 1982 were 243 000 (95% confidence limits 217 247-281 278) and 580 t. Cochrane (1985) estimated that the potential long-term yield of carp in Hartbeespoort Dam was 375 tons yr^{-1} compared to 170 tons yr^{-1} for *Oreochromis mossambicus* and 25 tons yr^{-1} for *Clarias gariepinus*, the most important indigenous species. Carp also constituted 74% of catches by anglers in Hartbeespoort Dam (Cochrane 1986). Cadieux (1980) reported that the fish most commonly caught by anglers in the Transvaal were *Oreochromis mossambicus* (33,6% of total catch) and common carp (28,9% or 2,74 million fish). In Lake Midmar, an impoundment in central Natal, alien fishes constituted 97,3% of angler catches, with bluegill comprising 65,2% and carp 30,8%. Small numbers of bass and brown trout were also caught (Breen 1983). Carp were also common in impoundments on the Orange River. They comprised the largest portion of the ichthyomass in Lake Verwoerd in five of six sampling years and were frequently caught by anglers (Hamman 1980). Carp were relatively uncommon in Lake le Roux possibly due to the steep sides of this impoundment, but were often caught by anglers (Allanson and Jackson 1983). Brown trout occur in Lakes Midmar, le Roux and Verwoerd but are rare (Hamman 1980; Allanson and Jackson 1983; Breen 1983).

Origin and dates of introduction

The origin of all the alien species has not been firmly established, but a combination of facts and educated guesses indicates that numerous species came from Europe (19) with lesser numbers from South America, Africa to the north of the Cunene/Zambezi catchments, North America and Asia (Table 4). Most alien species were introduced in the 20th century (20), of which eight were introduced in the past 35 years. The majority of the European and North American species were introduced in the 19th century, with all South American and African aliens being introduced in the 20th century. All three Asian species were introduced in the 18th century.

Reasons for the introductions

Aquatic animals have been introduced into or translocated within southern Africa for a variety of reasons (Table 5). While some invasions were caused accidentally by man, or cannot be assigned to a particular cause, the causes of others can be clearly identified. The sequence of invasions in South Africa is typical of that of Commonwealth or ex-Commonwealth countries ie initial introduction of goldfish *Carassius auratus* for ornament and common carp *Cyprinus carpio* for food, followed by introductions of sport and forage fish such as trout, bass and bluegill, then of mosquitofish *Gambusia affinis* for biocontrol and, most recently, of various species associated with the aquaculture and aquarium trade industries and the control of aquatic weeds.

The most frequent reasons for the intentional in-

Table 4. Summary of the origin and known or suspected year of introduction of alien aquatic species into southern Africa (from Bruton and de Moor in press).

	No of species				
	Year of introduction into southern Africa				
	1700 – 1799	1800 – 1899	1900 – 1949	1950 – 1985	Total
North America	0	5	1	0	6
South America	0	0	4	4	8
Europe	2	13	3	1	19
Asia	3	0	0	0	3
Africa	0	0	4	3	7
Total	5	18	12	8	43

troduction of aliens are sport fishing and biological control. Although numerous alien species have been introduced for aquaculture (Bruton and Merron 1985), only three are known to have established breeding populations in natural waters (black-brown mussel *Mytilus galloprovincialis*, which may also have been imported accidentally on ships' hulls, common carp *Cyprinus carpio* and rainbow trout *Salmo gairdneri*). The latter two species have also been introduced for sport fishing. The most frequent reasons for the intentional translocation of indigenous species are sport fishing, the stocking of impoundments and increasing the range of rare species. Overall 66 species (48,5%) have been introduced or translocated unintentionally. Because of the high proportion of accidental introductions, it is clear that attempts to control the importation of invasive animals will have to concentrate not only on the organisms themselves but also on the methods of introduction.

Intercatchment connections represent an important means of translocating indigenous and alien fishes as well as aquatic invertebrates in southern Africa. Almost all major river catchments are connected to adjacent catchments by tunnels, canals or pipes, and in some cases entirely different drainage basins are interconnected eg the Vaal to the Tugela and the Orange to the Fish River catchments. The Orange-Fish connection has introduced at least five invasive fish species (Cambray and Jubb 1977; Laurenson and Hocutt 1986) into the Great Fish River in the south coastal drainage system of the Cape Province, a separate zoogeographical subregion renowned for the high endemicity of its fauna (Skelton in press).

The Eastern National Water Carrier, currently under construction in South West Africa/Namibia, will link the Okavango River with man-made lakes near Windhoek. The connection may introduce fish species from the Okavango River to drainages in the central districts of South West Africa (Skelton and Merron 1984; Schrader 1985). Fortunately upstream migrations of fishes into the Okavango system are extremely unlikely as the northern part of the connection to the Okavango River consists of a pipe over 200 km long (Skelton and Merron 1984).

The Cuvelai and Cunene River systems in the north of South West Africa were linked artificially by the Ombalanta Canal in the mid-1970's and Schrader (1985) reports that at least 51 fish species could be exchanged between the two systems.

Thus far no alien aquatic organisms introduced into southern Africa for experimental purposes have escaped into the wild, as has occurred in Florida, United States of America (Courtenay and Robins 1973). The situation needs to be monitored, however, as no legislation exists which requires research organizations to ensure that holding tanks are situated so that accidental release into natural waters is impossible. The clawed toad *Xenopus laevis*, which was extensively used for human pregnancy tests in the past, was previously cultured and exported (Hey 1945) and has been translocated in association with fish culture activities (Cape Provincial Administration 1944-1945).

Impact of invasive aquatic animals in southern Africa

One of the major impacts of invasive aquatic animals is through the introduction of alien parasites and diseases which threaten natural communities and aquaculture and aquarist activities. Common carp have been implicated in the introduction of at least

Table 5. Reasons for the introduction of alien and translocation of indigenous aquatic species in southern Africa. As some invasive species were introduced or translocated for more than one reason, or for different reasons to different places, the totals here do not correspond to those in Table 1. Species whose taxonomy and distribution are poorly known but which are thought to be alien invasives in southern Africa (Bruton and Merron 1985, Table 3) are also included here.

Reason for introduction or translocation	No of invasive species			
	Alien	Indigenous	Total	% of total
Unintentional				
Accidental	27 (14 parasites)	3	30	22,1
Unintentional via inter-catchment connections	0	29	29	21,3
Unintentional via aquarium trade	5	2	7	5,1
Intentional				
Sport fishing (including forage fish)	14	11	25	18,4
Aquaculture (including food for fish)	3	9	12	8,8
Stocking man-made lakes and farm dams	1	11	12	8,8
Increasing the range of rare or desirable species (other than those above)	0	11	11	8,1
Biological control	6	3	9	6,6
Totals	56	79	135	

seven alien or suspected alien fish parasites into southern Africa (*Ichthyobodo necator*, *Chilodonella cyprini*, *C hexasticha*, *Apiosoma piscicola*, *Trichodina acuta*, *T nigra* and *Trichodinella epizootica*). Grass carp *Ctenopharyngodon idellus*, introduced into the eastern Transvaal for fish culture, were responsible for introducing the cestode *Bothriocephalus acheilognathi* from West Germany in 1975 (Brandt et al 1981). *Bothriocephalus acheilognathi* has since spread to most parts of this province and to the Cape (Boomker et al 1980; van As et al 1981). *Bothriocephalus acheilognathi* is particularly dangerous as it can infect a wide range of cyprinid fishes.

Furthermore, its intermediate hosts are planktonic copepods, to which it is not species-specific, with the result that the parasite will continue to spread throughout the subcontinent via local copepods. The fishlouse *Argulus japonicus*, an ectoparasitic crustacean which was probably introduced on goldfish, affects a broad spectrum of indigenous fish (Kruger et al 1983). During the summer of 1983 *A japonicus* occurred in very large numbers in some impoundments on the Vaal River (Kruger et al 1983). Both *B acheilognathi* and *A japonicus* have a significant impact on fish condition and Brandt et al (1981) regard bothriocephalosis as a threat to the aquaculture industry in South Africa.

Ichthyophthirius multifiliis, the pathogen responsible for white spot disease, has been recorded to cause mortalities in wild as well as pond and aquarium fish in South Africa (du Plessis 1952; van As and Basson 1984; van As et al 1984). Low infestations of the encysted cells of *I multifiliis* have little or no effect on the host fish, are not easily detectable and cannot be eradicated with the chemicals usually applied as treatment for fish infested with ectoparasites, such as formalin or malachite green. This parasite is therefore frequently translocated with aquarium fishes, which is its probable mode of importation. Another alien ciliophoran which owes its worldwide distribution to the importation of carp is *Trichodina acuta*. This parasite is now widespread in southern Africa and was one of the trichodinids responsible for mass mortalities of cultured *Oreochromis mossambicus* and *Cyprinus carpio* fry in the Transvaal (van As et al 1984).

Examples of the impact of alien fish parasites on fishes in southern Africa are given in Table 6. Parasites carried by aquatic invasives may also infect terrestrial mammals. *Lymnaea columella* is the intermediate host of the liver flukes *Fasciola hepatica* and *F gigantica* which infest domestic stock and, rarely, man. *Lymnaea columella* occupies a part of the habitat which is not colonized by the indigenous species *L natalensis* (the water:land interface) and may as a result come in closer contact with the primary host of the liver flukes (mammals) than *L natalensis*. In the mangrove swamp at Sodwana Bay in Zululand the alien species occurred on the trunks of mangrove trees whereas the indigenous species was recorded underwater in beds of *Utricularia* species (Bruton and Appleton 1975). In midland Natal and on the Pongolo floodplain *L columella* occurred on the water fringe whereas *L natalensis* was always found underwater (C C Appleton personal communication). The red-eared terrapin *Chrysemys scripta* (and other pet terrapins) has been implicated in over 300 000 cases of human salmonellosis in the United States of America (Wells et al 1973), but no incidences have as yet been reported in southern Africa.

A further alien invertebrate that has been found in South Africa is the hydrozoan medusa *Craspedacusta sowerbyi*. This species was recorded in zooplankton samples from a large artificial impoundment in the Natal midlands (Lake Midmar) in 1978 by Rayner (1981), but has not been reported since. *Craspedacusta sowerbyi* is indigenous to the Amazon basin in Brazil and is believed to have been introduced in its polyp form with aquatic plants. The impact of this species is not as yet known.

Seventeen of the 20 alien fish species which have established populations in southern Africa are internationally regarded as either pests, ie they are known to cause damage to natural communities (five species), or as nuisance animals, ie their disadvantages outweigh their advantages (12) (Welcomme 1984; Bruton and Merron 1985; Table 7). Several species are beneficial in one circumstance but harmful in another. For example, the rainbow trout supports an important aquaculture industry and recreational fishery, but has been implicated, with largemouth bass, in the reduction or local extinction

Table 6. Examples of the impact of alien parasites on indigenous and alien fishes in southern Africa.

Parasite	Impact	References
Chilodonella hexasticha	mortalities of various cichlid and cyprinid fishes	du Plessis (1952); Lombard (1968); Paperna and van As (1983); van As et al (1984)
Ichthyophthirius multifiliis	mortalities of various cultured and aquarium fishes	du Plessis (1952); van As and Basson (1984); van As et al (1984)
Trichodina acuta	mass mortalities of *Oreochromis mossambicus* and *Cyprinus carpio* fry	van As et al (1984)
Argulus japonicus	mortalities of various indigenous fish species	Kruger et al (1983)
Argulus japonicus	mortalities of *Salmo gairdneri*	van As (in press)
Bothriocephalus acheilognathi	mortalities of *Cyprinus carpio* and *Barbus kimberleyensis*	Brandt et al (1981)

Table 7. Established invasive aquatic animals (excluding parasites) in southern Africa which are internationally regarded as pests or nuisance animals. A = alien, T = translocated indigenous species (data from Paperna 1980; Newbery 1984; Taylor et al 1984; Welcomme 1984; O Bourquin personal communication and W R Courtenay personal communication).

International status categories and species in each category	Original reason for introduction or translocation
Pests	
Chrysemys scripta (A)	aquarium trade
Oreochromis mossambicus (T)	fish culture
Poecilia reticulata (A)	mosquito control
Tilapia rendalli (T)	weed control
Xiphophorus helleri (A)	aquarium trade
Nuisance animals	
Carassius auratus (A)	aquarium trade
Cyprinus carpio (A)	fish culture
Gambusia affinis (A)	mosquito control
Lepomis macrochirus (A)	sport fishing
Micropterus dolomieu (A)	sport fishing
Micropterus punctulatus (A)	sport fishing
Micropterus salmoides (A)	sport fishing
Perca fluviatilis (A)	sport fishing
Salmo gairdneri (A)	sport fishing
Salmo trutta (A)	sport fishing
Salvelinus fontinalis (A)	sport fishing
Tinca tinca (A)	sport fishing

of eight minnows (*Barbus* and *Oreodaimon* species, family Cyprinidae) and one kneria (*Kneria auriculata*, Kneriidae), which are regarded as rare or endangered in South Africa (Skelton 1977). The range of several stenotypic endemics in the western Cape, eg *Barbus burgi*, has been drastically reduced by the smallmouth bass (Gaigher et al 1980). A list of documented examples of alien fish predators affecting indigenous fish populations is given in Table 8.

Alien fishes have also been reported to alter habitats. They may decrease water quality, as in the carp which disturbs bottom sediments during feeding and increases the turbidity (Bruton 1985; Schrader 1985), but may at the same time release phosphates into the water (Welcomme 1984). Translocated *Tilapia rendalli swierstrae* have been shown to decimate communities of submerged aquatic macrophytes in Zimbabwe and Natal (Junor 1969; Wager and Rowe-Rowe 1972).

Aliens may compete with indigenous species for food and space and disrupt breeding patterns and parental care of offspring. An example of the latter occurred at Molapo Oog, a dolomitic sinkhole in the northern Cape. This waterbody was previously dominated by the indigenous cichlids, *Tilapia sparrmanii* and *Pseudocrenilabrus philander*, which nested on lightly vegetated and open substrata respectively. The introduction of black bass has resulted in a substantial reduction of cichlid populations, and nesting now occurs in a suboptimal habitat among dense vegetation (A J Ribbink personal communication). Noble and Hemens (1978) reported that overpopulation by common carp has reduced the abundance of the indigenous moggel *Labeo umbratus* in the Vaal River system.

Pike and Tedder (1973) and Gephard (1977) found that rainbow trout both prey upon and compete for food with the rare indigenous minnow *Oreodaimon quathlambae* in Lesotho. The National Institute for Water Research (1985) recorded a considerable dietary overlap between carp and the indigenous species *Barbus mattozi* and *B marequensis* in Hartbeespoort Dam. Mayekiso (1986) demonstrated an overlap between the diet of the eastern Cape rocky *Sandelia bainsii* and rainbow trout and largemouth bass in the Tyume River, Eastern Cape Province. He concluded further that habitat alterations and competition for food and space from alien fish species jeopardized the survival of *S bainsii*, a rare and threatened species (Skelton unpublished). *Sandelia bainsii* is also threatened by competition for food with the translocated omnivore *Clarias gariepinus* (Mayekiso 1986).

The sharptooth catfish *Clarias gariepinus*, which has been translocated into the Cape coastal drainage systems by the Orange-Fish tunnel, is an omnivorous predator capable of intensive predation on invertebrate and fish prey (Bruton 1979). The intensity of predation by sharptooth catfish on aquatic insects was shown by Weir (1972), who compared the invertebrate fauna of two similar man-made pools, one with and one without the predator, in Zimbabwe. Insect species diversity was reduced by 78% (31-7) in the Coleoptera and by 66% (15-5) in the Hemiptera.

Hybridization may be a serious threat posed by invasive fishes. In contrast to many other vertebrates, interspecific breeding in fishes often produces viable offspring. Hybridization occurs among wild fish species in some countries (eg United States of America, Courtenay and Stauffer 1984) but has not as yet been reported in southern Africa. Schrader (1985) has warned that there is a strong possibility that translocated *Oreochromis mossambicus* may interbreed with indigenous *O macrochir* and *O ander-*

Table 8. Examples of the impact of predation by alien fishes on indigenous fish species in southern Africa.

Alien species	Impact	River system	References
Micropterus salmoides	local extinction of *Sandelia capensis*	Paardevlei (western Cape)	Harrison (1936)
Micropterus dolomieu, M salmoides	local extinction or reduction of *Barbus calidus, Barbus phlegethon, Austroglanis gilli*	Olifants (west coast)	Barnard (1943); Jubb (1965, 1967); Gaigher (1973); Skelton (1977); Gaigher et al (1980)
M dolomieu, Salmo gairdneri, Salmo trutta	local extinction of *Barbus burgi*	Berg	Skelton (1977); Gaigher et al (1980)
M dolomieu, S gairdneri	local extinction of *Barbus burgi*	Eerste	Skelton (1977); Gaigher et al (1980)
S gairdneri	local extinction of *Oreodaimon quathlambae*	Umkomazana	Jubb (1966); Skelton (1977); Welcomme (1984)
M salmoides Micropterus punctulatus, S gairdneri	reduction of *Barbus trevelyani*	Buffalo, Keiskamma	Jubb (1965, 1967); Gaigher (1975) Skelton (1977); Gaigher et al (1980)
M dolomieu, S gairdneri	reduction of *Barbus treurensis*	Blyde, Limpopo	Skelton (1977); Kleynhans (1984, 1985)
S gairdneri, S trutta	reduction of *Kneria auriculata*	Incomati	Skelton (1977), Kleynhans (1979, 1984, 1985)
M dolomieu, M salmoides, S gairdneri	reduction of *Barbus burchelli*	Bree	Skelton (1977); Gaigher et al (1980); Cambray and Stuart (1985)

sonii in the northern rivers of South West Africa/ Namibia. There is also a risk of interbreeding between translocated and local stocks of *Tilapia* species and *Oreochromis* species. Hybrid tilapia or alien tilapia which may hybridize with local species have already been released into two small dams in Natal (T Pike personal communication). *Oreochromis aureus* and *O niloticus* were transported from the Fisheries Development Corporation hatchery at Amatikulu to a small dam in northern Natal in 1980. This stock was, however, destroyed by Natal Parks Board when the fishes showed poor growth. In 1982 the Tongaat Sugar Company released *O aureus* and *O niloticus* and their interhybrids into the reservoir of the Dudley Pringle Dam in the Wewe River catchment after an aquaculture venture had failed. The indigenous species *O mossambicus* occurs in this system but it is not known whether hybridization has occurred. Fortunately the Wewe system is very small and it is unlikely that the hybrids will distribute themselves more widely without man's assistance. Red tilapia (*O aureus* x *mossambicus* x *niloticus* hybrids) are kept at the Natal Parks Board Warm Water Fish Hatchery on the Umgeni River in Natal, but very strict safeguards are enforced to prevent their escape into the wild.

The mallard duck *Anas platyrhynchos* was introduced into southern Africa in the 1940's (Liversidge 1985). It is now widespread as a domestic animal in parks and on farms but has failed to establish viable feral populations in southern Africa. They are, however, capable of hybridizing with two indigenous South African duck species (*Alopochen aegyptiaca* and *Anas undulata*) in captivity, with only the *A platyrhynchos* x *A undulata* cross being fertile (Gray 1958).

Alien organisms rarely affect indigenous species in isolation, but are usually part of an overall environmental change which causes stress to local animal species. Predation by trout on minnows is therefore likely to be exacerbated by factors such as decreased water quality and manipulated water flows. The long-term solution to the problem of predation (or of parasitism, hybridization or competition) by invasive organisms is both to reduce their populations and to restore the aquatic system to a healthy state.

Effects of invasives on nature conservation values

It is useful to assess the likely impact of aquatic invasive animals in southern Africa on the achievement of the aims of the world conservation strategy. These are the maintenance of essential ecological processes and life support systems, the preservation of genetic diversity and the sustainable utilization of species and ecosystems (IUCN, 1979).

Aquatic invasive animals may have an impact on all three of these objectives. Changes to habitats, such as the removal of plants (Junor 1969) or increased turbidity (Bruton 1985), may affect essential ecological processes such as photosynthesis, carbon production, feeding and breeding by invertebrates and fishes, and predation rates on invertebrates and vertebrates. Genetic diversity is affected by invasives if predation leads to the local extinction of aquatic animal species (Table 8). The sustainable utilization by man of indigenous fish species in southern Africa has been detrimentally affected by alien fish parasites (Boomker et al 1980; van As et al 1981; Kruger et al 1983). In some cases indigenous species have been almost entirely replaced by predatory alien species, eg by bass in the Olifants River (Gaigher et al 1980) and in Paardevlei (Harrison 1936) in the western Cape. On the other hand, new fisheries have been created by stocking invasive pelagic species in large artificial impoundments ie Lake Kariba (Jackson 1966), or by stocking cold-tolerant fishes such as trout in high altitude streams which, for palaeohistorical reasons, have no large predatory fishes except *Sandelia bainsii* in the eastern Cape (Bruton and de Moor in press).

Management of faunal introductions

There are two aspects to the management of aquatic invasives (a) prevention of the further spread and amelioration of the impact of species already in the subcontinent, and (b) prevention of the importation or translocation, either intentional or unintentional, of further invasives.

The task of reducing the distribution and ecological impact of existing invasives will fall largely to the regional nature conservation authorities, and the priority here should be to determine which invasives are likely to cause the greatest harm and should therefore be most strictly controlled. Translocation through intercatchment connections needs to be controlled through the installation of grids and filters in canals and pipes designed to prevent the entrainment of fish eggs, juveniles and adults. The escape or release of aquarium fishes could be reduced by stricter legislation and a public education campaign. The introduction of aquatic animals for recreational fishing, aquaculture, stocking of impoundments or biocontrol should only be conducted after a careful evaluation of the potential impact of the candidate species has been made (see below).

Once an invasive species has become established, it is difficult to reduce its impact, but this may be attempted by selective removal of the invasive species and, possibly, stocking of indigenous species. In the Cape Province it has been proposed that the strict regulations protecting trout should be withdrawn, which may cause feral trout populations to decline if restocking does not take place (Skelton and Davies in press). Some invasive fishes thrive in disturbed habitats (Bruton this volume). Where possible, the obvious control method in this case is to restore the habitat to its original state so as to favour the indigenous species.

Controlling the future introduction of invasives will fall under the auspices of those bodies controlling imports and it is clear that considerable improvements can be made here. We have received numerous complaints from aquarium trade dealers and fish farmers that the import laws are vague, not based on scientific information, and are variably enforced. There is a clear need for the authorities to produce lists of species to distinguish between species which may be imported, those which should be prohibited and those which should be investigated before a decision is made.

Existing practices to prevent the introduction or translocation of fish parasites have generally failed, not only in southern Africa but also elsewhere. The only effective way in which the further spread of fish pathogens can be prevented is to control the importation and translocation of their hosts, especially alien fish species and their shipping waters.

A simple protocol for evaluating the desirability of introducing alien aquatic animals needs to be developed, based on that proposed for fishes by the International Council for the Exploration of the Sea, the European Inland Fisheries Advisory Commission, by Kohler and Stanley (1984) for the United States of America, and recently by the Committee for Inland Fisheries of Africa (CIFA 1985). This would provide a formal procedure for systematically making decisions about introductions on the basis of pertinent knowledge.

Because of the global movement of aquatic organisms and their established economic importance, it would be impractical to impose inflexible guidelines. Consequently, the protocol proposed by Kohler and Stanley (1984) has a tiered approach with several levels of review, each requiring progressively more information (Figure 1).

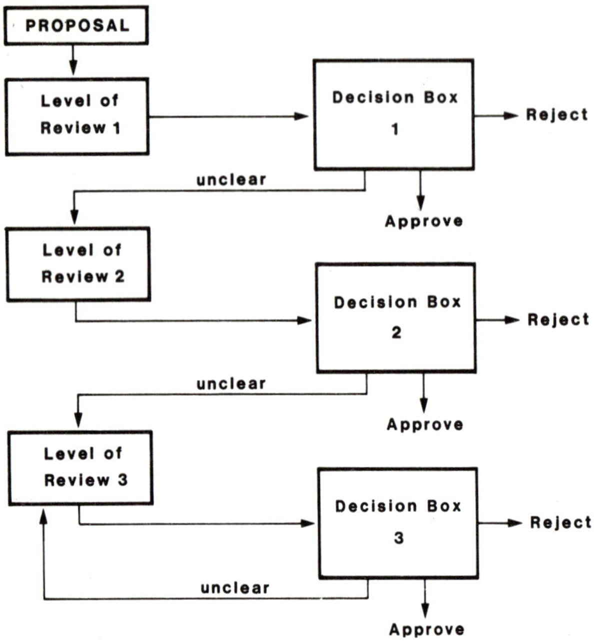

Figure 1. Diagrammatic representation of the review and decision model for evaluating proposed alien introductions (after Kohler and Stanley 1984).

This review and decision model involves the establishment of a committee of scientists who are qualified to evaluate proposals for the introduction of alien organisms. A person or organization intending to introduce an alien species is required to prepare a proposal which provides certain basic information (Table 9). On the basis of this proposal the members of the evaluation board answer a questionnaire (Table 10) which, when used with the model aids the decision on the case at hand. A computerized artificial intelligence system would lend itself to applications of this kind and its use in this connection is currently under investigation (Bruton unpublished).

The success of any evaluation system will depend ultimately on the quality of the database on the species concerned. It is essential, therefore, to build up a technical and bibliographic database on alien invasive species, as well as a reference collection, so as to assist in this process. There will always be considerable problems in evaluating large, multispecific consignments of aquatic organisms for the aquarium trade, but it is essential that more stringent controls are introduced.

Conclusions

Invasive animals provide both problems and opportunities for the managers of aquatic systems. In man-made or highly manipulated systems, or in areas where the provision of food or recreation is a higher priority than maintaining natural communities, the introduction of invasives would be justified on condition that the possibility of invasives deleteriously affecting adjacent natural systems has been ruled out. However, in natural or seminatural systems, any introduction, no matter how inconspicuous, is likely to cause changes. These changes are often difficult to detect or measure and we tend to see only the end product, eg a local extinction, of a long sequence of more subtle changes.

There are likely to be sharply increased demands on aquatic resources in southern Africa in future (Alexander 1985) and we cannot afford to allow invasive species to decrease the quality of water or of aquatic communities. Controls over invasive animals should be just as strict as those over chemical pollutants or physical alterations to aquatic systems.

Table 9. Questions which should be answered by a person or organization intending to introduce an alien aquatic organism into southern Africa (adapted from Kohler and Stanley 1984).

1. What alien species do you propose to introduce (common and scientific name)?
2. Where is its native range? What is the present range?
3. What is the purpose of the introduction?
4. Where, how many, and into what type of system would this alien species be introduced?
5. What precautions have or will be taken to ensure that the alien species is not harbouring diseases and parasites?
6. If the alien species is to be maintained in a closed system, what measures would be taken to guard against accidental escape to open water?
7. What is your knowledge of the acclimatization potential of the species?
8. What are the habitat and niche requirements of the species?
9. What is the reproductive potential and life history style of the species?

A bibliography of pertinent literature should be appended to the proposal and a preserved specimen should be sent for confirmation of identification to an approved research institution.

Table 10. Questionnaire for appraisal of introductions of alien aquatic organisms. Each member of an evaluation board circles the number most nearly matching his/her opinion about the probability of the occurrence of the event. If information is unavailable or too uncertain: "do not know" is marked (from Kohler and Stanley 1984; CIFA 1985).

	Response					
	No	Unlikely	Possibly	Probably	Yes	Do not know
1. Is the need valid	1	2	3	4	5	X
2. Is there no native species available that could serve the stated need?	1	2	3	4	5	X
3. Is the organism safe from overexploitation in its native range?	1	2	3	4	5	X
4. Are safeguards adequate to guard against importation of disease/parasites?	1	2	3	4	5	X
5. Would the introduction be limited to closed systems?	1	2	3	4	5	X
6. Would the organism be unable to establish a self-sustaining population in the range of habitats that would be available?	1	2	3	4	5	X
7. Would the organism have only positive ecological impacts?	1	2	3	4	5	X
8. Would all consequences of the introduction be beneficial to humans?	1	2	3	4	5	X
9. Is there a species synopsis and is it complete?	1	2	3	4	5	X
10. Does the database indicate desirability for introduction	1	2	3	4	5	X
11. Would benefits exceed risks?	1	2	3	4	5	X

It is also important to ensure that single interest groups, eg anglers or aquaculturalists, are not allowed to make introductions which are detrimental to the user community as a whole. It is essential that we set aside certain aquatic systems (or parts of them) as genetic reserves free of introduced and translocated species. Finally, it is necessary for in-depth conservation education programmes to be launched which highlight the potential dangers of invasive aquatic animals.

Inland waters have evolved over thousands or millions of years to reach their present state, and it is presumptuous of human society to judge that they can be 'improved' by introductions. Decisions on the introduction of aliens are often based on anticipated short-term benefits. There are sufficient examples of the disastrous consequences of unwise introductions in rivers and lakes around the world (eg Barel et al 1985) for us to adopt a cautious approach when evaluating future imports. There is also an urgent need for more field and experimental studies to be conducted on invasive aquatic animals, both in the original and the invaded country, so that we can avoid further mistakes.

Acknowledgements

We are grateful to P J Ashton, O Bourquin, R K Brooke, M M Coke, W R Courtenay jnr, J Deacon, I J de Moor, F J Kruger, I A W Macdonald, P R Miller, T Pike, A J Ribbink, P H Skelton, R E Stobbs and H Tomlinson for assistance with the preparation of the manuscript. This research was funded by the Invasive Biota Working Group of the National Programme for Ecosystem Research of the CSIR. We are particularly grateful to A A Ferrar for help and guidance.

Alexander W J R 1985. Hydrology of low latitude Southern Hemisphere landmasses. Hydrobiologia 125, 75-83.

Allanson B R and Jackson P B N 1983. Limnology and fisheries potential of Lake le Roux. South African National Scientific Programmes Report No 77. CSIR, Pretoria.

Barel C D N, Darit R, Greenwood P H, Fryer G, Hughes N, Jackson P B N, Kawanabe H, Lowe-McConnell R H, Nagoshi M, Ribbink A J, Trewavas E, Witte F and Yamaoka K 1985. Destruction of fisheries in Africa's lakes. Nature 315, 19-20.

Barnard K H 1943. Revision of the indigenous freshwater fishes of the S W Cape region. Annals of the South African Museum 36, 101-262.

Bell-Cross G 1972. The fish fauna of the Zambezi River System. Arnoldia, Rhodesia 5, 1-19.

Bell-Cross G 1976. The fishes of Rhodesia. Trustees of the National Museums and Monuments of Rhodesia, Salisbury.

Boomker J, Huchzermeyer F W and Naude T W 1980. Bothriocephalosis in the common carp in the eastern Transvaal. Journal of the South African Veterinary Association 51, 263-264.

Bowmaker A P, Jackson P B N and Jubb R A 1978. Freshwater fishes. In: Werger M J A and Van Bruggen A C (eds) Biogeography and ecology of southern Africa. W Junk, The Hague, pp 1181-1230.

Brandt F de W, Van As J G, Schoonbee H J and Hamilton-Atwell V L 1981. The occurrence and treatment of Bothriocephalosis in the common carp, *Cyprinus carpio* in fish ponds with notes on its presence in the largemouth yellowfish, *Barbus kimberleyensis* from the Vaal Dam, Transvaal. Water SA 7, 35-42.

Breen C M (ed) 1983. Limnology of Lake Midmar. South African National Scientific Programmes Report No 78. CSIR, Pretoria.

Bruton M N 1979. The food and feeding behaviour of *Clarias gariepinus* (Pisces, Clariidae) in Lake Sibaya, South Africa, with emphasis on its role as a predator. Transactions of the Zoological Society of London 35, 47-114.

Bruton M N 1985. The effects of suspensoids on fish. Hydrobiologia 125, 221-241.

Bruton M N and Appleton C C 1975. Survey of Mgobezeleni lake-system in Zululand, with a note on the effect of a bridge on the mangrove swamp. Transactions of the Royal Society of South Africa 41, 283-294.

Bruton M N and De Moor I J (in press). Invasive aquatic animals in southern Africa: an atlas and detailed assessment of the problem. South African National Scientific Programmes Report.

Bruton M N and Kok H M 1980. The freshwater fishes of Maputaland. In: Bruton M N and Cooper K H (eds) Studies on the ecology of Maputaland. Rhodes University, Grahamstown, pp 210-244.

Bruton M N and Merron S V 1985. Alien and translocated aquatic animals in southern Africa: a general introduction, checklist and bibliography. South African National Scientific Programmes Report No 113. CSIR, Pretoria.

Cadieux J 1980. Freshwater angling in the Transvaal: its environmental and economic impact. South African Journal of Science 76, 492-493.

Cambray J A and Jubb R A 1977. The Orange-Fish tunnel. Piscator 99, 4-6.

Cambray J A and Stuart C 1985. Aspects of the biology of a rare redfin minnow *Barbus burchelli* (Pisces: Cyprinidae) from South Africa. South African Journal of Zoology 20, 155-165.

Cape Provincial Administration 1944-1945. Inland Fisheries Department Reports No 1-2.

CIFA 1985. Introduction of species and conservation of genetic resources. Report of the Committee for Inland Fisheries of Africa, Report 85/13, 1-18.

Cochrane K L 1985. The population dynamics and sustainable yield of the major fish species in Hartbeespoort Dam. PhD Thesis, University of the Witwatersrand, Johannesburg.

Cochrane K L 1986. Distribution of the three major species of fish in the Hartbeespoort Dam in relation to some environmental factors. South African Journal of Zoology 21, 177-182.

Courtenay W R and Robins C R 1973. Exotic aquatic organisms in Florida with emphasis on fishes: a review and recommendations. Transactions of the American Fisheries Society 102, 1-12.

Courtenay W R and Stauffer J R (eds) 1984. Distribution, biology and management of exotic fishes. Johns Hopkins University Press, Baltimore.

Crass R S 1964. Freshwater fishes of Natal. Shuter and Shooter Pietermaritzburg.

Davies B R and Day J A 1986. The biology and conservation of South Africa's vanishing waters. Centre for Extra-mural Studies, University of Cape Town, Cape Town.

Du Plessis S S 1952. Fish diseases in Transvaal. Symposium on African Hydrobiology and Inland Fisheries. Publication 6, 128-130.

Gaigher C M 1973. The Clanwilliam River: it is not yet too late. Piscator 88, 75-78.

Gaigher I G 1975. The ecology of a minnow, *Barbus trevelyani* (Pisces: Cyprinidae) in the Tyume River, eastern Cape. Annals of the Cape Provincial Museums (Natural History) 11, 1-19.

Gaigher I G, Hamman K C D and Thorne S C 1980. The distribution, conservation status and factors affecting the survival of indigenous freshwater fishes in the Cape Province. Koedoe 23, 57-88.

Gephard S R 1977. Status of trout populations of Lesotho. Piscator 100, 71-74.

Gray A P 1958. Bird hybrids. Commonwealth Agricultural Bureaux, Bucks.

Hamman K C D 1980. Post-impoundment trends in the fish populations of the Hendrik Verwoerd Dam, South Africa. Journal of the Limnological Society of Southern Africa 6, 101-108.

Harrison A C 1936. Black bass in the Cape Province: a second report on the progress of American largemouth black bass. Department of Commerce and Industries, Fisheries and Marine Biological Survey Division, Union of South Africa. Investigational Report No 7, 1-119.

Hey D 1945. A preliminary report on the artificial cultivation of *Xenopus laevis* at the Jonkershoek Inland Fish Hatchery. Inland Fisheries Department, Cape Provincial Administration. Report No 2, 7-8.

IUCN 1979. World Conservation Strategy. IUCN, Morges.

Jackson P B N 1966. The establishment of fisheries in man-made lakes in the tropics. In: Lowe-McConnell R H (ed) Man-made lakes. Academic Press, London, pp 53-73.

Jubb R A 1965. Freshwater fishes of the Cape Province. Annals of the Cape Provincial Museums 6, 1-72.

Jubb R A 1966. *Labeo quathlambae*, a rare freshwater fish now feared to be extinct. Piscator 66, 89-90.

Jubb R A 1967. Freshwater fishes of southern Africa. Balkema, Cape Town.

Junor F J R 1969. *Tilapia melanopleura* Dum., in artificial lakes and dams in Rhodesia with special reference to its undesirable effects. Rhodesian Journal of Agricultural Research 7, 61-69.

Kleynhans C J 1979. The distribution and status of *Kneria auriculata* (Pellegrin) (Pisces: Kneriidae) in the Transvaal. Journal of the Limnological Society of Southern Africa 5, 27-29.

Kleynhans C J 1984. Die verspreiding en status van sekere seldsame vissoorte van die Transvaal en die ekologie van sommige spesies. DSc Thesis, University of Pretoria, Pretoria. 267 pp.

Kleynhans C J 1985. The future of rare fish species in the Transvaal. Fauna and Flora 42, 30-32.

Kohler C C and Stanley J G 1984. A suggested protocol for evaluating proposed exotic fish introductions in the United States. In: Courtenay W R and Stauffer J R (eds) Distribution, biology and management of exotic fishes. Johns Hopkins University Press, Baltimore. pp 387-406.

Kruger I, Van As J G and Saayman J E 1983. Observations on the occurrence of the fish louse *Argulus japonicus* Thiele, 1900, in the western Transvaal. South African Journal of Zoology 18, 408-410.

Laurenson L J B and Hocutt C H 1986. Colonization theory and invasive biota: the Great Fish River, a case history. Environmental Monitoring and Assessment 6, 71-90.

Liversidge R 1985. Alien bird species introduced into South Africa. In: Bunning L J (ed) Proceedings of the symposium on birds and man, Johannesburg 1983. Witwatersrand Bird Club, Johannesburg. pp 31-34.

Lombard G L 1968. A survey of fish diseases and parasites encountered in Transvaal. Limnological Society of Southern Africa Newsletter 11, 23-29.

Mayekiso M 1986. Some aspects of the ecology of the Eastern Cape rocky *Sandelia bainsii* (Pisces: Anabantidae) in the Tyume River, eastern Cape, South Africa. MSc Thesis, Rhodes University, Grahamstown.

Midgley D C and Pitman W V 1969. Surface water resources of South Africa. University of the Witwatersrand, Johannesburg. HRU 2/69.

National Institute for Water Research 1985. The limnology of Hartbeespoort Dam. South African National Scientific Programmes Report No 110. CSIR, Pretoria.

Newbery R 1984. The American red-eared terrapin in South Africa. African Wildlife 38, 186-189.

Noble R G and Hemens J 1978. Inland water ecosystems in South Africa - a review of research needs. South African National Scientific Programmes Report No 34. CSIR, Pretoria.

Paperna I 1980. Parasites, infections and diseases of fish in Africa. CIFA Technical Paper 7, 1-216.

Paperna I and Van As J G 1983. Epizootology and pathology of *Chilodonella hexasticha* (Kiernik, 1909) (Protozoa, Ciliata) infections in cultured cichlid fishes. Journal of Fish Biology 23, 441-450.

Pike T and Tedder A J 1973. Rediscovery of *Oreodaimon quathlambae* (Barnard). Lammergeyer 19, 9-15.

Rayner N A 1981. Studies on the zooplankton of Lake Midmar. MSc Thesis, University of Natal, Pietermaritzburg.

Roberts T R 1975. Geographical distribution of African freshwater fishes. Zoological Journal of the Linnean Society of London 57, 249-319.

Schrader H J 1985. Invasive alien fishes in South West Africa/Namibia. In: Brown C J, Macdonald I A W and Brown S E (eds) Invasive alien organisms in South West Africa/Namibia. South African National Scientific Programmes Report No 119. CSIR, Pretoria pp 35-40.

Scott W B and Crossman E J 1973. Freshwater fishes of Canada. Fisheries Research Board of Canada, Ottawa.

Skelton P H 1977. South African Red Data Book – Fishes. South African National Scientific Programmes Report No 14. CSIR, Pretoria.

Skelton P H 1980. Systematics and biogeography of the redfin *Barbus* species (Pisces: Cyprinidae) from Southern Africa. PhD Thesis, Rhodes University, Grahamstown.

Skelton P H (in press). South African Red Data Book – Fishes (revised edition). South African National Scientific Programmes Report.

Skelton P H and Davies M T T (in press). Proceedings of a colloquium: Trout in South Africa. ICHTHOS Special Edition No 1.

Skelton P H and Merron G S 1984. The fishes of the Okavango River in South West Africa with reference to the possible impact of the Eastern National Water Carrier on fish distribution. JLB Smith Institute of Ichthyology, Investigational Report No 9, 1-32.

Taylor J N, Courtenay W R and McCann J A 1984. Known impacts of exotic fishes in the continental United States. In: Courtenay W R and Stauffer J R (eds) Distribution, biology and management of exotic fishes. Johns Hopkins University Press, Baltimore. pp 322-373.

Van As J G (in press). Parasites and diseases of fish. In: Hecht T (ed) Aquaculture in South Africa. Haum, Johannesburg.

Van As J G and Basson L 1984. Checklist of freshwater fish parasites from southern Africa. South African Journal of Wildlife Research 14, 49-61.

Van As J G, Basson L and Theron J 1984. An experimental evaluation of the use of formalin to control trichodiniasis and other ectoparasitic protozoans on fry of *Cyprinus carpio* L. South African Journal of Wildlife Research 14, 42-48.

Van As J G, Schoonbee H J and Brandt F de W 1981. Further records of the occurrence of *Bothriocephalus* (Cestoda: Pseudophyllidea) in the Transvaal. South African Journal of Science 77, 343.

Wager V A and Rowe-Rowe D T 1972. The effects of *Tilapia rendalli* and *T mossambica* on aquatic macrophytes and fauna in five ponds. South African Journal of Science 68, 257-260.

Weir J C 1972. Diversity and abundance of aquatic insects reduced by introduction of the catfish *Clarias gariepinus* to pools in Central Africa. Biological Conservation 4, 169-175.

Welcomme R L 1984. International transfers of inland fish species. In: Courtenay W R and Stauffer J R (eds) Distribution, biology and management of exotic fishes. Johns Hopkins University Press, Baltimore. pp 22-40.

Wells J G, Clark G McC and Morris G K 1973. Evaluation of methods for isolating *Salmonella* and *Arizona* organisms from pet turtles. Applied Microbiology 27, 8-10.

Chapter 5

Alien and translocated terrestrial vertebrates in South Africa

R K Brooke*, P H Lloyd† and A L de Villiers†

*University of Cape Town and †Cape Department of Nature and Environmental Conservation.

Introduction

This review covers alien species of terrestrial amphibians, reptiles, birds and mammals which have bred in the wild at least once. Aquatic alien vertebrate species are covered by Bruton and van As (this volume). The area covered is that of continental South Africa including the Homelands, the Independent States and the Kingdoms of Lesotho and Swaziland, enclave and virtual enclave of South Africa respectively, but excluding the offshore islands dealt with by Cooper and Brooke (this volume). Major translocations of indigenous species outside their historic range are noted, particularly where this has led to hybridization between subspecies. We have not considered range expansions due to man-made modifications of the environment. Had we done so, we would have had to consider the majority of terrestrial vertebrate species in South Africa. Thus this review cannot be directly compared with the aquatic aliens review of Bruton and van As (this volume). Introductions, whether intentional or accidental, which have not led to breeding in the wild are not considered. It is not introduction that is of significance in conserving natural ecosystems. It is breeding in the wild which may lead to the establishment of self-sustaining populations capable of adversely affecting elements of natural ecosystems which is important for conservation and ecosystem management.

To produce this review or synthesis, the relevant scientific and some popular and semipopular literature has been scanned and combined with our personal knowledge and that of our colleagues. It must be emphasized, as Siegfried (1962) pointed out, that South African zoologists have, in general, paid little attention to terrestrial alien animals at large in South Africa in their writings. Unlike the position in some other countries such as Australia and New Zealand, there were no acclimatization societies here in the last century actively fostering terrestrial introductions and publishing newsletters to tell their constituents of their triumphs and failures. Hence the early data on introductions, whether deliberate or accidental, are meagre and it is not possible to discuss successful and unsuccessful introductions in the way that it is in Australia and New Zealand.

Address for correspondence: R K Brooke, Percy FitzPatrick Institute of African Ornithology, University of Cape Town, Rondebosch, 7700.

Amphibians and reptiles

There are no terrestrially breeding invasive alien amphibian or reptile species in South Africa. The tropical house gecko *Hemidactylus mabouia* from tropical Africa has established localized populations at East London and Port Elizabeth (W R Branch personal communication), both in the Cape Province and both major harbours. They seem to have arrived accidentally, presumably with seaborne cargo. Two other introductions of this or a closely related lizard from Sierra Leone to Simonstown and Gordon's Bay, south-western Cape Province, where they may now be established, were made deliberately in 1962 and again in 1976 (J R Nortier personal communication). While these populations may well survive indefinitely in urban areas they seem unlikely to expand into natural ecosystems.

A more interesting alien is the flowerpot snake *Ramphotyphlops braminus* which was presumably brought to Cape Town in the soil of potted plants from Indonesia before 1800 when the Dutch still controlled both that country and the Cape. It was collected in Cape Town by Dr Sir Andrew Smith in the 1820's or 1830's. He believed that it was indigenous and named it *Onycocephalus capensis* in 1846. It is still present in Cape Town and south to Plumstead on the Cape peninsula (McLachlan 1978) and it has recently been found at Edgemead, a north-eastern suburb of Cape Town (Visser 1979). It has also been found in Durban, obviously a discrete introduction (O Bourquin personal communication). Since it can reproduce parthenogenetically, it is easily introduced (Broadley 1983) and may occur in other localities in South Africa. It has been widely introduced into other countries (Broadley 1983) but it is not known to threaten any indigenous animal.

Indigenous mountain tortoises *Geochelone pardalis* were translocated to the Cape of Good Hope Nature Reserve before 1930 and the population is flourishing (G Wright personal communication). The species was also translocated to the De Hoop Nature Reserve, southern Cape Province, before 1966 where they also flourish (M de V Jordaan personal communication). The common dwarf gecko *Lygodactylus capensis* of Natal, the eastern Transvaal and northwards has been recorded in the Addo National Park, eastern Cape Province (Branch 1981). It is thought to have arrived in a crate from the Kruger National Park, eastern Transvaal. This record of an apparent accident would not be worth mentioning if the species had not recently been found breeding in nearby Port Elizabeth (W R Branch personal communication), allegedly for over 30 years. There have been two translocations of Bibron's gecko *Pachydactylus bibronii* indigenous to Little Namaqualand and Namibia into the Cape Peninsula. The first was to Kommetjie on the west coast before 1950 (Rose 1950; Siegfried 1962) but this population has died out (S Rose personal communication). During the 1960's several small releases were made near Fish Hoek on the east coast by John Wood, a dealer in reptile pets who did not retain scarred or defective specimens.

This population was found in 1985 to have maintained itself and bred (de Villiers unpublished). Hewitt and Power (1913) report that in loads of firewood imported from the north into Kimberley in the early days (1875-1910) of the diamond mining industry Jones's girdled lizard *Cordylus jonesii* and spotted harlequin snake *Homoroselaps lacteus* were translocated there but we do not know that they have persisted. A number of other reptiles have been translocated within southern Africa but there is no evidence for successful breeding in the new areas.

Birds

Alien bird species in South Africa have been briefly reviewed by Godfrey (1934), Bigalke (1937), Gebhardt (1959), Siegfried (1962), Winterbottom (1966), Liversidge (1979, 1985) and Long (1981). In addition, Brooke (in preparation) has compiled a bibliography of alien birds in southern and south-central Africa which has greatly assisted in compiling this section. Scientific nomenclature follows Howard and Moore (1980) as modified for local use by Clancey (1980). The published data on alien birds in South Africa is fuller than it is for mammals but seldom cover the exact details of an introduction or how many birds were involved. No alien bird species has managed to invade natural habitats as opposed to areas of human dwellings or crop cultivation or intensive animal husbandry.

Widespread alien birds

Pigeons *Columba livia* were first received in Cape Town in 1652 as domestic stock from Holland though they did not breed well until the next year (Thom 1952), presumably due to the change in seasons between the Northern and Southern Hemispheres. There have been many importations since, chiefly from north-west Europe, including Britain, but these introductions have not been documented in detail. It is not known when birds first abandoned their owners (or were abandoned) and made their homes in towns among the larger buildings as feral pigeons. By the 1960's most towns in South Africa had feral populations (Brooke 1981a) and these are occasionally augmented by lost homing pigeons (distinguished by being ringed) or abandoned private flocks. Feral pigeons share many of the larger towns with the indigenous and closely related speckled rock pigeon *C guinea*. The two species hybridize fairly easily in captivity and very occasionally in the wild (Rowan 1983), particularly where male feral pigeons have difficulty in finding conspecific mates as in the De Hoop vlei area (C J Uys personal communication). Speckled rock pigeons habitually travel long distances to feed in open country whereas feral pigeons feed in the nearest open areas available to them. Thus the two species rarely compete for food. Feral pigeons occasionally breed on sea cliffs in South Africa (Brooke 1981a,b) whereas speckled rock pigeons do so commonly but there is no evidence for a shortage of nest sites for pigeons on cliffs.

European starlings *Sturnus vulgaris* were deliberately introduced to Cape Town by Cecil Rhodes, probably in April 1897, in order to improve the amenities of the Cape by diversifying the bird fauna damned as poverty-stricken by Shelley (1875). Meinertzhagen (1952) gives the date on which he delivered the starlings to Rhodes as 14 January 1899, the day Rhodes left Britain for the Cape. However, we learn from Michell (1910) that on that day Rhodes arrived in Britain, leaving about 1 July 1899. Rhodes had previously left Britain for the Cape on 3 April 1897 and 19 or 20 May 1898 (Michell 1910). Meinertzhagen states that he trapped winter visitors to Britain which he could have done in March 1897 but hardly in May 1898 (see also van der Merwe 1984). The

European starling was one of many species introduced (Le Sueur 1913) but the principal one to establish itself and the only one to expand its range markedly (Winterbottom and Liversidge 1954; Liversidge 1962; Quickelberge 1972). They now occur throughout the southern parts of the Cape Province and up the west coast to Oranjemund in extreme south-western Namibia (Brown 1985) (see distribution map in van der Merwe 1984). They also occupy parts of the Karoo but abandon such areas during prolonged droughts (Winterbottom 1966, 1978). As Meinertzhagen (1952) pointed out, the birds he supplied were migrants from Fenno-Scandia, not British bred, and physiologically adapted to breeding in the short summers of northern Europe. It appears from Johnson and Cowan (1974) that European starlings are better adapted to breeding successfully in cool summers in Vancouver, western Canada, than Chinese jungle mynas *Acridotheres cristatellus* also introduced there. It would seem that by heredity European starlings in South Africa are not adapted to the warmer parts of the country and that further major range expansion is not likely to take place. European starlings are occasionally pests in soft-fruit growing areas since they are messy feeders on ripening fruit but their depredations are more significant for household gardeners with a few fruit trees or vines than for most bulk growers. When feeding in the countryside they associate freely and apparently amicably with flocks of indigenous starlings. The presence of European starlings does not appear to diminish indigenous numbers: pied starlings *Spreo bicolor* no longer breed on buildings in Cape Town and the Cape Peninsula (Craig 1985) because they feed by gleaning on open dry ground and the expansion of the metropolitan area has made commuting to distant feeding areas uneconomic. Pied starlings and redwinged starlings *Onychognathus morio* are larger than European starlings and their individuals can dominate European starlings when seeking the same food item (Brooke unpublished). In the country districts pied starlings and European starlings use the same nest sites, European starlings preferring buildings and pied starlings preferring holes in banks, but there is no evidence that pied starling populations are limited by nest site availability.

There have been two successful introductions of Indian mynas *Acridotheres tristis* into South Africa. The first took place in Durban in 1902 and was due to escapees from the caged bird trade (Kent 1927; van Nierop 1958). Cuthbertson (1952) and Calder (1953) quote an obituary in the Natal Mercury for 25 November 1932 which states that three pairs were brought from Mauritius and released in Durban in 1888. However, there is no evidence that this introduction was successful: the species was not mentioned by Woodward and Woodward (1899) in their account of Natal birds. Brooke (1976) showed that modern Natal populations appear to be derived from birds originating in Burma or Assam in north-east India. The third introduction is less well documented but took place in the Johannesburg area in the late 1930's (Roberts 1945; Rees 1949), the stock coming from outside South Africa of birds of standard Indian facies (A J Baker personal communication; Brooke unpublished). The early spread of the second introduction of Indian mynas in Natal has not been documented. They had, however, reached Pietermaritzburg, 65 km inland from Durban, before 1930 (Godfrey 1931). Expansion continued in western and northern Natal during the 1950's and 1960's, cf the distribution maps in Calder (1953) and Cyrus and Robson (1980). Expansion in the Transvaal seems to have been slower than in Natal, cf the distribution statements in Tarboton (1968) and Kemp et al (1985) who also provide a map. There was an unsuccessful introduction in Kimberley, northern Cape, in the 1950's (Bigalke 1964; Liversidge 1985). Although nonbreeding individuals have been recorded from time to time in the Transkei, south-west of Natal and even in the eastern Cape Province (eg Every 1979), there is no evidence that the breeding range is still moving southwards. If Johnson and Cowan's (1974) work on the reproductive physiology of Chinese jungle mynas in Vancouver also applies to Indian mynas in Natal, it may be that in moving south they are entering a climatic area to which they are less well adapted. This may also be the explanation for their slow rate of spread in the Transvaal. Like European starlings, Indian mynas are sometimes a pest of household fruit trees and like them may bring itch-causing mites into buildings when nesting in the eaves (Gilges and van der Spuy 1945). They are alleged to drive indigenous birds away from gardens but, as Clancey (1960) pointed out, there is no critical evidence for this. They can compete successfully with indigenous hole-nesters for holes (McCulloch 1965) as can European starlings (van der Merwe 1984). They interact less frequently with indigenous starlings than do European starlings (T B Oatley personal communication; Brooke unpublished).

The house sparrow *Passer domesticus* is believed to have reached Natal from India as pets of indentured sugar cane labourers in the 1880's or 1890's (Clancey 1964; Harwin and Irwin 1966). Introduction dates, 1893 and 1897, in Mackworth-Praed and

Grant (1963) refer to nominate race birds from Britain (Gebhardt 1944) which are not the source of Natal populations (eg Clancey 1980). Its spread in Natal is poorly documented though it had reached Pietermaritzburg by circa 1912 (Gebhardt 1944) and was common there by 1930 (Godfrey 1931). In 1948 it was first reported to have crossed the Drakensberg into the interior of South Africa and thereafter spread very rapidly throughout the country (Winterbottom 1961, 1972; Harwin and Irwin 1966; Vierke 1970). The Natal population has also spread through most adjacent countries. It is widespread in SWA/Namibia, Botswana, Zimbabwe, Mozambique, Malawi and Zambia, reaching the far north of the last (Conant 1980). While India provided the bulk of the stock which has overrun southern and south-central Africa, there have been other minor introductions by way of escaped aviary birds in East London (Godfrey 1934; Courtenay-Latimer 1942, 1955), Harare, Zimbabwe (Harwin and Irwin 1966), Maputo, Mozambique (Pinto 1959) and doubtless elsewhere. The introduction in Pretoria, Transvaal (Sclater 1905), did not lead to the establishment of a population. The East London birds came from Britain (Godfrey 1934) and the Maputo birds from Portugal in 1955 (Pinto 1959) and were nominate *domesticus* as opposed to *P d indicus* in Natal. However, a phenetic analysis of southern African house sparrows (Crowe and Brooke 1985) has shown that eastern and southern birds are *indicus* sensu lato; that nominate *domesticus* has had no marked effect on African phenotypes; and that birds from the dry west (eg those in Kimberley and Windhoek) have assumed a plumage type unknown elsewhere in the nearly worldwide range of the species. The nests and faeces of house sparrows in sheds may contaminate stored food but otherwise they cause little harm. However, it has been suggested that expansion of house sparrow populations is detrimental to Cape wagtail *Motacilla capensis* populations (Masterson 1976). This may well be so as the Cape wagtail is a scavenger as well as a generalized insectivore (Winterbottom 1964). House sparrows sometimes take over swallow nests (eg Patten 1957; Earle 1985) but this does not seem to affect the size of swallow populations. Their relationships with the larger Cape sparrow *P melanurus* are complex. In some small towns house sparrows occupy the centre whereas the peripheries with gardens are occupied by Cape sparrows. In other small towns and most large ones they coexist and it is hazardous to predict which one will be seen next though areas with large numbers of trees and shrubs seldom support many house sparrows (Harwin and Irwin 1966; Brooke unpublished).

Localized alien birds

The chaffinch *Fringilla coelebs* was among the species introduced to the Cape from Britain by Rhodes (Le Sueur 1913). It expanded along the eastern side of the Table Mountain chain from Groote Schuur estate where all Rhodes's species were liberated. However, in the last thirty years its range has been contracting and numbers decreasing (eg Ferguson and Tongue 1968; Connor 1972; Winterbottom 1979). It would seem to be headed for local extinction.

The ringnecked parakeet *Psittacula krameri* bred ferally in Cape Town in the 1850's (Layard 1867) but has not been found breeding there since although birds have been seen again in the 1980's (Murray and Murray 1984). It started to breed at the end of the 1960's at Sodwana Bay in Zululand (Vincent 1972) and now also breeds to the south in Durban North (Clancey 1976) and just west of Johannesburg, Transvaal (Weissenbacher and Allan 1985; Kemp et al 1985). All four populations are presumably based on escaped cage or aviary birds as there is no evidence for a southward expansion of their northern Afrotropical range. Whether they will become permanently established and start spreading is as yet unknown. Long (1981) lists many introductions of ringnecked parakeets but few have resulted in invasive status (chiefly in Egypt and Mauritius). He also points out that in India they are a pest both of growing and stored grain. Agricultural and nature conservation officials should monitor South African populations of the ringnecked parakeet and exterminate those that seem likely to become pests of grain crops. The plumheaded parakeet *P cyanocephala* established a breeding population in the Pretoria area in the 1970's (Kemp et al 1985) and the comments made on the ringnecked parakeet also apply here.

The rosycheeked lovebird *Agapornis roseicollis* is indigenous to SWA/Namibia and some parts of the Cape Province north of the Orange River. It has been introduced, apparently successfully, at Fish Hoek on the Cape Peninsula (Brooke 1984b), Middelburg, Transvaal (Liversidge 1985 but unknown to Kemp et al 1985), and Durban (O Borquin personal communication). It seems unlikely to present problems even if these introductions lead to the establishment of permanent populations. After all, the first introduction of the ringnecked parakeet died out (see above) as did an introduction of the greyheaded lovebird *A cana* (Woodward and Woodward 1899; Clancey 1964) and of the greenrumped parrotlet *For-*

pus passerinus (Brooke 1986), both in Natal. Escaped budgerigars *Melopsittacus undulatus* and cockatiels *Nymphicus hollandicus* are fairly common in the metropolitan areas of South Africa but have not been known to breed. The former may have done so at Umbogintwini, southern Natal, where a flock of about 50 birds was present from about 1947 onwards (Bowles 1966). Other parrots that have escaped but not bred, as far as is known, include the lesser sulphurcrested cockatoo *Cacatua sulphurea*, Nanday conure *Nandayus nenday*, brownthroated conure *Aratinga pertinax* (all Liversidge 1985), Jandaya conure *A jandaya*, duskyheaded conure *A weddellii* (both O Bourquin personal communication) and blackcheeked lovebird *Agapornis nigrigenis* (Kemp et al 1985). It may be that the niche for small avian seedeaters is now so effectively filled by weavers *Ploceus* species and allies and canaries *Serinus* species that small parrots have little chance to establish themselves. Early Pleistocene southern Africa had a much richer small parrot fauna than it has now but *Ploceus* and *Serinus* were hardly represented (Pocock 1971).

An alien species that has made use of human facilities to reach South Africa and establish a population in Durban is the Indian house crow *Corvus splendens* which has long used steam and motor vessels to hitch lifts around the Indian Ocean (Long 1981). They apparently made use of the increased traffic down the east African coast while the Suez Canal was closed between 1967 and 1980 to invade Durban in the early 1970's (Oatley 1973; Sinclair 1974). They have been recorded in Cape Town (eg Sinclair 1980) and East London (Liversidge 1985) but they are not established at either port. There is no indigenous crow of similar size (all are larger) in southern Africa and their principal competitors in Durban seem to be the smaller but abundant Indian mynas. It would seem that this population will become a permanent one, even if it is supported by occasional ship-assisted immigrants; and that it will spread along the densely settled Natal coasts. Long (1981) reports that it is regarded neutrally in India but an unsuccessful eradication campaign was instituted by the British administration in Zanzibar because of the general nuisance it was thought to be (Pakenham 1979).

Feral chickens *Gallus gallus* maintain a population at Gravelotte in the north-eastern Transvaal lowveld (Liversidge 1985). Long (1981) reports a Natal population but neither Liversidge (1985) nor ourselves have been able to find out anything about it. Chickens have been kept in South Africa since before 800 AD (Voigt and von den Driesch 1984) and have presumably had plenty of opportunity to establish permanent feral populations if they had the ecological ability to do so. Ringnecked pheasants *Phasianus colchicus*, presumably from Britain, were bred so successfully at Rooipoort estate near Kimberley in the 1930's that stock from there was established at Bredasdorp and Knysna in the southern Cape Province (Siegfried 1962) and also at Ceres, Elgin and Stellenbosch (Liversidge 1985). Populations persisted for a while but have died out. The same happened at Howick and the Hluhluwe Game Reserve in Natal (Macdonald and Birkenstock 1980). It appears that the number and variety of indigenous reptilian, mammalian and avian predators in South Africa is sufficient to make the establishment of alien Phasianidae very difficult on the continental land mass. The relative success of two phasianids, the peacock *Pavo cristatus* and the chukar partridge *Alectoris chukar*, on Robben Island (Cooper and Brooke this volume; Brooke and Prins in press) supports this view.

Among Rhodes's other importations of British birds to Cape Town were the blackbird *Turdus merula* and the song thrush *T philomelos*. The former bred for a few years at the beginning of the century and the latter persisted for more than 40 years as a breeding species (Gill 1936, 1952; McCauseland 1952). The closely related indigenous olive thrush *T olivaceus* probably provided sufficient competitive pressure to make permanent establishment of populations very difficult.

Escaped aviary birds of the American emberizine genus *Paroaria* have established temporary breeding populations: the redcowled cardinal *P dominicana* in Durban in the 1960's (Lawson 1971); the redcrested cardinal *P coronata* at Hermanus, southern Cape, in the 1950's (Winterbottom 1968) and at Randburg in the Transvaal in the 1970's (Kemp et al 1985; Liversidge 1985). It is not known why they were unsuccessful but the small size of the populations would have made them likely candidates for stochastic extinctions. Similarly, escaped strawberry finches *Amandava amandava* breed ferally at Rosherville, Transvaal (Liversidge 1985 but unknown to Kemp et al 1985). The European goldfinch *Carduelis carduelis* bred at Heidelberg, Transvaal, at the beginning of the century (Butler 1903; Brooke in press) but soon died out.

Translocation and subspecific hybridization

The indigenous race of the ostrich is *Struthio camelus australis* and wild-taken birds formed the original basis of the ostrich farming industry. To improve the feather qualities of the local birds they were bred to a few imported cocks of *S c camelus* from Algeria and Nigeria and of *S c syriacus* from Lebanon (Smit 1963, 1984). These hybrids form the stock farmed in the Little Karoo (D Swart personal communication). Outside this area ostriches living more or less ferally in the farming areas are essentially genetic *australis*. Obvious hybridization characters, particularly skin colour, bald headedness, white collar at the base of the neck, smooth eggshell (Liversidge 1978) have not been reported in modern South African ostriches outside the Little Karoo though they do not seem to have been searched for.

Domestic helmeted guineafowl derived from *Numida meleagris galeata* of west Africa have been kept on South African farms for well over a century (Layard 1867) and some hybridization with indigenous populations of *N m mitrata* and *N m coronata* has occurred. The most obvious sign is the presence of white or pale grey instead of dark grey in the plumage. In the 1890's eastern Cape Province *N m coronata* were translocated to the western Cape to improve the gamebird shooting (Sclater 1906; Skead 1962) and domestic birds were added to these. The result is that in the western Cape today the species is widespread, numerous and manifest hybrids are seen fairly often (Brooke unpublished). It seems likely that the development of agriculture along the southern Cape coast this century would have permitted a natural expansion of eastern Cape populations until they had reached the west coast.

Redeyed doves *Streptopelia semitorquata semitorquata* from Beira, Mozambique, (Winterbottom 1956) were bred very successfully in an aviary at Elgin, south-western Cape, in the 1920's and 1930's and all were eventually released there in 1933 at a time when the tree planting campaign initiated by Rhodes in the 1890's had greatly improved the habitat for the species. The result was that the indigenous race *S s australis* which had become very sparse through forest destruction formed a subspecific hybrid swarm with the newcomers and their joint numbers have now greatly increased (Brooke 1984a).

A number of other alien species have been seen at large in South Africa, having escaped from cages, aviaries or zoos or having been deliberately liberated. In most cases nothing is known about numbers involved, whether they bred successfully or why they did not succeed. In addition, indigenous species have been found outside their normal ranges in areas where escapes are likely and some may constitute failed translocations.

Mammals

Alien mammal species in South Africa have been briefly reviewed by Bigalke (1937) and Siegfried (1962). Scientific nomenclature follows Honacki et al (1982) as modified for local use by Smithers (1983). Comparatively little has been published on alien mammals in South Africa and much of what is known is in departmental reports and the personal knowledge of mammal workers in conservation departments.

Widespread alien mammals

The house rat *Rattus rattus* is now found virtually wherever people gather in large numbers in South Africa (Smithers 1983). It is recorded archaeologically from before 800 AD (Plug et al 1979) and probably reached South Africa via Arab traders moving down the east coast of Africa. Later arrivals from Europe after 1600 are very probable but apparently undocumented. The brown rat *R norvegicus* presumably reached South Africa from Europe after 1700 when it became the dominant species carried in boats (Atkinson 1985). It was certainly present in Cape Town by 1830 (Avery 1985). It is now found in and around most ports (Smithers 1983). The two *Rattus* species are the economically most harmful alien mammal species in South Africa, judging by work in other countries (eg Bajomi 1983; McDonald 1983) although no recent estimates of their damage to agriculture, stored foods and human health appear to have been made locally. However, Siegfried (1962) cites a figure of R1 million for grain destroyed by rats in 1945. The house mouse *Mus musculus* presumably also reached South Africa from Europe in the 17th century but its depredations do not seem to be nearly as serious economically as those of the two rats.

Many mammals have been imported as domestic stock and some have established temporary feral populations. Sheep *Ovis aries* were first introduced to South Africa over 2 000 years ago (Deacon this volume) and other species later. It is noteworthy that when whites first explored the interior of South Africa they did not report feral populations of stock mammals. Their value to mankind coupled, perhaps, with relative inability to cope with a large number of indigenous predators and pathogens has ensured that such populations usually die out fairly soon after establishment. At present, the most invasive are the donkey *Equus asinus*, the dog *Canis familiaris* and the cat *Felis catus*. The principal threat to nature conservation posed by feral donkeys is that of hybridization with the fragmented small populations of the

Cape mountian zebra *Equus zebra zebra*. The combination male donkey x female zebra produces mules, thus reducing the productivity of small zebra populations. Feral dogs freely kill both small stock and similarly sized indigenous mammals (Siegfried 1962) even in conserved areas. It is not known if populations persist by feral breeding or by recruiting stray household dogs, probably by both methods. Particularly in the rural areas, household dogs are kept out of doors under loose control and the loss of one or more animals is not thought to be a source of concern since they can readily be replaced. Feral cats come into existence in much the same way as dogs but they do not form packs and their hunting and kills are far less obvious. However, their predatory impact on small animals is probably considerable in inhabited areas and thus part of the total conservation pressure that increasing human numbers present to all wildlife. In addition, feral cats hybridize with the indigenous wild cat *F lybica* (Stuart 1981; Smithers 1983; Griffin and Panagis 1985). Pienaar (1964) instances a case in the Kruger National Park, apparently as long ago as the 1950's. However, the general statements cited above do not seem to be based on detailed published findings other than the one hybrid in 92 specimens from Botswana (Smithers 1971). This makes it difficult to assess the real threat posed by feral cats. Two furbearing mammals, the nutria *Myocastor coypus* and the North American mink *Mustela vison*, have been imported as stock for commercial fur production and have escaped to found feral populations in the eastern Cape in the 1950's (Siegfried 1962) but both appear to have since died out.

The history of introduction of the European fallow deer *Cervus dama* has been fully documented by Chapman and Chapman (1980). The species is widely kept as an ornamental and, since it breeds freely, little control is kept of them (Chapman and Chapman 1980). As a result feral populations are now widespread except in Natal (Smithers 1983). The impact of their browsing in indigenous ecosystems has not been studied. They were present in Cape Town before 1869. More were introduced there in 1897 as part of Rhodes's programme to improve the amenities of the Cape (Le Sueur 1913; Chapman and Chapman 1980). Many translocations and at least one further introduction from Britain to Volksrust, Transvaal, have taken place since then (Chapman and Chapman 1980; Smithers 1983).

Localized alien mammals

The grey squirrel *Sciurus carolinensis* was brought from the United States of America to Britain in the last century and then from Britain to the Cape, probably in the late 1890's (Le Sueur 1913) rather than in the early 1900's (Bigalke 1937; Davis 1950 who chronicles its early spread). The introduction was part of Rhodes's programme to improve the amenities of the Cape (Le Sueur 1913). It has persisted in areas where the alien trees *Pinus pinea*, *P pinaster* and *Quercus robur* occur in numbers but it only reaches new towns by being carried by people (Millar 1980). Its most easterly population is at Swellendam (Smithers 1983).

The Himalayan tahr *Hemitragus jemlahicus* provides an interesting case. A pair were sent to Groote Schuur Zoo, Cape Town, in 1935 from stock in the National Zoological Gardens, Pretoria (Siegfried 1962; Bigalke 1977). They or some of their progeny soon escaped and spread over Table Mountain creating an obvious conservation problem by their browsing and trampling. There were 330 animals by 1972 (Table 1; Lloyd 1975) when they were first studied and the population growth rate is estimated at 23% per annum (Lloyd 1985). A culling programme was instituted in 1973 which is still nominally in force (Table 1). It appears that despite heavy culling for a short period, the numbers of Himalayan tahr on Table Mountain are probably now greater than ever before due to culling not being sustained at a level higher than the population growth rate, a point of great importance in managing invasive

Table 1. Population growth and culling figures for the Himalayan tahr *Hemitragus jemlahicus* introduced with one pair to Table Mountain in 1935. Population and growth estimates are based on one substantial total count (on foot, 1972) and two supplementary counts (1973, 1974). All other estimates are calculated.

Year	Estimated population growth at 23% pa	Population estimated at 1 January	Number culled (year round off-take)	Calculated population estimate at 31 December
1972	–	330	–	330
1973	76	406	31 (7,6%)	375
1974	86	461	31 (6,7%)	430
1975	99	529	10 (1,9%)	519
1976	119	638	228 (35,7%)	410
1977	94	504	191 (37,9%)	313
1978	72	385	79 (20,5%)	306
1979	70	376	48 (12,8%)	328
1980	75	403	11 (2,7%)	393

Post 1980 figures not available

animals. The Himalayan tahr, at least, is also markedly invasive in New Zealand (Caughley 1977).

It appears that extermination could only be achieved at considerable expense and difficulty because of the steepness of the mountain slopes. Fortunately, Table Mountain (to which they have been restricted by the organization of the culling programme) and the peninsula chain running south from it is an ecological island and expansion into the main mountain chains of the southern Cape Province is very unlikely without translocation. The manifest effect of the Himalayan tahr on the flora of the mountain (Lloyd 1975) is evidence enough that it or any similar bovid should never be allowed free on any mountain in South Africa.

The only alien terrestrial vertebrate species to have been introduced as a form of biological control is the European wild boar *Sus scrofa*. It was introduced to pine plantations in the south-western Cape to control larvae of the pinetree emperor moth *Imbrasia cytherea* in the 1920's (Thomas and Kolbe 1942). Botha (1985) surveyed these introductions and found that a few have stabilized but that most have decreased or become extinct. The European wild boar would probably not become established in the warmer parts of South Africa since it is not immune to African swine fever and it would face competition from the indigenous bushpig *Potamochoerus porcus*.

Alien and translocated mammals on game ranches

The last 30 years have seen the establishment of game ranching as a form of commercial agriculture. This has led to the importation of alien mammals and the translocation of indigenous ones to areas in which they never occurred in the historical period. These animals are husbanded for their meat and for trophy hunting. As Chapman and Chapman (1980) point out in discussing the European fallow deer, the size of most South African game farms is equal to or larger than that of many nature reserves and that stock is not controlled intensively. Introductions to such farms are thus of concern to nature conservation officials and others since relatively little effort is made to prevent escapes. The alien species involved are: capybara *Hydrochaeris hydrochaeris*, llama *Lama glama* (van Ee 1962; Chapman and Chapman 1980; Lloyd and Millar 1983), dromedary *Camelus dromedarius* (Green 1948; Siegfried 1962; van Ee 1962; Lloyd and Millar 1983); sika *Cervus nippon* (Bigalke and Bateman 1962; Siegfried 1962), sambar *C unicolor* (Siegfried 1962), red deer *C elaphus* (Bigalke and Bateman 1962; Siegfried 1962; Lloyd and Millar 1983), hog deer *C porcinus* (van Ee 1962; Chapman and Chapman 1980; Lloyd and Millar 1983), Pere David's deer *Elaphurus davidianus*, Indian water buffalo *Bubalus bubalis* (Siegfried 1962; van Ee 1962; Chapman and Chapman 1980; Lloyd and Millar 1983), nilgai *Boselaphus tragocamelus* (van Ee 1962; Chapman and Chapman 1980), blackbuck *Antilope cervicapra*, addax *Addax nasomaculatus*, lechwe *Kobus leche* (Bigalke and Bateman 1962; van Ee 1962; Lloyd and Millar 1983), scimitarhorned oryx *Oryx dammah*, mouflon *Ovis musimon*, Barbary sheep *Ammotragus lervia* (van Ee 1962; Chapman and Chapman 1980). Where no authority is cited in the foregoing list or that in the next paragraph, the statement is based on unpublished data of the Cape Department of Nature and Environmental Conservation. Of the species listed, the scimitarhorned oryx can hybridize with the indigenous gemsbok *Oryx gazella* (Gray 1972) and is thus a possible conservation threat. The feeding ecology and behaviour of the mouflon and Barbary sheep are such that, like the Himalayan tahr discussed above, they could easily become numerous and difficult to eradicate if they escaped into mountainous country.

The following South African species have been translocated outside their historical range: warthog *Phacochoerus aethiopicus* (Howard and Marchant 1984), giraffe *Giraffa camelopardalis* (van Ee 1962; Howard and Marchant 1984), Lichtenstein's hartebeest *Alcelaphus lichtensteinii* (Lloyd and Millar 1983), suni *Neotragus moschatus* (Lloyd and Millar 1983), sable antelope *Hippotragus niger* (C Cawood 1983; M Cawood 1983; Howard and Marchant 1984), waterbuck *Kobus ellipsiprymnus* (Rautenbach et al 1981; Howard and Marchant 1984), gemsbok *Oryx gazella* (van Ee 1962; Lloyd and Millar 1983), red duiker *Cephalophus natalensis* (Howard and Marchant 1984), blue duiker *C monticola* (Brown and Gubb this volume), nyala *Tragelaphus angasii* (Bigalke and Bateman 1962; Bourquin and Sowler 1980; Howard and Marchant 1984; Brown and Gubb this volume), bushbuck *T scriptus* (van Ee 1962), blue wildebeest *Connochaetes taurinus* (van Ee 1962; Bourquin and Sowler 1980; Rautenbach et al 1981; Lloyd and Millar 1983; Howard and Marchant 1984), black wildebeest *C gnou* (Lloyd and Millar 1983), springbok *Antidorcas marsupialis* (de Graaff and Penzhorn 1976; Bourquin and Sowler 1980; Lloyd and Millar 1983; Howard and Marchant 1984), impala *Aepyceros melampus* (van Ee 1962; Bourquin and Sowler 1980; Lloyd and Millar 1983; Howard and Marchant 1984), bontebok *Damaliscus dorcas dorcas* (Penzhorn 1971; Lloyd and Millar 1983), blesbok *D d phillipsi* (Bourquin and Sowler 1980; Lloyd and Millar 1983; Howard and Marchant 1984), Cape mountain zebra *Equus zebra zebra*, Hartmann's mountain zebra *E z hartmannae* (Lloyd and Millar 1983).

Some of these translocations present more manifest conservation problems than the alien introductions. The nyala has taken very well to the dense thickets of the northern Cape, including those of alien *Prosopis* species, (Brown and Gubb this volume) where it is difficult to control in the dense vegetation. The two *Connochaetes* species are said to hybridize freely when translocated (their natural ranges are allopatric) and are also carriers of malignant catarrhal fever to cattle (de Vos 1973). There have been so many translocations of springbok that many demes are now contaminated genetically and it appears that the impala is headed in the same direction. Translocation of subspecies of *Damaliscus dorcas* has already led to the development of subspecifically hybrid populations (Lloyd and Millar 1983) and, if unfettered, would lead to the disappearance of the two subspecies. A similar threat is posed to the subspecies of *Equus zebra* unless effective control of translocations is implemented.

Discussion

Certain points arise from this summary of the available information on alien and translocated vertebrate species in South Africa. Unlike the position in Australia and New Zealand, only one species, the Himalayan tahr, has shown itself to be invasive of reasonably undisturbed natural ecosystems but it is confined to the ecological island of the Table Mountain chain on the Cape peninsula. Experience of it there has shown that, if introduced to other mountain areas, at least in the Cape Province, it would rapidly build up populations which would devastate the indigenous vegetation, be difficult to control and exceedingly difficult to exterminate. Invasion by ecologically similar bovids such as the mouflon and the Barbary sheep would be likely to cause similar problems. None of these species should be permitted on game farms.

No terrestrial amphibian or reptile can be considered invasive, even in the loosest sense. A number of alien bird and mammal species adapted to man-modified environments have invaded such areas in South Africa (cf Winterbottom 1966). The feral pigeon and the house sparrow have nearly countrywide ranges, the former probably by many deliberate translocations and the latter by its own dispersal efforts which greatly accelerated from 1948 onwards (Harwin and Irwin 1966), some 50 years after its original introduction from India to coastal Natal. The European starling from Scandinavia has spread widely in the Cape Province but has probably reached the limits of its South African range due to its relative intolerance of the heat and droughts to the north. The Indian myna has been introduced twice, first to Natal where it is now very widely distributed and second to Johannesburg where it is still locally distributed. The Natal Indian mynas came from Burma or Assam and have probably reached their southern breeding limits due to their relative intolerance of cool wet conditions to the south-west. The ringnecked parakeet has only been breeding ferally for 15-20 years and it is too early to say whether it will become invasive of man-modifed environments. Not surprisingly, the house rat and the house mouse have nearly countrywide ranges and the brown rat is found in and about most ports. The European fallow deer is very widespread but largely by deliberate translocation by those who fancy it as an addition to their properties.

It is clear from the small proportion of introduced species that have become invasive that South Africa does not provide a hospitable environment for alien terrestrial vertebrates outside man-modified areas. While the southern Cape Province has as temperate a climate as parts of Australia and New Zealand, it is also the temperate extension of a large tropical continent with a great suite of indigenous species making use of its resources, including many reptilian, mammalian and avian predators, not to mention pathogens. Their presence seems to make it very difficult for aliens to establish themselves. The success of European rabbits *Oryctolagus cuniculus*, peacocks and chukar partridges on Robben Island where there are few predators (Cooper and Brooke this volume; Brooke and Prins in press) strongly supports this view.

It is generally held that size of introduction has a considerable effect on survival and later capacity to invade a country. Unfortunately, there is little data on the size of South African introductions and what there is does not unequivocally support the view. The Himalayan tahr on Table Mountain are descended from one pair (Bigalke 1977). The European starlings are descended from 18 birds (Meinertzhagen 1952). It seems unlikely that there were more than 20-30 Indian mynas originally released (Kent 1927; van Nierop 1958). Yet an introduction of some 200 rooks *Corvus frugilegus* was unsuccessful (Le Sueur 1913) even though South Africa (like the whole Afrotropical Region) has few species of crows and the Indian house crow has hitched lifts on boats in sufficient numbers to establish a breeding population in Durban (Clancey 1980; Long 1981). We leave to others to offer explanations as to why some alien plants have become exceedingly invasive in South Africa whereas vertebrate animals have not.

Atkinson I A E 1985. The spread of commensal species of *Rattus* to oceanic islands and their effects on island avifaunas. International Council for Bird Preservation Technical Publication 3, 35-81.

Avery D M 1985. The dispersal of brown rats *Rattus norvegicus* and new specimens from 19th century Cape Town. Mammalia 49, 573-576.

Bajomi D 1983. Biological basis of the fight against rats. International Pest Control 25, 182-185.

Bigalke R 1937. The naturalisation of animals, with special reference to South Africa. South African Journal of Science 33, 46-63.

Bigalke R 1977. The Himalayan tahr on Table Mountain. Zoologica Africana 12, 504.

Bigalke R C 1964. Indian mynas (*Acridotheres tristis*) in Kimberley. Ostrich 35, 60.

Bigalke R C and Bateman J A 1962. On the status and distribution of ungulate mammals in the Cape Province, South Africa. Annals of the Cape Provincial Museums 2, 85-109.

Botha S A 1985. The feral pig: an unsuccessful alien in Africa. Poster presentation, National Synthesis Symposium on the Ecology of Biological Invasions, Stellenbosch 1985. Unpublished.

Bourquin O and Sowler S G 1980. The vertebrates of Vernon Crookes Nature Reserve: 1. Lammergeyer 28, 20-32.

Bowles Mr 1966. Wild budgerigars at Umbogintwini. Natal Bird Club News Sheet 134, 2.

Branch W R 1981. An annotated checklist of the lizards of the Cape Province, South Africa. Annals of the Cape Provincial Museums (Natural History) 13, 141-167.

Broadley D G 1983. FitzSimons' snakes of southern Africa. Delta Books, Johannesburg.

Brooke R K 1976. Morphological notes on *Acridotheres tristis* in Natal. Bulletin of the British Ornithologists' Club 96, 8-13.

Brooke R K 1981a. The feral pigeon - a 'new' bird for the South African list. Bokmakierie 33, 37-40.

Brooke R K 1981b. The breeding birds of Baboon Point cliff. Promerops 149, 6-7.

Brooke R K 1984a. A history of the redeyed dove in the southwestern Cape Province, South Africa. Ostrich 55, 12-16.

Brooke R K 1984b. South African Red Data Book - Birds. South African National Scientific Programmes Report No 97. CSIR, Pretoria.

Brooke R 1986. An apparent introduction of the greenrumped parrotlet to Natal. Albatross 285, 14.

Brooke R K (in press). A failed introduction of the European goldfinch in the Transvaal. Witwatersrand Bird Club News.

Brooke R K and Prins A J (in press). Review of alien species on South African offshore islands. South African Journal of Antarctic Research.

Brown C J 1985. Invasive alien birds in South West Africa/Namibia. In: Brown C J, Macdonald I A W and Brown S E (eds) Invasive alien organisms in South West Africa/Namibia. South African National Scientific Programmes Report No 119. CSIR, Pretoria.

Butler E A 1903. Goldfinch (*Carduelis elegans*) in South Africa. Zoologist, series 4, volume 7, 227-228.

Calder D R 1953. The distribution of the Indian myna. Bokmakierie 5, 4-6.

Caughley G 1977. Analysis of vertebrate populations. John Wiley and Sons, Chichester.

Cawood C C E 1983. Introducing the sable antelope to the eastern Cape. Redwing 1983, 31-33.

Cawood M 1983. Sable in the east Cape. Pelea 2, 106-107.

Chapman N G and Chapman D I 1980. The distribution of fallow deer: a world-wide review. Mammal Review 10, 61-138.

Clancey P A 1960. The Indian myna *Acridotheres tristis* (Linnaeus): is it guilty of driving the indigenous birds away? Natal Bird Club News Sheet 64, 1-2.

Clancey P A 1964. The birds of Natal and Zululand. Oliver and Boyd, Edinburgh.

Clancey P A 1976. Roseringed parakeet in Natal and Zululand. Albatross 241, 5.

Clancey P A (ed) 1980. S.A.O.S. checklist of southern African birds. Southern African Ornithological Society, Johannesburg.

Conant P 1980. House sparrow at Kalabo. Zambian Ornithological Society Newsletter 10, 77.

Connor M A 1972. House sparrows and chaffinches. Cape Bird Club Newsletter 101, 2.

Courtenay-Latimer M 1942. Letter to editor on introduced sparrows at East London. Ostrich 13, 181.

Courtenay-Latimer M 1955. The English sparrow at East London. Bokmakierie 7, 32.

Craig A J F K 1985. The distribution of the pied starling, and southern African biogeography. Ostrich 56, 123-131.

Crowe T M and Brooke R K 1985. Adaptive morphological variation in introduced house sparrows (summary). In: Bunning L J (ed) Proceedings of the Symposium on Birds and Man, Johannesburg 1983. Witwatersrand Bird Club, Johannesburg. p 30.

Cuthbertson M B 1952. Meet the mynahs. African Wildlife 6, 64-66.

Cyrus D and Robson N 1980. Bird atlas of Natal. University of Natal Press, Pietermaritzburg.

Davis D H S 1950. Notes on the status of the American grey squirrel, *Sciurus carolinensis* Gmelin, in the southwestern Cape (South Africa). Proceedings of the Zoological Society of London 120, 265-268.

De Graaff G and Penzhorn B L 1976. The re-introduction of springbok *Antidorcas marsupialis* into South African national parks - a documentation. Koedoe 19, 75-82.

De Vos V 1973. Common infectious and parasitic diseases of captured wild animals. In: Young E (ed) The capture and care of wild animals. Human and Rousseau, Cape Town.

Earle R A 1985. Predators, parasites and symbionts of the South African cliff swallow *Hirundo spilodera* (Aves: Hirundinidae). Navorsinge van die Nasionale Museum, Bloemfontein 5, 1-18.

Every B 1979. An annotated list of rare and poorly documented birds recently recorded in the eastern Cape Province XI. Bee-eater 30, 24.

Ferguson W T and Tongue P 1968. Chaffinches. Cape Bird Club Newsletter 86, 2.

Gebhardt E 1944. *Passer d domesticus* in Suedafrika. Ornithologische Monatsberichte 52, 54-55.

Gebhardt E 1959. Europaeische Voegel in ueberseeischen Laendern. Bonner Zoologische Beitraege 10, 310-341.

Gilges W and van der Spuy A J 1945. Indian mynah responsible for carrying scabies infection. Ostrich 16, 144-145.

Gill E L 1936. A first guide to South African birds. Maskew Miller, Cape Town.

Gill E L 1952. Comments on the September News Sheet. Cape Bird Club News Sheet 11, 2-3.

Godfrey R 1931. My first visit to Natal. Blythswood Review 8, 117-118.

Godfrey R 1934. Alien birds in South Africa. Blythswood Review 11, 57-58.

Gray A P 1972. Mammalian hybrids: a check-list with bibliography. Commonwealth Agricultural Bureaux, Slough.

Green L G 1948. To the river's end. Howard Timmins, Cape Town.

Griffin M and Panagis K 1985. Invasive alien mammals, reptiles and amphibians in South West Africa/Namibia. In: Brown C J, Macdonald I A W and Brown S E (eds) Invasive alien organisms in South West Africa/Namibia. South African National Scientific Programmes Report No 119. CSIR, Pretoria.

Harwin R M and Irwin M P S 1966. The spread of the house sparrow *Passer domesticus*, in south central Africa. Arnoldia (Rhodesia) 2(24), 1-17.

Hewitt J and Power J H 1913. A list of South African Lacertilia, Ophidia, and Batrachia in the McGregor Museum, Kimberley; with field notes on various species. Transactions of the Royal Society of South Africa 3, 147-176.

Honacki J H, Kinman K E and Koeppel J W (eds) 1982. Mammal species of the world. Association of Systematics Collections, Lawrence.

Howard P C and Marchant A N 1984. The distribution and status of some large mammals on private land in Natal. Lammergeyer 34, 1-58.

Howard R and Moore A 1980. A complete checklist of the birds of the world. Oxford University Press, Oxford.

Johnson S R and Cowan I McT 1974. Thermal adaptation as a factor affecting colonizing success of introduced Sturnidae (Aves) in North America. Canadian Journal of Zoology 52, 1559-1576.

Kemp M I, Kemp A C and Tarboton W R 1985. A catalogue of the birds of the Transvaal. Transvaal Museum/Transvaal Nature Conservation Division, Pretotia.

Kent C C 1927. The Indian myna. South African Journal of Natural History 6, 127-129.

Lawson W J 1971. Checklist of the birds of Durban. South African Avifauna Series 73, 1-80.

Layard E L 1867. The birds of South Africa. J C Juta, Cape Town.

Le Sueur G 1913. Cecil Rhodes, the man and his work. J Murray, London.

Liversidge R 1962. The spread of the European starling in the eastern Cape. Ostrich 33(3), 13-16.

Liversidge R 1978. Ostriches. Bokmakierie 30, 104.

Liversidge R 1979. Birds naturalised in South Africa. Bokmakierie 31, 43-44.

Liversidge R 1985. Alien bird species introduced into southern Africa. In: Bunning L J (ed) Proceedings of the symposium on birds and man, Johannesburg 1983. Witwatersrand Bird Club, Johannesburg. pp 31-44.

Lloyd P H 1975. A study of the Himalayan tahr (*Hemitragus jemlahicus*) and its potential effects on the ecology of the Table Mountain range. Unpublished report, Cape Department of Nature and Environmental Conservation.

Lloyd P H 1985. The Himalayan tahr on Table Mountain. Poster presentation, National Synthesis Symposium on the Ecology of Biological Invasions, Stellenbosch 1985. Unpublished.

Lloyd P H and Millar J C G 1983. A questionnaire survey (1969-1974) of some of the larger mammals of the Cape Province. Bontebok 3, 1-49.

Long J L 1981. Introduced birds of the world. David and Charles, London.

Macdonald I A W and Birkenstock P J 1980. Birds of the Hluhluwe-Umfolozi Game Reserve complex. Lammergeyer 29, 1-56.

Mackworth-Praed C W and Grant C H B 1963. Birds of the southern third of Africa, Vol 2. Longmans, London.

Masterson A N B 1976. Possible competition of house sparrow to the detriment of the Cape wagtail. Honeyguide 85, 48.

McCauseland D E 1952. Exotic species at the Cape. Cape Bird Club News Sheet 9, 2.

McCulloch D 1965. Indian mynas vs black-collared barbets. Natal Bird Club News Sheet 127, 4.

McDonald D 1983. Rodent control in the tropics. International Pest Control 25, 86-87.

McLachlan G R 1978. A population of *Typhlops braminus* (Daudin) on the Cape Peninsula (Reptilia: Typhlopidae). Zoologica Africana 13, 353-354.

Meinertzhagen R 1952. An historical note on the starling *Sturnus vulgaris* at the Cape of Good Hope. Bulletin of the British Ornithologists' Club 72, 47.

Michell L 1910. The life of the Rt Hon Cecil John Rhodes 1853-1902, Vol 2. Edward Arnold, London.

Millar J C B 1980. Aspects of the ecology of the American grey squirrel *Sciurus carolinensis* Gmelin in South Africa. MSc Thesis, University of Stellenbosch, Stellenbosch.

Murray J and Murray N 1984. Ringnecked parakeet in Cape Town. Promerops 164, 17.

Oatley T B 1973. Indian house crow - first SA sightings. Bokmakierie 25, 41-42.

Pakenham R H W 1979. The birds of Zanzibar and Pemba. British Ornithologists' Union Check-list 2, 1-134.

Patten G 1957. House sparrows occupying lesser striped swallow's nest. Witwatersrand Bird Club News Sheet 23, 6.

Penzhorn B L 1971. A summary of the re-introduction of ungulates into South African national parks (to 31 December 1970). Koedoe 14, 145-159.

Pienaar U de V 1964. The small mammals of the Kruger National Park a systematic list and zoogeography. Koedoe 7, 1-25.

Pinto A A da R 1959. Alguns novos "Records" de aves para o Sul do Save e Mocambique, incluindo o de um genero novo para a sub-regiao da Africa do Sul, com a descricao de novas subespecies. Boletim da Sociedade de Estudos da Provincia de Mocambique 118, 15-28.

Plug I, Dippenaar N J and Hanisch E O M 1979. Evidence of *Rattus rattus* (house rat) from Pont Drift, an Iron Age site in the northern Transvaal. South African Journal of Science 75, 82.

Pocock T N 1971. Pleistocene bird fossils from Kromdraai and Sterkfontein. Ostrich supplement 8, 1-6.

Quickelberge C D 1972. Status of the European starling at its present approximate eastern limits of spread. Ostrich 43, 179-180.

Rautenbach I L, Nel J A J and Root G A 1981. Mammals of Itala Nature Reserve, Natal. Lammergeyer 31, 21-37.

Rees J P 1949. Notes on the changes in bird life in a Johannesburg garden. Bokmakierie 1, 65-66, 79.

Roberts N 1945. Indian mynah and red-chested cuckoo in Johannesburg. Ostrich 16, 236-237.

Rose W 1950. The reptiles and amphibians of southern Africa. Maskew Miller, Cape Town.

Rowan M K 1983. The doves, parrots, louries and cuckoos of southern Africa. David Philip, Cape Town.

Sclater W L 1905. Check list of the birds of South Africa. Annals of the South African Museum 3, 303-387.

Sclater W L 1906. The birds of South Africa. Vol 4. R H Porter, London.

Shelley G E 1875. Three months on the coast of South Africa. Ibis, series 3, Vol 5, 59-87.

Siegfried W R 1962. Introduced vertebrates in the Cape Province. Cape Department of Nature Conservation Annual Report 19, 80-87.

Sinclair J C 1974. Arrival of the house crow in Natal. Ostrich 45, 189.

Sinclair J C 1980. House crow in Cape Town. Promerops 144, 7-8.

Skead C J 1962. A study of the crowned guinea fowl *Numida meleagris coronata* Gurney. Ostrich 33(2), 51-65.

Smit D J v Z 1963. Ostrich farming in the Little Karoo. Department of Agricultural Technical Services Bulletin 358, 1-103.

Smit D J v Z 1984. Russel Thornton's ostrich expedition to the Sahara 1911-1912. Karoo Agric 3(3), 19-27.

Smithers R H N 1971. The mammals of Botswana. National Museums of Rhodesia Memoir 4, 1-340.

Smithers R H N 1983. The mammals of the southern African subregion. University of Pretoria, Pretoria.

Stuart C T 1981. Notes on the mammalian carnivores of the Cape Province, South Africa. Bontebok 1, 1-58.

Tarboton W R 1968. Check list of birds of the south central Transvaal. Witwatersrand Bird Club, Johannesburg.

Thom H B (ed) 1952. Journal of Jan van Riebeeck, Vol 1, 1651-1655. Balkema, Cape Town.

Thomas A D and Kolbe F F 1942. The wild pigs of South Africa: their distribution and habits, and their significance as agricultural pests and carriers of disease. Journal of the South African Veterinary Medical Association 13, 1-11.

Van der Merwe F 1984. Europese spreeus verdring gryskopspegte. African Wildlife 38, 152-157.

Van Ee C A 1962. The distribution of ungulates in the Orange Free State. Annals of the Cape Provincial Museums 2, 53-55.

Van Nierop F 1958. The Indian mynah. Witwatersrand Bird Club News Sheet 27, 12.

Vierke J 1970. Die Besiedlung Suedafrikas durch den Haussperling (*Passer domesticus*). Journal fuer Ornithologie 111, 94-103.

Vincent J 1972. A new addition to the list of South African birds. Ostrich 43, 234-235.

Visser J 1979. New and reconfirmed records for the Cape Province with notes on some "rare" species (Sauria, Serpentes and Anura). Journal of the Herpetological Association of Africa 21, 40-50.

Voigt E A and von den Driesch A 1984. Preliminary report on the faunal assemblage from Ndondondwane, Natal. Annals of the Natal Museum 26, 95-104.

Weissenbacher D K H and Allan D 1985. Roseringed parakeet breeding attempts in the Transvaal. Ostrich 56, 169.

Winterbottom J M 1956. Red-eyed dove in the western Cape. Ostrich 27, 184.

Winterbottom J M 1961. Expansion of the range of the house sparrow. Cape Department of Nature Conservation Annual Report 16, 22-24.

Winterbottom J M 1964. Notes on the wagtails *Motacilla* of southern Africa. Ostrich 35, 129-141.

Winterbottom J M 1966. Some alien birds in South Africa. Bokmakierie 18, 61-62.

Winterbottom J M 1968. A check list of the land and fresh waterbirds of the western Cape Province. Annals of the South African Museum 53, 1-276.

Winterbottom J M 1972. Report on the house sparrow enquiry. Bokmakierie 24, 38, 37.

Winterbottom J M 1978. Birds. In: Werger M J A and van Bruggen A C (eds) Biogeography and ecology of southern Africa. W Junk, The Hague. pp 951-976.

Winterbottom J M 1979. A guide to the birds of the SW Cape. Cape Bird Club, Cape Town.

Winterbottom J M and Liversidge R 1954. The European starling in the south west Cape. Ostrich 25, 89-96.

Woodward R B and Woodward J D S 1899. Natal birds. P Davis and Sons, Pietermaritzburg.

The invasion of southern Africa's biomes

The study of alien organisms has advanced to markedly different levels in each of the biomes of the region. Initially most research and management was concentrated on the invasions by *Opuntia* in the savanna and karoo biomes but more recent research has tended to concentrate on the fynbos and the wide range of invaders that have established in it. Alien species have hardly been investigated in the grassland and forest biomes, with most of the information on the invasions occurring there being obtained incidental to other studies. Exceptions to this are the intensive research programmes on *Stipa trichotoma* in grasslands and *Acacia melanoxylon* in forests. The alien invasions occurring on the offshore islands of South Africa have been little studied but reasonably complete inventories of their biotas exist and allow for certain ecological conclusions to be drawn at the species richness level. These together with the results of some more detailed studies of aliens on the Prince Edward Islands are summarized by **Cooper and Brooke**. The invasions occurring in the karoo and desert biomes have been studied the least. The chapter of **Brown and Gubb** which covers the northern karoo and the desert and arid savannas of the northern Cape and South West Africa/Namibia is the first attempt to summarize information on alien invasions in these areas.

Macdonald and Richardson focus attention on a remarkable set of alien woody plants invading the fynbos. Australian species of *Hakea*, *Acacia*, *Albizia* and *Leptospermum,* and *Pinus* species from the Mediterranean region and North America constitute an invasive flora unique to southern Africa, even when compared with the other Mediterranean-climate regions of the world. It is an intriguing theoretical and practical question as to why herbaceous invaders have proven so much less of a problem in the fynbos than in the analogous environments of California, Chile and Western Australia. Opposed to this are the alien animals that have succeeded in the fynbos, which are the same species of the world's fauna that have thrived elsewhere.

By contrast, in the mesic savanna, grassland and forest mosaic of the eastern seaboard (**Henderson and Wells; Geldenhuys and co-authors**), cosmopolitan alien plants such as *Chromolaena odorata*, *Lantana camara*, *Melia azedarach*, *Psidium guajava* and *Opuntia* species thrive. Similarly, *Acacia mearnsii* is one of the dominant invaders in the upland grassland areas, a phenomenon apparent in montane grasslands all the way to East Africa. Much of the experience gained of these invasions in southern Africa might usefully be extrapolated both to elsewhere in Africa and further afield. As a corollary, we may predict that species such as partheneum weed *Parthenium hysterophorus*, which is invasive in analogous biomes in Asia, are likely to invade these southern African biomes.

Possibly it is the environmental singularity of our deserts and semi-desert shrublands that has given rise to the situation where almost all the major alien species are restricted to the river courses traversing these biomes. The major problem apparently looming as a result of the incipient widespread invasion of these by *Prosopis* species has its analogy in Australia. This genus might yet provide candidates for the list of the world's worst weeds.

Certainly the alien invaders of our islands are not uniquely southern African. This chapter is included here to provide an insular contrast to the continental invasions which dominate the rest of the book. The considerable experience gained by our scientists working in the Prince Edward islands is included, hopefully to assist those responsible for the management of similar oceanic islands throughout the globe.

The single common denominator of all these chapters is that there is always at least one alien species capable of extensive invasions in a biome, albeit sometimes only in specialized habitats. Riparian habitats appear particularly heavily invaded throughout southern Africa. This finding has ominous implications for the remainder of Africa, where the study of alien invasions is even less advanced than in southern Africa, and the resources to combat them severely limited.

Chapter 6

Alien species in terrestrial ecosystems of the fynbos biome

I A W Macdonald* and D M Richardson†

*University of Cape Town and †South African Forestry Research Institute.

Introduction

In this chapter we review alien invasions of the fynbos biome and address the following questions: (a) which species have invaded? (b) what are the characteristics of the invasions? and (c) do any of these characteristics appear to be unique to the biome? An outline of the salient features of the environment of the fynbos, the biotic communities which constitute it, and the history of human activies in the biome serve as background to the review.

Geographic extent

The fynbos biome is defined by Moll and Bossi (1984) and is here taken to also include all the Mountain Renosterveld (Acocks 1975). As such it occupies 77 172 km^2. The biome lies at the southern tip of Africa and includes the Cape Fold Belt Mountains and the coastal lowlands lying to the west and south of these mountains. The biome has a latitudinal range from 31° S on the west coast to 34° 50′S at Cape Agulhas to 33° 15′S near Grahamstown in the east. The longitudinal range is from 17° 51′E to nearly 27° E near Grahamstown. Various outlier patches occur in the mountains of the Karoo and further north in Transkei and Natal but their extent is small and their identity as true fynbos communities disputed (Kruger 1979).

Geology and soils

The predominant geological features are the quartz arenites of the Table Mountain and Witteberg Groups of the Cape Supergroup which form the main mountain and ridge crests. Shales of the Malmesbury Series of the Nama System and Bokkeveld Series of the Cape System form outcrops on the valley floors and coastal lowlands. Extensive areas of Quaternary and Tertiary sands are found along the southern and western coasts. Soils derived from sandstone rock types are predominantly acid and poor in nutrients. Those derived from shales are relatively fertile and often saline. Granites occur locally, especially along the west coast at lower altitudes (Lambrechts 1979) and limestones and calcareous rock are largely restricted to the south coast.

Climate

The biome has climatic gradients from west to east, south to north, coast to interior and also a pronounced altitudinal gradient (Campbell 1983). These gradients, superimposed on the mountainous topography, give rise to a mosaic of spatially diverse climates (Fuggle and Ashton 1979). In the west, the lower mountain slopes experience a true Mediterranean-type climate with cold, wet winters and dry, hot summers (Köppen's climatic type Cs); summer drought is less severe on the upper peaks (Campbell 1983). Further east, rain falls throughout the year, with maxima in spring and autumn (Köppen's climatic type Cf; Fuggle and Ashton 1979). The biome is comparatively well watered and 19% of the major catchments of southern Africa fall within the biome (van der Zel 1981), an area comprising only about four per cent of the area of the subregion (Goldblatt 1978). Persistent high speed winds are an important factor in most fynbos ecosystems (Taylor 1984; Van Wilgen and Kruger 1985).

Fire regime

Possibly the most important aspect of the climate is that, in combination with the low-nutrient soils, it gives rise to patterns of plant production and combustibility that favour a fire regime of high intensity fires at intervals varying from 10-40 years depending on site conditions and sources of ignition (Kruger and Bigalke 1984). This fire regime is the single factor that most clearly differentiates the fynbos biome from the adjacent biomes. Fynbos plant species have characteristic adaptations that enable them to survive these periodic high-intensity fires (eg serotiny, myrmecochory and the ability to resprout).

Address for correspondence: I A W Macdonald, Percy FitzPatrick Institute of African Ornithology, University of Cape Town, Rondebosch, 7700.

Biogeography

The combination of topography, geology and climate has given rise to environments that are ecologically unique in southern Africa, so much so that the area has been classified as a separate floral kingdom, distinct from the rest of Africa (Goldblatt 1978). The flora of the biome is renowned for its high alpha, gamma and delta species diversities (Kruger and Taylor 1979; Bond 1983). There is a high degree of endemism in several taxonomic groups: the flora (Goldblatt 1978); archaic invertebrates (Stuckenberg 1962); terrestrial molluscs (Van Bruggen 1978); Coleoptera (Endrödy-Younga 1978); Lepidoptera (Cottrell 1978); amphibia (Poynton 1960) and fishes (Bowmaker et al 1978). This reflects the antiquity and quasi-insular nature of the fynbos biome. Although a distinct biotic zone is recognized for the mammalian fauna (Rautenbach 1978), the species richness and endemism in this and other higher vertebrate classes does not parallel that of the flora (Bigalke 1978; Siegfried and Crowe 1983).

Vegetation and fauna

The vegetation of the biome is usually characterized by the presence of proteoid shrubs which form an upper storey of highly variable density. Ericoid shrubs and restioid hemicryptophytes together usually form the understorey. In mid-high sparse to tall mid-dense shrublands (Campbell et al 1981), woody shrubs make up the greatest part of the above-ground living phytomass but the contribution of this component is very variable (van Wilgen 1984). Trees are generally restricted to ravines and riparian habitats and seldom dominate the landscape. Towards the east, grasses become more important in the understorey. Populations of the dominant plant species are generally even-aged with recruitment occurring directly post-fire (Kruger 1979). Plants are often protected by secondary chemicals and the level of herbivory is generally low except for the period directly after fire.

In general, the lowland environment comprises a few different plant communities, usually covering relatively large areas (Siegfried and Crowe 1983) whereas in the montane environment, a much greater variety of communities, often widely dispersed, occur as relatively small entities over the landscape. The great diversity of plant communities in this zone reflects the influence of geographic factors and of a complex landscape.

Populations of small mammalian granivores are of the same order as in other biomes (Cody et al 1983), but avian granivores are scarce except in some instances after fire when they take advantage of serotinous seed releases. Avian communities are generally sparse. Nectarivores, feeding mainly on nectar of Proteaceae and Ericaceae and the associated insect fauna, and small insectivores are the most important guilds represented (Siegfried and Crowe 1983). Low bird density in the biome is ascribed to the low insect density (Siegfried 1979).

Patterns of land use and disturbance

Overall impacts of man

Man has been present in the fynbos biome for at least 600 000 years and the implications of the activities of prehistoric man for alien invasions have been extensively reviewed (Deacon this volume). The impacts on the biome of the burning, hunting and stock grazing of these peoples, although profound, have probably been smaller than those arising from man's activities since European colonization. Within 330 years, settled agriculture and the cultivation of a wide variety of alien plants has led to the complete transformation of at least 68% of the 28 508 km^2 of lowland ecosystems (Moll and Bossi 1984).

Transformation of the montane ecosystem has been less extensive, only 10% of the 42 064 km^2 occurring in the main block having been transformed by 1981 (Moll and Bossi 1984), and 10 823 km^2 (or 26%) currently being protected in State Forests or as proclaimed Mountain Catchment Areas (NAKOR 1984).

Simplification of faunal communities

Indigenous large mammals have been virtually eliminated throughout the biome (Skead 1980) and domestic livestock now graze much of the remaining areas of natural vegetation on the lowlands. Large mammalian predators (Stuart et al 1985) and a few of the large avian predators (Boshoff et al 1983) have been eliminated or reduced in numbers within the biome.

Alteration of fire regimes

One of man's main impacts throughout the biome, as it relates to alien invasions, has been the alteration of the natural fire regime (Macdonald 1984). We suggest the main changes have been: (a) increased fire frequency; (b) altered seasonal incidence (Van Wilgen and Richardson 1985a) and (c) reduced fire size. Lower mean fire intensities would have resulted from

(a) and (b). As in the precolonial era, the altered fire regime is virtually the only direct human impact on large tracts of the montane environment. Intensive grazing of recently burnt veld by domestic stock has, in places, led to severe degradation of natural communities (Wicht 1945). The degree to which fire regimes have been altered is greatest around urban areas (Kruger and Bigalke 1984) where alien organisms are also most prolific.

Afforestation with alien trees

One of the first actions of the European colonists was the planting of alien tree species (eg Krüger 1966) and afforestation was strongly advocated by authorities in the Cape of Good Hope in the late nineteenth and early twentieth century (Shaughnessy 1980 and this volume). Trees were planted for fuel and timber, to stabilize driftsands and for aesthetic reasons. Afforestation in the biome was on a limited scale up to the time of the First World War. It reached a peak in the period between the First and Second World Wars and again after the Second World War. Mainly *Pinus radiata*, *P pinaster*, *P canariensis* and *Eucalyptus* species were planted for commercial use. *Acacia cyclops* and *A saligna* were planted widely for driftsand reclamation and by 1938, 102 km^2 of lowland fynbos had been thus treated (King 1943).

History of introduced species

The role of prehistoric man

The early Khoi and San inhabitants of the area apparently introduced few species, if any, and not even commensals such as the house rat *Rattus rattus* and domestic cat *Felis catus* (Skead 1980; Deacon this volume). The introduction of alien species began with European settlement in 1652, although there is a putative record of *Rattus rattus* from Robben Island in 1614 (Skead 1980).

Plant introductions

Shaughnessy (1980 and this volume) has documented the history of the more important invasive alien trees and shrubs in the Cape Town area. Most were intentionally introduced for shade, timber and sand-binding over the period 1830 to 1900, although *Pinus pinaster* was introduced as early as 1680. Many of the species that have spread were extensively propagated and cultivated throughout the biome (Shaughnessy 1980; Stirton 1978). Notable exceptions are *Hakea sericea* and *Sesbania punicea* where the extent of the invasions are disproportionately greater than the extent of their cultivation (Kruger et al this volume).

Faunal introductions

Most vertebrate introductions have been intentional. Examples are the grey squirrel *Sciurus carolinensis* (Millar 1980), European starling *Sturnus vulgaris* and chaffinch *Fringilla coelebs* (Siegfried 1962). However, in cases such as the Himalayan tahr *Hemitragus jemlahicus* and *Felis catus*, the species have established from individuals which escaped from captivity (Siegfried 1962). Still other introductions have been inadvertent. The slow worm *Ramphotyphlops braminus*, for example, was possibly introduced to the Kirstenbosch Botanical Gardens with potted plants (McLachlan 1978). By contrast almost all invertebrate introductions have been unintentional. These include the numerous alien molluscs, the termite *Cryptotermes brevis*, the ant *Iridomyrmex humilis* and the wasp *Vespula germanica* which are thought to have been introduced in cargo or in ships' ballast (Skaife 1955; van Bruggen 1964; Whitehead and Prins 1975; Coaton 1981).

Introductions from elsewhere in southern Africa

Several southern African species have been introduced into the biome and have established free-living populations here, eg the guineafowl *Numida meleagris* (Skead 1962), the gecko *Pachydactylus bibroni* (Siegfried 1962), several species of trees and shrubs (Moll and Scott 1981; de Villiers and McDowell 1982) and grasses (Adamson and Salter 1950). Within southern Africa, the fynbos biome seems to be uniquely susceptible to invasion by species from elsewhere in the subcontinent. Possibly this is a reflection of the quasi-insular nature of the biome, a situation mirrored by the mesic south-west of Australia which is also being invaded by intracontinental aliens (Macdonald 1985a,b).

Patterns of invasions by some successful alien species

Figure 1 shows patterns of invasion by four alien organisms in the fynbos biome and serves to illustrate differences in patterns and rates of spread of plants, molluscs, birds and mammals.

Hakea sericea has spread to occupy most mountain veld around Cape Town, Franschhoek and George. Spread was rapid in all directions and was not significantly aided by human dissemination, in marked contrast to most of the other important woody plant species (Shaughnessy this volume). Invasion by the snail *Theba pisana*, after its introduc-

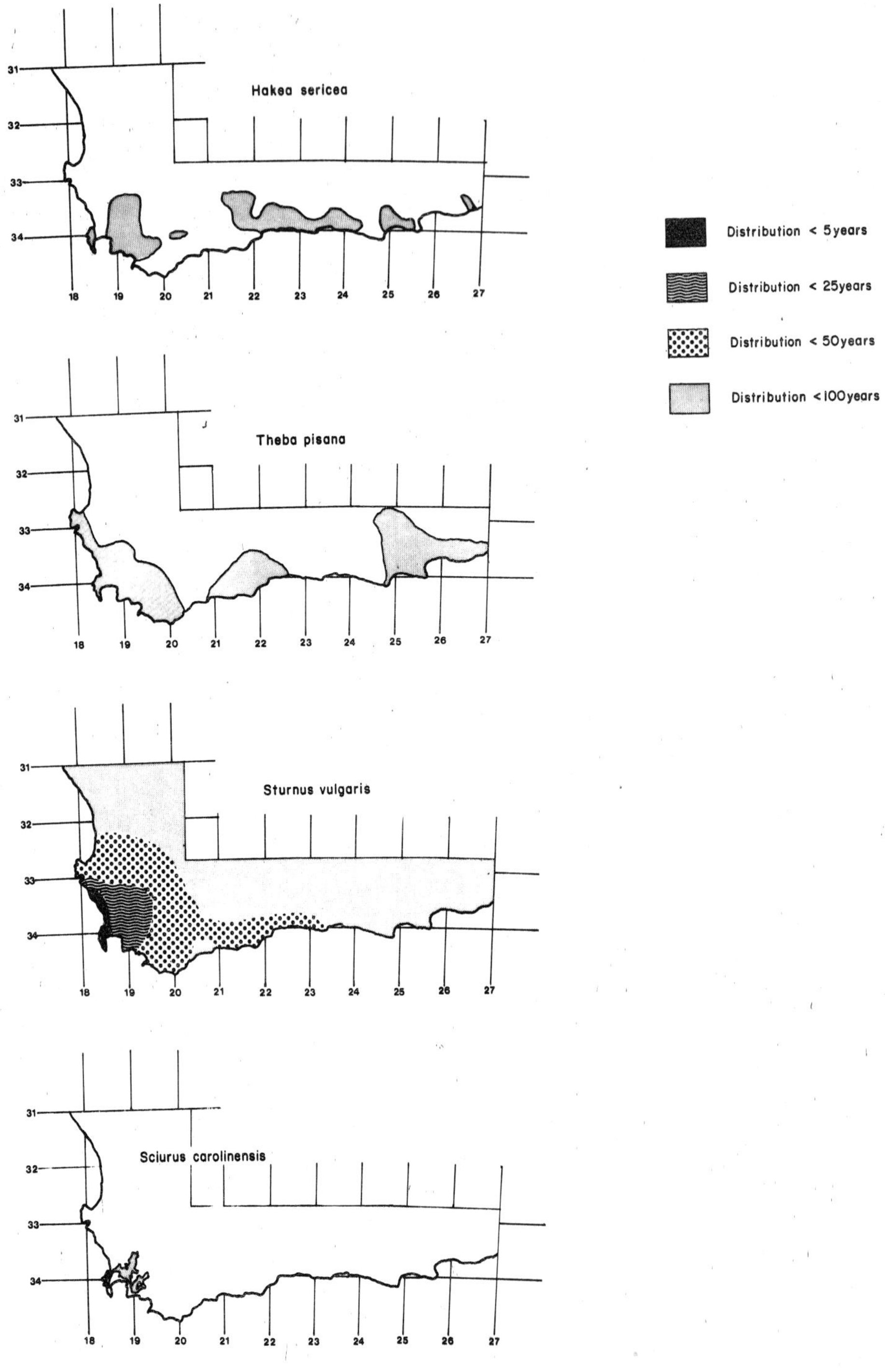

Figure 1. Distributions of some successful alien invaders of the fynbos biome.

tion to ports was facilitated by artificial dispersal by rail, road or ship though natural dispersal over short distances is apparently also implicated (Van Bruggen 1964). The starling *Sturnus vulgaris*, introduced in 1899, spread along a front in all directions from Cape Town to occupy the entire biome within 70 years of introduction. Spread was most rapid eastwards, where the availability of alien trees and man-made structures for roosting and nesting facilitated rapid colonization (Winterbottom and Liversidge 1954). The squirrel *Sciurus carolinensis* spread rapidly to occupy all potential habitat around Cape Town but further spread in the biome was prevented by the absence of suitable habitat. The slow increase in this species range has apparently been facilitated by the increase and maturation of stands of its favoured alien tree species (Millar 1980).

The nature of the invasions

Invasions of Mountain Fynbos by trees and shrubs

The most important alien invaders of the biome are numerous tree and shrub species, most of which have spread from cultivation and now cover small portions of the biome entirely. Coastal mountain ranges throughout the biome are invaded by *Hakea sericea*, *Pinus pinaster* and *Acacia longifolia* which form dense impenetrable thickets. *Hakea sericea*, introduced to the biome in around 1830 and planted on a very limited scale around Cape Town, Franschhoek and George, spread rapidly eastwards from Cape Town and Franschhoek and in all directions from George to cover almost 4 800 km^2 or 14% of the area of Mountain Fynbos about 130 years after its introduction (Kluge and Richardson 1983; Macdonald 1984). The present distribution of this species is positively associated with the occurrence of soils derived from quartzite and sandstone and its spread eastwards from Cape Town and northwards from George has been restricted to some extent by barriers of unsuitable nutrient-rich substrates (Fugler 1979; Richardson 1984). Typical stands of this species consist of three to four metres tall shrubs at a density of around 9 000 stems per ha ten years after a fire (Van Wilgen and Richardson 1985b). The other major invader of Mountain Fynbos, *P pinaster*, has been found to be the most widely dispersed species in the Cape Peninsula mountains where it was present in 82% of 98 circular quadrats (10,5 ha) systematically located throughout the northern Peninsula mountains (Hall 1961). The spread of *P pinaster* on Stellenboschberg has been analysed from successive aerial surveys (Smit and De Kock 1984). The area covered by "dense infestations" (more than 50 plants km^{-2}) increased from four per cent in 1938 to 36% in 1977. The most recent estimate of the area invaded by *Hakea* and *Pinus* species throughout the biome is 7 592 km^2 (Macdonald et al 1985a).

Hakea sericea and *P pinaster* have the following factors in common which may have contributed to their success: both are dispersed by wind; both have winged seeds which are held in heat-resistant serotinous cones and which are often released after fire has killed the parent plants. However, the degree of serotiny is much higher in *H sericea* than in *P pinaster* (F J Kruger personal communication).

Acacia longifolia is a comparatively recent arrival in most Mountain Fynbos areas. However, it has recently shown rapid increases in both its distribution and density (McLachlan et al 1980; Taylor et al 1985) and is currently ranked the second most important alien plant invader in the biome (Macdonald and Jarman 1984). Its seeds are spread both by birds and water and it often forms dense thickets following fire, alien clearing operations or soil disturbance (McLachlan et al 1980; Pieterse 1984; Macdonald and Richardson unpublished).

Invasions of lowland areas and riparian communities by trees and shrubs

In contrast to the situation in Mountain Fynbos, *Acacia cyclops* and *A saligna* are the most important species in the Strandveld and Coastal Fynbos. A survey of the western lowlands between the Berg River and False Bay revealed that by 1971, 17% of the remaining area of natural vegetation was invaded by alien woody plants (Boucher 1981). The seeds of the most important woody plant invaders of the lowlands are dispersed by birds and mammals (Middlemiss 1963; Knight and Macdonald 1985), man and water (Macdonald and Jarman 1984; Macdonald et al 1985a). Studies of sequential air photos have shown that certain *Acacia* species invade most rapidly along watercourses. The direction and rate of spread suggest that seed dispersal in water is the major cause of these patterns (Brownlie 1982). Possibly as a consequence of the efficacy of waterborn seed dispersal, riparian habitats throughout the biome are heavily invaded by numerous species, eg *A longifolia*, *A saligna*, *A mearnsii*, *Albizia lophantha* and *Sesbania punicea* (Macdonald and Jarman 1984). A helicopter survey of the lowlands in the western Cape showed that *Sesbania punicea* was present in 64% of the 532 km of rivercourse surveyed (Bruwer 1983). *Acacia* and other thicket-forming alien woody plant species in the biome were estimated to have invaded some 8 962 km^2 in 1984 (Macdonald et al 1985a). The extent of invasion by herbaceous alien plants has not been studied.

Vertebrate invasions

Of the vertebrate invaders, only the starling *Sturnus vulgaris* and the house sparrow *Passer domesticus* have spread throughout the biome. *Fringilla coelebs* is confined to the Cape Peninsula and is declining in numbers (Cape Bird Club 1979). Of the introduced mammals only the Himalayan tahr *Hemitragus jemlahicus* has been able to invade unmodified fynbos ecosystems and even it has not spread from Table Mountain because of intensive shooting since the early 1970's and because there are no corridors to other suitable habitats. The house mouse *Mus musculus* and *Rattus rattus* both have extensive ranges in the biome but are thought to occur mainly in areas of human habitation especially in drier areas (Smithers 1983). However, instances are known where both species have invaded areas of indigenous fynbos vegetation (Macdonald et al this volume). The brown rat *R norvegicus* is mainly restricted to ports and coastal towns (Smithers 1983) but has been recorded invading seminatural areas within the biome (Middlemiss 1975; Stuart 1976). The squirrel, *Sciurus carolinensis*, introduced to Cape Town in 1900, rapidly became naturalized but is entirely dependent on introduced tree species (*Pinus* species and *Quercus robur*) and has not invaded natural ecosystems (Millar 1980).

Invertebrate invasions

A possibly important alien invertebrate that has invaded natural communities in the biome is the Argentine ant *Iridomyrmex humilis* (Skaife 1955; Donnelly 1983). The extent of invasion by this species is, however, related to the extent of disturbance in the form of roads and human habitation (De Kock 1984). *Iridomyrmex humilis* was found at 42% of 83 sites sampled within the fynbos between the Cederberg and the De Hoop Nature Reserve during 1983/84 (De Kock 1984). *Theba pisana* and *Helix adspersa* are the two most widely distributed alien molluscs in the biome, although at least 17 terrestrial species are known from the Cape Peninsula alone (Van Bruggen 1964). *Theba pisana*, first noted in the biome in 1881, is now found along most of the coastal strip south of St Helena Bay and has extended its range on the lowlands inland around Cape Town and Port Elizabeth (Van Bruggen 1964).

The case of *Phytophthora cinnamomi*

The pathogenic root fungus *Phytophthora cinnamomi* was thought to have been introduced to the biome and was considered to pose a considerable threat to its flora (Knox-Davies 1975). More recently, analyses of the current distribution of *P cinnamomi* in the river systems of the biome (Von Broembsen 1984a), its host range (Von Broembsen 1984b) and the patterns of mortality in indigenous vegetation (Von Broembsen and Kruger 1985) have led to the conclusion that the species is indigenous. However, the possibility that one of the two genetic strains of the species present in the biome is alien requires further investigation.

Ecological impacts

Effects of woody plants on soil erosion rates

The only known impacts of invasive alien organisms on ecosystem processes in the fynbos biome are those of woody plant species. Soil erosion and deposition are considered to be severely affected in several ways. River bank erosion has been accelerated following the invasion of riparian vegetation by species such as *Acacia mearnsii*, *A longifolia*, *A saligna*, *Sesbania punicea* and *Pinus pinaster*. These species establish in and grow up through the indigenous vegetation which is well adapted to the flash floods that occur in most fynbos catchments. The alien trees, however, are not able to withstand these floods and are ripped out, often dislodging mats of indigenous vegetation such as those of *Prionium serratum* (Beyers 1959; Heydorn and Tinley 1980; Macdonald unpublished). The exposed mineral soil is then subjected to accelerated erosion.

Soil erosion is also considered to be accelerated under dense stands of certain woody alien species such as *Pinus* species which characteristically have a very sparse ground cover (Cowling et al 1976; Jacot Guillarmod 1980). However, other writers consider the establishment of alien woody plants to have decreased erosion rates, both of water erosion, as on Table Mountain (Lückhoff 1951), and of wind erosion, as on the Cape Flats (Wicht 1945). Following the clearing of long-established infestations of these species, particularly where the plants are felled, allowed to dry out and then burned, soil erosion is often accelerated on steep slopes (Cowling et al 1976; Richardson and Van Wilgen in press).

Effects of woody plants on coastal sand movements

Alien trees and shrubs have had a marked effect

on the movement of coastal dune sands, thus altering sediment dynamics. The species mainly responsible for this are the Australian wattles *Acacia cyclops* and *A saligna*. These have been extensively planted by the state forestry authorities to stabilize areas of windblown sand (Walsh 1968; Shaughnessy 1980; Shaughnessy this volume). Although some of this area was caused by man-induced changes in the vegetal cover (King 1943), recent research has shown that there are naturally occurring unvegetated dunes within the biome (McLachlan et al 1982; Lord et al 1985; Lubke and Avis 1985). Where the dunes that have been stabilized form part of a headland bypass dunefield, the reduction in sand movement can have profound effects on the distribution and extent of sandy beaches in the vicinity (Heydorn and Tinley 1980; Lord et al 1985; Lubke 1985). The ability of alien plants to establish and grow more successfully than indigenous plants on coastal sand dunes has been observed in southern and eastern Australia and has been recorded as having had profound effects on coastal sediment dynamics in these areas as well (Heyligers 1985; Sauer 1985).

Effects of woody plants on hydrology and fire regime

Although it is widely contended that dense stands of alien woody plants reduce the flow of water from fynbos catchments (eg Jacot Guillarmod 1980) there are no data available on this point. Extrapolation from plantations of alien tree species indicates that this reduction does occur (Van Wyk 1977; Bosch and Hewlett 1982).

Fuel loads are greater in dense stands of aliens than in fynbos but fires are generally more easily ignited in fynbos where there is an abundance of fine material in the herbaceous layer. Only under extreme weather conditions is fire intensity much higher in invaded areas. Fires under these conditions are more difficult to contain and are potentially more damaging to ecosystems than fires in indigenous vegetation (Van Wilgen and Richardson 1985b).

Effects of woody plants on geochemical cycling

Dense infestations of alien *Acacia* species affect geochemical cycling. *Acacia cyclops* has been shown to fix nitrogen (Roux and Warren 1963) and soil nitrogen levels have been shown to be increased by an order of magnitude following dune stabilization with this species (McLachlan et al 1982). Carbon cycling is also radically changed with above-ground biomass and mean litterfall rates being approximately three times higher in thickets of *A cyclops* and *A saligna* than in indigenous fynbos vegetation (Milton 1981; Milton and Siegfried 1981). Soil organic matter content under a thicket of *A cyclops* has been found to be twice that under indigenous dune thicket on the Alexandria dunefield east of Port Elizabeth (McLachlan et al 1982).

Milton (1981) has calculated from litterfall data and information on the chemical composition of alien and indigenous plants, that annual inputs of nitrogen and phosphorus to the soil under a dense thicket of *Acacia* would be about nine times greater than those under indigenous fynbos vegetation. Subsequent investigations of soil phosphorus levels in Strandveld have shown that both total and available phosphorus are approximately 50% greater under *Acacia* thickets than under fynbos (E. Witkowski and D.T. Mitchell unpublished). No work has yet been conducted on the relative rates of mineral cycling in alien and indigenous communities. Because fynbos plants are generally adapted to conditions of low soil nutrients the eutrophication of fynbos soils possibly has important implications for the survival or re-establishment of indigenous species on invaded sites. In the comparable heathlands of South Australia Heddle and Specht (1975) showed that three years' application of phosphate fertilizer resulted in the elimination or reduction in density of several indigenous species and their replacement by a range of herbaceous alien plants as measured twenty years after the fertilization.

Effects of woody plants on species richness of indigenous plant communities

Impacts on community properties have been more extensively documented than have those on ecosystem properties. Once again it is the impacts of alien woody plants that are best known. Plant species richness has been shown to be reduced in communities dominated by alien plants relative to uninvaded indigenous communities (Figure 2). This reduction in species richness has been best documented in a 'before and after' study of the fynbos in an area planted to *Pinus radiata* (Richardson and Van Wilgen 1986). The alien trees were planted in small pits in the fynbos with minimal disturbance to the indigenous vegetation. After 35 years the mean number of species per 0,1 m^2 quadrat was 2,1 (1,8 if alien species were excluded) compared with 8,5 prior to planting. The cover of the indigenous vegetation had decreased from 74 to 19% and plant density from 260 to 78 plants m^{-2}. Different life forms showed varying degrees of resistance to suppression by alien plants. Proteaceous shrubs with canopy-stored seeds and limited dispersal capabilities were particularly susceptible (Richardson and van Wilgen 1986).

All the studies where paired estimates had been made of indigenous plant diversity, density or cover show these parameters to be reduced by dense or moderately dense infestations of alien plants (Cowling et al 1976; Behrens 1985). The poor development of understorey vegetation in dense stands of alien trees is similar to that shown in indigenous forest vegetation in the south-western Cape (Bester 1976). Allelopathy, documented for *Pittosporum undulatum* in Australia (Gleadow 1982), has become evident where this species has invaded scrub and forest patches in parts of the biome (Richardson and Brink 1985).

Measures applied to control invasive trees and shrubs sometimes reduce the rate of recovery of indigenous elements relative to recovery following normal disturbance. Burning of slashed *H sericea* shrubs under severe weather conditions has resulted in very intense fires which kill subterranean rootstocks and seeds on or in the soil (Richardson and Van Wilgen, in press). Following the felling of an infestation, with or without subsequent burning, species richness is still depressed. Species richness has not been monitored in the long term so that it is not known if it ever returns to preinfestation levels and, if it does, how long it takes to do so. In eastern Cape grassy fynbos it has been shown that, six years after the eradication of a 15 to 20-year old multispecies alien infestation (primarily *Pinus* species), the fynbos had regenerated sufficiently for a fire to be followed by normal post-fire regeneration of the vegetation (Jacot Guillarmod 1980, 1983; Richardson et al 1984). This agrees with the unsubstantiated assertion of Lückhoff (1951) that if *Pinus pinaster* infestations up to 20 years old are removed, recovery of the fynbos is still possible.

If infestations are not removed there is little doubt that local populations of at least certain indigenous species will become extinct. An analysis of a sample of 70 plant species in the south-western Cape which were evidently threatened with extinction showed that 33 were affected thus by invasive alien woody plants (Hall et al 1980). Similarly, of the 58 threatened plant species occurring on the Cape Peninsula, 31 are jeopardized by alien woody plants (Hall and Ashton 1983). As the extent of the remaining areas of natural vegetation decreases and alien plant invasions increase, more indigenous species will become en-

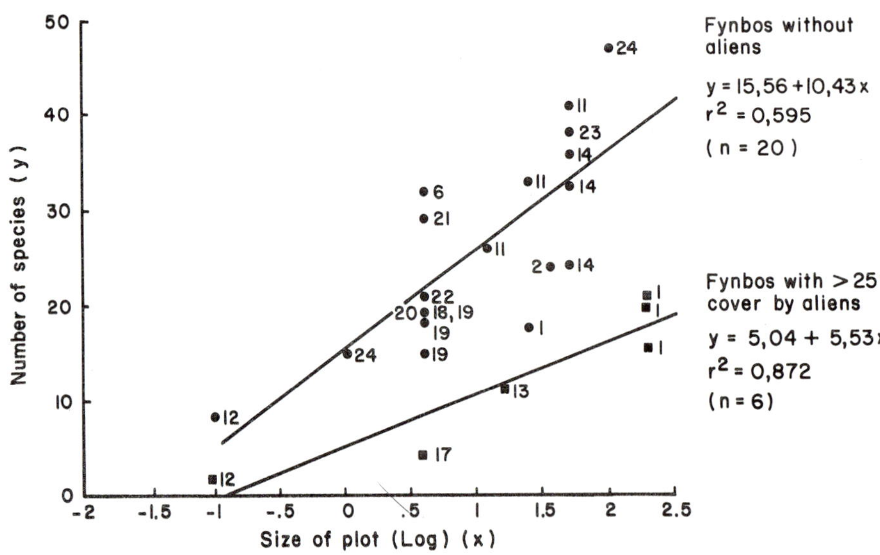

Figure 2. The relationships between the number of indigenous vascular plant species and plot size in fynbos areas densely infested (■) by alien woody plants and in unifested areas (●). Each number indicates a different site. Numbering system relates to an unpublished study of the authors which is being prepared for publication. Sources of data are: (1) Cowling et al 1976; (2) Behrens 1985; (6) Richardson and Van Wilgen in press; (11) Van Wilgen and Kruger 1985; (12) Richardson and Van Wilgen 1986; (13) Milton 1976; (14) Van Wilgen 1981; (17) Richardson and Forsyth unpublished; (18) Richardson and Van Wilgen unpublished; (19) (20) and (21) le Maitre unpublished; (22) Richardson and Forsyth unpublished; (23) Taylor 1984; and (24) Bond 1983.

dangered. Several authors have predicted the almost total disappearance of the fynbos flora if the spread of alien plants is not checked (Wicht 1945; Taylor 1977).

Effects of woody plants on the indigenous fauna

The impacts of alien woody plants on the fauna of the biome have also been considerable. Thickets of *Acacia cyclops* and *A saligna* have been shown to have a depauperate avifauna (total species number 86 with 37 occurring on five per cent or more of the bird lists made in this habitat) compared to the coastal fynbos they had replaced (total = 146 with 56 occurring on five per cent or more of lists) (Winterbottom 1970). The avifauna shows a reduced percentage of insectivores and an increase in granivores. The species that decrease most in frequency of occurrence in these thickets are those dependent on the smaller ericoid and graminoid plants within the indigenous vegetation eg Karoo robin *Erythropygia coryphaeus*, grassbird *Sphenoeacus afer*, greybacked cisticola *Cisticola subruficapilla* and lesser doublecollared sunbird *Nectarinia chalybea*. Of the 14 species most frequently recorded in these thickets 10 are generalist species that are also amongst the 14 species most frequently recorded in gardens eg Cape turtle dove *Streptopelia capicola*, laughing dove *S senegalensis*, fiscal shrike *Lanius collaris* and Cape sparrow *Passer melanurus* (Winterbottom 1970). The alien woody plants are not merely having local effects on the avifauna; studies have shown that at least two species, the hadeda ibis *Bostrychia hagedash* and pied barbet *Lybius leucomelas*, which are dependent on trees for roosting and nesting, have invaded the biome at least partly in response to the spread of alien trees (Macdonald 1986; Macdonald et al 1986). Historical analysis of the avifauna of terrestrial habitats in the south-western portion of the biome indicates that 36 species have invaded the area as a result of alien vegetation. A further 27 species have markedly expanded their ranges within the biome. Thus 63 of the current total avifauna of 224 species (28%) have been favoured by these alien plants (Macdonald et al 1985b).

Small mammal communities are reduced in species richness in some instances. In the Alexandria dune field, pure stands of *Acacia cyclops* were found to have only a single species of rodent, the four-striped fieldmouse *Rhabdomys pumilio*, while five species were recorded on adjacent littoral dunes having an indigenous plant cover (McLachlan et al 1982). Biomass of *R pumilio* in mixed thickets of *A cyclops* and *A saligna* on the Cape Flats has been estimated at 2,5-9,4 kg^{-ha} (David 1980). This is approximately five to 10 times higher than the total rodent biomass in fynbos (David 1981). Its diet in the alien thicket was primarily *Acacia* seed (Shelton 1975). On the other hand, in *Hakea sericea* stands of varying density (1%, 35% and 95% of canopy cover) communities of indigenous small mammals, birds and arthropods were found to be similarly rich, both in species and in numbers (Breytenbach this volume). The effects of these faunal communities on their indigenous food plants were, however, much greater in the densely infested stand to the extent that recruitment of these plants following a fire could be jeopardized.

Impacts of other alien groups

Ecological impacts have not been investigated for herbaceous alien plants which, within this biome, are generally restricted to modified ecosystems, such as along railway lines (Raitt 1983), or are present only briefly following fire in areas of natural vegetation.

The browsing and trampling of *Hemitragus jemlahicus* is considered to have given rise to a reduction in plant cover and a concomitant increase in soil erosion (Lloyd 1975).

Iridomyrmex humilis displaces indigenous ant species (Skaife 1955; Donnelly 1983; De Kock 1984). The implications of this for the regeneration of the many myrmecochorus indigenous plant species has been stressed by Bond and Slingsby (1984). Such displacement has changed the pattern and reduced the abundance of seedling recruitment in *Mimetes cucullatus*, for example (Bond and Slingsby 1984).

History of control

The evolution of control programmes

Invasive trees and shrubs have long been a cause of concern in the biome. Phillips (1938) reported that farmers in the Bathurst district first met to discuss the spread of *Hakea sericea* in about 1863. Shaughnessy (1980) defines three phases of opinion by authorities concerning alien trees around Cape Town between 1884 and 1936: i) a determination to achieve total cover of trees (1884-1905); ii) defined plantations plus areas of abandoned plantations beyond which a policy of laissez faire applied (1906-1935); iii) areas beyond demarcated plantations were to be actively cleared of pines (1936 onwards). This is an accurate reflection of developments in the fynbos biome as a whole.

The control of *Hakea* and *Pinus* species in Mountain Fynbos areas

In the case of *Hakea* species and *Pinus pinaster*, the Weeds Act of 1937 did little to stop their spread. The lack of a definite mountain catchment management policy, together with the general opposition to the use of controlled burning as a management tool, made the attainment of control objectives impossible. A biological control programme for *H sericea* was initiated in 1962, resulting in the release of two species of host-specific seed-attacking insects (Neser and Annecke 1973; Moran and Neser this volume; Neser et al this volume). The other important development in the control of *H sericea* and other alien plants in mountain land has been the initiation of an active control policy by the Forestry Branch (Fenn 1980). With the initiation of regular prescribed burning in the late 1970's, proclaimed catchments were divided into management units. Each unit was to be burned at approximately 15-year intervals. All woody alien plants are felled prior to a burn and the area is burned about 18 months later to kill seedlings (Kruger 1977; Fugler 1983). This programme has been very successful and of the 7 592 km^2 once infested by *Hakea* and *Pinus* species, 1 579 km^2 (21%) had been successfully cleared by the end of 1984 (Macdonald et al 1985a). Assuming that the current rate of progress is maintained, most areas for which the Forestry Branch is responsible will have been cleared by around 1995 (Kluge and Richardson 1983).

The relatively recent emergence of the hard-seeded *Acacia* species and *Albizia lophantha* as important invaders of Mountain Fynbos is complicating control programmes. Prolific regeneration of these species after fire is now having to be treated with arboricides which is both expensive, often ineffective, and potentially damaging to indigenous plant regeneration (Macdonald et al 1985a).

The control of woody aliens in the lowlands

Acacia saligna and *A cyclops*, the most important aliens in lowland vegetation of the biome, were, until very recently, widely planted for driftsand reclamation and a variety of other purposes. The use of these species has been strongly discouraged by the Forestry Branch since 1980 and official management instructions currently stipulate that indigenous species should be used for reclamation. In contrast to the montane environments of the biome, where some control is exercised over land use in proclaimed catchments, most land invaded by *A saligna* and *A cyclops* falls outside the area of jurisdiction of the Forestry Branch. Control of these species is carried out by a number of smaller organizations, municipalities and volunteer groups. In total it is estimated that of the 8 962 km^2 infested by alien plants other than *Hakea* and *Pinus* species, ie mainly *Acacia* species, only 870 km^2 (10%) had been successfully cleared by 1984 (Macdonald et al 1985a). The ability of several of these species to coppice after felling and the density and longevity of the soil-stored seed bank makes control of these species difficult and expensive (Milton and Hall 1981; Macdonald et al 1985a). The control of these species, particularly on privately owned farm land, poses one of the most significant challenges to fynbos conservation.

The control of herbaceous alien plants

The large-scale control of herbaceous alien plant invaders in areas of natural vegetation has not yet been undertaken within the biome. The only exceptions have been the subshrub *Hypericum perforatum* and the grass *Stipa trichotoma*, both of which are confined mainly to areas of modified vegetation within the biome. The former has been the subject of a successful biocontrol programme (Kluge and Gordon 1983) while the latter is being controlled manually in at least one of the small areas currently known to be infested (Hall and Bulley 1980). Both these species are being controlled primarily on account of their potential threat to agriculture.

The control of alien fauna

The control of alien animals has not yet been accorded a high priority. Early attempts to limit the spread of *Sciurus carolinensis* through the application of a bounty system were abandoned as being futile (Millar 1980). The control of *Hemitragus jemlahicus* on Table Mountain is possibly the only example of the control of an alien animal for nature conservation reasons in the biome. The population density has been reduced to and kept at low levels through a programme of shooting initiated in the early 1970's and no increase in distribution range has been observed since then (Lloyd 1975). Research is being conducted on the ant *Iridomyrmex humilis*, to define the extent of its distribution, to analyse its effects and explore methods of control; recommendations have been made for limiting its further spread into fynbos areas (Macdonald and Jarman 1984; De Kock 1984). Until the ecological impacts of faunal aliens such as *Theba pisana* and *Vespula germanica* have been investigated, no rational basis exists for evaluating the importance of controlling these species.

Control priorities within the biome

The obvious impacts of alien woody plants has led to their control being accorded overriding significance in the management of natural and seminatural vegetation throughout the fynbos biome. This situation will not change unless major progress is made in the control techniques available for these species. In this respect, the biological control of the Australian *Acacia* species should be given the highest priority.

Conclusions

No attempt is made to draw conclusions regarding the assertion that the fynbos biome is especially susceptible to alien plant invasions, as this matter is dealt with in detail elsewhere (Macdonald 1984; Macdonald et al this volume). Instead conclusions will be made on the three questions posed at the start of this review.

Which species have invaded?

The most important alien invaders, both in terms of the extent of their infestations and their impacts, are undoubtedly the numerous species of woody trees and shrubs. However, the biome has been invaded by species from most taxonomic groups, with normally only a few species proving to be important in each of the faunal groups.

What are the characteristics of the invasions?

The invasions by woody plant species have been widespread and have often resulted in dense stands which radically alter biotic and abiotic features of the invaded landscape. These species have been able to invade relatively undisturbed ecosystems and the regular occurrence of fire in these ecosystems has tended to favour their spread and proliferation.

The impact of these woody plant invasions on the indigenous flora has been marked. Most indigenous plants in the biome appear to be intolerant of shade and decline in vigour or die once overtopped by invading tree species.

The invasions of herbaceous species and alien animals have generally been much less successful, with many of the alien species being restricted to sites of human disturbance.

What appears to be unique about alien invasions of the fynbos?

The dominance of the biome's invasive alien flora by trees and shrubs, at least in terms of the more important species, is possibly one of the features unique to the biome. Another is the large reduction in plant species diversity that follows the invasion of fynbos vegetation by alien woody plants.

The apparent high susceptibility of the biome to invasion by species from elsewhere in the subcontinent is also possibly unique. That more alien animal species have been able to establish populations in this biome than elsewhere in the subcontinent might also be unique (see Macdonald et al this volume). However, this could simply be the result of the extensive modification to ecosystems and impoverishment of the fauna that has characterized the biome in recent times.

The apparent occurrence of a "vacant" tree niche in the biome (Campbell et al 1979), the alledged occurrence of a "vacant" fish niche in the biome's rivers (Bruton this volume) and its susceptibility to invasion from elsewhere in the subcontinent all agree with the hypothesis that this small biome is in several ways akin to an island. The high levels of endemism in its indigenous biota strengthen this hypothesis. As in many "insular" environments throughout the world, alien organisms have wreaked havoc in the fynbos biome and will continue to do so until effectively controlled.

Acocks J P H 1975. Veld types of South Africa. Memoirs of the Botanical Survey of South Africa 40, 1-128.

Adamson R S and Salter T M (ed) 1950. Flora of the Cape Peninsula. Juta, Cape Town.

Behrens T 1985. *Pinus pinaster* clearance from the Back Table of Table Mountain. BSc Project, University of Cape Town, Cape Town. 20 pp.

Bester A P 1976. A conservation and management report of Newlands Forest. BSc Project. University of Cape Town, Cape Town. 21 pp.

Beyers A P 1959. Foreword. In: The control of alien vegetation committee (eds) The green cancers in South Africa. Botanical Society of South Africa, Kirstenbosch, p 3.

Bigalke R C 1978. Mammals. In: Werger M J A (ed) Biogeography and ecology of southern Africa. W Junk, The Hague. pp 981-1048.

Bond W 1983. On alpha diversity and the richness of the Cape Flora: A study in southern Cape Fynbos. In: Kruger F J, Mitchell D T and Jarvis J U M (eds) Mediterranean-type Ecosystems: The role of nutrients. Ecological Studies 43. Springer-Verlag, Berlin. pp 337-356.

Bond W and Slingsby P 1984. Collapse of an ant plant mutualism - the Argentine ant, *Iridomyrmex humilis* and myrmecochorous Proteaceae. Ecology 65, 1031-1037.

Bosch J M and Hewlett J D 1982. A review of catchment experiments to determine the effect of vegetation changes on water yield and evapotranspiration. Journal of Hydrology 55, 3-23.

Boucher C 1981. Western Cape Province lowland alien vegetation. In: Moll E (ed) Proceedings of the Symposium on coastal lowlands of the Western Cape. University of the Western Cape, Cape Town. p 67.

Boshoff A F, Vernon C J and Brooke R K 1983. Historical atlas of the diurnal raptors of the Cape Province. Annals of the Cape Provincial Museums 14, 173-297.

Bowmaker A P, Jackson P B N and Jubb R A 1978. Freshwater fishes. In: Werger M J A (ed) Biogeography and ecology of southern Africa. W Junk, The Hague. pp 1181-1230.

Brownlie S F 1982. The effects of recent landuse on a fynbos site - final report. Unpublished report. School of Environmental Studies, University of Cape Town, Cape Town. 158 pp.

Bruwer J P 1983. Besmetting van *Sesbania punicea* en ander onkruide in die lope van sekere riviere in Wes-Kaap. Unpublished report. Department of Agriculture, Elsenburg. 5 pp.

Campbell B M 1983. Montane plant environments in the fynbos biome. Bothalia 14, 283-298.

Campbell B M, Cowling R M, Bond W and Kruger F J 1981. Structural characterization of vegetation in the fynbos biome. South African National Scientific Programmes Report No 52. CSIR, Pretoria.

Campbell B M, MacKenzie B and Moll E J 1979. Should there be more tree vegetation in the mediterranean climate region of South Africa. Journal of South African Botany 45, 543-557.

Cape Bird Club 1979. A guide to the birds of the south western Cape. Cape Bird Club, Cape Town.

Coaton W G H 1981. The West Indian dry-wood termite, *Cryptotermes brevis*. Pamphlet 273, Directorate of Forestry, Pretoria. 6 pp.

Cody M L, Breytenbach G J, Fox B, Newsome A E, Quinn R D and Siegfried W R 1983. Animal communities: diversity, density and dynamics. In: Day J A (ed) Mineral nutrients in Mediterranean ecosystems. South African National Programmes Report No 71. CSIR, Pretoria. pp 91-110.

Cottrell C B 1978. Aspects of the biogeography of southern African butterflies. Zambezia (Supplement). 87 pp.

Cowling R M, Moll E J and Campbell B M 1976. The ecological status of the understorey communities of pine forests on Table Mountain. South African Forestry Journal 99, 13-24.

David J H M 1980. Demography and population dynamics of the striped field mouse *Rhabdomys pumilio* in alien *Acacia* vegetation on the Cape Flats, Cape Province, South Africa. PhD Thesis, University of Cape Town, Cape Town. 349 pp.

David J H M 1981. The mammals of the coastal lowlands of the western Cape. In: Moll E (ed) Proceedings of a symposium on coastal lowlands of the Western Cape. University of the Western Cape, Cape Town. pp 53-56.

De Kock A E 1984. Verslag oor Fynbos-bioom projek: Argentynse mier-opname. Unpublished Report, Jonkershoek Forestry Research Centre, Stellenbosch. 62 pp.

De Villiers A L and McDowell C R 1982. The indigenous exotic vegetation of Table Mountain. African Wildlife 36, 120.

Donnelly D 1983. Comparison of community structure of epigaeic invertebrate fauna with particular reference to ants and beetles between fynbos sites under different management practices. MSc Thesis, University of Stellenbosch, Stellenbosch. 113 pp.

Endrödy-Younga S 1978. Coleoptera. In: Werger M J A (ed) Biogeography. and ecology of southern Africa. W Junk, The Hague. pp 797-822.

Fenn J A 1980. Control of hakea in the Western Cape. In: Neser S and Cairns A C P (eds) Proceedings of the 3rd National Weeds Conference of South Africa. Balkema, Cape Town. pp 167-173.

Fuggle R F and Ashton E R 1979. Climate. In: Day J, Siegfried W R, Louw G N and Jarman M L (eds) Fynbos ecology: a preliminary synthesis. South African National Scientific Programmes Report No 40. CSIR, Pretoria. pp 7-15.

Fugler S R 1979. Some aspects of the autecology of three hakea species in the Cape Province, South Africa. MSc Thesis, University of Cape Town, Cape Town. 295 pp.

Fugler S R 1983. The control of silky hakea in South Africa. Bothalia 14, 977-980.

Gleadow R M 1982. Invasion by *Pittosporum undulatum* of the forests of Central Victoria. 2. Dispersal, germination and establishment. Australian Journal of Botany 30, 185-198.

Goldblatt P 1978. An analysis of the flora of southern Africa: its characteristics, relationships and origins. Annals of the Missouri Botanical Garden 65, 369-436.

Hall A V 1961. Distribution studies of introduced trees and shrubs in the Cape Peninsula. Journal of South African Botany 27, 101-110.

Hall A V and Ashton E R 1983. Threatened plants of the Cape Peninsula. Threatened plants Research Group, University of Cape Town, Cape Town.

Hall A V and Bulley S M 1980. Nassella tussock in the south western Cape. In: Neser S and Cairns A L P (ed) Proceedings of the 3rd National Weeds Conference of South Africa. Balkema, Cape Town. pp 33-38.

Hall A V, De Winter M, De Winter B and Van Oosterhout S A M 1980. Threatened plants of southern Africa. South African National Scientific Programmes Report No 45. CSIR, Pretoria.

Heddle E M and Specht R L 1975. Dark Island heath (Ninetymile Plain, South Australia). VIII. The effects of fertilizers on composition and growth, 1950-1972. Australian Journal of Botany 23, 151-164.

Heydorn A E R and Tinley K L 1980. Estuaries of the Cape Part 1. Synopsis of the Cape Coast. Research Report 380. National Research Institute for Oceanology, CSIR, Stellenbosch.

Heyligers P C 1985. The impact of introduced plants on foredune formation in south-eastern Australia. Proceedings of the Ecological Association of Australia 14, 23-41.

Jacot Guillarmod A F M 1980. Exotic weeds and macchia vegetation in the Eastern Cape. The Eastern Cape Naturalist 69, 21-23.

Jacot Guillarmod A F M 1983. Recovery of eastern Cape heathland after fire. Bothalia 14, 701-704.

King N L 1943. Historical sketch of the development of forestry in South Africa. Journal of South African Forestry Association 1, 4-16.

Kluge R L and Gordon A J 1983. The control of *Hypericum perforatum* by the beetle *Chrysolina quadrigemina* in a field at Tulbagh. Paper presented at the 5th National Weeds Conference of South Africa, Stellenbosch. 9 pp.

Kluge R L and Richardson D M 1983. Progress in the fight against hakea. Veld and Flora 69, 136-138.

Knight R S and Macdonald I A W 1985. Seed dispersal of alien acacias by the Black Korhaan. Poster paper. 7th annual Fynbos Biome Research Meeting, Stellenbosch.

Knox-Davies P S 1975. Decline disease of silver trees and other indigenous species. Veld and Flora 61(2), 20-21.

Krüger B 1966. The pear tree blossoms: a history of the Moravian Mission Stations in South Africa 1737-1869. Moravian Mission Press, Genadendal.

Kruger F J 1977. Invasive woody plants in the Cape Fynbos with special reference to the biology and control of *Pinus pinaster*. Proceedings of the 2nd National Weed Conference of South Africa. Balkema, Cape Town. 47-55.

Kruger F J 1979. South African heathlands. In: Specht R L (ed) Heathlands and related shrublands of the world. A. Descriptive studies. Elsevier, Amsterdam. pp 19-80.

Kruger F J and Bigalke R C 1984. Fire in fynbos. In: Booysen P de V and Tainton N M (eds) Ecological effects of fire in South African ecosystems. Ecological studies 48. Springer-Verlag, Berlin. pp 67-114.

Kruger F J and Taylor H C 1979. Plant species diversity in Cape Fynbos: Gamma and delta diversity. Vegetatio 41, 85-93.

Lambrechts J J N 1979. Geology, geomorphology and soils. In: Day J, Siegfried W R, Louw G N and Jarman M L (eds) Fynbos ecology: A preliminary synthesis. South African National Scientific Programmes Report No 40. CSIR, Pretoria. pp 16-26.

Lloyd P H 1975. A study of the Himalayan Tahr (*Hemitragus jemlahicus*) and its potential effects on the ecology of the Table Mountain Range. Unpublished Report. Cape Department of Nature and Environmental Conservation. 80 pp.

Lord D A, Illenberger W K and McLachlan A 1985. Beach erosion and sand budget for the Port Elizabeth beachfront. University of Port Elizabeth, Institute for Coastal Research, Report 1, 1-48.

Lubke R A 1985. Erosion of the beach at St Francis Bay, Eastern Cape, South Africa. Biological Conservation 32, 99-127.

Lubke R A and Avis A M 1985. Succession on the coastal dunes and dune slacks at Kleinemonde, Eastern Cape, South Africa. Paper presented at 1985 AETFAT Congress.

Lückhoff C A 1951. Table Mountain. Balkema, Cape Town.

Macdonald I A W 1984. Is the fynbos biome especially susceptible to invasion by alien plants? A re-evaluation of available data. South African Journal of Science 80, 369-377.

Macdonald I A W 1985a. Birding down under. Bokmakierie 37, 24-28.

Macdonald I A W 1985b. The invasion of Australian ecosystems by alien organisms: some observations by a South African. Unpublished report. University of Cape Town, Cape Town. 17 pp.

Macdonald I A W 1986. Range expansion in the Pied Barbet and the spread of alien tree species in southern Africa. Ostrich 57, 75-94.

Macdonald I A W and Jarman M L (eds) 1984. Invasive alien organisms in the terrestrial ecosystems of the fynbos biome, South Africa. South African National Scientific Programmes Report No 85. CSIR, Pretoria.

Macdonald I A W, Jarman M L and Beeston P M (eds) 1985a. Management of invasive alien plants in the fynbos biome. South African National Scientific Programmes Report No 111. CSIR, Pretoria.

Macdonald I A W, Powrie F J and Richardson D M 1985b. Avifaunal responses to the spread of alien trees in the fynbos. Poster paper. 7th Annual Fynbos Biome Research Meeting, Stellenbosch.

Macdonald I A W, Richardson D M and Powrie F J 1986. Range expansion of the Hadeda Ibis in southern Africa. South African Journal of Zoology 21, 331-342.

McLachlan A, Sieben P R and Ascaray C 1982. Survey of a major coastal dunefield in the Eastern Cape. University of Port Elizabeth Zoology Department Report Series 10, 1-48.

McLachlan D, Moll E J and Hall A V 1980. Resurvey of the alien vegetation in the Cape Peninsula. Journal of South African Botany 46, 127-146.

McLachlan G R 1978. A population of *Typhlops braminus* (Daudin) on the Cape Peninsula. Zoologica Africana 13, 353-354.

Middlemiss E 1963. The distribution of *Acacia cyclops* in the Cape Peninsula by birds and other animals. South African Journal of Science 59, 419-420.

Middlemiss E 1975. The Rondevlei Bird Sanctuary 1952-1974: A record of an environment. Unpublished Report, Divisional Council of the Cape, Cape Town. 425 pp.

Millar J C G 1980. Aspects of the ecology of the American Grey Squirrel. *Sciurus carolinensis* Gmelin in South Africa. MSc Thesis, University of Stellenbosch, Stellenbosch. 313 pp.

Milton S J 1976. Understorey vegetation of a pine plantation, Bosboukloof, Jonkershoek, S W Cape. Unpublished BSc Honours project, Forestry Faculty, University of Stellenbosch, Stellenbosch. 65 pp.

Milton S J 1981. Litterfall of the exotic acacias in the south west Cape. Journal of South African Botany 47, 147-155.

Milton S J and Hall A V 1981. Reproductive biology of Australian acacias in the south western Cape Province, South Africa. Transactions of the Royal Society of South Africa 44, 465-487.

Milton S J and Siegfried W R 1981. Aboveground biomass of Australian acacias in the southern Cape, South Africa. Journal of South African Botany 47, 701-716.

Moll E J and Bossi L 1984. A current assessment of the extent of the natural vegetation of the fynbos biome. South African Journal of Science 80, 355-358.

Moll E J and Scott L 1981. Trees and shrubs of the Cape Peninsula. Eco-lab, University of Cape Town, Cape Town.

NAKOR 1984. Register of conserved areas - October1984. Unpublished report. Department of Environment Affairs, Pretoria. 36 pp.

Neser S and Annecke D P 1973. Biological control of weeds in South Africa. Entomological Memoir 28. Department of Agricultural Technical Services, Pretoria.

Phillips E P 1938. The naturalized species of Hakea. Farming in South Africa 13, 424.

Pieterse P J 1984. Aspects of the seed production of *Acacia longifolia*. Paper presented at the 6th National Weeds Conference, Nelspruit.

Poynton J C 1960. Preliminary note on the zoogeography of the amphibia of southern Africa. South African Journal of Science 56, 307-312.

Raitt L M 1983. An analysis of the flora and floral phenology of a disturbed area of the Cape Flats. Paper presented at the 5th National Weeds Conference, Stellenbosch.

Rautenbach I L 1978. A numerical re-appraisal of southern African biotic zones. Bulletin of the Carnegie Museum of Natural History 6, 175-187.

Richardson D M 1984. A cartographic analysis of physiographic factors influencing the distribution of *Hakea* species in the south western Cape. South African Forestry Journal 128, 36-40.

Richardson D M and Brink M P 1985. Notes on *Pittosporum undulatum* in the south western Cape. Veld and Flora 71, 75-77.

Richardson D M and Van Wilgen B W (in press). The effects of fire in felled *Hakea sericea* and natural fynbos and implications for weed control in mountain catchments. South African Forestry Journal.

Richardson D M and Van Wilgen B W 1986. Effects of thi.ty five years of afforestation with *Pinus radiata* on the composition of mesic mountain fynbos near Stellenbosch. South African Journal of Botany 52, 309-315.

Richardson G R, Lubke R A and Jacot Guillarmod A 1984. Regeneration of grassy fynbos near Grahamstown (eastern Cape) after fire. South African Journal of Botany 3, 153-162.

Roux E R and Warren J L 1963. Studies in the autecology of Australian acacias in South Africa. 2. Symbiotic nitrogen fixation in *Acacia cyclops*. A. Cunn. South African Journal of Science 59, 294-295.

Sauer J C 1985. How and why is Australian seashore vegetation different. Proceedings of the Ecological Society of Australia 14, 17-22.

Siegfried W R 1962. Introduced vertebrates in the Cape Province. Cape Department of Nature Conservation, Annual Report for 1962. 80-87.

Siegfried W R 1979. Animal ecology. In: Day J, Siegried W R, Louw G N and Jarman M L (eds) Fynbos ecology: A preliminary synthesis. South African National Scientific Programmes Report No 52. CSIR, Pretoria. pp 127-132.

Siegfried W R and Crowe T M 1983. Distribution and species diversity of birds and plants in fynbos vegetation of the Mediterranean climate zone, South Africa. In: Kruger F J, Mitchell D T and Jarvis J U M (eds) Ecological studies, Vol 43: Mediterranean-type Ecosystems: The role of nutrients. Springer-Verlag, Berlin. pp 403-416.

Shaughnessy G L 1980. Historical ecology of alien woody plants in the vicinity of Cape Town, South Africa. PhD Thesis, University of Cape Town, Cape Town. 421 pp.

Shelton P 1975. Analysis of the stomach contents of the striped field mouse. *Rhabdomys pumilio* Sparrman. BSc Honours Project Report, Department of Zoology, University of Cape Town, Cape Town.

Skaife S H 1955. The Argentine ant, *Iridomyrmex humilis* Mayr. Transactions of the Royal Society of South Africa 34, 355-377.

Skead C J 1962. A study of the Crowned Guineafowl, *Numida meleagris coronata* Gurney. Ostrich 33, 51-65.

Skead C J 1980. Historical mammal incidence in the Cape Province. Vol 1. The Western and Northern Cape. Cape Department of Nature and Environmental Conservation, Cape Town. 903 pp.

Smit H A P and De Kock G L 1984. Ruimtelike diffusie van die indringer plant *Pinus pinaster* teen die hange van Stellenboschberg. South African Forestry Journal 129, 10-16.

Smithers R H N 1983. The mammals of the southern African Subregion. University of Pretoria, Pretoria.

Stirton C H (ed) 1978. Plant invaders - beautiful but dangerous. Department of Nature and Environmental Conservation of the Cape Provincial Administration, Cape Town.

Stuart C T 1976. Contributions to the biology and ecology of the rodents and insectivores of the Andries Venter Research Station, Jonkershoek Valley. Unpublished Report, Cape Department of Nature and Environmental Conservation. 36 pp.

Stuart C T, Macdonald I A W and Mills M G L 1985. History, current status and conservation of large mammalian predators in the Cape Province, Republic of South Africa. Biological Conservation 31, 9-17.

Stuckenberg B R 1962. The distribution of the montane palaeogenic element in the South African invertebrate fauna. Annals of the Cape Provincial Museums 2, 190-205.

Taylor H C 1977. The Cape floral kingdom, an ecological view. Proceedings of the 2nd National Weeds Conference of South Africa. Balkema, Cape Town. pp 19-33.

Taylor H C 1984. A vegetation survey of the Cape of Good Hope Nature Reserve. II. Descriptive Account. Bothalia 15, 259-291.

Taylor H C, Macdonald S A and Macdonald I A W 1985. Invasive alien woody plants in the Cape of Good Hope Nature Reserve. II. Results of a second survey from 1976 to 1980. South African Journal of Botany 51, 21-29.

Van Bruggen A C 1964. The distribution of introduced mollusc species in southern Africa. Beaufortia 11, 161-169.

Van Bruggen A C 1978. Land molluscs. In: Werger M J A (ed) Biogeography and ecology of southern Africa. W Junk, The Hague. pp 877-923.

Van der Zel D W 1981. Optimum mountain catchment management in southern Africa. South African Forestry Journal 116, 75-81.

Van Wilgen B W 1981. Some effects of fire frequency on fynbos plant community composition and structure at Jonkershoek, Stellenbosch. South African Forestry Journal 118, 42-55.

Van Wilgen B W 1984. Adaptation of the United States Fire Danger Rating System to fynbos conditions, Part I. A fuel model for Fire Danger Rating in the fynbos biome. South African Forestry Journal 131, 13-17.

Van Wilgen B W and Kruger F J 1985. The physiography and fynbos vegetation communities of the Zachariashoek Catchments, south western Cape Province. South African Journal of Botany 51, 379-399.

Van Wilgen B W and Richardson D M 1985a. Factors influencing burning by prescription in mountain fynbos catchment areas. South African Forestry Journal 134, 22-32.

Van Wilgen B W and Richardson D M 1985b. The effects of alien shrub invasions on vegetation structure and fire behaviour in South African fynbos shrublands: A simulation study. Journal of Applied Ecology 22, 955-966.

Van Wyk D B 1977. Die invloed van bebossing met *Pinus radiata* op die totale jaarlikse afvoer van die Jonkershoek strome. MSc Thesis, University of Stellenbosch, Stellenbosch. 206 pp.

Von Broembsen S 1984a. Distribution of *Phytophthora cinnamoni* in rivers of the south western Cape Province. Phytophylactica 16, 221-225.

Von Broembsen S 1984b. Occurrence of *Phytophthora cinnamoni* on indigenous and exotic hosts in South Africa, with special reference to the south western Cape Province. Phytophylactica 16, 227-229.

Von Broembsen S and Kruger. F J 1985. *Phytophthora cinnamoni* associated with mortality in native vegetation in South Africa. Plant Disease 68, 715-717.

Walsh B N 1968. Some notes on the incidence and control of driftsands along the Caledon, Bredasdorp and Riversdale Coastline of South Africa. Department of Forestry Bulletin 44, 1-79.

Whitehead V B and Prins A J 1975. The European Wasp *Vespula germanica* (F) in the Cape Province. Journal of the Entomological Society of southern Africa 38, 39-42.

Wicht C L 1945. Report of the committee on the preservation of the vegetation of the south west Cape. Special publication of the Royal Society of South Africa. 55 pp.

Winterbottom J M 1970. The birds of the alien acacia thickets of the south western Cape. Zoologica Africana 5, 49-57.

Winterbottom J M and Liversidge R 1954. The European starling in the south west Cape. Ostrich 25, 89-96.

Chapter 7

Invasive alien organisms in the Namib Desert, Upper Karoo and the arid and semi-arid savannas of western southern Africa

C J Brown* and A A Gubb†

*Department of Agriculture and Nature Conservation, SWA/Namibia and †McGregor Museum.

Introduction

Little information exists on alien organisms in arid and semi-arid biomes in southern Africa. The only publications which deal extensively with data from these regions are those of Brown et al (1985) and Macdonald and Nott (in press), although references to alien species were also found in a wide range of published literature. This chapter provides an inventory of all known invasive alien organisms recorded in those arid and semi-arid regions of southern Africa for which some information is available and attempts to explain the incidence and degree of invasion in terms of the major ecological factors governing arid and semi-arid environments.

The regions of South West Africa (SWA)/Namibia and the northern Cape Province of South Africa included here are the Namib desert, an escarpment zone intermediate between the true desert and the savannas, and a zone of arid and semi-arid savannas (Figure 1). The Northern Cape Savanna falls between the Kalahari Savanna in the north and the Upper Karoo in the south with the Cape-Transvaal/Orange Free State provincial boundary forming its eastern limit. Acocks's (1975) map is the basis for defining the zones in South Africa. Ecological regions in SWA/Namibia are based on the preliminary vegetation map by Giess (1971).

The area is bounded on the east and south by the 400 mm mean annual rainfall isohyct except for small portions in the north-east within both SWA/Namibia and the northern Cape Province.

Regional descriptions

The Namib Desert extends along the Atlantic coast from the Cunene River in the north to the Orange River in the south. The gravel plains of the Northern and Central Namib region, situated north of the Kuiseb River, rise to a mountainous scarp in the east and are transected by numerous episodic rivers. The Southern Namib, from the Kuiseb River south to the northern limit of the winter rainfall area, just north of Luderitz consists of a dune sea extending 320 km and about 120 km inland. The Desert and Succulent Steppe includes the zone of winter rainfall which extends to the Orange River. The barren desert plains with sand dunes of the coast grade inland into hills and mountains.

The Escarpment Zone runs north-south and separates the Namib Desert from the savanna regions. In the north the transition between desert and savanna is poorly defined. In the south sandstone and calcrete are the major geological materials while in the central regions schist and granite predominate. Episodic rivers in the south terminate in the sand dunes of the southern Namib, sometimes in large temporary lakes or marshes, while the central and northern areas are traversed by deep river valleys and canyons.

The Dwarf Shrub Zone occupies the southern central parts of SWA/Namibia. This consists of mountains and hills interspersed by extensive plains. Geologically it is dominated by sandstones and shale. The region is bisected by the Fish River system, the flow of which is episodic, but supplemented by releases of water from the Hardap Dam. Just prior to the confluence with the Orange River the Fish River forms a deeply eroded canyon.

The Upper Karoo is situated on the central plateau of South Africa. Inselbergs rise abruptly from extensive stony plains (Leistner 1979). The substrata consist of sedimentary rocks, mainly shales, mudstones and fine sandstones (Hamilton and Cooke 1960). Soils are generally shallow and classed as desert soils (van der Merwe 1962). Seasonal rivers drain either

Address for correspondence: C J Brown, Department of Agriculture and Nature Conservation, Private Bag 13306, Windhoek, 9000, South West Africa/Namibia.

into the Orange River or into large brackish, silty pans which are practically bare (Werger 1978).

Five savanna types occur in the area under discussion (Figure 1). The Mopane Savanna occupies the north-western region of SWA/Namibia. The area is mountainous in the west, but otherwise consists of plains. The western area is underlain by schist, metamorphic basement complexes and quartzites while in the east sandstone and calcrete predominate. The area is bounded in the north-west by the perennial Cunene River. All other rivers are episodic. Rivers rising in this region drain to the west (eg Huab, Uniab, Hoanib). In addition, the Cuvelai system drains from Angola southwards through this region into the Etosha Pan (van der Merwe 1983).

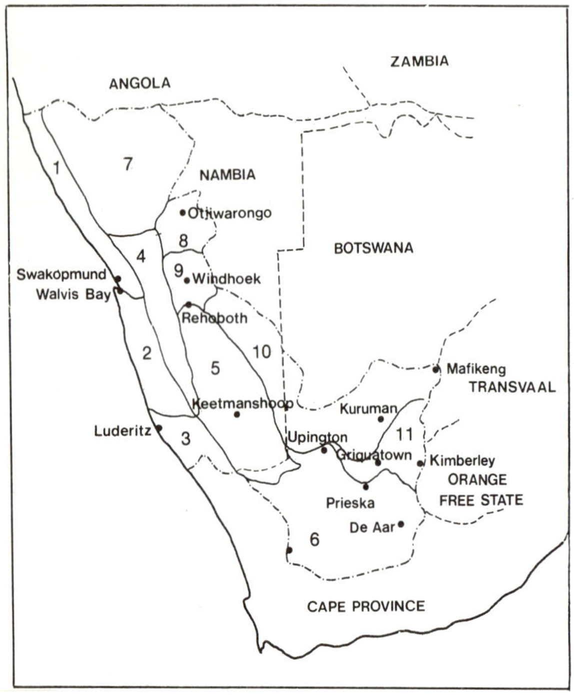

– – — – Boundary of area covered by this chapter.

Figure 1. The arid and semi-arid regions of SWA/Namibia and the Cape Province covered by this chapter. The regions in SWA/Namibia are based on Giess' (1971) preliminary vegetation map and those in the Cape are after Gubb (unpublished).

A. The Namib Desert, divided into the Northern and Central Namib (1); the Southern Dune Namib (2); and Desert and Succulent Steppe winter rainfall Namib (3).

B. An intermediate habitat between the true desert and the dry savannas, divided into the Escarpment Zone (4); the Dwarf Shrub Zone (5); and the Upper Karoo (6); this latter being defined as including Acocks's (1975) Orange River Broken Veld (Types 32 and 40), his Arid Karoo and Desert Grasslands (Types 29 and 35) and his Central Upper Karoo (Type 27).

C. The arid and semi-arid savannas, divided into the Mopane Savanna (7); the Thornbush Savanna (8); the Highland Savanna (9); the Kalahari Savanna (10); and the Northern Cape Savanna (11).

The Thornbush Savanna in the central part of SWA/Namibia north of Windhoek consists mainly of plains with scattered hills and inselbergs, but with some low mountains to the west. Geologically the area is complex being underlain mainly by granite in the west, and schist and sandstone with some unconsolidated sand cover in the east. The most important river course running through this region is the Omatako which runs north-east to the Okavango River, although a number of west-flowing rivers (eg Swakop, Omaruru and Ugab) rise here.

The Highland Savanna occupies the central area of SWA/Namibia around Windhoek. The area consists of low mountains and rolling hills and gives rise to numerous episodic streams and rivers. Rocks consist of Swakop schist of the Damara sequence.

The Kalahari Savanna extends from south-eastern SWA/Namibia into the northern Cape Province and consists mainly of longitudinal aeolian sand dunes, although these may be poorly developed or entirely absent in the north-eastern Cape. Traversing this region are the Auob, Nossob and Kuruman River systems which flow episodically into the Molopo River. There are pans throughout the Kalahari most of which are shallow depressions in exposed areas of underlying limestone (Leistner 1967).

The terrain of the Northern Cape Savanna, situated north of the Orange River, is generally rocky and broken. Prominent features are the Langeberg, Kuruman-Asbestos mountains and the extensive Ghaap Plateau. The geology and soil types are complex; the Langeberg and parallel mountain ranges are quartzitic, the Kuruman-Asbestos mountains consist mainly of banded ironstone and the Ghaap Plateau is covered by calcrete and dolomite. On the eastern boundary rocky ridges of andesite and dolerite intrusions are common. Soils are shallow, consisting of arenosols and lithosols (Harmse 1978). The Vaal, Harts and Riet Rivers with their wide valleys coincide approximately with the eastern boundary of this region.

Climate

The south-western region of SWA/Namibia

receives winter rainfall. Rainfall in much of the Upper Karoo and the southern part of the Dwarf Shrub Zone is equinoctial (Tyson 1978). The rest of the region has mainly summer rainfall. Mean annual rainfall decreases from east to west and from north to south and ranges from about 400 mm in the north-east to less than 20 mm on the west coast. Coastal regions are subject to frequent fogs which may extend up to 100 km inland. Fog precipitation at the coast (Swakopmund) is between 35-45 mm yr^{-1} (three times the annual rainfall) while 40 km inland it falls to about 20 mm yr^{-1} (Goudie 1972). Reliability of rainfall varies greatly and variability increases sharply to the west.

Structure of indigenous communities

In the Northern and Central Namib most of the vegetation is limited to the river courses and stony hills and mountains. The gravel plains are barren except after rains when they become covered by annual grasses, mainly *Stipagrostis* species. The dunes of the Southern Namib are sparsely vegetated by clumps of *S sabulicola*, often in association with *Trianthema hereroensis*. Succulents are typical of the winter rainfall region of the Namib with trees and shrubs being confined to river courses (Giess 1971).

The Escarpment Zone is characterized by a great variety of species, many of which are endemic. Typical of the area are succulent trees such as *Euphorbia guerichiana* and *Cyphostemma* species and several species of *Acacia* and *Commiphora*. The Dwarf Shrub Zone is dominated by karoo shrubs and grasses. Trees, such as *Acacia erioloba* and *Tamarix usneoides*, are found only along rivers. The plains of the Upper Karoo are regulated by an open, dwarf to low shrub formation with *Eriocephalus* and *Lycium* species dominating. Taller shrubs (eg *Rhus undulata*) occur on the slopes. The sandy flats of Bushmanland support low desert grasslands.

The Mopane Savanna is dominanted by *Colophospermum mopane* and ranges in structure from dense woodlands to sparse shrublands. The Thornbush Savanna is made up of grassland with trees and bigger shrubs in dense or open clumps of varying size. *Acacia* species are the dominant trees and encroachment of the grassland by species such as *A mellifera* subspecies *detinens* and *Dichrostachys cinerea* is widespread. The Highland Savanna region is dominated by stands of open, low *Acacia* species such as *A hereroensis*, as well as of *Combretum apiculatum* and other small trees with a grass understorey of *Anthephora pubescens* and *Brachiaria nigropedata* in areas not overgrazed. The Kalahari Savanna consists of sparse to open, low to high *Acacia* woodlands with *Terminalia sericea* trees being common in the north. Grasses such as *Stipagrostis amabilis* and *S uniplumis* are almost entirely replaced by *Schmidtia kalahariensis* in disturbed areas. The Northern Cape Savannas are dominated by shrubs such as *Tarchonanthus camphoratus*, *Euclea undulata* and *Rhus* species with trees such as *Olea europaea* subspecies *africana* and *Acacia karroo* on the Ghaap plateau and *A erioloba* and *A tortilis* subspecies *heteracantha* on the sandy loam plains of the eastern regions.

Patterns of land use

Much of the Namib falls within conservation areas, eg the Skeleton Coast and Namib-Naukluft Parks, or areas with mining concessions and restricted access. The coastal littoral is more or less disturbed by diamond mining. Substantial proportions of the Mopane and Kalahari Savannas are conserved by the Etosha National Park and the Kalahari Gemsbok National Park respectively. Other regions have little or no conserved areas. Land use in some of the communal areas in Damaraland and Kaokoland of SWA/Namibia is often of low intensity. Cattle and game ranching are the principal forms of land use in northern SWA/Namibia. The central region is mixed large- and smallstock farming with some game farming and the southern commercial farming areas are mainly devoted to small stock, as is the Upper Karoo. The eastern regions of the Northern Cape are primarily cattle farming areas. Intensive agricultural land use is restricted mainly to areas close to the Orange, Vaal and Riet Rivers and an area near the Hardap Dam (Figure 2).

Methods

There is little published information on alien organisms in the northern Cape Province. For SWA/Namibia information was drawn from Brown et al (1985) and Macdonald and Nott (in press). Much information was obtained from the Windhoek State Herbarium and the McGregor Museum Herbarium in Kimberley. National lists were used to search for alien plants (Willman 1946; Merxmuller 1966-1972; Macdonald 1983; Gibbs Russell 1984) but none of these lists are complete or agree on what is exotic or indigenous. Information was also obtained from colleagues in the respective herbaria, the Directorate of Nature Conservation and Recreation Resorts in

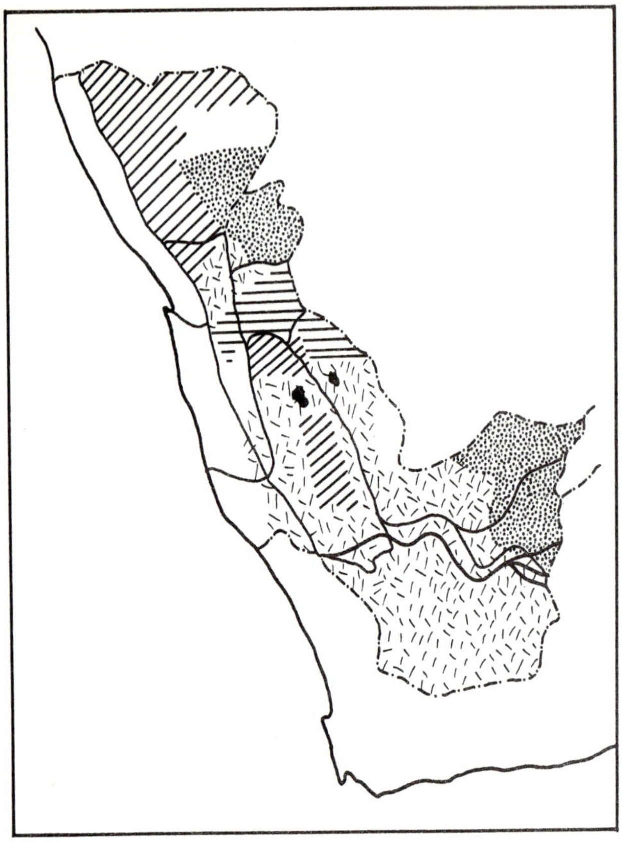

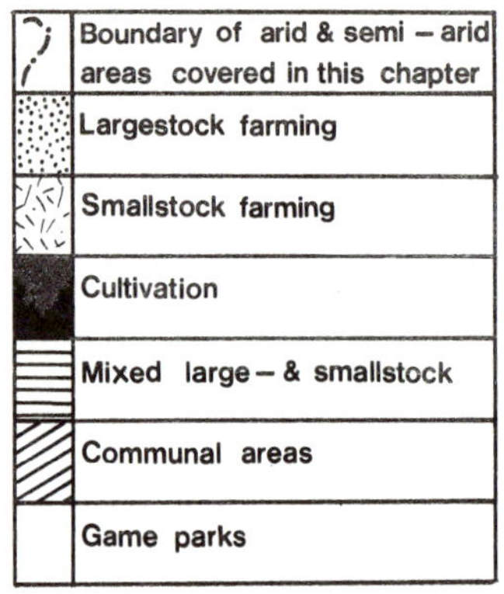

Figure 2. Patterns of land use in the arid and semi-arid areas of SWA/Namibia and the Cape Province covered by this chapter.

SWA/Namibia and the Cape Department of Nature and Environmental Conservation.

Alien organisms may establish in arid and semi-arid biomes only with careful nurturing by man, or may be highly successful and spread rapidly in all habitats encountered with no assistance from man. In the first instance, although alien, the species are not considered to be invasive. Between these two extremes are found a range of aliens of varying success. In this chapter invasive alien organisms have been classed as low, moderate or highly invasive based on the capacity of the species to invade large areas rapidly, to withstand fluctuations in climatic conditions, and to invade many different habitats.

Evaluations of these characteristics are based on herbarium records, personal and published observations (eg in Brown et al 1985; Macdonald and Nott in press). Appendices A and B list all the flora and fauna known to be alien and invasive in the arid and semi-arid biomes covered by this chapter, give an assessment of their success as invaders, and summarize pertinent observations.

Invasive alien plants

Biogeographic origins

Of the alien plant species introduced to the arid and semi-arid areas of southern Africa the most successful have been those species indigenous to other arid environments such as the western Americas, Asia and Australia. Most of the alien plants recorded originate from Europe (54 species), South and Central America (31 and 24 species respectively) and from Asia (25 species) (Table 1). Only eight species are from Australia and five from other parts of Africa. The percentages of the highly invasive ie most successful species from each source region vary markedly. Of the Australian species 38% are considered to be highly invasive as are 33% of the Central American species. Only 7% of the species from Europe and none of the species from other parts of Africa are highly invasive.

In total about 170 species of alien terrestrial plants have been recorded in seminatural and natural habitats in the area covered by this chapter (Appen-

Table 1. The regions of origin of alien plants recorded in the areas of SWA/Namibia and the Northern Cape Province. %H = percentage of highly invasive species from each place of origin.

Place of origin	Number of alien plant species and degree of invasiveness				
	Low	Moderate	High	Total	%H
North America	11	3	2	16	13
Central America	9	7	8	24	33
South America	18	9	4	31	13
Europe	37	13	4	54	7
Asia	13	6	6	25	14
Africa	2	3	0	5	0
Australia	3	2	3	8	38

dix A). However, only about 27 currently give cause for concern.

Incidence by regions

Table 2 lists the numbers of invasive alien plants for the different ecological regions. The Northern Cape Savanna has the highest number followed by the Upper Karoo and the Highland Savanna. The Namib Desert regions have low numbers of alien plants as do the Escarpment Zone, Dwarf Shrub Zone, Mopane Savanna and Thornbush Savanna. High numbers of alien species in the Northern Cape and Highland Savannas can be partially attributed to the relatively high intensity of collection in these areas. However, these also provide a greater diversity of habitats which allows a greater variety of species to become established.

The regions containing the greatest number of alien plants have certain features in common. They are almost completely utilized for commercial livestock production and receive a relatively high and reliable rainfall compared to the highly variable rainfall of the Kalahari and Namib Deserts. The topography is generally hilly and mountainous with numerous watercourses, many of which rise from large catchments in adjoining regions. They tend also to have a more highly developed infrastructure with centres of relatively large human populations (eg Kimberley, Upington, Griquatown, Windhoek, Rehoboth), compared to other regions and have a relatively long history of occupation by Europeans.

In contrast, the areas with few alien plants are characteristically very arid, often have a substratum which allows little runoff, eg in the Southern Dune Namib, have few watercourses, and little infrastructural development with low human population density or have only fairly recently been developed. Alien plants in these areas are usually confined to watercourses many of which rise in the more developed areas.

Table 2. The number of invasive alien plants, grouped according to growth form, recorded in each of the ecological regions.

Ecological Region	Number of species recorded in categories of growth form			
	Trees and shrubs	Forbs	Grasses	Total
Northern and Central Namib	4	6	0	10
Southern Dune Namib	3	5	0	8
Succulent Steppe Namib	2	2	0	4
Escarpment Zone	4	8	0	12
Dwarf Shrub Zone	2	6	0	8
Upper Karoo	19	23	1	43
Mopane Savanna	5	8	0	13
Thornbush Savanna	7	9	0	16
Highland Savanna	7	24	3	34
Kalahari Savanna	10	19	0	29
Northern Cape Savanna	27	95	21	143

Alien trees and large shrubs are more successful in invading indigenous vegetation than are other growth forms (Table 3). Thirty-six per cent of trees and large shrubs are classed as highly invasive while only 14% of forbs and no grasses are considered highly invasive.

Table 3. The degrees of invasion found for species of different growth forms. Values in brackets are percentages of total species within each growth form.

	Number of species according to degree of invasion			
Growth form	High	Moderate	Low	Total
Trees and large shrubs	13 (36)	10 (28)	13 (36)	36
Forbs	14 (13)	32 (29)	65 (59)	111
Grasses	– (0)	5 (22)	18 (78)	23
Total	27 (16)	47 (28)	96 (56)	170

Incidence by habitats

In the analysis of invasive organisms a distinction has been drawn between seminatural and natural vegetation. Seminatural vegetation is here taken as being vegetation which has been considerably modified by human activities and usually lies adjacent to transformed habitats such as gardens, urban areas, artificial watering points, and cultivated fields. Natural vegetation is taken to refer to vegetation which is still in pristine condition or has been disturbed only to a minimal extent by human activities. Alien species restricted to transformed habitats have not been included. The frequency of occurrence of alien plant species in seminatural and natural habitats is given in Table 4. It is evident that seminatural habitats have been more extensively invaded than have natural habitats, having a 43% greater incidence of alien occurrence. This trend is further emphasized by examining the incidence of invasion by those plants considered to be highly invasive (Table 5). These species occur twice as frequently in seminatural habitats than in natural habitats.

In the Cape Province the habitats most susceptible to invasion, particularly by forbs, are those of the banks of perennial rivers (Table 6). This can be attributed to the relatively dense human population near rivers, cultivation and other forms of disturbance, as well as waterborne seed. In addition, many of the species recorded in these habitats grow only under moist conditions and are able to survive in this region only because of the perennially moist conditions. This habitat type is not present in that part of SWA/Namibia under review; in this area farm homestead surroundings have the greatest number of invasive aliens. It is apparent that areas of high human disturbance, eg farm homestead and kraal surroundings, outskirts of towns, road verges, and mining areas are the habitats which support the greatest number and diversity of alien plants in desert and semidesert biomes. The second group of habitats apparently vulnerable to invasion are those disturbed episodically by natural events. These include channels and beds of ephemeral streams and dry floodplains. Rivers in semi-arid and arid areas flow and flood their banks at irregular intervals, destroying vegetation and depositing silt. These habitats are therefore unstable and are prone to invasion. Less disturbed natural habitats have a low incidence of invasion.

Some characteristics of successful invader species

Disturbed conditions also favour certain indigenous species, especially annuals, which establish well in disturbed sites. Recruitment is episodic, usually confined to periods of rainfall. The species which become established will depend upon, amongst other

Table 4. The total number of records of alien plant species in seminatural and natural habitats.

	Frequency by habitat type		
Habitat	Trees and large shrubs	Forbs	Grasses
Seminatural	96	237	31
Natural	67	171	16

Table 5. The number of species of alien plants considered to be highly invasive, recorded for seminatural and natural vegetation.

	Number of species recorded			
Habitat	Trees and large shrubs	Forbs	Grasses	Total
Seminatural	11	13	0	24
Natural	6	4	0	10

factors, the timing and amount of rainfall and the number of propagules present.

Invasive alien plants in arid and semi-arid biomes have a number of seasonal and phenological advantages over indigenous vegetation. Some species germinate more rapidly and at a slightly earlier stage, eg *Salsola kali*. Some are very early deciduous (*Melia azedarach*) or very late deciduous (*Prosopis* species). Many are xerophytic ephemerals, which do not always respond to wet periods, especially if the vegetation is in good condition and there is competition from the higher cover of indigenous species. If the vegetation has been overgrazed then these invasive ephemerals are able to germinate at moisture levels too low for the indigenous vegetation and dominate the community. Many aliens can withstand low moisture levels or have access to more water than indigenous species, eg *Prosopis* species (Vinjevold et al 1985). Many aliens produce huge crops of flowers, (eg *Melia azedarach*, *Prosopis* species), over a short period of time with a resultant high seed production and widespread dispersal opportunities compared to those of indigenous flora. The fruit is often highly palatable, (eg *Prosopis* species, *Opuntia* species, *Melia azedarach*), and is heavily utilized by domestic stock and game (Muller 1985; Tarr and Loutit 1985). In addition, seedlings of many alien plants germinate freely around and beneath the canopies of their adult plants, (eg *Argemone subfusiformis*, *Datura* species, *Tagetes minuta*, *Bidens* species, *Agave americana*,

Table 6. The number of invasive alien plants recorded from specific habitats in those areas of the Cape Province and SWA/Namibia covered by this chapter.

	Trees and large shrubs		Forbs		Grasses		Totals	
Habitat	Cape	SWA/ Namibia	Cape	SWA/ Namibia	Cape	SWA/ Namibia	Cape	SWA/ Namibia
Disturbed by human agency								
1. Farm homesteads	21	7	37	13	3	–	63	20
2. Kraals (traditional villages)	11	7	17	12	–	–	31	19
3. Outskirts of towns (xeric sites)	18	6	19	12	3	1	41	19
4. Timber plantations	5	–	–	–	–	–	7	–
5. Roadside verges and adjacent fields	8	3	28	10	7	1	44	14
6. Railway verges and adjacent fields	4	2	4	8	–	1	13	10
7. Quarries	6	2	7	6	–	–	15	8
8. Mining areas	9	3	14	7	–	–	25	10
9. Waterholes	5	3	7	4	–	–	14	7
10. Irrigated lands	1	–	36	–	11	–	48	–
11. Irrigation canals	1	–	21	–	3	–	25	–
12. Outskirts of towns (mesic sites)	4	3	31	6	3	2	37	9
Naturally disturbed								
13. Dry drainage lines and riverbeds	5	4	3	7	–	–	7	11
14. Dry river gravel	8	1	6	2	–	–	16	3
15. Dry river alluvium/floodplain	10	2	15	5	4	–	29	7
16. Moist drainage lines and riverbeds	4	2	15	5	2	–	20	7
17. Perennial river banks	8	–	50	–	3	–	64	–
18. Episodic river banks	2	4	5	6	1	–	7	10
Undisturbed								
19. Rocky habitats	10	4	15	1	1	–	26	5
20. Sandy habitats	6	2	20	1	1	–	32	3
21. Seasonal and perennial marshes	–	–	9	–	2	–	10	–
22. Seasonal pans	1	–	6	–	1	–	7	–
23. Springs and seeps	–	1	9	–	1	–	9	–
24. Islands in perennial rivers	4	–	–	–	–	–	4	–
25. In perennial rivers	–	–	–	–	–	–	–	–

Melia azedarach, *Opuntia* species, *Prosopis* species) thereby forming dense monospecific stands. Intraspecific competition within indigenous species would usually prevent the dominance of an area by a single species.

Accounts of invasions by individual species

This section concentrates on those species which have been most successful and are therefore of greatest cause for concern. Detailed information is available only for SWA/Namibia (Brown et al 1985; Macdonald and Nott in press).

The most successful alien plants for the whole region under discussion are *Prosopis* species, *Eucalyptus* species, *Nicotiana glauca*, *Melia azedarach*, *Opuntia* species, *Datura* species, *Argemone subfusiformis*, *Bidens biternata* and *Salsola kali*. In this portion of SWA/Namibia *Eucalyptus* species and *M azedarach* are not considered highly invasive, probably because there are no perennial rivers in the area. *Prosopis* species pose by far the greatest threat of invasion in arid and semi-arid biomes. *Datura* species are the most important invasive alien forbs in SWA/Namibia while *Salsola kali* is particularly widespread in the northern Cape Province. No alien grass species have been very successful in invading these biomes.

Prosopis species

Prosopis species were introduced into SWA/Namibia in 1912 as fodder and shade trees (Muller 1985). Since then, this genus has been spread widely throughout the drier regions of SWA/Namibia and South Africa and today is found in and around most towns and villages, around many farmsteads, along roadsides (where it is still being planted) and in the courses of episodic rivers, where it is most successful. There is confusion as to which species are present and little ecological information is available. *Prosopis* species have been observed to form monospecific stands. Indigenous species are apparently outcompeted, possibly for water (Vinjevold et al 1985). Annual seed production is prolific (960 seeds m^{-2} of canopy, Brown unpublished) and this may be advantageous at the establishment phase. Seed is distributed by water, domestic stock and a wide range of indigenous animals (Muller 1985; Brown and Gubb personal observations). Particularly dense infestations of *Prosopis* occur on municipal land near Windhoek, near Rehoboth (Macdonald and Nott in press), in the vicinity of the Orange River, and in the Swakop River in the Namib-Naukluft Park. The Swakop River region was intensively overgrazed by domestic stock until 1977 when the land was included in the park. *Prosopis* occurs in densities of 300-400 large, multistemmed trees per 100 m of river length in this area (R Davies personal communication).

Limited control methods have been attempted. The Windhoek Municipality has successfully controlled *Prosopis* with Tordon 125 and Tordon 225, but the environmental side-effects of the chemicals are not yet known. Control of *Prosopis* in the Swakop River commenced in July 1984 and continues. Trees are cut with a chain saw and the branches are packed about the stump. These are allowed to dry for at least four months and are then burnt. Clearing is at a rate of about 25 trees per 100 man hours; 12 km of the 63 km of infestation in the Swakop River have been successfully cleared. Less than two per cent survival of treated trees has been recorded. These trees are treated a second time and to date no further regrowth has been reported.

Prosopis species are cultivated by farmers and reinfestation from sources upstream occurs each time the river flows. Clearing operations should commence in river catchments and work downstream. This is true for all invasive species with waterborne seed. In order to obtain farmers' cooperation in the control of *Prosopis*, research into finding suitable indigenous replacement species should be a matter of priority.

Nicotiana glauca

This species is widespread over the drier western regions of SWA/Namibia and along the length of the Orange River. It is believed to have entered SWA/Namibia in horse fodder imported from South America at the turn of the century. Habitats especially susceptible to *N glauca* are river courses, road verges, and areas surrounding human habitation. Particularly dense infestations occur along the lower 20 km of the Ugab River in the Skeleton Coast Park. Seeds are dispersed by water and reinfestation from upstream occurs after flooding of the river; the rate of spread is rapid (Tarr and Loutit 1985). Reinvasion after flooding results in colonization of most of the open riverbed. Seedlings and immature plants are easily removed by hand provided they do not occur in compacted soils, dried silts or rocky areas. No large-scale control has been attempted and it is suggested that unless a national control campaign be initiated, starting in the catchment areas and working westwards, little would be achieved by clearing the coastal areas.

Opuntia species

Eight species of *Opuntia* have been recorded from the regions under discussion, four in SWA/Namibia. The most important species is *O ficus-indica* which was introduced as a fodder plant as well as for its edible fruit (Muller 1985). It has been recorded over much of SWA/Namibia but has not spread as rapidly as might have been expected even though it can propagate vegetatively as well as by seed. Biological control using cochineal *Dactylopius opuntiae* has been successful only in some areas.

Opuntia engelmannii occurs in only one area of the Windhoek district, having been introduced in 1922 to the farm Lichtenstein as a garden ornamental. Concern regarding this species is based on its very rapid rate of spread and the fact that it displaces indigenous vegetation. The species is spread by water along river courses and by baboons, which, in their attempts to reach the fruit, tear off cladodes and scatter them. These cladodes give rise to new plants. In addition, it is thought that, by eating the fruit, baboons disperse viable seeds. The entire Highland Savanna region is a suitable habitat for invasion by *O engelmannii*. No biological control has as yet been successful; introduction of *Cactoblastis cactorum* gave disappointing results and control is further hampered by the fact that the plants occur on cliffs. However, investigations are under way at the Uitenhage Weed Laboratory to find a suitable biocontrol agent (Muller 1985).

Datura species

Three species of *Datura* are recorded for the arid regions under discussion, all of which occur in SWA/Namibia. The most invasive of these is *D innoxia* although *D ferox* and *D stramonium* are also widespread (Muller 1985). Areas most susceptible to invasion by all three species are sandy river courses, damp drainage lines, the verges of artificial impoundments and to a lesser degree, areas of human disturbance.

Heavy infestations of *D innoxia* have been recorded from Avis Dam near Windhoek (Brown personal observations), sections of the Kuiseb River (Vinjevold et al 1985) and rivers in the Skeleton Coast Park (Tarr and Loutit 1985). Very dense stands occur in the Hoanib River floodplain. Dispersal of plants is probably mainly by means of waterborne seed. This makes control in the lower reaches of rivers almost impossible without coordinated action upstream. Test plots in the Hoanib River floodplain which were cleared have shown that complete reinvasion follows flooding. Furthermore, *D innoxia* plants in this area produce on average 15 pods per plant at any one time, with each pod containing an average of 650 seeds (Tarr and Loutit 1985). Of these, about 87% germinate, and seeds may remain viable in the soil for up to 20 years (Landsdell 1927). The seeds and leaves of *D innoxia* contain several poisonous alkaloids and there is no evidence that they are browsed by livestock and game. The absence of *D innoxia* from around water points in the Kuiseb River is thought to be due to trampling (Vinjevold et al 1985). While the seeds of this species have been fatal to young ostriches *Struthio camelus* in South Africa (Landsdell 1927), over 1 000 seeds were found in the crop and stomach of a doublebanded sandgrouse *Pterocles bicinctus* (Brown et al 1984) and both Cape turtle doves *Streptopelia capicola* and laughing doves *S senegalensis* have been observed feeding on these seeds (Brown personal observation).

It is believed that *D innoxia* is likely to constitute an ecological threat in the future, especially where it occurs in dense stands. Areas of dense indigenous vegetation appear not to be susceptible to invasion, but floods create channels through such areas which soon become invaded, thereby reducing the indigenous floral components (Tarr and Loutit 1985).

Argemone subfusiformis

This species is widespread over much of the arid and semi-arid regions, particularly in riverbeds, along roadways, and around human habitation. Dense infestations have been recorded in the Omaruru River in the vicinity of Omaruru town, in rivers near Rehoboth, and around dams in the Windhoek district. Little information is available on this plant, but it is not thought to pose a very serious threat to indigenous vegetation. No control has been attempted.

Invasive alien animals

Alien animals known or thought to have established wild, self-sustaining populations in the semi-arid and arid biomes of SWA/Namibia and the northern Cape Province are listed in Appendix B. About 12 species of parasites, five fishes, no amphibians or reptiles, six birds and 37 mammals have been recorded.

Parasites

The best known alien parasites are those which infest domestic animals and man. They are all fairly easily controlled, but total eradication is unlikely. Few of the parasites that infest domestic stock have been recovered from indigenous hosts in SWA/Namibia. Apparently, it is only when indigenous hosts are in poor condition that they become heavily infested with alien parasites (Gubb personal observations). The harsh climate of SWA/Namibia makes it relatively immune to invasion. Only species with close host associations for their whole life cycles, eg lice, or which are environmentally resistant during their free-living stages, eg tapeworms, or whose life cycle is synchronized to coincide with advantageous environmental conditions, eg *Oestrus ovis*, seem able to survive. In general, the effect of indigenous parasites on alien hosts appears to be far more dramatic than the effect of alien parasites on indigenous hosts (Biggs and Horak 1985).

Fishes

Three alien species of fishes can be considered as potentially highly invasive in SWA/Namibia. These are the carp *Cyprinus carpio*, the largemouth bass *Micropterus salmoides* and the Mocambique tilapia *Oreochromis mossambicus*. All three species are widespread in farm and state dams in SWA/Namibia. Concern regarding these species is centred around their possible entry into the Okavango River via the Omatako drainage system which rises in the Thornbush Savanna region.

Further possible routes of invasion of natural systems are provided by artificial water canals linking the Cunene and Cuvelai systems, and the Okavango River with the central regions of SWA/Namibia (Schrader 1985).

Both carp and Mocambique tilapia have been found in pools in the Namib desert and probably originated from artificial impoundments on farms further east. Both species are considered permanent inhabitants of the Namib Desert, even during periods of prolonged drought (Dixon and Blom 1974).

Birds

At present the only successful alien bird in the drier regions of southern Africa is the house sparrow *Passer domesticus*, which has colonized virtually the entire area under discussion. It invaded the Karoo in the early 1950's and was recorded in the Kalahari by 1959. First records for SWA/Namibia date from the late 1950's and early 1960's, in the extreme south of the country. The house sparrow was first recorded from Windhoek in 1962, Swakopmund in 1964, Luderitz in 1965 and it had reached the southern boundary of Etosha National Park by 1969 (Brown 1985). This extremely rapid, but irregular series of jumps has been reported for the species in other areas (see Harwin and Irwin 1966; Vierke 1970). The mean speed of advance in SWA/Namibia was about 150 km yr^{-1} but with the south being colonized more quickly than the north (260 km yr^{-1} and 100 km yr^{-1} respectively). The house sparrow is commensal with man, occurring in built-up areas, around farmyards and in gardens. As it is reliant on human habitation this species is not considered to be a threat to indigenous species.

The feral pigeon *Columba livia* is found in all towns and on many farms but the distribution of self-sustaining feral populations is probably restricted to larger urban areas. Hybridization between *C livia* and *C guinea* has been achieved in captivity, but the fertility rate and survival rate of the nestlings are extremely low. In addition, the behavioural characteristics of the two species in the wild state are too different for successful inter-breeding to be likely (Rowan 1983). *Columbia livia* is not considered a threat to indigenous birds in semi-arid and arid biomes.

The European starling *Sturnus vulgaris* has not invaded the semi-arid and arid biomes of southern Africa successfully. At present the species is resident at Victoria West and Oranjemund (since 1970). It has been suggested that the lack of fruit, hard compacted ground which prevents probing with the bill for invertebrates, and the large distances between Oranjemund and the next town, have prevented it from spreading further (see also Macdonald et al this volume). If this species were to establish a population further north it may be expected to spread rapidly (Brown 1985).

Mammals

Alien mammals fall into three main categories, namely large mammals intentionally introduced for domestic or game farming purposes, household pets, and rodents which have been accidentally introduced by man. No member of the first group is considered to be of cause for concern as they are all easily controlled. It is of interest to record, however, that feral populations of domestic donkeys *Equus asinus*, horses *E caballus* and goats *Capra hircus* occur in remote regions of SWA/Namibia. Some of these may interbreed with indigenous species, eg donkeys with Hartmann's mountain zebra *E zebra hartmanae* (Brown and Jenkins in press).

In the second group, the domestic cat *Felis catus* which has been introduced to most parts of the semi-arid and arid biomes is said to interbreed freely with the indigenous African wild cat *Felis lybica* (Griffin and Panagis 1985), resulting in the loss of genetic integrity in the latter species.

In the third group, house mice *Mus musculus* and

house rats *Rattus rattus* have both established numerous urban populations and both have feral populations at Sandwich Harbour (Griffin and Panagis 1985).

Conclusion

It is apparent that alien plants are of far greater concern in arid and semi-arid biomes than are alien animals. Of the approximately 170 invasive alien plants recorded for the area under discussion 27 are considered to be very successful and thus of concern. By contrast, about 40 invasive alien vertebrate species have been recorded of which six are, or could become, the cause of concern.

Although most alien plants originate from Europe and South America the greatest number of highly successful species have come from Central America and Asia with the highest percentage of highly invasive species being for Australia. Most of the species listed are from semi-arid or arid areas of the continents of origin.

The ecological regions most invaded by alien plants are the Northern Cape Savanna, the Upper Karoo and the Highland Savanna. These regions have most or all of the following characteristics in common:

(1) a relatively large human population and a relatively long history of European settlement resulting in disturbance of natural habitats, and the introduction of large numbers of propagules;
(2) a large percentage of the areas used for farming;
(3) permanent water sources ie rivers and artificial impoundments, which allow mesophytic species to extend well beyond their normal arid tolerances;
(4) relatively high, reliable rainfall that has a season of at least four months; and
(5) hilly areas with numerous watercourses and good runoff.

Within the ecological regions the habitats containing the greatest number and diversity of alien plants can be grouped into two classes, namely (i) habitats disturbed by human agency, eg outskirts of towns, farm homesteads, and road verges, and (ii) naturally disturbed habitats, eg riverine habitats and floodplains of intermittant discharge systems. Undisturbed natural habitats had a low incidence of invasion.

The immediate implementation of stringent control measures is essential. At present invasion in these arid and semi-arid regions is less extensive than in areas of the subcontinent which receive a higher rainfall, and should therefore be easier to bring under control.

Principles of control are essentially the same as those employed in other ecosystems. However, there is an urgent need for information on the distribution, status and invasive potential of all alien organisms occurring in the arid regions of the Cape Province. Research into providing an alternative, indigenous fodder and shade tree to replace *Prosopis* species and others is needed. Wherever infestations of water dispersed alien plant species are to be controlled the possibility of reinfestation from headwater regions must be considered. In some cases reinfestation is such an important consideration that control should only be initiated where the upper reaches can be included in the control programme.

Legislation prohibiting the introduction and distribution of potentially invasive species should be introduced in SWA/Namibia, where such legislation is lacking at present, and should be enforced throughout the area. Any person wanting to import alien species of plants or animals to an area should be required to demonstrate that the species is unlikely to become invasive.

Increased public awareness of the undesirability of invasive alien organisms is of prime importance; the control of invasive aliens is ultimately the responsibility of all sectors of the population.

Acknowledgements

We wish to thank the following people for assistance with the preparation of this chapter: S E Brown, B H Erasmus, G E Gubb, H Kolberg, R Liversidge and S Lloyd.

B Whiting, R Brooke, L Henderson, I Macdonald, F Kruger and an unknown referee made useful comments on earlier drafts of this chapter.

Acocks J P H 1975. Veld types of South Africa. Memoirs of the Botanical Survey of South Africa 40. Department of Agricultural Technical Services, Pretoria.

Bailey L H 1951. Manual of cultivated plants. MacMillan Publishing Company Incorporated, New York.

Biggs H C and Horak I G 1985. Some invasive alien parasites in South West Africa/Namibia. In: Brown C J, Macdonald I A W and Brown S E (eds) Invasive alien organisms in South West Africa/Namibia. South African National Scientific Programmes Report No 119. CSIR, Pretoria. pp 41-43.

Brown C J 1985. Invasive alien birds in South West Africa/Namibia. In: Brown C J, Macdonald I A W and Brown S E (eds) Invasive alien organisms in South West Africa/Namibia. South African National Scientific Programmes Report No 119. CSIR, Pretoria. pp 31-34.

Brown C J and Jenkins A R (in press). Hybridization between a Hartmann's mountain zebra and a donkey. Madoqua.

Brown C J, Macdonald I A W and Brown S E (eds) 1985. Invasive alien organisms in South West Africa/Namibia. South African National Scientific Programmes Report No 119. CSIR, Pretoria.

Brown C J, Riekert B R and Vinjevold R 1984. *Datura innoxia* seeds eaten by Doublebanded Sandgrouse. Lanioturdus 20 (11), 1.

Dixon J E W and Blom M J 1974. Some aquatic vertebrates from the Namib Desert, South West Africa. Madoqua II. 3, 31-32.

Gibbs Russell G E 1984. List of species of Southern African plants. Memoirs of the Botanical Survey of South Africa No 48. Department of Agricultural Technical Services, Pretoria.

Giess W 1971. A preliminary vegetation map of South West Africa. Dinteria 4, 1-114.

Goudie A 1972. Climate, weathering, crust formation, dunes and fluvial features of the Central Namib Desert, near Gobabeb, South West Africa. Madoqua II. 1, 15-31.

Griffin M and Panagis K 1985. Invasive alien mammals, reptiles and amphibians in South West Africa/Namibia. In: Brown C J, Macdonald I A W and Brown S E (eds) Invasive alien organisms in South West Africa/Namibia. South African National Scientific Programmes Report No 119. CSIR, Pretoria. pp 44-47.

Hamilton G N G and Cooke H B S 1960. Geology for South African students. Central News Agency, South Africa.

Harmse H J von M 1978. Schematic soil map of Southern Africa south of latitude 10 deg 30 min S. In: Werger M J A (ed) Biogeography and ecology of Southern Africa. W Junk, The Hague.

Harwin R M and Irwin M P S 1966. The spread of the house sparrow, *Passer domesticus*, in south-central Africa. Arnoldia 2 (24), 1-17.

Henderson M and Anderson J G 1966. Common weeds of South Africa. Memoirs of the Botanical Survey of South Africa 37. Department of Agricultural Technical Services, Pretoria.

Holm L G, Plucknett D L, Pancho J V and Herberger J P 1977. The world's worst weeds; distribution and biology. The University Press of Hawaii, Honolulu.

Landsdell K A 1927. Weeds of South Africa. Department of Agriculture, Pretoria. Bulletin 18, 1-6.

Leistner O A 1967. The plant ecology of the southern Kalahari. Memoirs of the Botanical Survey of South Africa 38, 1-178.

Leistner O A 1979. Southern Africa. In: Perry R A and Goodall D W (eds) Arid-land Ecosystems: Structure and Management Vol 1, IBP 16, 109-143.

Macdonald I A W 1983. Invasive alien plants in southern Africa. List accompanying questionnaire survey of invasive alien plants in southern African protected areas. Percy FitzPatrick Institute, University of Cape Town, Cape Town.

Macdonald I A W and Nott T B (in press). Invasive alien organisms in central South West Africa/Namibia: results of a reconnaisance survey conducted in November 1984. Madoqua.

Merxmuller H 1966-1972. Prodromus einer flora von Sudwestafrika. Verlag von J. Cramer, Lehre/Germany.

Muller M A N 1985. Invasive alien plants in commercial grazing areas, diamond areas and Hereroland. In: Brown C J, Macdonald I A W and Brown S E (eds) Invasive alien organisms in South West Africa/Namibia. South African National Scientific Programmes Report No 119. CSIR, Pretoria. pp 1-5.

Rowan M K 1983. The doves, parrots, louries and cuckoos of Southern Africa. David Philip, Cape Town.

Schrader H J 1985. Invasive alien fishes in South West Africa/Namibia. In: Brown C J, Macdonald I A W and Brown S E (eds) Invasive alien organisms in South West Africa/Namibia. South African National Scientific Programmes Report No 119. CSIR, Pretoria. pp 35-40.

Staff of the L H Bailey Hortarium, Cornell University 1978. Hortus Third; a concise dictionary of plants cultivated in the United States and Canada. MacMillan Publishing Company Incorporated, New York.

Tarr P W and Loutit R 1985. Invasive alien plants in the Skeleton Coast Park, western Damaraland and western Kaokoland. In: Brown C J, Macdonald I A W and Brown S E (eds) Invasive alien organisms in South West Africa/Namibia. South African National Scientific Programmes Report No 119. CSIR, Pretoria. pp 19-23.

Tyson P D 1978. Rainfall changes over South Africa during the period of meteorological record. In: Werger M J A (ed) Biogeography and ecology of Southern Africa. W Junk, The Hague.

Van der Merwe C R 1962. Soil groups and sub-groups of South Africa. Science Bulletin No 356, Government Printer, Pretoria. pp 1-355.

Van der Merwe J H (ed) 1983. National Atlas of South West Africa. Unifoto, Cape Town.

Vierke J 1970. Die Besiedlung Sudafrikas durch den Haussperling (*Passer domesticus*). Journal fur Ornithologie 111, 94-103.

Vinjevold R D, Bridgeford P and Yeaton D 1985. Invasive alien plants in the Namib-Naukluft Park. In: Brown C J, Macdonald I A W and Brown S E (eds) Invasive alien organisms in South West Africa/Namibia. South African National Scientific Programmes Report No 119. CSIR, Pretoria. pp 24-27.

Werger M J A 1978. Biogeographical division of Southern Africa. In: Werger M J A (ed) Biogeography and ecology of Southern Africa. W Junk, The Hague.

Willman M 1946. Preliminary check list of the flowering plants and ferns of Griqualand West (Southern Africa). Deighton, Bell and Company Ltd, Cambridge.

APPENDIX A

The alien flora recorded in arid and semi-arid habitats in the areas of SWA/Namibia and the Northern Cape Province covered by this chapter, the habitats and ecological regions in which they have been recorded, their invasiveness in semi-natural and natural habitats (L = low, M = moderate, H = high) and their places of origin. Habitats are as follows: 1 = farm homesteads, 2 = kraals, 3 = outskirts of towns (xeric sites), 4 = timber plantations, 5 = roadside verges, 6 = railway verges, 7 = quarries, 8 = mining areas, 9 = waterholes, 10 = irrigated lands, 11 = irrigation canals, 12 = outskirts of towns (mesic sites), 13 = dry drainage lines and river beds, 14 = dry river gravel, 15 = dry river floodplains, 16 = moist drainage lines and riverbeds, 17 = perennial river banks, 18 = episodic river banks, 19 = rocky habitats, 20 = sandy habitats, 21 = seasonal and perennial marshes, 22 = seasonal pans, 23 = springs and seeps, 24 = islands in perennial rivers, 25 = in perennial rivers.
Ecological regions are as follows: NN = Northern and Central Namib, SN = Southern Dune Namib, WN = Desert and Succulent Steppe (winter rainfall), EZ = Escarpment Zone, DS = Dwarf Shrub Zone, UK = Upper Karoo, MS = Mopane Savanna, TS = Thornbush Savanna, HS = Highland Savanna, KS = Kalahari Savanna, NS = Northern Cape Savanna. The places of origin of the alien species were obtained from Bailey (1951), Henderson and Anderson (1966), Merxmuller (1966–1972), Holm et al (1977) and Staff of the L H Bailey Hortarium, Cornell University (1978).

Alien flora	Habitats recorded	Ecological regions recorded	Degree of invasion: Seminatural	Degree of invasion: Natural	Region of origin
Trees and large shrubs					
Agave spp	1,2,3	UK,MS,KS,NS	L	–	North and South America
Caesalpinia gilliesii	1,3	UK,NS	L	–	South America
Cannabis sativa	1,2,17	NS	L	L	Asia
Casuarina equisetifolia	1,3	NS	L	–	Australia
Cereus peruvianus	1,2,3,8,13,14,19	UK,NS	M	M	South America
Crotalaria juncea	6,9,15	NS	L	L	Asia
Eucalyptus camaldulensis	1-8,15,18-20,24	UK,KS,NS	H	H	Australia
Eucalyptus microtheca	1-8,15,18-20,24	UK,KS,NS	H	M	Australia
Eucalyptus sideroxylon	4	UK,NS	L	–	Australia
Eucalyptus sp	4,15,17	UK	L	L	Australia
Gleditsia triacanthos	1,3	UK,KS,NS	L	–	North America
Lantana camara	1,3,5	TS	L	–	South America
Melia azedarach	1-3,5,8,13,15-18,20	UK,TS,KS,NS	H	H	South West Asia
Nicotiana glauca	1-3,5-8,13-16,18,19	All regions	H	H	South America
Opuntia aurantiaca	3,14,19	NN,HS,NS	L	M	South America
Opuntia engelmannii	1,18,19	HS	L	M	Central America
Opuntia ficus-indica	1-3,7,8,13-16,19	EZ,UK,MS,TS,HS,KS,NS.	M	H	Central America
Opuntia imbricata	1,3,8,14,15,19	UK,NS	M	H	Central America
Opuntia rosea	10,19,20	NS	–	H	Central America
Opuntia stricta	19	HS	–	L	Central America
Opuntia sp (bl-gr cultivar)	1-3,5,6	UK,NS	M	–	–
Opuntia sp (yellow spines)	1,3,19	NS	L	L	–
Parkinsonia aculeata	1,3	UK,KS,NS	L	–	Central America
Physalis peruviana	11,15,17	NS	M	L	South America
Populus deltoides	12,17,24	UK,NS	L	M	Central America
Prosopis spp (5 possible spp)	1-9,12-20	All regions	H	H	North and South America
Ricinus communis	1,9,14,15,23	NN,SN,EZ,MS,TS,UK,NS	L	M	Africa and Asia
Salix babylonica	1,9,12,16-18	UK,NS	M	M	Asia
Schinus molle	1-3,5,7-9,14,20	UK,KS,NS	H	M	Central America
Sesbania punicea	12,17,24	TS,HS,NS	L	M	South America
Tamarix ramosissima	22	NS	–	M	Europe and Asia
Forbs					
Acanthospermum australe	1,2	NS	L	–	South America
Acanthospermum hispidum	3	HS	L	–	South America
Alhagi camelorum	14	NS	–	L	–
Alternanthera pungens	1-3,5,8,19,20	UK,KS,NS	H	M	South America
Alternanthera sessilis	15	NS	–	L	Central America
Amaranthus blitoides	3,5	NS	L	–	Central and South America
Amaranthus cruentus	3	HS	L	–	Asia
Amaranthus hybridus	1,2,5,19,20	NS	M	M	North America
Amaranthus viridus	3,5	NS	L	–	Central America
Anagallis arvensis	1	NS	L	–	Europe
Apium graveolens	10,11,16,17	NS	M	L	–
Apium leptophyllum	12	HS	L	–	Central America
Argemone subfusiformis	1-9,12-16,18	All regions	H	H	Central America
Aster subulatus	10-12,17	UK,NS	H	M	Australia
Atriplex nummularia	1-3,5,15,20,22	UK,KS,NS	M	M	Australia

Alien flora	Habitats recorded	Ecological regions recorded	Degree of invasion		Region of origin
			Seminatural	Natural	
Forbs continued					
Atriplex semibaccata	5,8,19,20,22	UK,NS	L	M	North America
Bidens biternata	1-3,5,6,9-12,16-18,20,23	EZ,UK,MS,TS,HS,KS,NS	H	H	North America and Asia
Bidens pilosa	1,9-12,17,20	UK,HS,KS,NS	H	L	Asia
Blackiella inflata	1,8,15,20	UK,NS	L	M	Australia
Boerhavia diffusa	10,19	NS	–	M	–
Boerhavia pterocarpa	1,2,5,6	NS	M	–	Central America
Brayulinea densa	1-3,5,7,8,19,20	UK,HS,KS,NS	H	M	Central America
Capsella bursa-pastoris	10,11,17	NS	M	L	Europe
Cardiospermum grandiflorum	1,16	TS	L	L	South America and Africa
Carduus marianus	10,16,21	NS	L	L	Europe, North Africa, Asia
Chenopodium album	1-3,10,17,20	UK,KS,NS	M	L	Europe and Asia
Chenopodium ambrosioides	1-3,10,12-16,18	NN,SN,WN,DS,MS,TS,HS, NS	M	–	Central America
Chenopodium carinatum	5,17,20	NS	L	L	–
Chenopodium glaucum	17,19,22	UK,KS,NS	–	L	–
Chenopodium multifidum	5	NS	L	–	South America
Chenopodium murale	3,5,17	NS	L	L	Europe
Chrysanthellum americanum	21	NS	–	L	North America
Cichorium intybus	10,11	NS	M	–	Europe and Asia
Conyza bonariensis	5,8,11,17	NS	M	M	–
Conyza canadensis	8,15,17	HS,NS	L	L	South America
Conyza floribunda	3	HS	L	–	–
Coronopus integrifolius	17,19,20	NS	–	M	South America
Datura ferox	1-3,5-10,13-16,18,20,22	All regions except WN	H	H	Europe and Asia
Datura innoxia	1-3,5,6,12-18	NN,SN,EZ,DS,MS,TS,HS,NS	H	H	Central America
Datura stramonium	1-3,5,6,8-10,13-16,18,20	NN,SN,EZ,DS,UK,MS,KS,NS	H	H	Central America
Dodonaea viscosa	3	TS	L	–	–
Eclipta prostrata		HS	–	–	Asia
Euphorbia chamaesyce	5	HS,NS	L	–	Central America
Euphorbia peplus	5	HS,NS	L	–	Europe
Fagopyrum esculentum	19	NS	–	L	Asia
Flaveria bidentis	1-3,5,10,13,14,16	NN,EZ,MS,TS,HS,NS	M	–	Central America
Fumaria officinalis	1,10,17	NS	L	L	Europe
Galinsoga parviflora	20	NS	–	L	South America
Glycyrrhiza glabra	10	UK	L	–	Europe
Gnaphalium luteo-album	12,16,17,20,21	UK,KS,NS	L	M	Europe
Gomphrena celosioides	1,2,5,9,17,19	HS,NS	M	L	South America
Hibiscus trionum	10,17	NS	L	L	Europe
Ipomoea purpurea	12,17	NS	L	L	South America
Lactuca serriola	12,17,21,23	NS	L	M	Europe
Lamium amplexicaule	1,10,12,17	NS	M	L	–
Lappula echinata	1,3,20	UK,NS	L	L	Europe
Lepidium sp cf *bonariense*	1,5,10,17,20,23	NS	L	L	South America
Lithospermum arvense	8	NS	L	–	–
Malva parviflora	1,10-12,17,19	UK,KS,NS	H	L	Europe
Medicago aschersoniana	1,10-12,15-17	NS	M	L	–
Medicago denticulata	11,15,17	NS	L	L	Europe and Asia
Medicago nigra	1,10-12,15-17	NS	L	L	–
Medicago sativa	11,16	NS	L	L	Europe, South West Asia
Melilotus alba	1,10-12,15-17	NS	M	L	Europe and Asia
Melilotus indica	1,10-12,15-17	NS	H	M	Europe and Asia
Modiola caroliniana	10-15,17	NS	L	L	–
Nicandra physaloides	10,11,17	NS	L	L	South America
Nothoscordium inodorum	1,10,21	NS	L	L	North America
Oenothera indecora	1,5,17	NS	L	L	Central America
Oenothera rosea	10,11,17	NS	M	M	Central America
Orobanche ramosa	3	HS	L	–	Europe and North America
Oxalis corniculata	12,17,19	KS,NS	L	L	North and South America
Oxalis latifolia	12	NS	L	–	Central America

Alien flora	Habitats recorded	Ecological regions recorded	Degree of invasion		Region of origin
			Seminatural	Natural	
Forbs continued					
Papaver hybridum	3	NS	L	–	–
Papaver rhoeas	10	NS	L	–	Europe and Asia
Physalis angulata	–	HS	L	L	North and South America
Physalis peruviana	10,11,15,17	HS,NS	M	L	South America
Plantago lanceolata	5,12,17,18,20	NS	L	L	Europe
Plantago major	10-11,16-18,23	NS	L	M	Europe
Polygonum aviculare	1,12,13,15,17,19	UK,KS,NS	L	M	North and South America
Polygonum convolvulus	1,3	NS	L	–	Europe
Portulaca oleracea	1,3	NS	L	–	Europe
Rumex angiocarpus	3,5,12	NS	L	–	Europe
Salsola kali	1-3,5-10,13-16	DS,UK,HS,KS,NS	H	H	Asia
Schkuhria pinnata	1,5,10,20	UK,HS,KS,NS	L	L	South America
Silene gallica	7,8,12,17	NS	L	L	Europe
Solanum mauritianum	12	NS	L	–	Asia
Solanum nigrum	1,2,17,19,20	NS	L	M	Europe
Sonchus asper	1	NS	L	–	Europe
Sonchus maritimus	1,8,14	NS	–	L	Europe
Sonchus oleraceus	5-8,10-12,17,21,23	UK,NS	M	M	Europe
Spergularia media	17,22	NS	–	L	Europe
Spergularia rubra	12	NS	L	–	Europe
Stellaria media	12,17	NS	L	L	Europe
Tagetes minuta	1-3,5-8,12-18,23	EZ,UK,MS,TS,HS	H	M	South America
Taraxacum dens-leonis	17,19,21,23	KS,NS	–	M	Europe
Tragopogon porrifolius	10,12	NS	M	–	Europe
Urtica urens	17	NS	–	L	Europe
Verbena officinalis	17	NS	–	L	Europe
Verbesina encelioides	1,10	UK,KS,NS	L	–	North America
Vicia atropurpurea	12	NS	L	–	Europe
Vicia sativa	12,17	NS	L	L	Europe and Asia
Xanthisma texanum	3	NS	L	–	North America
Xanthium spinosum	1-3,9,13-16,18	EZ,HS,KS,NS	M	M	Central America
Xanthium strumarium	10	NS	L	–	Central America
Zinnia peruviana	7,8,15,19	NS	L	M	South America
Grasses					
Arundo donax	21	NS	–	L	Europe
Avena sativa	5	NS	L	–	Europe
Briza maxima	10	NS	L	–	Europe
Bromus diandrus	16,17	NS	–	L	Europe
Bromus japonicus	10,11	NS	L	–	Europe and Asia
Bromus unioloides	11	NS	L	–	South America
Cenchrus incertus	10,15	NS	L	L	Central America
Digitaria sanguinalis	1,11,12,19	NS	M	L	Europe
Eragrostis pilosa	–	HS	L	L	Europe
Eragrostis tef	5,10	NS	M	–	North East Africa
Eragrostis tef × *E* spp	5,10	NS	M	–	–
Eragrostis virescens	5,10	UK,NS	L	L	South America
Hordeum murinum	10	NS	L	–	–
Lolium multiflorum	1,10	NS	L	–	Europe
Lolium temulentum	1,10,15	NS	L	L	Europe
Paspalum dilatatum	5,17	NS	L	L	North America
Pennisetum setaceum	3,5,12	HS	M	–	Africa
Phalaris arundinacea	10	NS	L	–	Europe and North America
Phalaris minor	3	NS	L	–	Europe
Poa annua	3	NS	L	–	Europe
Polypogon monspeliensis	12,16-18,21-23	HS,NS	L	H	Europe
Sorghum almum	5,10,15	NS	L	L	–
Sorghum halepense	5,10,15	NS	L	L	Europe

APPENDIX B

The alien fauna recorded for the semi-arid and arid biomes of the Cape Province and SWA/Namibia covered by this chapter, the region from which they have been recorded (CP = Cape Province, N = SWA/Namibia) and their invasive potential (L = low, M = moderate, H = high). Under "Notes" is recorded whether the species is free-ranging (F) or confined to a game farm etc (C). If the species is known to hybridize with indigenous species, this is indicated (H). Species which occurred historically in the areas under discussion, or have a very limited range, and have been reintroduced (R) are indicated.

	Region	Potential aggression	Notes
Parasites (SWA/Namibia only)			
Cysticercus bovis (man-cattle measles)			
Cysticercus cellulosae (man-pig measles)			
Damalinia spp (biting lice)			
Gasterophilus intestinalis (bot fly of horses)			
Haemmatopinus spp (blood-sucking lice)			
Linognathus spp (blood-sucking lice)			
Oestrus ovis (nasal bot fly of sheep)			
Taenia saginata (tape worm of man; measle in cattle)			
Taenia solium (tapeworm of man, measle in pig)			
Fishes (SWA/Namibia only)			
[*Clarias gariepinus* (sharptoothed catfish, a strain)]		M	
Cyprinus carpio (carp)		H	
Micropterus salmoides (largemouth bass)		M	
Oreochromis mossambicus (Mocambique tilapia)		H	
[*Tilapia sparrmanii* (banded tilapia, a strain)]		M	
Birds			
Acridotheres tristis (Indian myna)	CP	H	
Columba livia (feral pigeon)	CP & N	L	
Melopsittacus undulatus (budgerigar)	CP & N	M	
Passer domesticus (house sparrow)	CP & N	H	
Struthio camelus (ostrich)	CP	–	R (hybrids)
Sturnus vulgaris (European starling)	CP & N	M	
Mammals			
Addax nasomaculatus (addax)	CP	L	C
Aepyceros melamptus (impala)	N	L	C
Ammotragus lervia (barbary sheep)	CP	M	F
Axis porcinus (hogdeer)	CP	L	F
Bos spp (domestic cattle)	CP	M	F
Boselaphus tragocamelus (nilgai)	CP	M	F
Bubalus arnee (Indian water buffalo)	CP	L	F
Bubalis bubalis (water buffalo)	N	L	C
Camelus dromedarius (dromedary camel)	CP	M	F
Canis familiaris (domestic dog)	CP & N	M	F, H (*C. mesomelas*)
Capra hircus (domestic goat)	CP & N	M	F
Capra ibex (ibex)	N	L	C
Cephalophus monticola (blue duiker)	CP	L	C
Ceratotherium simum (white rhino)	N	–	F, R (from Zululand)
Cervus dama (fallow deer)	CP & N	M	F
Cervus elaphus (red deer)	CP	L	F
Connochaetes gnou (black wildebeest)	N	L	C
Connochaetes taurinus (blue wildebeest)	N	–	F, R
Damaliscus dorcas (bles- and bontebok)	CP & N	M	C, R (blesbok-bontebok)
Equus asinus (domestic donkey)	N	M	F, H (*E. zebra* and *E. burchelli*)
Equus caballus (domestic horse)	N	L	F, H (*E. burchelli*)
Equus zebra subspp	CP	M	F, H (*E. z. zebra – E. z. hartmannae*)
Felis catus (domestic cat)	CP & N	H	F, H (*F. lybica*)
Hemitragus jemlahicus (Himalayan thar)	CP	M	F
Kobus leche (red lechwe)	CP	L	F
Mus musculus (house mouse)	CP & N	H	F
Oryctolagus cuniculus (domestic rabbit)	CP	?	F (Possesion Island)
Ovis spp (domestic sheep)	CP	M	F
Rattus rattus (house rat)	CP & N	H	F
Rattus norvegicus (brown rat)	N	M	F
Redunca arundinum (reedbuck)	N	L	C
Sus scrofa (European wild boar)	CP	M	F
Syncerus caffer (African buffalo)	N	–	R (from Addo, CP)
Tragelaphus angasii (nyala)	CP & N	L	F
Tragelaphus scriptus (bushbuck)	CP	L	F

Chapter 8

Alien plant invasions in the grassland and savanna biomes

L Henderson and M J Wells
Botanical Research Institute.

Introduction

We review alien plant invasions in the grassland and savanna of southern Africa, and in particular South Africa. The semi-arid savanna of South West Africa/Namibia and the northern Cape of South Africa are covered more fully elsewhere (Brown and Gubb this volume).

The following descriptions of the grassland and savanna biomes in southern Africa (south of 22° S) have been derived from Rutherford and Westfall (in press). The savanna biome is largest and comprises 46,16% (about 959 000 km^2) of southern Africa, while the grassland biome is third largest and covers 16,52% (about 343 000 km^2) of southern Africa.

The grassland biome

The grassland biome is situated mainly on the central plateau of South Africa, inland areas of the seaboard of Natal and mountain areas in the south-eastern Cape Province. Topography is mainly flat to rolling but can also be mountainous. Altitude ranges from 300 m on the coastal plateau to 2 850 m in the Drakensberg. Rain falls mainly in summer and the mean annual rainfall is mostly between 400 and 2 000 mm. Moist grassland, in areas with a mean annual rainfall greater than about 625 mm, constitutes by far the greater part of the biome. Frosts are common and the mean lowest minimum temperature for the coldest month is generally below one degree centigrade. Lightning flash densities to the ground, the average number of days with thunder and hail frequency at a given point are higher than for all other biomes. The frequency of fire depends mainly on plant production (rate of fuel accumulation) and is least frequent in drier areas but in the moister areas can occur annually.

The indigenous vegetation of most of the biome is open grassland with trees and shrubs being associated with local specialized niches. A climatically marginal variation of moist grassland is found east of the escarpment which experiences higher minimum temperatures than the rest of the biome. Grassland here is maintained by fire and persistent wood removal.

The main land uses are the cultivation of crops, particularly maize and dairy, beef and wool production. Silviculture is limited largely to the regions receiving more than 800 mm of rain per year, in the eastern sector of the biome. The biome incorporates the greatest urban concentration in southern Africa, namely the Witwatersrand complex. Approximately 1,12% of the area of the biome is formally conserved.

The savanna biome

Composition of the savanna varies locally. The topography consists mainly of extensive plains but is also rugged along mountain scarps and on valley slopes. Altitude ranges from sea level to a few hundred metres in the south and to about 2 000 m in the north. The mean annual rainfall is mostly in excess of 235 mm. Much of the biome in South Africa receives more than 400 mm of rain per year. Savanna is excluded from high rainfall areas with low winter minimum temperatures but not from low rainfall areas with relatively severe winter frosts. Frost is absent or very rare along the east coast and northwards through the lowveld of Natal, Mozambique and Transvaal. Annual fires are rare in arid savanna where plant productivity is low but are possible in higher rainfall areas with woody plant canopy cover usually less than 40%.

The indigenous vegetation at biome scale is characterized by the codominance of grasses and trees or shrubs. The trees and shrubs are between three and seven metres tall and often multistemmed. Local dominance of trees or shrubs may occur, for example, where overgrazing leads to bush encroachment.

The main land use is extensive grazing by stock animals especially cattle. Where rainfall is adequate (usually more than 500 mm per year) and the soils are more fertile, agricultural crops are cultivated.

Address for correspondence: L Henderson, Botanical Research Institute, Private Bag X101, Pretoria, 0001.

Silviculture is limited to the high rainfall areas bordering on the grassland biome. Urbanization is limited mainly to the coastal areas and the higher altitude inland areas. In South Africa, approximately 8,5% of the biome is formally conserved.

Alien species and their history

The following analysis of aliens is based on 711 species of naturalized alien plants which are now included in the flora of the grassland and savanna biomes in southern Africa. The total species count was obtained largely from Wells et al (1986). Other sources were Gibbs Russell et al (1984) and Macdonald (1983a). Using information on specimen labels in the National Herbarium we compiled lists of alien plants that are known to be present in areas of natural and seminatural grassland and savanna vegetation.

The aliens belong to 98 plant families. The largest family, with 101 species, is the Poaceae, followed by the Asteraceae with 91 species and the Fabaceae with 77 species. The fourth largest family, the Solanaceae, has 34 species. Trees, shrubs and climbers contribute 225 of the species or 32%, the remaining 486 being herbs of which 401 are perennials. Of these species 167 or 23% are particularly rapid invaders (Macdonald 1983a).

Europe and Asia have provided almost as many species (313 or 44%) as the rest of the world. Most species came from Europe which does not have a very rich flora but which has had a long history of traffic with southern Africa. The Americas have provided 251 species or 35% (160 or 23% from South America, 61 or nine per cent from North America and 19 or three per cent from Central America). Australia has only provided 43 species or six per cent. The rest of Africa has provided 45 species or six per cent and less than one per cent came from islands in the Pacific, Indian and Atlantic oceans. The remaining 56 species (8%) are of uncertain origin.

It is significant that although Europe and Asia have provided the largest number of species only 18% of these are particularly aggressive invaders. In contrast, 36% of the species from South America and 47% from Australia have become aggressive invaders. These results suggest that new invaders are more likely to come from the southern continents, with similar environments to southern Africa, than from the northern regions.

The earliest recorded dates of 151 species that are known to be particularly aggressive invaders were obtained from the literature and the National Herbarium. Based on the earliest recorded dates for the country as a whole and not just for the grassland and savanna biomes, approximately 90% of all species arrived between 1820 and 1950 (Figure 1). The greatest rate of arrival occurred between 1820 and 1910. The graph suggests that the rate of arrival of invasive aliens is decreasing.

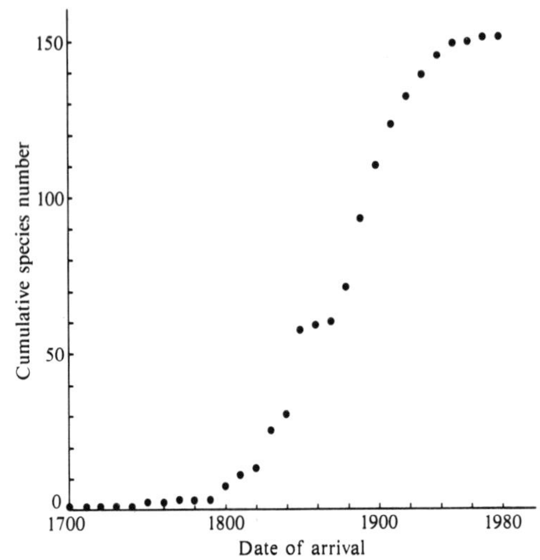

Figure 1. Cumulative number of species against date of arrival in time-course of introduction of 151 invasive aliens in the grasslands and savannas in South Africa.

The invasion of the grassland and savanna biomes by alien species spans a period of about 200 years. *Opuntia ficus-indica* was first recorded invading savanna vegetation in the eastern Cape in the 1770's (Macdonald 1891). For a long time, however, it appears to have remained largely within the confines of cultivation as it is not included in Drege's list of plants found in those parts between 1826 and 1834 (Macdonald 1891). Two other species are also known to have been naturalized in the grassland and savanna biomes before 1850 and these are *Opuntia vulgaris* (Neser and Annecke 1973) and *Arundo donax* (National Herbarium). Species naturalized before 1900 include *Acacia dealbata*, *A mearnsii*, *Opuntia aurantiaca*, *Melia azedarach* and *Solanum mauritianum* (Table 4). Most invasion of alien pest plants, however, has taken place only within the past 115 years (Table 4).

The nature of the invasions

Correlation of invasion with environmental factors

The distribution of aliens is clearly associated with disturbance, particularly roads, railways, fence lines, abandoned lands, human habitation and plantings (Wells et al 1980; Henderson and Musil 1984; Gubb 1985). Some species, however, are found far from plantings or apparent disturbance, for example *Lantana camara*, *Melia azedarach* and *Opuntia ficus-indica* (Henderson and Musil 1984). Streambanks which are subjected to periodic disturbance through flooding, are often heavily infested (Macdonald 1983b; Henderson and Musil 1984; Gubb 1985; Macdonald and Jarman 1985).

Table 1. The number of naturalized alien species in the climatic regions of the grassland and savanna biomes (National Herbarium and Wells et al 1986).

Climatic region	Biome	Number of species
Subtropical and temperate (summer and all-year rainfall)	grassland, savanna	711
Subtropical (summer rainfall)	savanna	583
Temperate (summer and all-year rainfall)	grassland, savanna	532
Temperate (summer rainfall)	grassland, savanna	483
Temperate (all-year rainfall)	mainly savanna	298
Exclusive subtropical	savanna	153
Exclusive temperate	grassland, savanna	103

The greatest concentration of species (583 species or 82%) occurs in the subtropical region of the savanna biome (Table 1). Moist situations (ie swamps, marshes, streambanks and lakeshores), which occupy a very small proportion of the total land surface area, have been invaded by many species (Table 2), with 51% of all species being recorded in moist situations and 50% of all tree, shrub and climbing species occurring along streambanks. In the Transvaal, Henderson and Musil (1984) recorded 40 of the 61 species (66%) of naturalized alien trees and shrubs in the streambank habitat. In a roadside survey of the Pretoria-Witwatersrand-Vereeniging complex in the central Transvaal aliens were recorded at all river crossings, 77% of perennial stream crossings and at 16% of all temporary stream crossings. Streambanks in savanna and grassland were almost equally invaded (Wells et al 1980).

Table 2. The number of naturalized alien species in habitats classified by moisture regime in the grassland and savanna biomes (National Herbarium and Wells et al 1986).

Moisture regime	Number of species: Total	Perennials	Trees, shrubs and climbers
Terrestrial dry	594	315	195
Streambanks	261	151	112
Terrestrial moist (swamps, vleis etc)	211	103	40
Aquatic	21	19	– (herbs)
Streambanks, terrestrial moist and aquatic	364	214	118

Factors contributing to the vulnerability of streambanks to alien invasion include periodic natural disturbances compounded by human-related disturbances, the often perennial availability of moisture and reliable dispersal by water. Rivers and streams may play a role in reducing water stress in invading species and thereby allowing acclimatization, particularly in the more arid regions. Water has been of particular significance in transporting the otherwise rather immobile seeds of several alien species belonging to genera such as *Acacia*, *Caesalpinia*, *Sesbania* and *Prosopis*. Rivers and streams not only provide routes for the invasion of alien species but their banks also serve as a seed reservoir from which further spread can take place.

In the Transvaal the greatest invasion of alien species occurs in those areas of the grassland and savanna biomes adjacent to the eastern escarpment. Grassland and savanna in the western Transvaal have been subject to much less invasion by aliens than the eastern parts. Factors contributing to this contrast include less disturbance, the absence of extensive plantations of invader species, a drier climate and a landscape that offers a smaller range of habitats (Henderson and Musil 1984).

Similar invasion trends appear to hold true for the rest of the grassland and savanna biomes in southern Africa. The distributions of the most important invaders (Table 4) overlap to the greatest extent in a relatively narrow belt extending from the eastern Cape in the south through the Transkei, Natal and Swaziland to the northern Transvaal. This belt extends from the coast to the escarpment and includes grassland, savanna and forest. It also includes the whole of the summer rainfall subtropical zone. West of this belt fewer alien species are found but dense stands of one or a few species are encountered, for example, *Prosopis* species in the northern Cape and SWA/Namibia, and *Opuntia imbricata*, *O ficus-indica* and *O aurantiaca* in the arid parts of the eastern and northern Cape.

Important invaders and the extent of their infestations

At least 32 alien species are known to form moderately dense to dense stands in the grassland and eastern savanna biomes (Table 3). The information in Table 3 was obtained largely from a survey of alien trees and shrubs in the Transvaal (Henderson and

Table 3. Alien invaders that occur in moderately dense to dense stands in major habitat classes in the grassland and eastern savanna biomes.

	Grassland biome		Savanna biome	
Species and (life form*)	Terrestrial habitats	Streambanks	Terrestrial habitats	Streambanks
Acacia dealbata (T)	XY	XY		X
A decurrens (T)	X	XY		
A mearnsii (T)	XY	XY		X
A melanoxylon (T)		XY		
Arundo donax (T/S)				X
Caesalpinia decapetala (S/C)		X		XY
Cassia bicapsularis (T/S)				X
C didymobotrya (T/S)				X
+ *Cereus peruvianus* (T/S)			X	
Cestrum laevigatum (T/S)				X
Chromolaena odorata (S/C)			XY	XY
Eucalyptus spp (T)		X		
Eupatorium adenophorum (H)	X			
Jacaranda mimosifolia (T)			X	X
Leucaena leucocephala (T/S)				X
Lantana camara (S/C)			XY	XY
Mangifera indica (T)				X
Melia azedarach (T)		X	X	XY
Morus alba (T)		X		X
+ *Opuntia aurantiaca* (S)	X		XY	
+ *O ficus-indica* (T/S)	X		XY	X
+ *O imbricata* (S)	X		X	
+ *O rosea* (S)			X	
+ *O vulgaris* (T/S)	X			
+ *Pereskia aculeata* (S/C)			X	
Pinus patula (T)	X			
Populus canescens (T)		XY		XY
Psidium guajava (T/S)			X	
Rubus cuneifolius (S)	XY	XY		
Sesbania punicea (S)		X		XY
Solanum mauritianum (T/S)		XY	X	X
Stipa trichotoma (H)	X			

XY signifies dense stands
*T = tree, S = shrub, C = climber, H = herb/grass, + = succulent

Musil 1984) and the proceedings of a workshop on invasive alien plants in Natal (Macdonald and Jarman 1985). In the western savanna regions *Prosopis* species form moderately dense to dense stands (Brown and Gubb this volume).

The distributions of some important plant invaders are summarized in Table 4. Distribution is expressed in terms of quarter-degree squares and degree squares occupied by 1985. The information was obtained from the National Herbarium, Henderson and Musil (1984 and unpublished), Stirton (1978) and Liggitt (1983). On the whole, distribution records are very fragmentary and the information in Table 4 should be used with caution when comparing distributions. The inferred present distribution of a species is given in the last column in the table.

Table 4. Age and area: years since the first record of naturalization and number of quarter-degree squares and degree squares occupied by some important alien plants in the grassland and savanna biomes.

		Quarter-degree squares				Degree squares	
Species	Earliest record of naturalization	Years	Total	Grass-lands	Savannas	Actual records	Probably occupied
Opuntia ficus-indica	1770's (E Cape) (Macdonald 1891)	215*	195	44	151	45	78
Acacia dealbata	1870 (Natal) (Smith 1966)	115	100	80	20	24	31
Acacia mearnsii	1870's? (Natal, Lesotho)	115?	118	86	32	28	36
Opuntia aurantiaca	1892 (E Cape) (Zimmermann and van de Venter 1981)	93	113	52	61	27	50
Melia azedarach	1894 (Natal) (Medley-Wood 1894)	91	163	50	113	40	53
Solanum mauritianum	1894 (Natal) (Medley-Wood 1894)	91	60	30	30	22	27
Rubus cuneifolius	early 1900's (Natal) (Phillips et al 1939)	85	29	23	6	11	18
Populus canescens	1903 (Tranvaal) (Burtt Davy 1904)	82	93	63	30	28	52
Salix babylonica	1904 (Transvaal) (Burtt Davy 1904)	81	84	67	17	22	45
Lantana camara	1907 (Natal) (Medley-Wood 1907)	78*	140	30	110	35	36
Psidium guajava	1919? (Natal?) (Sim 1919)	66?	39	3	36	15	16
Sesbania punicea	after 1912 (Transvaal and Swaziland) (Burtt Davy and Pott-Leendertz 1912)	60?*	66	22	44	30	35
Stipa trichotoma	1928 (E Cape) (Wells 1974)	57	12	12	–	5	7
Chromolaena odorata	1940's (Natal) (National Herbarium)	45	49	8	41	9	13?

In general there is a weak correlation between date of introduction and area colonized but *Opuntia ficus-indica*, the oldest invasive plant species, has also achieved the widest distribution. By 1932, before the implementation of biocontrol measures, infestations were spread over roughly 80 degree squares in the whole of South Africa (Anonymous 1932) and 64 degree squares in the grassland and savanna biomes, with dense infestations in 24 degree squares. Although biocontrol has greatly reduced *O ficus-indica* it is still present in the wild state in most of these areas and has possibly increased its range. A survey of the Transvaal (Henderson and Musil 1984) revealed that *O ficus-indica* is the most widespread species in this province and occurred in 61% of all quarter-degree squares surveyed.

The next most widespread species is *Melia azedarach*. In the Transvaal it was recorded in 52% of all quarter-degree squares surveyed (Henderson and Musil 1984). The next most widespread species are *Lantana camara*, *Acacia mearnsii*, *Opuntia aurantiaca* and *A dealbata*. However, since *Populus canescens* and *Salix babylonica* are widely planted in the grassveld biome these species could be a lot more widespread than actual records show. In the Transvaal both species were recorded in more than 30% of all quarter-degrees surveyed and *S babylonica* was recorded at 62% of all the watercourse crossings which were sampled in the grassland areas (Henderson and Musil 1984). *Opuntia aurantiaca*, a poorly collected species, is also suspected of having a much wider distribution than records show.

The species that appear to have infested the greatest land area in South Africa are *O aurantiaca* (in excess of 15 000 km^2: Zimmermann and Moran 1982) and *O ficus-indica* (formerly about 10 000 km^2 densely infested but now infesting less than 3 500 km^2: Zimmermann and Moran 1982). For both species the total area given includes the area invaded within the karoo biome.

Chromolaena odorata was estimated in 1982 to infest about 8 000 km^2 in Natal (Liggitt 1983), including forest communities. *Stipa trichotoma* was estimated in 1983 to infest about 600 km^2 in the eastern Cape (Myburgh 1985). *Acacia mearnsii* was estimated in 1972 to infest 479 km^2 in Natal and the Transvaal (Stubbings 1977) and in 1976 it was estimated to infest 800 km^2 in Natal and the Cape (Downing 1978). These values, however, include its extent in parts of the forest and fynbos biomes.

Little information exists on actual densities achieved by individual species. Macdonald (1983b) reports that *Melia azedarach* was present at a density of 120 trees per hectare (0,01 km^2) in a riverine community in Natal where all the indigenous trees had a combined density of only 105 trees ha^{-1}. *Chromolaena odorata* was present at a density in excess of 48 000 plants ha^{-1} in a riverine infestation in Natal (Macdonald 1983b). *Cereus peruvianus* has been recorded at a maximum density of 39 333 plants ha^{-1} in the northern Transvaal (Taylor and Walker 1984).

Comparisons with indigenous invasions

Alien invaders are not the only plants to have spread in the grassland and savanna biome areas. Wells et al (1983) state that: 'three main types of invasion (of grassland) by indigenous species take place.

(1) On its higher, drier margins the grassland is being invaded by small xerophytic 'karroid' shrubs. This is locally referred to as 'the spread of the karoo'.
(2) On its lower, moister margins grassland is being invaded by the adjacent savanna communities.
(3) Within the grassland, particularly on the higher, wetter ridges, unpalatable grassland species including tussocks and poisonous plants are on the increase.'

All these changes and to a lesser extent the invasion of grassland by indigenous fynbos species, have been documented by Acocks (1975). The question arises as to whether alien invaders are contributing to or acting independently of these changes. No studies have been made but the following observations are relevant.

(1) Very few low shrubs of alien origin are present at the interface between karoo and grassland. However, arborescent *Prosopis* species and *Opuntia ficus-indica*, which have spread from farm plantings, are invading some of these areas. *Opuntia ficus-indica* occurs mainly on rocky ridges and valley sides, whilst the *Prosopis* species 'favour watercourses or areas with deep sediments and groundwater near the surface' (Harding in press). Both are sources of supplementary forage and the possible influence of this on the carrying capacity and grazing pressure on adjacent grassland, especially during drought, needs to be investigated.
(2) Aliens are present throughout the invading savanna, but do not appear to increase the extent or efficiency of this community's invasion of grassland, except in a few instances such as

Lantana camara in the eastern Transvaal. The dramatic invasions by woody aliens such as Australian *Acacia* species sometimes parallel bush encroachment, by spreading from streams. Generally, however, they spread, as do *Pinus* and *Eucalyptus* species, directly from plantations, gardens and roadside plantings. In this they seem to be hardier or more efficient than their indigenous counterparts — perhaps a question of differing climatic histories in the areas where they evolved (Brain 1986).

(3) *Stipa trichotoma* is the only alien grass that has yet become a troublesome invader of South African grassland. Other alien *Stipa* species already occur in the country and together with numerous species not yet introduced from South America, might in future invade the grasslands. Their reaction to selective grazing appears to be very similar to that of the indigenous *Aristida* species which are increasing.

Ecological impacts

An impact is defined here as a change to ecosystem, community or population processes caused by aliens once they have invaded the landscape. Processes which are irreversibly affected, such as a decrease in species richness, are most important. The most obvious impact of alien species on both the grassland and savanna biomes is their partial or complete replacement of indigenous communities, thereby reducing species richness and altering the appearance and character of the land (Henderson and Musil 1984). Alien trees and shrubs appear to have a greater impact than herbaceous plants in transforming habitats (Wells et al this volume).

Possibly the greatest impact of aliens in either of the biomes has been in the moist subtropical grasslands east of the escarpment which have been greatly exploited for silviculture. Where there used to be open grassland with scattered trees there are now extensive plantations, some of which have been abandoned and have formed thickets. This applies particularly to *Acacia mearnsii* which, according to Beard (1956), may provide a less efficient watershed cover than ungrazed natural grassland. Net infiltration of water into the soil is reduced by higher rainfall interception and runoff. Water runoff is highest in spring when the soil beneath the plantations is resistant to wetting.

The increased woody element in the grassland biome particularly in the moist subtropical grasslands is likely to have an impact on the fire regime of these areas. Increased total fuel load, a reduction in dead fine material and a reduction in ground cover is likely to increase the probability of intense fires and to increase the intervals between fires.

The degradation of the riverine habitat caused by alien species has been documented in several studies (eg Wells et al 1980; Macdonald 1983b; Henderson and Musil 1984; Brown and Gubb this volume; Macdonald and Jarman 1985). The indigenous streambank communities in both biomes are being replaced by a spectrum of alien species some of which form pure stands, jeopardizing the survival of the indigenous flora and fauna and the water resources of the biomes. Alien woody species such as *Populus canescens*, *Acacia dealbata* and *Arundo donax* tend to spread from the riparian habitat beyond the zone previously occupied by indigenous woody species. The resulting dense thickets must have considerable effects on total transpiration losses and on the dynamics of the stream channel, especially where thickets extend for several kilometres along watercourses (Wells et al 1980).

Population genetic processes may be interfered with through the hybridization of alien pest plants with indigenous species. Stirton (1979) reports that some alien *Rubus* species hybridize with indigenous *Rubus* species, and that this can give rise to new invasive plant species.

The history of management and control

The history of the management and control of alien invasive plants in the grassland and savanna biomes dates back to the campaigns against *Xanthium spinosum* (1860) (Smith 1966) and *Opuntia ficus-indica* (1883) (Stirton 1978). The campaign against *X spinosum* has been largely ineffective and today *Xanthium* is still a serious weed.

Various methods of mechanical and chemical control were employed to suppress *Opuntia ficus-indica* but with little success. Then in 1932 a biological control campaign was started which has been partially successful. An estimated 75% of the original 10 000 km^2 of fairly dense prickly pear was reclaimed through the action of *Cactoblastis cactorum* and *Dactylopius opuntiae* (Zimmermann and Malan 1980). The residual infestations, however, have increased considerably since the end of the biological control campaign in the early 1950's. With no further introductions of new natural insect enemies for improved biological control of *O ficus-indica* envisaged, much more reliance will have to be placed on improved chemical control efforts (Zimmermann and Malan 1980).

The most costly campaign against an alien plant has been that waged against *Opuntia aurantiaca* since 1933. Despite a biocontrol campaign started in 1935 and efforts to combat the plants mechanically and chemically it has continued to spread. Since 1957 a herbicidal control programme has been implemented at an annual expenditure estimated at about 2,5 million US dollars (Zimmermann and Moran 1982). By 1978 the total infestation in the Cape (ie including the Karoo) covered an area of 16 000 km^2 and it was increasing at a rate of 70 $km^2 yr^{-1}$ (Stirton 1978). By 1983 20 eastern Cape districts were infested totalling 5 460 km^2 (with 75% under control). In Natal nearly 90 km^2 (with 67% under control) and in the Orange Free State 12 km^2 (with 69% under control) are reported to be infested with this alien (Anonymous 1984).

An intensive eradication campaign against *Stipa trichotoma* was initiated in 1976 after a survey in the eastern Cape revealed that 490 km^2 of land in six districts were infested. Despite efforts at mechanical and chemical control the extent of the infestations increased to 600 km^2 in 1983. The herbicides which have been used are extremely costly and largely ineffective as they do not prevent the emergence of large numbers of seedlings after treatment. There is, however, renewed hope for the chemical control of this species with an aerially applied herbicide which is absorbed by the roots and which is long-lasting, selective, comparatively inexpensive and poses little danger to fish and mammals (Myburgh 1985).

Apart from *Opuntia ficus-indica* and *O aurantiaca* several other species have been subjected to biological control methods. These include *Opuntia imbricata*, *O rosea*, *O lindheimeri*, *O vulgaris* (almost totally eradicated - this was the first attempt at biocontrol in 1913 - and was completely successful (Neser and Annecke 1973)), *O spinulifera*, *Lantana camara* (so far only successful in very localized areas, for example, at Margate on the Natal coast; the enemies introduced so far are not effective against all cultivars (Anonymous 1985)), *Sesbania punicea* (all indications are that the campaign will be very successful (C J Cilliers personal communication)), *Cirsium vulgare*, *Eupatorium adenophorum*, *Solanum elaeagnifolium* and *Harrisia martinii* (S Neser personal communication). Species which are being considered as targets for biocontrol include *Rubus cuneifolius*, *Solanum mauritianum*, *Chromolaena odorata* and *Acacia* species (S Neser personal communication). In this respect *Melia azedarach* deserves attention (Macdonald 1983b).

Of the 700 species of naturalized aliens known to occur in the grassland and savanna biomes, legislation has been passed against 53 species, 43 species have been declared weeds and 10 have been declared invader plants. Despite this legislation the uncontrolled spread of alien invaders in the veld and the presence of declared weeds in urban and rural gardens is commonplace. This seems to be because the general public has the attitude that the authorities should carry out or subsidize all weed control, and because the laws cannot be adequately enforced.

Cost of eradication, intangible or unperceived benefits of control, insufficient man-power, difficulty in finding all the plants or parts as in eg *Opuntia aurantiaca*, the inability to recognize alien species, large and long-lived seed banks in the soil (eg *Acacia* species, *Stipa trichotoma*) and the ineffectiveness of available control measures are probably additional factors contributing to the lack of control.

While much emphasis has been placed on the eradication and control of aliens after they have invaded, less attention has been given to preventing initial invasion, particularly through improved veld management. In both savanna and grassland the maintenance of a vigorous plant cover, particularly the grass layer, is seen as an important factor in the prevention of alien invasions. Equally important is the effective management of primary infestation routes and sources, such as roads, railways and rivers. The management of rivers is greatly complicated by their many different uses, the many landowners involved and, in places, by the ruggedness of the terrain.

Acocks J P H 1975. Veld types of South Africa. Memoirs of the Botanical Survey of South Africa 40, 1-128.

Anonymous 1932. Streke met turksvye besmet, 1932. Map supplied by Division of Entomology, Biological Control Section, Port Elizabeth in 1961.

Anonymous 1984. Ploegvore van gister. Litjieskaktus duur voort. Agricultural News, 26 October 1984. 8 p.

Anonymous 1985. Vordering met biologiese beheer van indringerplante. Agricultural News, 29 March 1985. 5 p.

Beard J S 1956. Results of the mountain home rainfall interception and infiltration project in black wattle, 1953/54. Journal of the South African Forestry Association 27, 72-85.

Brain C K 1986. Prospects for a southern African continental record of Cainozoic temperature changes. South African Journal of Science 82, 94-96.

Burtt Davy J 1904. Alien plants spontaneous in the Transvaal. Report of the South African Association for the Advancement of Science. The Argus, Johannesburg. pp 252-299.

Burtt Davy J and Pott-Leendertz R 1912. A first check-list of the flowering plants and ferns of the Transvaal and Swaziland. Annals of the Transvaal Museum 3(3), 119-182.

Downing B H 1978. Environmental consequences of agricultural expansion in South Africa since 1850. South African Journal of Science 74, 420-422.

Gibbs Russell G E and the Staff of the National Herbarium 1984. List of species of southern African plants. Memoirs of the Botanical Survey of South Africa 48, 1-144.

Gubb A A 1985. A preliminary ecological synthesis of invasive alien biota in the northern Cape and Great Karoo. Unpublished manuscript, McGregor Museum, Kimberley, 45 pp.

Harding G B (in press). The status of *Prosopis* as a weed. Applied Plant Science.

Henderson L and Musil K J 1984. Exotic woody plant invaders of the Transvaal. Bothalia 15 (1,2), 297-313.

Liggitt B 1983. The invasive alien plant *Chromolaena odorata*, with regard to its status and control in Natal. Institute of Natural Resources, Pietermaritzburg. Unpublished.

Macdonald A C 1891. Prickly pear in South Africa. Agricultural Journal 4(2), 21-25.

Macdonald I A W 1983a. Invasive alien plants in southern Africa. Being list A accompanying the questionnaire survey of invasive alien plants in southern African protected areas. Percy FitzPatrick Institute of African Ornithology, University of Cape Town, Cape Town. 101 pp.

Macdonald I A W 1983b. Alien trees, shrubs and creepers invading indigenous vegetation in the Hluhluwe-Umfolozi Game Reserve Complex in Natal. Bothalia 14 (3,4), 949-959.

Macdonald I A W and Jarman M L (eds) 1985. Invasive alien plants in the terrestrial ecosystems of Natal, South Africa. South African National Scientific Programmes Report No 118. CSIR, Pretoria.

Medley-Wood J 1894. Preliminary catalogue of indigenous Natal plants. Natal Mercury, Durban.

Medley-Wood J 1907. Handbook to the flora of Natal. Bennett and Davis, Durban.

Myburgh G 1985. Nou kan ons Nasella uitklop. Landbouweekblad 401, 8-10.

Neser S and Annecke D P 1973. Biological control of weeds in South Africa. Entomology Memoir 28, 1-27.

Phillips E J, Harding J J and du Toit R 1939. American Bramble. Farming in South Africa 14, 272-274.

Rutherford M C and Westfall R H (in press). The biomes of southern Africa — an objective categorization. Memoirs of the Botanical Survey of South Africa.

Sim T R 1919. Flowering trees and shrubs for use in South Africa. The Specialty Press, Johannesburg.

Smith C A 1966. Common names of South African plants. Memoirs of the Botanical Survey of South Africa 35, 1-642.

Stirton C H (ed) 1978. Plant invaders — beautiful but dangerous. The Department of Nature and Environmental Conservation of the Cape Provincial Administration, Cape Town.

Stirton C H 1979. Taxonomic problems associated with invasive alien trees and shrubs in South Africa. In: Kunkel G (ed) Proceedings of the IX Plenary Meeting of AETFAT, Las Palmas, Gran Canaria, pp 218-229.

Stubbings J A 1977. A case against controlling introduced Acacias. Proceedings of the 2nd National Weeds Conference of South Africa, Stellenbosch. pp 89-107.

Taylor S E and Walker B H 1984. Autecology of an invading population of the cactus *Cereus peruvianus* (queen of the night) in the central Transvaal. South African Journal of Botany 3(6), 387-396.

Wells M J 1974. *Nassella trichotoma* (Nees) Hack. in South Africa. Unpublished duplicated papers presented at the 1st South African National Weeds Conference of South Africa. Pretoria. pp 125-137.

Wells M J, Balsinhas A A, Joffe H, Engelbrecht V M, Harding G and Stirton C H 1986. A catalogue of problem plants in southern Africa incorporating the national weed list of South Africa. Memoirs of the Botanical Survey of South Africa, 53. Government Printer, Pretoria.

Wells M J, Duggan K J and Henderson L 1980. Woody plant invaders of the central Transvaal. In: Neser S and Cairns A L P (eds) Proceedings of the 3rd National Weeds Conference of South Africa. Balkema, Cape Town. 11-23.

Wells M J, Engelbrecht V M, Balsinhas A A and Stirton C H 1983. Weed flora of South Africa 3: more power shifts in the veld. Bothalia 14 (3,4), 967-970.

Zimmermann H G and Malan D E 1980. Present status of prickly pear control in South Africa. In: Neser S and Cairns A L P (eds) Proceedings of the 3rd National Weeds Conference of South Africa. Balkema, Cape Town. 79-86.

Zimmermann H G and Moran V C 1982. The ecology and management of cactus weed in South Africa. South African Journal of Science 78, 314-320.

Zimmermann H G and Van de Venter H A 1981. Jointed cactus. Weed Series, Division of Agricultural Information AI. Farming in South Africa.

Chapter 9

Alien invasions in indigenous evergreen forest

C J Geldenhuys*, P J le Roux† and K H Cooper‡

*South African Forestry Research Institute, †Directorate of Forestry and ‡Wildlife Society of Southern Africa.

Introduction

Indigenous forests occur sporadically in the eastern zones of the Republic of South Africa from the Soutpansberg inland at 22° 40'S and the Tongaland coast at 27° S to the Cape Peninsula at 34° S (Figure 1). Huntley (1984) distinguished the lowland forest biome of coastal dunes and lowlands from the Afromontane forest biome of the Drakensberg escarpment, Natal and eastern Cape midlands and southern Cape mountains and coastal plateau. In terms of floristic composition and stand structure the forests are related to lowland Guineo-Congolian and upland Afromontane Regions of tropical Africa (White 1983). Elements of both biomes intergrade and distinction is clear only at the extremes. The forests are tall (10-35 m) and closed and have multiple layers. The canopy is evergreen to semideciduous, depending on the moisture and nutrient conditions of the site. Afromontane forests include several genera and include *Ilex*, *Olea*, *Pittosporum*, *Podocarpus*, *Rapanea* and *Xymalos*. Epiphytes are often abundant in zones of fog, though lianes are generally rare. Lowland forests have an abundance of lianes but relatively few epiphytes. Tree genera include *Blighia*, *Celtis*, *Cola*, *Croton*, *Ficus*, *Harpephyllum*, *Protorhus*, *Strychnos* and several Afromontane genera (White 1983; Huntley 1984). Important frugivorous birds are the sombre bulbul *Andropadus importunus*, yellow-bellied bulbul *Chlorocichla flaviventris*, rameron pigeon *Columba arquatrix*, louries *Tauraco* species, tinker barbets *Pogoniulus* species and the white-eye *Zosterops pallidus*. Mammals important in the dispersal of plant propagules include fruit bats of the genera *Epomophorus* and *Rousettus*, samango monkey *Cercopithecus albogularis*, vervet monkey *C pygerythrus*, baboon *Papio ursinus* and bushpig *Potamochoerus porcus*.

The forests grow on a wide range of geological formations, including quartzitic sandstones, mudstones, shales, schists, aeolian sands, conglomerates, dolerite sand granite-gneiss. The derived soils vary in depth, water-holding capacity and nutrient status. Summer is the main rainy season in the eastern parts of the country. The southern Cape receives rain all year with maxima in early and late summer. A winter maximum occurs in the south-western Cape. Annual rainfall along the coastal lowlands ranges from 500 mm in the eastern Cape to 1 500 mm in Zululand. Coast-facing slopes along the eastern escarpment receive 600-1 700 mm on lower plateaux and 900-1 950 mm on higher mountain slopes, with additions through interception of fog. The southern and south-western Cape mountain foothills and coastal platform receive 650-1 200 mm and fog occurs at altitudes above 300 m (Geldenhuys 1983a).

The forests occupy less than one per cent (2 000-3 000 km^2) of the land area (Anonymous 1982; Huntley 1984). They occur as scattered patches of varying size and have a large ratio of forest margin to forest area. This accentuates the importance of forest margins in forest survival. Forest margins are well developed where regular fires and cultivation are absent (Phillips 1931; Granger 1984; Huntley 1984). The forest interior is subject to regular small-scale disturbances causing gaps about 0,005-0,1 ha in size. These arise in nature from single trees that die standing, from windfalls of single trees, and from occasional lightning strikes (Geldenhuys and Maliepaard 1983; Geldenhuys unpublished). Lightning fires, flooding and landslides on steep slopes during heavy rainstorms (Bosch and Hewlett 1980; Anonymous 1985; C J Geldenhuys unpublished) occasionally expose larger areas in the forests.

The forests play a role in the welfare of South African society which is disproportionately greater than their small extent and low potential for commercial exploitation. Heavy exploitation in the past, both by peasant communities and for commercial timber consumption, caused deterioration and destruction of forests (eg King 1938; Phillips 1963; Darrow 1973). Forests were often cleared for

Address for correspondence: C J Geldenhuys, Saasveld Forestry Research Centre, Private Bag X6515, George, 6530.

Figure 1. Forestry zones in South Africa in relation to annual rainfall ≥800 mm. Indigenous forests occur mostly in areas within this rainfall zone (after Anonymous 1984).

agriculture, and grazing by cattle opened up the forest interior (Taylor 1962; Geldenhuys 1983b). The forests enhance the landscape, contribute to the potential for tourism and recreation in the high-rainfall areas of the country, and protect soil and water in catchments and along streams (Geldenhuys 1983b). The forest biome supports a large number of interdependent plants and animals (Geldenhuys 1983b). The lists of rare and endangered vertebrate species in South Africa indicate that none of 46 reptiles and amphibians, 13 of 102 birds and eight of 71 mammals live in evergreen forests (Meester 1976; Skinner et al 1977; McLachlan 1978; Brooke 1984). They are the sources of products important in traditional tribal remedies.

The purpose of this paper is to review information relating to the potential for aliens to invade the indigenous forest under natural conditions and under current land-use practices.

History of forest disturbance by man and of alien introductions

Early forestry and the rehabilitation of forests in South Africa

The initial exploitation and destruction of forests was followed by forest protection and the evolution of management systems (Phillips 1963). Establishment of alien trees in all parts of the country began immediately after European colonization. Fast-growing alien trees were planted in forest gaps and in plantations along the margins of forests to provide timber, fibre and firewood needs. Valuable introduced timber species, in particular easily raised, fast-growing, light-demanding species from rain-forest or other sources were selected. Planting trees for their aesthetic value was a secondary objective. Jan van Riebeeck planted the first alien tree, *Quercus robur*, in South Africa in 1656. Initially this species was planted for its timber, but later for shade and shelter (Donald 1978). Commercial afforestation of the Cape Peninsula, mostly with species of *Pinus* and *Eucalyptus*, began at Tokai in 1884 (Spilhaus 1950; Campbell and Moll 1977).

Alien trees were planted in overexploited portions of the eastern Grootvadersbosch forest near Swellendam during 1901-1913 (Von Breitenbach 1976). *Eucalyptus* species, *Pinus radiata* and *P taeda* were planted in an area of 100 ha where forest was cleared. *Acacia melanoxylon*, *Cedrela toona*, *Cinnamomum camphora*, *Fraxinus angustifolia*, *Quercus canariensis*, *Q robur*, *Q suber* and *Sequoia sempervirens* were planted in gaps, individually or in small groups. *Acacia mearnsii* was sown in the shrublands on surrounding slopes. These plantings have had various consequences. Windthrows of large *Eucalyptus* trees occasionally caused large gaps. *A melanoxylon* and *C toona* have spread into gaps and along the forest margin and have been noted to suppress regeneration of indigenous trees.

Phillips (1963) summarized the history of introduction of aliens into the southern Cape. *Quercus robur* was planted around George by 1780. Captain C Harrison, Conservator of Forests, introduced *Q robur*, *Populus* species and *Eucalytpus globulus* to the Tsitsikamma in 1856 and several *Pinus* species and *Araucaria excelsa* between 1856 and 1876. *Acacia melanoxylon* occurred as large trees around George and Knysna by 1876. From 1890 burnt areas, glades, and clearings in the forest were planted mainly with *Acacia dealbata*, *A mearnsii*, *Eucalyptus camaldulensis* and other *Eucalyptus species*, *Pinus pinaster* and other *Pinus* and *Quercus* species. In 1891, forest gaps arising from the felling of indigenous trees, totalling 7,5 ha, were planted with alien trees. On average, eight alien plants were used to replace each of the 12 000 trees felled under licence that year. From 1892 belts of *Pinus canariensis*, *P halepensis*, *P pinaster* and *Q robur* were planted to protect forest margins against fire. In 1907 establishment of plantations was favoured. Planting of aliens in forest gaps was temporarily stopped although establishment of fire belts continued. Since 1930 several aliens were planted in the Diepwalle forest in gaps resulting from heavy logging or burning (Robertson unpublished). The objectives were to aid regeneration, to build up the humus content of the soil, to assist in raising the forest canopy and to select species for more extensive planting in South Africa. Many of the original gaps were considerably enlarged by the removal of indigenous regeneration to obtain areas large enough to warrant the trials. Gap sizes varied between 0,02 and 0,08 ha. Trees were planted in pure groups in cultivated planting spots and regularly weeded. Numerous introduced species were noted to have failed shortly after establishment, or to have shown poor growth, including Australian rain forest species such as *Agathis australis*, *Cedrela odorata*, *Elaeocarpus kirtonii*, *Flindersia australis*, *F brayleyana* and *F oxleyana*.

Darrow (1973) described the major events of forest destruction and afforestation in the Amatole mountains and other parts of the eastern Cape during 1850-1910. First attempts at afforestation with *Pinus pinea*, *Quercus robur*, *Salix* spp and *Tamarix gallica* took place at Cathcart in 1877, but these were largely failures. During 1882-1884 successful attempts were made near Katberg with one *Acacia* species, two *Casuarina* species, two *Cedrus* species, a *Fraxinus* species, six *Juniperus* species, 13 *Pinus* species, and species of *Quercus*, *Sequoia*, *Taxodium* and *Thuja*, but these were later destroyed by fire. At Cata and Isidenge a large number of species, including 21 *Eucalyptus* species, eight *Pinus* species, five *Acacia* species, one *Quercus* species and an assortment of conifers were planted. In 1884 1,4 ha of *Acacia mearnsii* was planted at Fort Grey near East London. By the turn of the century 3 500 ha of alien trees were established in 27 plantations throughout the eastern Cape. Almost every alien tree species introduced into South Africa up to that time could be found in these plantations.

The subtropical climate of the coasts of Natal and Zululand allowed people to establish a large variety of alien plants for timber, fruit or ornament (Sim 1905). Growth was rapid compared to other parts of the country. Ward (1985) listed 26 alien species from Ngoye forest, 18 of which were introduced timber species planted for test purposes during 1912-1922. The trees were planted on a total area of 5,7 ha in the grasslands adjacent to the forest and/or in clearings within the forest (Huntley 1965). Most afforestation in Zululand started after 1910. *Acacia* species, *Cedrela toona*, *Cinnamomum camphora*, *Eucalyptus saligna*, other *Eucalyptus* species, *Grevillea robusta* and *Pinus palustris* were favoured. At Qudeni *Acacia melanoxylon* and *Cupressus lusitanica* have been planted in the forest since 1911 and these species grew well. Afforestation with *Pinus patula* and sand dune stabilization with *Casuarina equisetifolia* started after World War II (Marwick 1973).

The first plantations on the north-eastern Transvaal escarpment were established early this century on Woodbush State Forest. *Eucalyptus grandis*, *E saligna*, other *Eucalyptus* species, *Pinus patula*, *P pseudostrobus* and other *Pinus* species were used. Westfalia Estate on the slopes below the escarpment at 1 250 m was afforested with *Eucalyptus* species during 1931-1940 (Scheepers 1978). Since 1906 various tropical and subtropical hardwoods ie three *Cedrela* species, *Chukrasia tabularis*, two *Gmelina* species and *Khaya nyasica*, have been planted in the eastern and northern Transvaal in the vicinity of indigenous forest to meet the demand for furniture and joinery timber (Von dem Bussche 1982).

Present extent of afforestation

In 1960 the area of commercial timber plantations in South Africa was 900 000 ha. This increased to 1,1 million hectares in 1983, of which 52% was in the Transvaal, 36% in Natal, seven per cent in the southern Cape, three per cent in the western Cape, and two per cent in the eastern Cape (Figure 1). *Pinus* species and other coniferous species covered 54% of the total area, *Eucalyptus grandis* covered 26%, tan wattles (mainly *Acacia mearnsii*) 12%, other *Eucalyptus* species 7%, *Populus* species 0,5%, and other broad-leaved species 0,2% (Anonymous 1985). New afforestation has decreased from an annual rate of 28 170 ha yr^{-1} in 1971 to one of 10 034 ha yr^{-1} in 1982 (Figure 2).

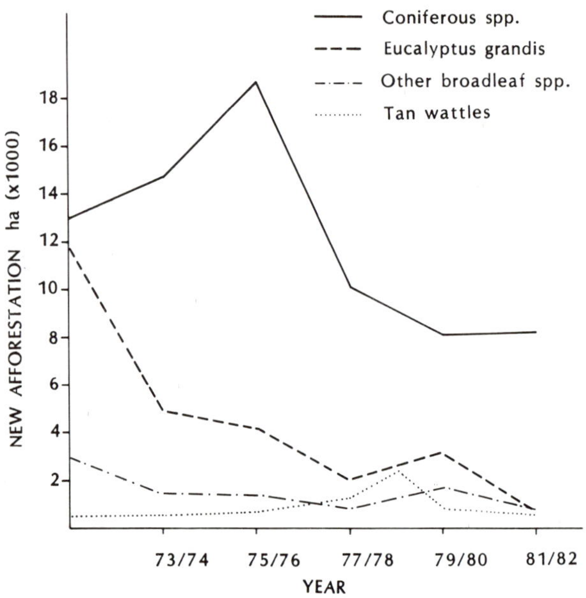

Figure 2. New afforestation in South Africa since 1971, mostly in shrublands and grasslands close to indigenous forests.

Introduction of ornamental plants

During 1870-1910 over 300 species of trees and shrubs, mostly aliens, were sold from forestry nurseries. *Pittosporum undulatum* was planted as early as 1901 in the Tokai Arboretum, Cape Peninsula, and was almost certainly widely planted in the early part of the century especially in Cape Town (D M Richardson personal communication). In Jonkershoek valley, Stellenbosch, this species has spread from ornamental trees planted about 1930 (Richardson and Brink 1985).

Introduced animals

Feral pigs *Sus scrofa* were introduced to forestry plantations in the western and southern Cape from domestic stock introduced for the control of the indigenous moth *Imbrasia cytherea* whose larvae defoliate pine trees (Smithers 1983). The pigs did not naturalize in the adjoining forest, which is in contrast to the situation on other continents (S A Botha personal communication). The Natal Parks Board has introduced nyala antelope *Tragelaphus angasii* into several areas south of its natural range at northern Lake St Lucia (Brooke and Macdonald 1983). Introductions of alien animals specifically to indigenous forest habitats, if there have been any, have clearly not succeeded.

Distribution of invasive species in forest habitats

Little has been published on alien species in forest habitats. A questionnaire survey was therefore used to collect information on invasive alien plants currently occurring in forest margins and gaps and closed forest. The objectives were to analyse the distribution, conditions and naturalization and characteristics of potential invader plants in forests. The offices of eight state forestry regions supplied information. Respondents were knowledgeable foresters and were requested to report the alien species known to have naturalized in forest habitats, to indicate which habitats were involved and to identify which alien species were regarded as invaders. Distinction was made between passive invaders, which grew and reproduced successfully in the adopted habitat but were not yet widely dispersed, and active invaders, which were widely dispersed. This information was supplemented with information from Henderson and Anderson (1966), Stirton (1978b), Scheepers (1978), Moll (1981), Macdonald (1983), Henderson and Musil (1984) and Ward (1985) (Tables 1 and 2). A total of 82 species was reported or had been recorded in the literature. Of these, 77% were reported as being confined to forest margins and larger gaps in the forests. Eighty-eight per cent were trees, shrubs and woody lianes. Less than 44% were regarded as invaders. Ten species of herbs and soft shrubs were reported, probably because they are less well known or less conspicuous. Eighty-four per cent of the aliens recorded from closed forest were regarded as passive or active invaders, compared to only 32% of the aliens from open forest. This suggests that alien plants which are able to reproduce and establish in the closed forest are more likely to become problem plants in the forest biome. However, some active invader plants of closed forest are not able to establish under those conditions, such as *Acacia melanoxylon* (Geldenhuys this volume), and probably also *A longifolia*, *A mearnsii*, *Cedrela toona*, *Grevillea robusta*, *Jacaranda mimosifolia*, *Pinus patula*, *Populus canescens* and *Rubus cuneifolius*. Their present occurrence in closed forest is most probably the result of their establishment in what used to be large gaps. A greater variety of successful invaders is found in Natal and the Transvaal than elsewhere. Henderson and Musil (1984) attributed the greater degree of alien invasion in Transvaal forests along roadsides and streambanks, when compared to the drier bushveld types, to the more favourable growing conditions throughout the year and the history of greater relative disturbance in the forests.

Processes of invasion and ecological impacts

Processes of invasion deal with the movement of propagules, their fate until establishment and their growth to maturity. Impacts are defined here as the positive or negative effects of the alien organism on the structure and processes of the forest ecosystem. The degree and direction of the impact often depends on the frequency, intensity and size of disturbance. The topic is treated here by selecting several different consequences of the introduction of alien plants, to illustrate the range of impacts evident.

Indigenous trees establish under stands of aliens

There are several cases where indigenous trees have expanded their local distributions by colonizing the understorey in stands of aliens. Large numbers of indigenous seedlings established under a 40-year-old monospecific stand of *Acacia melanoxylon* near Hogsback in the Amatole mountains of the eastern Cape (Seagrief 1965; Geldenhuys this volume). Wilson (1982) noted the destruction of the regeneration during subsequent clear-felling of the stand. She also noted the establishment of *Podocarpus* species under *Acacia mearnsii*, *A melanoxylon*, *Pinus pinaster*, *P radiata* and *Rubus (cuneifolius?)* in the same area, even on relatively dry sites. Several indigenous tree species were recorded in stands of *Cupressus lusitanica*, *Eucalyptus saligna* and *Pinus patula* on the edge of Ngoye forest in Zululand (Huntley 1965).

Alien forest species evidently differ in their effects on recruitment of indigenous species. For example, very few shrubs or herbs establish in *Pinus patula*

Table 1. The occurrence and status of alien plants in indigenous forests in South Africa as determined from a questionnaire survey of state forestry officials and from published information. The category, **Widespread**, indicates that the plant occurs throughout the region, otherwise only very localized. The letters W, S, E, N, T indicate the regions western Cape, southern Cape, eastern Cape, Natal and Transvaal, and u and c indicate upland and coastal zones within a region. A passive invader (+) is able to grow and reproduce but is not dispersed widely. The active invader (*) is able to grow, reproduce and is dispersed widely.

(a) Aliens occurring only along forest margins and in large gaps.		
Herbs and soft shrubs		
Agave spp. Nc T	*Arundo donax* T	*Bidens pilosa* N T
Cenchrus brownii Nc	*Cirsium vulgare* W Ec N	*Coix lachryma-jobi* Nc
Commelina benghalensis Nc	*Crotalaria agatiflora* T	*Opuntia aurantiaca* (Eu Nc) +
Rivina humilis Nc+		
Woody shrubs and lianes		
Acacia cyclops (W S Ec)+	*Bambusa balcooa* T	*Carica benghalensis* Nc
Cassia coluteoides Nc	*Cassia didymobotrya* T	*Cestrum laevigatum* (Ec NC)+
Hakea sericea S Eu Nc	*Lantana camara* (S Ec) + (N T)*	*Leptospermum laevigatum* S
Leucaena leucocephala Nc	*Ligustrum japonicum* T	*Montanoa bipinnatifida* Nc
Montanoa hibiscifolia Nc	*Nicotiana glauca* T	*Phytolacca spp* (S Nu T)+
Pyracantha angustifolia T	*Rosa* sp T	*Rubus niveus* T+ (Stirton 1981)
Sesbania punicea W S	*Solanum mauritianum* W Eu	*Tecoma stans* T
Vitis sp T		
Widespread		
Caesalpinia decapetala (Eu Nu T)*	*Cassia bicapsularis* Nc	*Cassia floribunda* Nc T
Ricinus communis (Nc T)*	*Sesbania punicea* Nc* T+	*Solanum mauritianum* (N T)*
Trees		
Acacia dealbata T+	*Acacia decurrens* T	*Acacia pycnantha* (W Ec)+
Acacia saligna (W S Ec)+	*Agathis australis* S	*Albizia lophantha* (W S)+
Anacardium occidentale Nc	*Araucaria cunninghamii* S	*Bauhinia variegata* T
Casuarina spp Ec Nc T	*Chamaecyparis lawsoniana* S	*Cupressus* sp T
Ensete spp Nc	*Eucalyptus* spp S W Eu N T	*Litsea sebifera* Nc
Mangifera indica Nc	*Melia azedarach* (Nc T)*	*Morus alba* T+
Pinus spp W S Eu N T	*Populus* spp T+	*Psidium cattleianum* Nc
Prunus persica T	*Quercus robur* W S Eu	*Salix babylonica* T+
Schinus terebinthifolius Nc+	*Sequoia sempervirens* W S	*Syncarpia glomulifera* S
(b) Aliens occurring along forest margins, in small and large gaps and in closed forest.		
Herbs and soft shrubs		
Opuntia ficus-indica (Eu Nc T)+		
Woody shrubs and lianes		
Homalanthus populifolius W	*Passiflora edulis* S E N T	*Rubus cuniefolius* (W S Eu Nu T)*
Widespread		
Chromolaena odorata N*	*Pereskia aculeata* N*	*Solanum seaforthianum* Nc*
Trees		
Acacia longifolia (W S Eu N)+	*Cedrela toona* (W T)+	*Cinnamomum camphora* (W S Nc)+
Citrus spp S Eu Nu T	*Grevillea robusta* T+	*Jacaranda mimosifolia* Nc (T)*
Pinus patula T*	*Pittosporum undulatum* W+	*Populus canescens* W+
Widespread		
Acacia mearnsii (W S E N T)*	*A melanoxylon* (W S Eu Nu T)*	*Psidium guajava* (Nc T)*

stands whereas well developed understoreys of indigenous shrubs and trees are found in *Eucalyptus* plantations. It appears that this succession could eventually result in a montane forest climax (Van der Schijff and Schoonraad 1971). Dense stands of various indigenous forest canopy species have been observed to establish and grow in older stands of *Eucalyptus* and *Pinus* species in the southern Cape.

Table 2. Number of species and status of alien plants in forests in different zones of South Africa. *Symbols indicate status of the alien plants: N = no threat is posed; P = passive and A = active successful invader.

	Herbs and soft shrubs			Woody shrubs and lianes			Trees			All taxa		
Zone	N*	P	A	N	P	A	N	P	A	N	P	A
Forest margin and large gaps only												
Western Cape	1			2	1		7	3		10	4	
Southern Cape				3	3		14	2		17	5	
Eastern Cape coastal	1				3		1	2		2	5	
Eastern Cape upland		1		2		1	6			8	1	1
Natal coastal	6	2		8	1	4	9	1	1	23	4	1
Natal upland	2				1	3	4			6	1	3
Transvaal	4			9	3	4	10	2	1	23	5	5
Forest margins, small and large gaps and closed forest												
Western Cape				1		1		4	2	1	4	3
Southern Cape				1		1	1	2	2	2	2	3
Eastern Cape coastal				1					1	1		1
Eastern Cape upland		1		1		1	1	1	2	2	2	2
Natal coastal		1		1		3	1	2	2	2	3	5
Natal upland				1		3	1	1	2	2	1	5
Transvaal		1		1		1	1	3	4	2	4	5
All regions												
Open forest	7	2		18	4	5	18	8	1	43	14	6
Open and closed forest		1		2		4	1	7	4	3	8	8

In general, *Eucalyptus* stands have a better developed understorey vegetation of indigenous tree species. *Pterocelastrus tricuspidatus* and *Rapanea melanophloeos* are most abundant with moderate numbers of *Podocarpus* species, *Olea capensis* subspecies *capensis* and *Ocotea bullata*. In the western Cape understorey vegetation of plantations is generally sparse or absent, although dense stands of indigenous trees, in particular *Kiggelaria africana*, occur under stands of various alien tree species on the eastern slopes of Table Mountain on the Cape Peninsula (Geldenhuys unpublished).

Dependence by aliens on disturbance in the forests

There are numerous examples of species that have not spread into indigenous forest unless aided by man, or which have naturalized only in disturbed forest and other modified ecosystems.

Several alien species were planted in gaps in the Diepwalle forest near Knysna at 450 m altitude, but only *Araucaria cunninghamii*, *Chamaecyparis lawsoniana*, *Sequoia sempervirens* and *Syncarpia glomulifera* have grown rapidly in both diameter and height (Geldenhuys unpublished). Maximum tree height of 27-33 m was greater than the 24-28 m of the tallest trees of the surrounding, relatively high indigenous forest canopy. Stand basal area was 50-184 m^2ha^{-1} compared to the 30-40 m^2ha^{-1} of the forest. Mean annual basal area increment was 1,3-3,9 $m^2ha^1yr^{-1}$ compared to 0,3-0,4 $m^2ha^{-1}yr^{-1}$ for the forest. Litter mass within the alien stands was 61% (11-101%) higher than that of the forest. Despite their successful growth, no seedling regeneration of the alien trees was observed within or outside these stands.

Where the forest is modified, many aliens may establish themselves. Thirteen arborescent aliens established in indigenous vegetation at higher altitudes on the Cape Peninsula. Only *P pinaster*, denoted as rare, occurred in the broad-leaved forest in ravines. The broad-leaved bushland on moist talus slopes near cliffs, at forest margins and in more exposed ravines contained abundant *Acacia saligna* and *Hakea sericea*, occasional *Albizia lophantha*, *Eucalyptus* species and *Pinus pinaster* and rare

Acacia longifolia, *A melanoxylon*, *P pinea* and *P radiata* (Hall 1961). In coastal Natal the improved site conditions for germination of forest species following fire are exploited by aliens such as *Acacia* species, *Asimina* species, *Cestrum* species, *Chromolaena odorata*, *Euphorbia* species, *Lantana camara*, *Melia azedarach*, *Psidium guajava* and *Solanum* species (Macdonald 1978; Granger 1984). In the Ntendeka Wilderness Area in Natal areas exposed by landslides on steep slopes during cyclone Demoina in January 1984, were soon colonized by several indigenous pioneer tree species as well as by the aliens *Rubus cuneifolius* and *Solanum mauritianum* (J Huisamen personal communication). *Acacia melanoxylon* is establishing widely in many vegetation types, including modified ecosystems. The hard-coated seed is widely dispersed by birds and germinates in abundance after fire, logging or general soil disturbance. However, even established plants of *A melanoxylon* cannot compete successfully in the closed forest (Geldenhuys this volume).

Shade-tolerant alien plants establish in closed forest

Regeneration under monospecific plantations of three species on Witfontein State Forest in the southern Cape demonstrates markedly different patterns of invasion. The stands were adjacent on level ground near a perennial stream.

The species were the indigenous *Curtisia dentata* and the aliens *Eucalyptus diversicolor* and *Pinus radiata*. Age, relevant stand variables and standing litter for the three stands are summarized in Table 3. Individuals of the aliens *Acacia melanoxylon*, *Cinnamomum camphora*, *Eucalyptus globulus*, *Pinus canariensis* and *P radiata* that had invaded the *C dentata* stand were removed during 1926-1928, but a grove of trees of *C camphora* was left on its southern boundary. A recent survey in the *C dentata* stand indicated that *C camphora* occurred throughout this stand, up to 140 m from the remaining mature trees on the southern boundary. There were 531 stems ha^{-1} for trees greater than 20 mm DBH, 3,1 stems m^{-2} for saplings of 0,3-3,0 m height and 23 stems m^{-2} for seedlings of 0,1-0,3 m height. Other than the trees originally planted, only small seedlings of *C dentata* were found scattered throughout the stand although the records indicate that these trees had already started profuse fruiting by 1926. Eleven indigenous species were recruited to the plantations, of which only *Rapanea melanophloeos* and *Clutia pulchella* were successful (Table 3). All the species recruited to the plantations were bird-dispersed except *E diversicolor* and *P radiata*. The shade-tolerant *C camphora* was disseminated into the stands of aliens, but only established successfully in the *C dentata* stand where there was sufficient shade, moisture and a thin litter layer. This indicates the potential for the shade-tolerant *C camphora* to spread into indigenous forest. Scattered plants of *A mearnsii* and *A melanoxylon*, up to five metres tall, grew in the *Eucalyptus* stand. The *Acacia* species had not established under the closed canopy of the *C dentata* stand (Geldenhuys and Van der Merwe unpublished).

Table 3. The stand variables and understorey vegetation of three plantations in Witfontein State Forest near George.

	Plantation species		
Stand variables	*Curtisia dentata*	*Eucalyptus* diversicolor*	*Pinus* radiata*
Planting date	1899	1945	1967
Age, years	86	40	18
area, ha	1,8	5,0	0,7
Stand density, stems ha^{-1}	1020	380	340
Diameter breast height, mm	182	415	318
Canopy height, m	17	37	28
Crown cover, %	91	60	57
Litter, g m^{-2}	791	1818	815

	Density (d, stems m^{-2}) and frequency (f, %) of understorey vegetation of 0,3 – 3,0 m height					
Understorey species	d	f	d	f	d	f
Rapanea melanophloeos	0,8	36	1,4	47	1,0	21
Clutia pulchella	1,2	52	0,8	59	0,3	2
**Acacia mearnsii*	–	–	0,9	16	–	–
**Acacia melanoxylon*	–	–	0,6	25	–	–
**Cinnamomum camphora*	3,1	52	–	–	–	–
*Eucalyptus diversicolor	–	–	0,8	16	0,3	2
**Pinus radiata*	–	–	–	–	0,3	2

*Alien species

Pittosporum undulatum has invaded undisturbed natural communities in the western Cape. These include tall scrub dominated by *Olea europaea* subspecies *africana*, *Kiggelaria africana*, *Rhus angustifolia* and *Maytenus oleoides* and riparian forest patches in mountain ravines and along the Eerste River and its tributaries near Stellenbosch. The large seedling to adult ratio indicates that the invasion of *P undulatum* is still in its early stages (Richardson and Brink 1985).

Creepers overtop indigenous trees

The spiny creepers *Caesalpinia decapetala* and *Pereskia aculeata* are important invaders in subtropical areas. They grow to form dense masses in the canopy and cause trees to collapse as well as shading out the understorey (Bruton and Cooper 1980).

Scramblers increase flammability of the forest edge

Chromolaena odorata (known as *Eupatorium odoratum* in the tropics) is one of the most successful alien plants in coastal Natal and occurs on the margins and within almost every indigenous forest, as well as in intervening grassland and savanna (Macdonald and Jarman 1985). This plant has been found in single-tree windfall gaps in the middle of patches of pristine forest in several reserves and under closed forest at numerous places in Hluhluwe Game Reserve (I A W Macdonald personal communication). It does not tolerate dense shade and is generally absent from mature forest. It burns vigorously, causing severe denudation of the forest edge when fires sweep up to a forest (Macdonald 1983; Cooper unpublished). *Lantana camara* has a similar behaviour, is confined to the forest margin and disturbed forest but is less aggressive and burns less vigorously. It is noted for its rapid invasion of natural veld, waste ground, forest margins and derelict or cultivated lands (Stirton 1978a).

Changed food plant preferences in frugivores potentially affect forest reproduction

A wide range of birds and some mammals eat the fruit of introduced plants such as *Acacia melanoxylon*, *Cestrum laevigatum*, *Cinnamomum camphora*, *Melia azedarach*, *Pittosporum undulatum*, *Psidium guajava*, *Rubus cuneifolius* and *Solanum mauritianum* (see also Kruger et al and Dean this volume). The extensive spread of fruiting alien plants in forest habitats must in the long term have profound consequences to the original plant-animal relationships. Firstly, vector competition could arise between indigenous and introduced fruiting plants which are dispersed by indigenous animals (Macdonald 1983; Oatley 1984; Dean 1985). The rameron pigeon *Columba arquatrix* normally feeds on fleshy fruits of indigenous trees in genera such as *Celtis*, *Ocotea*, *Olea* and *Prunus* and was primarily a species associated with evergreen forest (Phillips 1927; Oatley 1984). Modern populations feed extensively on the fruits of introduced plants. Oatley collected 109 pigeons in Natal of which 94,5% had fed exclusively on *S mauritianum* fruit, 4,6% had fed on *S mauritianum* and other fruit and 0,9% had fed exclusively on *Morus alba*. Although some pigeons were observed to feed on fruits of indigenous trees such as *Celtis africana*, *Prunus africana* and *Rauvolfia caffra*, many more pigeons were feeding on *S mauritianum* fruit at the forest edge nearby. Furthermore, stands of *S mauritianum* attracted at least 41 bird species either feeding on fruits, foraging on insects and other arthropods in the canopy or understorey, or nesting (Dean 1985). Secondly, the more regular and widespread availability of fruit of alien plants could increase the density of frugivores. Oatley (1984) suggested that the heavy exploitation of the forests since the turn of the century markedly reduced the density of frugivorous birds. He attributed the present high density of *C arquatrix* in Natal to its ability to exploit the new habitats of extensive *Pinus*, *Eucalyptus* and *Acacia* plantations for nesting and roosting and into which it dispersed the seed of *S mauritianum*. This plant is rapidly colonizing forest edges and disturbed sites within forest. Thirdly, the indigenous pioneer and other tree species with fleshy fruits could gain from the widespread occurrence of alien plants and the higher density of frugivores (Dean 1985). Indigenous bird-dispersed canopy trees are found in stands of alien trees (Huntley 1965; Seagrief 1965; Wilson 1982). Very often the stands of aliens were established in grassland or shrubland where forest species would not readily establish. Macdonald (1983) noted that indigenous forest-edge plants such as *Trema orientalis* and *Dombeya burgesiae* were absent from stands of *S mauritianum*. However, it is not normal for pioneer plants to establish in dense stands of other pioneer plants.

Introduced antelope displaced native antelope

The nyala *Tragelaphus angasii*, that had been introduced to areas south of northern Lake St Lucia markedly reduced the forest understorey through their browsing, are thought to have displaced bushbuck *Tragelaphus scriptus* and caused the disappearance of the small forest duiker *Cephalophus monticola* (Brooke and Macdonald 1983; Cooper unpublished; Natal Parks Board unpublished). The reduction in the plant cover along forest margins enabled the establishment of *Chromolaena odorata* in some areas (Macdonald and Jarman 1985).

Management and control

The control of alien plants in forests started early

in this century when foresters ceased to plant alien trees in forest gaps and established these trees in plantations outside the forests instead (Phillips 1963). Phillips (1928, 1931) claimed that *A melanoxylon* and species of *Eucalyptus* and *Pinus* threatened the forest through their spread, establishment and rapid growth in disturbed forest and their suppression of the regeneration of indigenous species.

That efforts to control alien invasions in indigenous forests are relatively limited is probably due to several factors. Firstly, the area of indigenous forest is small relative to that of agricultural lands, timber plantations and mountain catchment areas. Several alien invader plants of indigenous forest are more serious problems in commercial plantations. *Caesalpinea decapetala*, *Chromolaena odorata*, *Lantana camara*, *Pereskia aculeata*, *Phytolacca* species, *Rubus* species and *Solanum mauritianum* obstruct plantation operations and increase the fire hazard (Le Roux 1980a,b, 1982; Maree 1982; Hinze 1983; Anonymous 1984). Secondly, current methods of alien plant control in plantations such as the use of herbicides and the mowing and hoeing out the plants, are expensive. Landowners managing forests as part of a larger property have to minimize their costs. Direct financial benefits from the forests are minimal. Thirdly, several invader plants such as *Acacia melanoxylon* provide direct economic benefits (Geldenhuys this volume). Finally, the general notion is held that alien plants are unable to penetrate closed forest. The control of alien plant invasions into forests is primarily dependent on the maintenance of a closed forest canopy and a well-developed forest margin.

The current policy of the Department of Environment Affairs (Forestry Branch) makes provision for the systematic removal of alien plants from the intensively managed indigenous forests in the southern Cape (Von Breitenbach 1968). Removal of utilizable aliens from the forest interior is done with minimal damage to the forest canopy and existing natural regeneration or minimal exposure of the soil surface. Accessible trees of *A melanoxylon* are utilized to supplement the supply of furniture timber from indigenous species. Such trees are removed by lopping the crown before felling to minimize disturbance, as is also done for utilizable indigenous trees (Geldenhuys 1982). Utilizable trees in inaccessible or sensitive sites and nonutilizable trees of all alien species are ring-barked and/or poisoned so as to die standing and to allow for a gradual breakdown of their crowns. However, single trees or small groups of aliens of exceptional form are retained for their historical and/or dendrological value if the species involved is not invasive (Von Breitenbach 1976).

Many alien invaders in forests are pioneer plants establishing in disturbed sites. They often form dense, pure stands or thickets inhibiting the establishment of indigenous early secondary species. They sometimes increase the fire hazard and decrease the biotic diversity associated with the forest margin. When these stands mature, they become less dense and can allow the establishment of some indigenous regeneration. Pioneer tree stands, however, take several years to develop to the point where shade-tolerant indigenous tree species can establish. Ring-barking, bark-stripping, poisoning and handpulling of aliens on the forest margin to reduce their density, carried out with minimal damage to indigenous species, could assist the establishment and growth of the forest species. Several examples exist to indicate that careful manipulation of such stands of aliens, rather than their eradication, contributes to the establishment and growth of indigenous trees and to successful and less expensive control of the alien plants. This is in sharp contrast to the control of woody invaders in other vegetation types and in modified ecosystems where the aliens must be eradicated, or nearly so, lest they replace the natives. It is common practice in the southern Cape to reconstruct forest on narrow strips of aliens, for example between a road and the forest. Aliens are removed and the area planted to indigenous species, or killed standing to allow indigenous tree regeneration to develop if present, or hand-pulled from the developing stand of indigenous regeneration (Von dem Bussche 1975; Geldenhuys 1982). Stands established in this manner have developed a canopy height of eight metres within six years (J C van Daalen personal communication). Recent removal of the overstorey of *Cupressus lusitanica*, *Eucalyptus saligna*, *Pinus patula* and other alien species on the edge of Ngoye forest (M C Ward personal communication) has stimulated the growth of the indigenous tree understorey recorded by Huntley (1965).

Some species, however, require special treatment. The cactus *Pereskia aculeata* propagates vegetatively from any other small fragment of tissue. Plant parts can survive for more than six months without water or contact with the soil. The only known solution is total destruction of the vegetation in infested areas by repeated slashing, treatment with herbicide and burning (le Roux unpublished). Chemical, mechanical and cultural control methods have been applied to *Lantana camara* but with limited success, and only biological control methods have any real promise (Stirton 1978a).

Some indigenous plants may be unwittingly eradicated when they are confused with alien invaders of similar appearance. The indigenous *Solanum giganteum*, which has smaller reddish berries and a spinescent stem, is often confused with *S mauritianum* and eradicated (Geldenhuys unpublished). Infestations of *Caesalpinia decapetala* have been overlooked due to confusion with indigenous thorny creepers such as *Entada spicata* and *Acacia schweinfurthii* (Macdonald 1983).

Synthesis

Most alien invader plants in the forest establish only in disturbed forest margins or large gaps in heavily exploited forest. Nevertheless, the forest biome is particularly resistant to alien invasion and shows remarkable recovery potential following infrequent disturbance such as results from timber harvesting. Currently, indigenous timbers are harvested on a commercial scale only from the southern Cape forests, the largest forest complex in the country (60 500 ha). About 10% of the forest area is involved. Management is by a version of the single-tree selection system, where compartments are cut over on a 10-year-cycle with the aim of removing only overmature trees or those at the point of death, or the thinning of overdense patches. Special attention is given to the manipulation and rehabilitation by this means of previously heavily logged areas. In other areas, of the southern Cape and elsewhere, dead or dying trees are sporadically harvested singly. Most of the forests outside Natal, which has many privately owned forests (Cooper unpublished), are protected on State Forest land (Geldenhuys 1982, 1983b). Afforestation with alien timber species has furthermore provided forests with protection against fire and other disturbance and has contributed to their recovery (Lister 1902; Henkel et al 1936; King 1938; Taylor 1962; Phillips 1963; Van der Schijff and Schoonraad 1971; Darrow 1973; Grut 1977; Weisser and Marques 1979; Le Roux 1981).

Analysis of alien plants occurring in the forest indicate a total of 36 passive and active invaders (Table 2). With a few exceptions, alien plants invading forests are characterized by shade intolerance, propagule dispersal by birds and mammals, the accumulation of seed stores due to regular production of hard-coated seeds and the ability to reproduce vegetatively. Only eight of these were introduced as timber trees. *Acacia mearnsii* and *A melanoxylon* are the only timber trees which have successfully invaded modified forest throughout the country (Table 1). To date, no alien animal has been able to naturalize in the indigenous forest.

Most South African indigenous forest trees can be characterized as shade-tolerant species of middle to late successional stages. Many of the shade-intolerant alien invader trees substitute for or supplement the few indigenous pioneer forest trees such as *Virgilia oroboides* (Phillips 1926), *Trema orientalis* (Scheepers et al 1968) and *Acacia karroo* (Weisser and Marques 1979). *Acacia mearnsii*, *A melanoxylon* and *Eucalyptus* species are important pioneer trees in rain forest succession in eastern Australia (Hopkins et al 1977; Geldenhuys 1980). Bazzaz (1979) discussed the relative performance of species of early and late successional stages in broad-leaved deciduous forest. Seedlings of shade-tolerant species are adapted to less variable and less extreme environments in the forest with respect to temperature, humidity and wind. Maximum fluctuation in these variables occurs at or near the soil surface in open, early successional habitats. In general photosynthesis, respiration and transpiration rates of early successional trees fall between the high rates of "sun-adapted", early successional herbaceous plants and the low rates of late successional trees. The latter are photosynthetically more efficient at low light intensities than early successional herbs, and early successional trees have intermediate abilities. Theoretically the early successional alien trees would ameliorate the more variable and extreme conditions of regularly disturbed shrubland and grassland and should allow the establishment of the shade-tolerant forest species. Comparison of alien invasions in plantations and forest suggest that the dense shade of the mixed, multilayered forest provide little growing space for the majority of the light-demanding aliens. Afforestation with alien trees generally took place on sites suitable for forest. These stands ameliorated the environment and allowed forest succession to proceed. Similarly, only relatively open stands of indigenous forest provide growing space for a range of the less light-demanding alien species, but species which would not be able to establish in dense forest. A few species, probably those belonging to the later stages of succession such as *Cestrum laevigatum*, *Cinnamomum camphora*, *Psidium guajava* and *Pittosporum undulatum*, however, have shown potential to invade closed forest.

Frugivores are important in the dissemination of the seed of most alien invaders in forests. The effect of changes in the food plants of bird and mammal dispersers on the reproduction of indigenous plants is an important subject for study if we are to understand the impact of alien invasions on forest. It should also form an important component of investigations into control measures for plant invaders in timber plantations.

Acknowledgements

This report was funded through the Conservation Forestry Research Programme of the Department of Environment Affairs (Forestry Branch). The assistance of Forestry personnel in the different regions in collecting information on present invaders, the assistance of Messrs C J van der Merwe and C Jacobs in the collection of field data, the preparation of figures by Mr J Dobson and the useful comments of colleagues and referees are acknowledged with thanks.

Anonymous 1982. Forestry guide plan for South Africa. Department of Environment Affairs, Pretoria.

Anonymous 1984. The forestry industry in South Africa. Department of Environment Affairs (Forestry Branch), Pretoria.

Anonymous 1985. Annual report of the Department of Environment Affairs 1983-84. Government Printer, Pretoria.

Bazzaz F A 1979. The physiological ecology of plant succession. Annual Review of Ecology and Systematics 10, 351-371.

Bosch J M and Hewlett J D 1980. Sediment control in South African forests and mountain catchments. South African Forestry Journal 115, 50-55.

Brooks P M and Macdonald I A W 1983. The Hluhluwe-Umfolozi Reserve - an ecological case history. In: Owen-Smith R N (ed) Management of large mammals in African conservation areas. HAUM, Pretoria. pp 51-77.

Brooke R K 1984. South African Red Data Book - Birds. South African National Scientific Programmes Report No 97. CSIR, Pretoria.

Bruton M N and Cooper K H (eds) 1980. Studies on the ecology of Maputaland. Rhodes University, Grahamstown and Wildlife Society, Durban.

Campbell B M and Moll E J 1977. The forest communities of Table Mountain, South Africa. Vegetatio 34, 105-115.

Darrow W K 1973. Forestry in the Eastern Cape Border Region. Bulletin 51, Department of Forestry, Pretoria.

Dean W R J 1985. The utilization of *Solanum mauritianum* woodlands by birds in the north-eastern Transvaal. In: Bunning L J (ed) Proceedings of the Symposium on Birds and Man. Witwatersrand Bird Club, Johannesburg. pp 89-97.

Donald D G M 1978. The oak at the Cape. Veld and Flora 64, 98-102.

Geldenhuys C J 1980. Report on a study tour of rainforests in Australia, New Zealand and Chile. Unpublished report. Saasveld Forestry Research Centre, George. 45 pp.

Geldenhuys C J 1982. The management of the southern Cape forests. South African Forestry Journal 121, 1-7.

Geldenhuys C J 1983a. Distribution and classification of the indigenous evergreen forest. In: Odendaal P B, Owen D L, Bosman D L, Schönau A P G, von Gadow K and van der Zel D W (eds) South African Forestry Handbook. South African Institute of Forestry, Pretoria. pp 262-268.

Geldenhuys C J 1983b. A critical evaluation of indigenous forest management. Proceedings of the Jubilee Symposia, University of Stellenbosch, Stellenbosch. Foris 98, 497-524.

Geldenhuys C J and Maliepaard W 1983. The causes and sizes of canopy gaps in the Southern Cape forests. South African Forestry Journal 124, 50-55.

Granger J E 1984. Fire in forest. In: Booysen P de V and Tainton N M (eds) Ecological effects of fire in South African Ecosystems. Springer-Verlag, Berlin. pp 177-197.

Grut M 1977. Notes on the history of forestry in the western Cape, 1652-1872. South African Forestry Journal 100, 32-37.

Hall A V 1961. Distribution studies of introduced trees and shrubs in the Cape Peninsula. Journal of South African Botany 27, 101-110.

Henderson L and Musil K J 1984. Exotic woody plant invaders of the Transvaal. Bothalia 15, 297-313.

Henderson M and Anderson J G 1966. Common weeds in South Africa. Memoirs of the Botanical Survey of South Africa 37, 1-440.

Henkel J S, Ballenden S St C and Bayer A W 1936. An account of the plant ecology of the Dukuduku Forest Reserve and adjoining areas of the Zululand coast belt. Annals of the Natal Museum 8, 95-125.

Hinze W H F 1983. *Solanum mauritianum* as a forest invader plant. In: Neser S and Cairns A L P (eds) Proceedings of the 5th National Weeds Conference of South Africa, Stellenbosch. pp 327-331.

Hopkins M S, Kikkawa J, Graham A W, Tracey J G and Webb L J 1977. An ecological basis for the management of rain forest. In: The Border Ranges: a land use conflict in regional perspective, Royal Society, Brisbane. pp 57-66.

Huntley B J 1965. A preliminary account of the Ngoye Forest Reserve, Zululand. Journal of South African Botany 31, 177-205.

Huntley B J 1984. Characteristics of South African Biomes. In: Booysen P de V and Tainton N M (eds) Ecological effects of fire in South African Ecosystems. Springer-Verlag, Berlin. pp 2-17.

King N L 1938. Historical sketch of the development of forestry in South Africa. Journal of the South African Forestry Association 1, 4-16.

Le Roux P J 1980a. Encroaching plant species in state plantations in South Africa with special reference to their occurrence in the afforested area of the Transvaal. In: Neser S and Cairns A L P (eds) National Proceedings of the 3rd Weeds Conference in South Africa. Balkema, Cape Town. pp 181-186.

Le Roux P J 1980b. The control of encroaching plant species on state forests: a management approach. In: Neser S and Cairns

A L P (eds) Proceedings of the 3rd National Weeds Conference in South Africa. Balkema, Cape Town. pp 187-194.

Le Roux P J 1981. Supply of fuel-wood for rural populations in South Africa. South African Forestry Journal 117, 22-27.

Le Roux P J 1982. Plant invader species in pine plantations in South Africa. In: Van de Venter H A and Mason M (eds) Proceedings of the 4th National Weeds Conference of South Africa. Balkema, Cape Town. pp 155-168.

Lister J S 1902. Report on forestry in Natal and Zululand. Times Printing and Publishing Company, Pietermaritzburg.

Macdonald I A W 1978. Proposed monitoring projects for the dune forest in Umlalazi Nature Reserve with special reference to measuring the effects of the accidental forest fire. Unpublished report, Natal Parks Board. 22 pp.

Macdonald I A W 1983. Alien trees, shrubs and creepers invading indigenous vegetation in the Hluhluwe-Umfolowi Game Reserve Complex in Natal. Bothalia 14, 949-959.

Macdonald I A W and Jarman M L 1985. Invasive alien plants in the terrestrial ecosystem of Natal, South Africa. South African National Scientific Programmes Report No 118. CSIR, Pretoria.

Maree H B 1982. The economics of the control of invader plants in forestry. In: Van de Venter H A and Mason M (eds) Proceedings of the 4th National Weeds Conference of South Africa. Balkema, Cape Town. pp 145-154.

Marwick C W 1973. Kwamahlati, the story of forestry in Zululand. Bulletin 49, Department of Forestry, Pretoria. 72 pp.

McLachlan G R 1978. South African Red Data Book - reptiles and amphibians. South African National Scientific Programmes Report No 23. CSIR, Pretoria.

Meester J A J 1976. South African Red Data Book - small mammals. South African National Scientific Programmes Report No 11. CSIR, Pretoria.

Moll E J 1981. Trees of Natal. Eco-Lab Trust Fund. University of Cape Town, Cape Town.

Oatley T B 1984. Exploitation of a new niche by the rameron pigeon *Columba arquatrix* in Natal. Proceedings of the 5th Pan-African Ornithological Congress. pp 323-330.

Phillips J F V 1926. *Virgilia capensis* Lamk. ('Keurboom'): A contribution to its ecology and sylviculture. South African Journal of Science 23, 435-454.

Phillips J F V 1927. The role of the "Bushdove" *Columba arquatrix* T. & K. in fruit-dispersal in the Knysna Forests. South African Journal of Science 24, 435-440.

Phillips J F V 1928. The behaviour of *Acacia melanoxylon* R. Br. ('Tasmanian blackwood') in the Knysna forests. Transactions of the Royal Society of South Africa 16, 31-43.

Phillips J F V 1931. Forest succession and ecology in the Knysna Region. Memoirs of the Botanical Survey of South Africa 14, 1-327.

Phillips J 1963. The forests of George, Knysna and the Zitzikama - a brief history of their management 1778-1939. Government Printer. Pretoria.

Richardson D M and Brink M P 1985. Notes on *Pittosporum undulatum* in the south-western Cape. Veld and Flora 71, 75-77.

Scheepers J C 1978. The vegetation of Westfalia Estate on the north-eastern Transvaal escarpment. Memoirs of the Botanical Survey of South Africa 42, 1-230.

Scheepers J C, Van der Schijff H P and Keet J D M 1968. An ecological account of the *Trema* plantations of Westfalia Estate. Tydskrif vir Natuurwetenskappe 8, 105-120.

Seagrief S C 1965. Establishment of *Podocarpus latifolius* in Blackwood plantations at the Hogsback. South African Journal of Science 61, 433-437.

Sim T R 1905. Tree planting in Natal. P Davis and Son, Pietermaritzburg.

Skinner J D, Fairall N and Bothma J du P 1977. South African Red Data Book - large mammals. South African National Scientific Programmes Report No 18. CSIR, Pretoria.

Smithers R H N 1983. The mammals of the Southern African subregion. University of Pretoria, Pretoria.

Spilhaus M W 1950. Indigenous trees of the Cape Peninsula. Juta, Cape Town.

Stirton C H 1978a. Tickberries on the march. Veld and Flora 64, 8-11.

Stirton C H (ed) 1978b. Plant invaders - beautiful but dangerous. Department of Nature and Environmental Conservation of the Cape Provincial Administration, Cape Town.

Taylor H C 1962. A report on the Nxamalala forest. Forestry in South Africa 2, 29-51.

Van der Schijff H P and Schoonraad E 1971. The flora of the Mariepskop complex. Bothalia 10, 461-500.

Von Breitenbach F 1968. Southern Cape indigenous forest management manual Volume 2. Department of Environment Affairs, Pretoria.

Von Breitenbach F 1976. Bewaringsplan vir Grootvadersbosch. Unpublished report, Western Cape Forestry Region, Cape Town. 66 pp.

Von dem Bussche G H 1975. Indigenous forest conservation management. South African Forestry Journal 93, 25-31.

Von dem Bussche G H 1982. The establishment of hardwood plantations for the production of furniture and joinery timber in the Transvaal. South African Forestry Journal 121, 11-23.

Ward M C 1985. The current extent of alien infestations: KwaZulu areas. In: Macdonald I A W and Jarman M L 1985 (eds) Invasive alien plants in the terrestrial ecosystems of Natal, South Africa. South African National Scientific Programmes Report No 118. CSIR, Pretoria. pp 7-14.

Weisser P J and Marques F 1979. Gross vegetation changes in the dune area between Richards Bay and the Mfolozi River, 1937-1974. Bothalia 12, 711-721.

White F 1983. The vegetation of Africa: a descriptive memoir to accompany the UNESCO/AETFAT/UNO vegetation map of Africa. Natural Resources Research 20, UNESCO, Paris.

Wilson M 1982. The re-establishment of yellowwood forest. Veld and Flora 68, 111-112.

Chapter 10

Alien plants and animals on South African continental and oceanic islands: species richness, ecological impacts and management

J Cooper and R K Brooke
University of Cape Town.

Introduction

South African marine islands can be conveniently classified into two categories: continental and oceanic. Continental (offshore) islands are situated on the African continental shelf. Oceanic islands are the Prince Edward islands (Marion Island and Prince Edward Island) situated in the southern Indian Ocean. The presence and history of nonmarine aliens at the two island groups has recently been reviewed in detail (Brooke and Prins in press; Watkins and Cooper in press). This paper, after briefly describing the two island groups, compares and contrasts the continental and oceanic islands in terms of their present nonmarine plant and animal biotas, both indigenous and alien. Ecological impacts of aliens are then considered, as well as management implications and priorities. Mention is made of those few alien species that occur at both island groups. By alien species on continental islands we mean any species there which is not indigenous or native to the adjacent mainland, as well as any mainland species deliberately translocated to one or more islands (cf Brooke and Prins in press). Alien species on the Prince Edward islands are any species known or thought not to be indigenous. We recognize four ways by which alien species may reach islands: by deliberate introduction; by accidental introduction; by escape of domestic stock or pets; and by natural dispersion after introduction to the adjacent mainland in the case of continental islands. In the great majority of cases there is no information into which category an alien species on an island falls. Exceptions occur among woody plants, birds and mammals.

Address for correspondence: J Cooper, Percy FitzPatrick Institute of African Ornithology, University of Cape Town, Rondebosch, 7700.

Description of the islands

Continental islands

South African continental islands extend from Hollamsbird Island in the north to Bird Island, Algoa Bay, in the east (Figure 1). All but one island situated off the coast of South West Africa/Namibia are South African territory. The farthest offshore is only 10,7 km from the mainland. The continental islands are relatively small, the largest, Robben Island, being 507 ha (Table 1). Those smaller than one hectare are usually classified as 'offshore rocks' and are not considered further. Following this classification there are 24 islands for which some information on aliens is available for 20 (Brooke and Prins in press; Table 1). Two South African continental islands are now connected to the mainland by artificial causeways (Cooper et al 1985; Brooke and Prins in press). Following Brooke and Prins (in press), animals known to have reached Marcus Island via its causeway are not included in this paper (see Cooper et al 1985 for mammals introduced in this manner).

The continental islands (see descriptions in Rand 1963a,b; Brooke and Crowe 1982) have a warm temperate climate, are low lying and are not well vegetated, especially the smaller ones. The continental islands can be subdivided into three groups on rainfall pattern: virtually no rain falls on the islands (Hollamsbird to Sinclair, Figure 1) off South West Africa/Namibia though heavy fogs are of regular occurrence; rainfall is virtually confined to the winter months on the islands (Bird, Lambert's Bay, to Seal, False Bay) off the western coast of the Cape Province; rain may fall at any time of the year on islands (Dyer to Bird, Algoa Bay) off the southern and eastern coasts of the Cape Province.

Most continental islands are breeding localities for large numbers of a community of 13 seabird species (eg Rand 1963a,b) and a few support colonies of

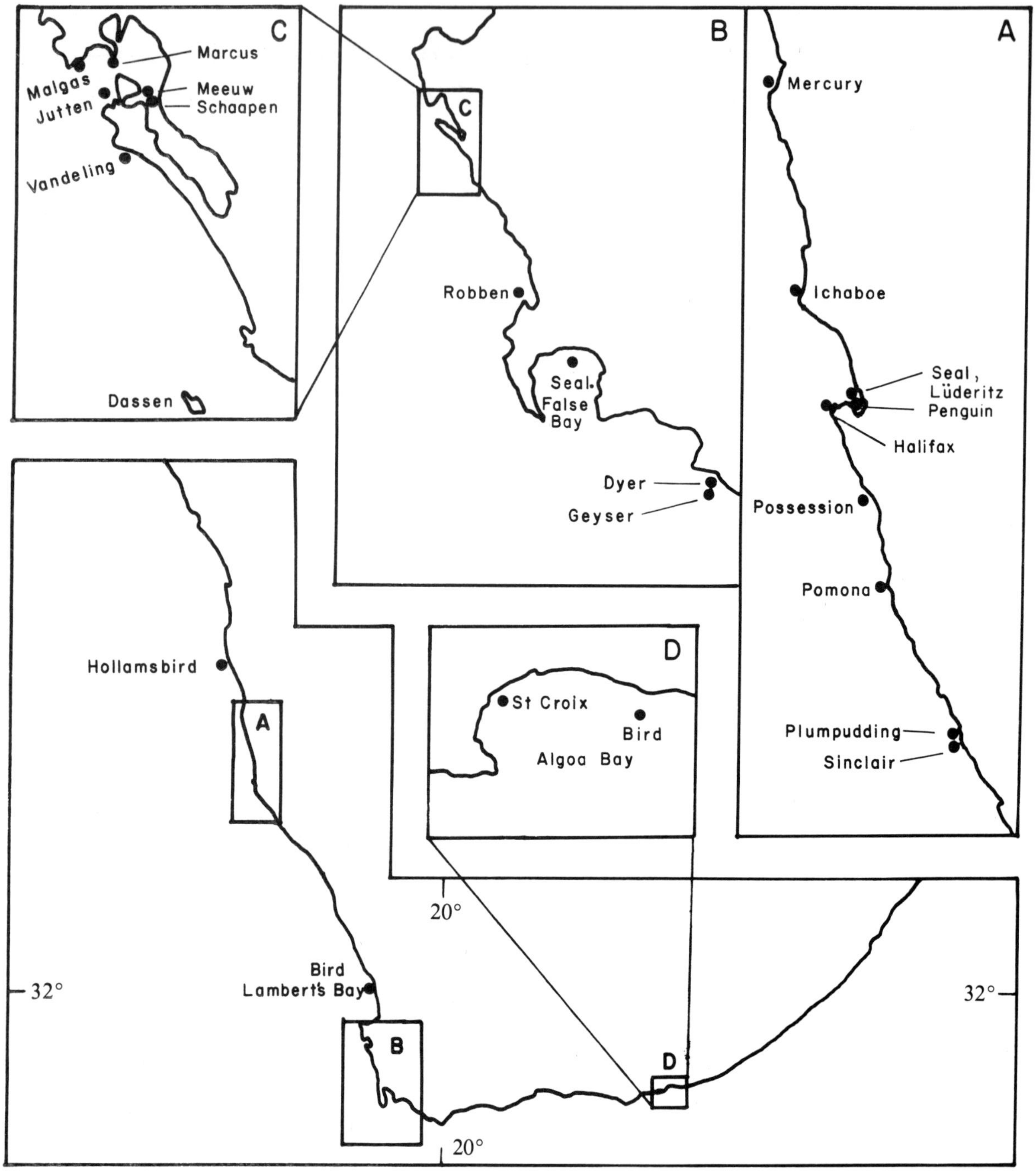

Figure 1. The South African continental islands.

Table 1. Sizes of South African continental islands and their present numbers of known indigenous and alien taxa, after Adamson (1934); Kriel et al (1980); Cooper (1981); Randall et al (1981); Brooke et al (1982); Brooke and Crowe (1982); Cooper et al (1982); Crawford et al (1982a,b); Crawford et al (1983); Hockey (1983); Randall and Randall (1984); Shaughnessy (1984); Shelton et al (1984); Cooper et al (1985); Brooke and Prins (in press); C W Heyl personal communication; FitzPatrick Institute unpublished.

Name	Size (ha)	Distance offshore (km)	Indigenous taxa		Alien taxa		Total
			Plants	Animals	Plants	Animals	
Hollamsbird	1	10,7	0	3	not studied		
Mercury	3	1,0	0	12		4	16
Ichaboe	6,5	1,3		28		6	34
Seal, Luderitz	44	1,0	3	14	0	2	19
Penguin	36	0,9	1	12	0	2	15
Halifax	46	0,2	2	10	0	1	13
Possession	90	2,7	2	42	0	1	44
Pomona	3	0,2	0	6	not studied		
Plumpudding	1	0,6	0	7	not studied		
Sinclair	3	0,3		8	not studied		
Bird, Lambert's Bay	3	0,2		22	2	3	27
Malgas	9	0,9	3	30	6	5	44
Marcus	11	1,4	10	51	12	9	82
Schaapen	41	0,5	12	41	5	4	62
Meeuw	7	0,1	19	26	8	1	54
Jutten	46	1,3	7	48	12	9	76
Vondeling	21	1,0	2	44	3	3	52
Dassen	222	9,0	16	90	15	14	135
Robben	507	6,7	101*	52	37*	16	206
Seal, False Bay	2	5,7		14		3	17
Dyer	20	3,5	6	51	8	10	75
Geyser	3	4,8	0	0	1	9	8
St Croix*	2,5	3,9	2	23	8	6	39
Bird, Algoa Bay*	19	8,4	6	12	3	6	27
All islands	–		134	208	53	35	430

*Pre-1971 records included due to inadequacy of recent collections, after Brooke and Prins (in press).

Cape fur seals *Arctocephalus pusillus* (eg Shaughnessy 1984).

The South African continental islands were first visited in the sixteenth century (Cooper et al 1985) and Robben Island, for example, has been continuously inhabited since 1652 (de Villiers 1971). Whereas the dates of introduction of some aliens to continental islands are quite accurately known, eg European rabbit *Oryctolagus cuniculus* in 1654/1658 on Robben Island (Cooper and Brooke 1982), very little is known about the history of most alien species, especially plants and invertebrates.

In a few cases it is possible to say how alien species reached continental islands. With the partial exception of *Acacia* species which can be bird dispersed, it may be assumed that all woody plants were deliberately introduced. Among birds the European starling *Sturnus vulgaris* and the house sparrow *Passer domesticus* reached continental islands by natural dispersion after introduction to the South African mainland (Brooke et al this volume) whereas all others were deliberately translocated to Robben Island to improve its amenities. Among mammals, the house rat *Rattus rattus* and the house mouse *Mus musculus* are assumed to be accidental introductions whereas, with the exception of the domestic cat *Felis catus*, the others were deliberately introduced or translocated, the European rabbit to 11 continental islands where it still occurs on seven (Cooper and Brooke 1982) and the rest to Robben Island. The domestic cat has become feral on two continental islands (Dassen and Robben) by escaping from domestic stock (Cooper et al 1985).

Oceanic islands

The Prince Edward islands consist of two islands, Marion and Prince Edward, situated 22 km apart at approximately 47°S, 38°E in the southern Indian Ocean and circa 1 800 km from the African continent, the nearest mainland. The islands are relatively large (Marion: 29 000 ha; Prince Edward: 4 400 ha). They are mountainous with peat bog and mire vegetation and have a subantarctic climate. Large populations of at least 27 seabird species breed on them as do three species of seals (van Zinderen Bakker et al 1971; Williams et al 1979). Detailed descriptions of a number of aspects of their environment are given in van Zinderen Bakker et al (1971). Abbott (1974) gives a summary of biogeographical and environmental facts for Marion and Prince Edward Islands, as well for 17 other subantarctic and southern temperate islands.

Man first set foot on the Prince Edward islands circa 1802 (Watkins and Cooper in press). Marion Island has been continuously occupied since 1948 whereas Prince Edward Island has never been permanently inhabited.

Little is known about the dates of introduction of alien species to the Prince Edward islands with the exception of the domestic cat which was introduced to Marion Island in 1949 to control house mice within the meteorological station. By 1951 cats had become feral. A number of plants and the house mouse are thought to have been introduced to Marion Island in the early 19th century by visiting sealers (Watkins and Cooper in press).

Species richness

Continental islands

At least 88 terrestrial alien taxa are present on 20 South African continental islands (Table 1). This total is made up of 53 vascular plants, 19 macro-invertebrates and 16 vertebrates (Table 2). Nearly all (16) of the macro-invertebrates are insects and only one vertebrate, the angulated tortoise *Chersine angulata*, is not a bird or mammal. Robben Island is the largest and supports the greatest number of aliens (53 species) even though the figure excludes its invertebrates which have not been studied (Table 3). Totals of aliens for islands off South West Africa/Namibia (Mercury to Possession, Figure 1) and on islands in the eastern Cape Province, South Africa (St Croix and Bird, Algoa Bay), are likely to be underestimates due to insufficient collecting, especially of plants and macroinvertebrates. However, the small totals of indigenous plants for the South West African/Namibian islands are presumably a reflection of their desert nature due to very low rainfall.

Table 2. Present numbers of known indigenous and alien taxa of 20 South African continental islands. Sources as in Table 1.

Taxa	Indigenous	Alien	Total	% Alien
Vascular plants	134	53	187	28,3
Gastropods	0	1	1	100,0
Earthworms	1	0	1	0,0
Centipedes	5	0	5	0,0
Insects	108	16	124	12,9
Crustaceans	3	1	4	25,0
Arachnids	27	1	28	3,6
Macroinvertebrates	144	19	163	11,7
Amphibians	1	0	1	0,0
Reptiles	8	1	9	11,1
Birds	53	7	60	11,7
Mammals	2	8	10	80,0
Vertebrates	64	16	80	20,0
Animals	208	35	243	14,4
Total	342	88	430	20,5

The alien taxa of the 20 continental islands studied form 20,5% of the total number (430) of taxa known to be present. Alien plants form a higher percentage than do alien animals, just as in mainland South Africa (Brooke et al this volume; Wells et al this volume). The percentage of the macroinvertebrate taxa made up of aliens is relatively low and similar to that for reptiles and birds (Table 2). Most of the alien invertebrates (84,2%) are insects. Robben Island supports 207 (48,1%) of all taxa presently recorded on the continental islands of which 25,6% are aliens (Table 3). This higher percentage reflects the more disturbed nature of Robben Island in comparison with the other smaller continental islands for which the equivalent figure is 18,6% for the 11 other south-western Cape islands (Table 4). The percentage would probably be even higher if the macroinvertebrates of Robben Island had been studied. Although Robben Island supports more alien plants than do the 11 smaller south-western Cape islands (Tables 3 and 4) they form a smaller proportion of the total flora due to the high number of indigenous plants on Robben Island (101) as opposed to a total of 46 on the other 11 islands. Robben Island supports relatively high numbers of alien birds and mammals, mostly deliberately introduced (Cooper et al 1985; Brooke and Prins in press).

Table 3. Present numbers of known indigenous and alien taxa of Robben Island, after Adamson (1934); Cooper et al (1985); Brooke and Prins (in press); C W Heyl personal communication; Fitz-Patrick Institute unpublished.

Taxa	Indigenous	Alien	Total	% Alien
Vascular plants	101	38	139	27,3
Amphibians	1	0	1	0,0
Reptiles	12	0	12	0,0
Birds	39	7	46	15,2
Mammals	1	8	9	88,9
Vertebrates	53	15	68	22,1
Total*	154	53	207	25,6

*No collections of macroinvertebrates have been made on Robben Island, otherwise the totals would probably have been substantially larger.

The relationship between numbers of indigenous and alien taxa and island size for the 11 south-western Cape islands (Table 1) was compared using two analyses: log-log regression and a generalized linear model. Use of log-log regression showed a more significant relationship between indigenous taxa and island area than between alien taxa and island area (Table 5). The slopes (z values) obtained were generally less than those obtained by Brooke and Crowe (1982) using a data set which included four south-western Cape offshore rocks, ie less than one hectare. Figures obtained for the 11 islands were similar to those expected for species-area relationships for very small islands (Williamson 1981).

A statistically more rigorous analysis is provided by fitting a generalized linear model using a logarithmic link function and Poisson error distribution (Nelder and Wedderburn 1972; McCullagh and Nelder 1983) to fit the numbers of indigenous and alien taxa using the logarithm of island size as an independent variable (Table 6). In all categories island size was found to be significantly related to the number of taxa present. The exponent to which the area is raised is almost identical for each category. The values of the multipliers and their levels of significance vary considerably. The values for alien plants and animals are smaller than those for indigenous plants and animals and are less significant (Table 6). This indicates that the relationship between island area and number of taxa is weaker for alien taxa than for indigenous taxa which have had a longer period in which to colonize the islands (Brooke and Crowe 1982).

Table 4. Present numbers of known indigenous and alien taxa of 11 south-western Cape islands, excluding Robben Island. Sources as in Table 1.

Taxa	Indigenous	Alien	Total	% Alien
Vascular plants	46	28	74	37,8
Gastropods	0	1	1	100,0
Earthworms	1	0	1	0,0
Centipedes	5	0	5	0,0
Insects	98	11	109	10,1
Crustaceans	3	1	4	25,0
Arachnids	27	1	28	3,6
Macroinvertebrates	134	14	148	9,5
Reptiles	6	1	7	14,3
Birds	28	2	30	6,7
Mammals	2	4	6	66,7
Vertebrates	36	7	43	16,3
Animals	170	21	191	11,0
Total	216	49	265	18,5

Table 5. Log-log regression analyses of numbers of indigenous and alien taxa and area of 11 south-western Cape islands.

Taxa	Slope (z)	Coefficient of determination (r^2)	Significance
All indigenous	0,45	0,72	P<0,001
All aliens	0,53	0,56	P<0,01
All taxa	0,47	0,70	P<0,01

Table 6. The relationship between numbers of indigenous and alien taxa and area in hectares (h) of 11 south-western Cape islands obtained by using the generalized linear model.

Taxa	Predicted number $a = h^{b}$	t-value of a	Significance	t-value of b	Significance
Indigenous plants	$2,23\ h^{38}$	2,74	$P<0,025$	4,71	$P<0,001$
Alien plants	$2,00\ h^{40}$	2,30	$P<0,025$	4,77	$P<0,001$
Indigenous animals	$13,75\ h^{35}$	21,67	$P<0,0005$	10,42	$P<0,0005$
Alien animals	$1,90\ h^{37}$	2,01	$P<0,05$	4,19	$P<0,0025$
All plants	$4,23\ h^{39}$	6,88	$P<0,0005$	6,70	$P<0,0005$
All animals	$15,66\ h^{36}$	24,31	$P<0,0005$	11,23	$P<0,0005$
All aliens	$3,90\ h^{38}$	6,19	$P<0,0005$	6,23	$P<0,0005$
All indigenous	$15,99\ h^{36}$	24,81	$P<0,0005$	11,44	$P<0,0005$
All taxa	$19,88\ h^{36}$	29,99	$P<0,0005$	13,06	$P<0,0005$

These results suggest that students of species richness-area relationships should consider whether the species or taxa included are indigenous or alien. It appears that the inclusion of aliens tends to reduce the significance of species richness-area relationships. However, Abbott and Grant (1976), from a study of Australasian islands, suggest that alien species are likely to be better colonizers of islands than indigenous species and, by implication, that the inclusion of aliens should result in steeper species richness-area curves. In this respect it should be noted that the slope value for all alien taxa on the 11 south-western Cape islands is higher than for all indigenous taxa, although, as stated above, the relationship is not as significant.

Our results also suggest, as Brooke and Crowe (1982) argued on other grounds, that the south-western Cape islands are not close to a state of species equilibrium since the establishment of alien species has not obviously been at the expense of indigenous species (Cooper and Brooke unpublished). Abbott and Grant (1976) claimed the same for most Australasian islands.

Distance to the nearest mainland (Table 1) was found not to be a significant independent variable in the generalized linear model. Brooke and Crowe (1982) also found no significant relationship between distance offshore and species richness for 15 south-western Cape islands. They considered that this was because all the islands were nearly equally accessible to potential colonists. It is clear from Williamson (1981) that the distances involved among the South African continental islands should not lead to a significant distance effect on species richness.

Oceanic islands

In all, 26 alien species presently occur at the Prince Edward islands, made up of 10 vascular plants, 13 macroinvertebrates and three vertebrates (Watkins and Cooper in press). Only three alien species, the grass *Poa annua*, the chironomid midge *Limnophyes pusillus* and the spider *Myro kerguelenensis*, have been recorded at uninhabited Prince Edward Island (Watkins and Cooper in press). This low number may be partly due to a low level of collecting compared with inhabited Marion Island. A further 20 species of aliens recorded at Marion Island are considered to be domestic and nonferal or are no longer present (Watkins and Cooper in press) and are not included here.

Aliens at oceanic Marion Island form 22,8% of the total biota of 114 taxa (Table 7). Alien plants form 31,3% of the total vascular flora of 32 species and alien invertebrates form 23,5% of the total macroinvertebrate fauna of 71 taxa (Table 7).

Abbott (1974) found that distance from the nearest land, whether island or mainland, was relatively unimportant in accounting for variation in the number of indigenous plant, insect and terrestrial bird species on 19 southern temperate and subantarctic islands, including Marion and Prince Edward Islands. He found that both island area and mean temperature of the coldest month were important in explaining variation in the numbers of plants but not of insects and terrestrial birds. The numbers of plant species themselves explained much of the variation in numbers of insects and birds. Comparable data on alien species were not analysed by Abbott (1974) and such an analysis remains an important field of study at subantarctic islands.

Table 7. Numbers of present indigenous and alien taxa on Marion Island, after van Zinderen Bakker et al (1971); Williams et al (1979); Gremmen (1981); Watkins and Cooper (in press); J E Crafford in litteris; Fitzpatrick Institute unpublished.

Taxa	Indigenous	Alien	Total	% Alien
Vascular plants	22	10	32	31,3
Gastropods	2	0	2	0,0
Earthworms	2	1	3	33,3
Insects	21	10	31	32,3
Arachnids	13	2	15	13,3
Macroinvertebrates	38	13	51	23,5
Fish	0	1	1	100,0
Birds	25	0	25	0,0
Mammals	3	2	5	40,0
Vertebrates	28	3	31	9,7
Animals	66	16	82	19,5
Total	88	26	114	22,8

Continental and oceanic islands compared

The relative remoteness, shorter period of human occupation and colder climate should all serve to explain the smaller list of alien taxa for the Prince Edward islands compared with that for the continental islands. Further collecting on the continental islands may increase the disparity in totals. However, an equally important reason may be that the two largest continental islands (Dassen and Robben) have been farmed or ranched in the past (Cooper et al 1985) and gardens have been established on some. This must inevitably have led to the introduction of aliens. Apart from keeping a few species of domestic animals and the presence of a greenhouse in the past, no farming, ranching or gardening has taken place at the Prince Edward islands (Watkins and Cooper in press).

Five alien species presently occur in common at both island groups. These are: *Poa annua*, chickweed *Stellaria media*, lesser house fly *Fannia canicularis*, feral cat and house mouse (Brooke and Prins in press; Watkins and Cooper in press). All five species are cosmopolitan, have high tolerances of different climatic conditions and are markedly associated with man-modified environments in their introduced ranges (eg Chippendall 1955; Zeuner 1963; Henderson and Anderson 1966; Smith 1973; Gleeson and van Rensburg 1982).

Alien vascular plants form 60,2% of the total alien biota of the continental islands compared with 38,5% for oceanic Marion Island. A comparison of the lists of alien plants (Brooke and Prins in press; Watkins and Cooper in press) shows that woody species, including intentionally planted trees, have become established only on continental islands. The climate of the oceanic Prince Edward islands is apparently too severe to support alien trees and there are no indigenous trees (van Zinderen Bakker et al 1971; Gremmen 1982; Watkins and Cooper in press). There are no fish on the continental islands through lack of permanent water. No reptiles occur at the oceanic islands which are presumably too cold. Perhaps the most significant difference is in the numbers of alien birds: seven (8% of all alien taxa), all land birds, occur at continental islands and none at the oceanic islands. The Prince Edward islands support only one nonseabird, the lesser sheathbill *Chionis minor* (Williams et al 1979). The paucity of land and shore birds at the Prince Edward islands has been attributed to lack of suitable food supplies and vegetation cover and to the consequences of intense avian predation (Burger et al 1980; Burger 1985). In comparison, the continental islands support 40 indigenous breeding nonseabirds (Brooke and Crowe 1982; C W Heyl personal communication; FitzPatrick Institute unpublished).

A comparison between the continental and oceanic islands of the proportions of the biota made up of alien taxa (Tables 2 and 7) show similar figures for vascular plants and for all taxa combined. The continental islands support smaller proportions of alien invertebrates and vertebrates (but higher absolute numbers) than does Marion Island. This is, of course, partially due to the indigenous biota of the continental islands containing nearly four times as many taxa as oceanic Marion Island. The continental islands are relatively rich in both alien and indigenous taxa whereas oceanic Marion Island has few species, notwithstanding its much greater area (cf Williamson 1981).

Ecological impact of aliens

Our knowledge of alien species, other than birds and mammals, present on South African continental and oceanic islands is usually based on collection lists containing taxa not always identified to species, and not on community studies. It is therefore seldom known how abundant any species or taxon may be and whether it has any impact on an island's biotic environment. We give below, examples where some knowledge does exist on these aspects.

At some continental islands, notably Robben Island, intentionally introduced trees, other plants and herbivorous mammals have greatly altered the habitat, although little is known about the original biota on South African continental islands (Brooke 1983; Brooke and Prins in press). Gillham (1963) showed that European rabbits had a marked effect on plant communities of the continental islands on which they were abundant due to their selective grazing. Whereas their elimination would be unlikely to restore the original plant communities, it would permit some recovery of the natural communities, particularly on Schaapen Island (Cooper and Brooke 1982).

Two plant species, the grass *Agrostis stolonifera* and the forb *Sagina apetala*, are widespread aliens that are presently spreading at Marion Island (Gremmen 1982; Watkins and Cooper in press). The former species is able to spread into undisturbed vegetation and replace the native dwarf shrub *Acaena magellanica* (Gremmen 1982; V R Smith personal communication).

On continental Dassen Island in April/May 1985 the population of feral cats was estimated at 65 adults. These were killing chicks of jackass penguins *Spheniscus demersus* at an estimated rate of 531 a year, ie about nine per cent of the annual production of chicks. The kill estimate was based on the stomach contents of 44 cats killed at that time. Very few other birds were being taken (Berruti in press). The jackass penguin is regarded as vulnerable (Brooke 1984). At oceanic Marion Island an estimated 2 139 ± 290 cats were responsible for killing 455 000 burrowing petrels (Procellariidae) of seven species in 1975 (van Aarde 1980). Cats are thought to have caused the extinction of the breeding population of the common diving petrel *Pelecanoides urinatrix* at Marion Island (Brooke 1984 and references therein).

House mice at Marion Island are predominantly insectivorous (Gleeson and van Rensburg 1982) on an island where most invertebrates are detritivores. House mice are therefore more likely to affect plant decomposition indirectly than plant production directly (Smith 1977). Nothing is known of the diet of house mice on continental islands where they seem to occur mainly in or around buildings (Cooper and Brooke unpublished). Nonetheless, a population on St Croix Island maintained itself for a number of years between the destruction of the first group of buildings and the erection of a second group (Brooke and Prins in press).

A feature of the continental island faunas is the absence of mammalian predators other than introduced feral cats. This has permitted the establishment of European rabbits on seven islands (Cooper and Brooke 1982). On Robben Island it has permitted the establishment of populations of peacocks *Pavo cristatus* and chukar partridges *Alectoris chukar*. None of these species has become established on the South African mainland with its great suite of predators and pathogens (Brooke et al this volume). Most alien vertebrates with wide ranges on the South African mainland also occur on various continental islands. This seems to be due to the species concerned (European starling, house sparrow, house rat, house mouse) having become adapted to man-modified environments before reaching South Africa (Brooke et al this volume). The other widespread alien vertebrates in South Africa are the feral pigeon *Columba livia* which lives in towns of a size not found on the continental islands and the Indian myna *Acridotheres tristis* which does not breed in the Cape Province or South West Africa/Namibia (Brooke et al this volume) off which all the continental islands lie.

Management implications and priorities

To date, control of alien biota at South African continental and oceanic islands has been attempted for not one plant and very few animal species.

Control of feral cats has already started at Dassen and Marion Islands with some success (Howell 1984; van Aarde 1984; Berruti in press). The feral cat population at Dassen Island was reduced substantially by day and night shooting in April/May 1985 and further control measures leading to elimination of the species are planned (Berruti in press). Control of feral cats at Marion Island has been attempted by the introduction, in March 1979, of a host-specific, contagious virus which causes feline panleucopaenia (Howell 1984). This introduction reduced the population from a level estimated in 1975 to be 2 139 ± 290 cats to 615 ± 107 cats in 1982 (van Rensburg in press). This reduced population is estimated to have killed 142 000 burrowing petrels of six species in 1982 (van Rensburg 1985). A planned secondary control measure entails night shooting (van Rensburg in press).

Sporadic control of house rats has recently been undertaken at continental Bird Island, Lambert's Bay, by poisoning and trapping. However, since the island is connected to the mainland by a causeway, it is unlikely that the species could be permanently eliminated (Cooper et al 1985).

Control of brown trout *Salmo trutta* has been undertaken at Marion Island in recent years. The species occurs in one river only and, indeed, may no longer be present (Watkins and Cooper in press).

The intentionally introduced large herbivorous mammals and the ostrich *Struthio camelus* on Robben Island which occur in small numbers (Cooper et al 1985; Brooke and Prins in press) could be easily removed by shooting. Control of European fallow deer *Cervus dama* has recently started at Robben Island (P M Norton personal communication).

The successful control of several species of mammalian predators at Marcus Island has been achieved by building a predator-proof wall and by poisoning and trapping (Cooper et al 1985). Since these predators were self-introduced via the causeway they have not been considered elsewhere in this paper.

Following Cooper and Brooke (1982), Cooper et al (1985), Brooke and Prins (in press) and Watkins and Cooper (in press), we recommend that the following management activities be undertaken, in approximate order of priority.

(1) Control of feral cats on Dassen, Robben and Marion Islands, with the aim of eventual extinction.
(2) Continuation of strict precautions (detailed in Watkins and Cooper in press) to avoid introducing aliens (especially *Agrostis stolonifera*, *Sagina apetala* and house mice) to Prince Edward Island from Marion Island.
(3) Removal of European rabbits from several continental islands.
(4) Drawing up management plans for each island group that include measures for halting further introductions of aliens.
(5) More detailed surveys for alien species on a number of continental islands, especially islands off South West Africa/Namibia, Robben Island and in the eastern Cape Province and on oceanic Prince Edward Island.

Acknowledgements

We acknowledge financial and logistical support from the Sea Fisheries Research Institute, the South African Departments of Transport and Environment Affairs, the South African National Committee for Oceanographic Research and the South African Scientific Committee for Antarctic Research. We thank J E Crafford, C W Heyl, P M Norton and V R Smith for supplying unpublished information and L G Underhill for undertaking a generalized linear model analysis.

Abbott I 1974. Numbers of plant, insect and land bird species on nineteen remote islands in the southern hemisphere. Biological Journal of the Linnaean Society 6, 143-152.

Abbott I and Grant P R 1976. Nonequilibrial bird faunas on islands. American Naturalist 110, 507-528.

Adamson R S 1934. The vegetation and flora of Robben Island. Transactions of the Royal Society of South Africa 22, 279-296.

Berruti A (in press). The predatory impact of feral cats *Felis catus* and their control on Dassen Island. South African Journal of Antarctic Research.

Brooke R K 1983. On the 17th century avifauna of Robben Island, South Africa. Cormorant 11, 15-20.

Brooke R K 1984. South African Red Data Book - Birds. South African National Scientific Programmes Report No 97. CSIR, Pretoria.

Brooke R K, Cooper J, Shelton P A and Crawford R J M 1982. Taxonomy, distribution, population size, breeding and conservation of the whitebreasted cormorant, *Phalacrocorax carbo*, on the southern African coast. Gerfaut 72, 188-220.

Brooke R K and Crowe T M 1982. Variation in species richness among offshore islands of the southwestern Cape. South African Journal of Zoology 17, 49-58.

Brooke R K and Prins A J (in press). Review of alien species on South African offshore islands. South African Journal of Antarctic Research.

Burger A E 1985. Terrestrial foodwebs in the sub-Antarctic: island effects. In: Siegfried W R, Condy P R and Laws R M (eds) Antarctic nutrient cycles and food webs. Springer-Verlag, Berlin. pp 582-591.

Burger A E, Williams A J and Sinclair J C 1980. Vagrants and the paucity of land bird species at the Prince Edward islands. Journal of Biogeography 7, 305-310.

Chippendall L K A 1955. A guide to the identification of grasses in South Africa. In: Meredith D (ed) The grasses and pastures of South Africa. Central News Agency, Johannesburg. pp 1-527.

Cooper J 1981. Biology of the bank cormorant, part 1: distribution, population size, movements and conservation. Ostrich 52, 208-215.

Cooper J and Brooke R K 1982. Past and present distribution of the feral European rabbit *Oryctolagus cuniculus* on southern African offshore islands. South African Journal of Wildlife Research 12, 71-75.

Cooper J, Brooke R K, Shelton P A and Crawford R J M 1982. Distribution, population size and conservation of the Cape cormorant *Phalacrocorax capensis*. Fisheries Bulletin of South Africa 16, 121-143.

Cooper J, Hockey P A R and Brooke R K 1985. Introduced mammals on South and South West African islands: history, effects on birds and control. In: Bunning L J (ed) Proceedings of the symposium on birds and man, Johannesburg 1983. Witwatersrand Bird Club, Johannesburg. pp 179-203.

Crawford R J M, Cooper J and Shelton P A 1982a. Distribution, population size, breeding and conservation of the Kelp Gull in southern Africa. Ostrich 53, 164-177.

Crawford R J M, Shelton P A, Brooke R K and Cooper J 1982b. Taxonomy, distribution, population size and conservation of the crowned cormorant *Phalacrocorax coronatus*. Gerfaut 72, 3-30.

Crawford R J M, Shelton P A, Cooper J and Brooke R K 1983. Distribution, population size and conservation of the Cape gannet *Morus capensis*. South African Journal of Marine Science 1, 153-174.

De Villiers S A 1971. Robben Island. C Struik, Cape Town.

Gillham M E 1963. Some interactions of plants, rabbits and seabirds on South African islands. Journal of Ecology 51, 275-294.

Gleeson J P and Van Rensburg P J J 1982. Feeding ecology of the house mouse *Mus musculus* on Marion Island. South African Journal of Antarctic Research 12, 34-39.

Gremmen N J M 1981. The vegetation of the Subantarctic islands, Marion and Prince Edward. Geobotany 3, 1-149.

Gremmen N J M 1982. Alien vascular plants on Marion Island. Comite National Francais des Recherches Antarctiques 51, 315-323.

Henderson M and Anderson J G 1966. Common weeds in South Africa. Botanical Survey Memoir of South Africa 37, 1-440.

Hockey P A R 1983. The distribution, population size, movements and conservation of the African black oystercatcher *Haematopus moquini*. Biological Conservation 9, 233-262.

Howell P G 1984. An evaluation of the biological control of the feral cat *Felis catus* (Linnaeus, 1758). Acta Zoologica Fennica 172, 111-113.

Kriel F, Crawford R J M and Shelton P A 1980. Seabirds breeding at Robben Island between 1949 and 1980. Cormorant 8, 87-96.

McCullagh P and Nelder J A 1983. Generalized linear models. Chapman and Hall, London.

Nelder J A and Wedderburn R W M 1972. Generalized linear models. Journal of the Royal Statistical Society A135, 370-384.

Rand R W 1963a. The biology of guano-producing seabirds. 4. Composition of colonies on the Cape islands. Division of Sea Fisheries of South Africa Investigational Report 43, 1-32.

Rand R W 1963b. The biology of guano-producing seabirds. 5. Composition of colonies on the South West African islands. Division of Sea Fisheries of South Africa Investigational Report 46, 1-26.

Randall B M and Randall R M 1984. Land and shore birds at St Croix Island. Bokmakierie 36, 51-55.

Randall R M, Randall B M, Batchelor A L and Ross G J B 1981. The status of seabirds associated with islands in Algoa Bay, South Africa, 1973-1981. Cormorant 9, 85-104.

Shaughnessy P D 1984. Historical population levels of seals and seabirds on islands off southern Africa, with special reference to Seal Island, False Bay. Investigational Report of the Sea Fisheries Research Institute of South Africa 127, 1-61.

Shelton P A, Crawford R J M, Cooper J and Brooke R K 1984. Distribution, population size and conservation of the jackass penguin *Spheniscus demersus*. South African Journal of Marine Science 2, 217- 257.

Smith K G V (ed) 1973. Insects and other arthropods of medical importance. Trustees of the British Museum (Natural History), London.

Smith V R 1977. A qualitative description of energy flow and nutrient cycling in the Marion Island terrestrial ecosystem. Polar Record 18, 361-370.

Van Aarde R J 1980. The diet and feeding behaviour of feral cats, *Felis catus* at Marion Island. South African Journal of Wildlife Research 10, 123-128.

Van Aarde R J 1984. Population biology and the control of feral cats on Marion Island. Acta Zoologica Fennica 172, 107-110.

Van Rensburg P J J 1985. The feeding ecology of a decreasing feral house cat, *Felis catus*, population at Marion Island. In: Siegfried W R, Condy P R and Laws R M (eds) Antarctic nutrient cycles and food webs. Springer-Verlag, Berlin. pp 620-624.

Van Rensburg P J J (in press). Control of the Marion Island cat (*Felis catus*) population: why and how. South African Journal of Antarctic Research.

Van Zinderen Bakker E M, Winterbottom J M and Dyer R A (eds) 1971. Marion and Prince Edward Islands, report on the South African biological and geological expedition/1965-1966. Balkema, Cape Town.

Watkins B P and Cooper J (in press). Introduction, present status and control of alien species at the Prince Edward islands, sub-Antarctic. South African Journal of Antarctic Research.

Williams A J, Burger A E, Berruti A and Siegfried W R 1979. The Prince Edward Islands: a sanctuary for seabirds in the Southern Ocean. Biological Conservation 15, 59-72.

Williamson M 1981. Island populations. Oxford University Press, Oxford.

Zeuner F E 1963. A history of domesticated animals. Hutchinson, London.

Towards understanding the ecology of biological invasions

The current approach to the management of invasive species has largely been an empirical one, based on individual prescriptions for particular species or areas. The section which follows serves to review our current understanding of local biological invasions in order to develop a predictive capability for their management and prevention. It represents an attempt to move from an entirely reactive, to at least a partially pro-active mode of operation.

We have already noted in the chapter by Shaughnessy, how important the history of introductions is in determining present-day patterns of invasion. Given the patterns of introduction and propagation, how do we explain subsequent invasion? We may hypothesize first, that the species which disperse most effectively will be the ones that invade most rapidly. Second, one might argue that the species with most abundant recruitment and most vigorous and thrifty offspring will be the most effective invaders. If we look beyond predictions of invasions to evaluation of control strategies then we find a knowledge of the ecophysiology of the target plant may be important in designing optimum biological control programmes. Elsewhere in this volume, Moran et al have documented the success achieved by one or two control insects on *Opuntia* species, whereas many are needed for *Lantana camara,* for example. Comparative studies of carbon metabolism such as those reviewed by **Whiting and co-authors**, could assist in such cases.

Which regions of the country and which habitats within regions are the ones where problems with invasive species will occur soonest, or with the most severe consequences? These chapters show that unambiguous answers to these questions are not yet at hand. It is not clear that any one biome within the region is distinctly more prone to invasion than any other, though **Macdonald and co-authors** have some evidence that ecosystems in the fynbos biome are more readily invaded than others. But there is consistent evidence from all the biomes that it is the riparian habitats that are invaded soonest and quickest. And wherever the case, it is the habitat disturbed by man which consistently accommodates the introduced species - though there are many cases where species are invading habitats where it is difficult to prove that human disturbance is a prerequisite for their success. An effective set of guidelines for preventing the introduction of invasive species and for managing those already present is beyond our capacity at this stage.

Chapter 11

Processes of invasion by alien plants

F J Kruger, D M Richardson and B W van Wilgen
South African Forestry Research Institute.

Introduction

A model to describe the process of colonization or invasion of a new habitat by a species has three principal elements (Roughgarden and Diamond 1985). Firstly, the colonizing species must overcome the barriers to dispersal between its native habitat and the new; secondly, on arrival it must withstand the rigours of the new habitat, and thirdly, if the habitat allows growth and reproduction, then the colonizer must survive the adverse interactions with resident species which are competitors, inhibitors, or predators, or establish new mutualistic relations (Figure 1). Such a model may conveniently be applied at two scales: to that of intercontinental transfer, and then to the subsequent processes involved in the spread within a region.

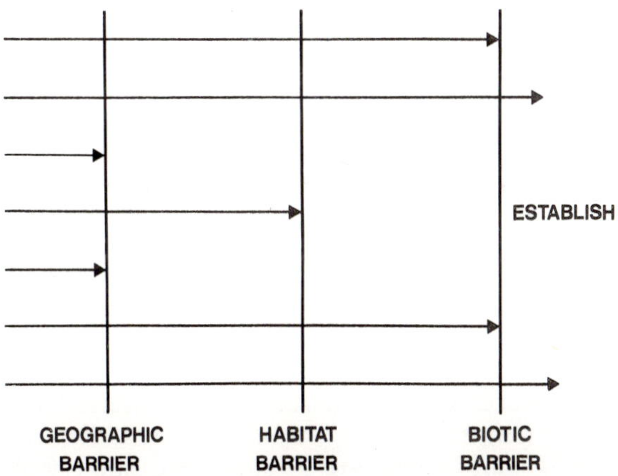

Figure 1. A model to describe the processes involved in the invasion of a new habitat by an alien species. The geographic barrier must be overcome by dispersal. The habitat barrier requires preadaptation to the conditions of the new environment or a genetic adjustment to it. The biotic barrier includes the forces of predation, herbivory, competition and interference which must be overcome in the new community, or the new mutualistic relationships which must develop. See text.

Address for correspondence: F J Kruger, South African Forestry Research Institute, P O Box 727, Pretoria, 0001.

In this chapter we examine the information on introduced plant species that are invasive in South Africa to find general explanations for their invasions. Few studies have been aimed directly at explaining the invasions so that it is often possible only to derive tentative hypotheses from the results reported. Because of this we often resort to the forestry literature, since many tree species have been introduced to South Africa and their introduction to and propagation in uncultivated land has been studied experimentally.

Entry of plant species to South Africa

Introductions

By definition, introduced species are those translocated by man. Species of the coastal littoral and adjacent habitats may arrive without the aid of man but no problematic species have dispersed to South Africa this way. There is evidently a steady arrival of exotic pollens on Marion Island, for example, but no seeds have been detected in this drift (Scott and Van Zinderen Bakker 1985).

Of the invasive species introduced by man, many have been inadvertent introductions. Baker (1986) lists four modes of such introductions, to which we may add one for South Africa, ie introduction of *Nicotiana glauca* with hay for horses (Tarr and Loutit 1985) (Table 1). Modes of entry of alien invasive plants to South Africa are discussed by Wells et al (this volume). Because species which are inadvertently introduced are seldom, if ever, propagated, there may be many such introductions which have failed to survive.

Species intentionally introduced would have been propagated deliberately and this would tend to favour a proportionately higher rate of success in establishment. At least 400 alien species of trees and shrubs are successfully cultivated in South Africa or are naturalized (Poynton 1984) and this is about 40% of the native tree and shrub flora (De Winter et al 1978). Among herbs, domesticated varieties are used more often than in the case of alien trees and shrubs, especially among ornamentals and crop plants. Because domesticated varieties of ornamentals have been selected for garden conditions and not for competitive ability, and are often sterile or nearly so (Baker 1986), they are seldom invasive. In South Africa, the relative rarity of invaders among introduced pasture plants is striking (Wells et al this volume). Thus, the frequency of invasive species may be higher among the woody species than among the herbaceous. In South Africa, the need to introduce woody plants to a landscape relatively devoid of trees (eg Shaughnessy this volume) would have motivated the importation of more such trees and shrubs than was the case in other regions of the world.

Preadaptation to the new habitat

Those who introduced species have often attempted to match the climates and substrata of the source regions with those of the target region in South Africa, and then selected appropriate species from such matched source regions (King 1938). Many introduced species were therefore preadapted to the new environment and some at least should have spread from cultivation. What is interesting is why all introduced species have not spread. Although bioclimatic matching may aid successful introductions, it does not necessarily result in invasive spread. *Pinus caribaea* achieves vigorous growth on the Zululand coast, but does not invade because the local climate does not allow adequate reproduction (Poynton 1979a). The many species of *Eucalyptus* introduced to South Africa were selected to match the new habitats. Over 170 species have been introduced to South Africa over the past two centuries (Poynton 1979b, 1984), but few rank high on any list of species invasive in unmodified habitats (Henderson and Musil 1984; Macdonald and Jarman 1984, 1985), though about 25% of the species introduced tend to spread (Poynton 1984) and they are often problem weeds in forest plantations (Le Roux 1980). The reproductive biology and early growth of *Eucalyptus* species seem to make them poor competitors in the natural environments of South Africa (see below).

Baker (1986) draws the distinction between major weeds, whose habitat requirements are general, and minor weeds, with special habitat requirements and distributional ranges which are actually and potentially limited. Poynton (1984) has summarized the data on cultivated trees and shrubs in South Africa and has divided the country into 48 silvicultural zones on the basis of climate, with relative aridity, seasonality of rainfall, and severity of frost being the most important factors. He has classified the performance of each species in each zone, as well as their tendency to invade the surrounding habitat. Among the 300 or so species of *Acacia*, *Eucalyptus* and *Pinus* introduced into South Africa there is great variation among species in the number of silvicultural zones to which each is suited (Poynton 1984; see below). Among those which are invasive this ranges from about six to 36 of the 48 zones. There is a poor correlation between the number of zones to which species are suited and the number in which they are classified as invasive ($r^2 = 0{,}177$, $P = 0{,}189$). Some species with wide habitat tolerances invade widely whereas others, with similar histories and tolerances, invade only in a few regions. *Acacia dealbata*, for example, is suited to about half the zones and is recorded as invading in all of these. By contrast, *Acacia melanoxylon*, which has about the same ecological amplitude, is recorded as an invasive only in half of its zones, ie the humid ones. *Acacia*

Table 1. Modes of entry of alien plants to a new continent (Baker 1986), with some examples of species introduced to South Africa.

Mode of entry	Example
Inadvertent introductions	
by ballast in ships	
by impure seed imports	*Hypericum perforatum* (Gordon 1978)
by adhesion to domesticated animals	*Medicago* spp (Deacon, this volume)
in soil surrounding the roots of nursery stock	
in fodder for horses	*Nicotiana glauca* (Tarr and Loutit 1985), *Stipa trichotoma* (Stirton 1978)
Deliberate introduction as	
forage plants	*Prosopis* spp (Stirton 1978)
fibre plants	—
medicinal plants	—
ornaments	*Sesbania punicea, Opuntia aurantiaca* (Stirton 1978)
erosion controls	*Acacia cyclops, A. saligna* (for stabilization of drift-sands, Stirton 1978)
forest plantation	*Pinus* spp (Poynton 1979a), *Eucalyptus* spp (Poynton 1979b), *Acacia meansii, A. melanoxylon* (Stirton 1978)

podalyriifolia invades in all the 18 zones to which it is suited but *A cultriformis*, a species suited to the same zones and used for similar horticultural purposes, is recorded as invasive in only two. Clearly, some introduced species in South Africa may be classified as major weeds in Baker's definition, and others as minor weeds. *Pinus pinaster* and *P radiata* are able to invade habitats within the winter- and all-year-rainfall regions of South Africa which vary widely in rainfall and soil conditions (eg Kruger 1977). These are major weeds. Among minor weeds in the same region one may count *Pittosporum undulatum*, whose distribution is severely limited by microclimatic conditions required for establishment (Gleadow 1982; Richardson and Brink 1985).

The genotype of *P pinaster* that has invaded most of the area concerned is the "Franschhoek" form, introduced in the seventeenth century (Kruger 1977). There is marked racial variation in *P pinaster* and provenance trials have shown which have the most desirable characteristics (Rycroft and Wicht 1948). The preferred form, from Portuguese provenances, apparently produces less cones and therefore less seed under South African conditions than the others (Rycroft and Wicht 1948). It remains to be seen whether this genotype will have general capabilities for invasion after its widespread introduction subsequent to 1948.

Intraspecific variation in tolerance of habitat factors may determine whether the introduced genotype will establish or which habitats within a bioclimatic region will be invaded. Frosts in the highveld grasslands of South Africa exclude most species of trees. Introduced species of frost-hardy *Eucalyptus* vary in their tolerance, both between and within species (Darrow 1984). In *Eucalyptus viminalis*, for example, there is pronounced variation among populations of different provenances in their survival of frost (Darrow 1984) and this is evidently related to native microclimatic conditions (Figure 2). *Eucalyptus viminalis* is a slow invader of grassland habitats (Poynton 1984). The habitat invaded locally would depend on which genotype had been introduced, at least in relation to frost. The overall degree of invasion would depend on the variety of genotypes selected for introduction. In the past, selection of varieties may not have been optimal in this respect, but formal programmes of provenance trials in forestry, for example, will ensure progressively better matching of genotypes and environment.

Preadaptation to the biota and communities

Invasive weeds are often plants which have evolved in long association with man in the source region, often the Mediterranean Basin (Baker 1986). Few of the grasses which have succeeded in overwhelming communities elsewhere (eg Mack 1986), and which often include the forms which Baker referred to, have penetrated substantially into native South African pastures. In these habitats we have seen invasions by plant forms not normally classified as weeds, ie trees and shrubs, although there are some exceptions such as *Stipa trichotoma* in the grasslands and *Bromus japonicus* in the winter-rainfall karoo. Macdonald (1985a) has drawn the contrast between the situation in South Africa and that in Australia, for example, where most invasive plants are herbs. He suggested that this was because the abundant African herbivores impaired the vigour of introduced herbs to the point that any advantage over equivalent native species was lost.

The role of such biotic factors in determining whether an introduced species invades successfully has received little attention. Many alien species of timber trees suffer severe pressure from native mammals (Bigalke 1980). Certain species and even races are often preferred, and these may be excluded from further propagation as a result. This is illustrated by

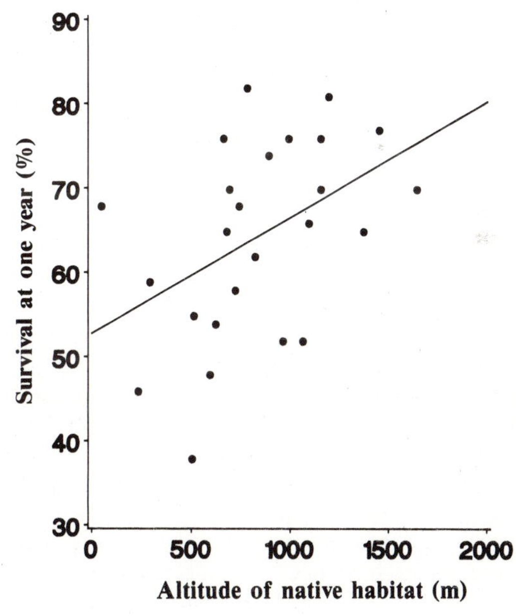

Figure 2. Variation in the survival of frost by *Eucalyptus viminalis* at one year after planting at Jessievale, eastern Transvaal. Survival was significantly related to altitude of native provenance (r = 0,53). Data from Darrow (1984).

the case of *Pinus oocarpa*, a potentially productive Central American timber tree. Trials have shown markedly different rates of casualty among plants of different provenances, with damage caused mainly by bushpigs *Potamochoerus porcus* and, to a lesser extent, baboons *Papio ursinus* (eg Steyn 1980). Termites are known to have prevented the successful establishment of *Eucalyptus* species in plantations in tropical regions prior to the advent of chemical pesticides (Poynton 1979b). It is thus likely that the establishment of some introduced species in South Africa has been prevented or at least delayed by the effects of the native biota.

Invasion of new habitat by the established alien

Evolution of invasions after introduction

Introduced species sometimes persist in small colonies, often in cultivation, for several generations before invasion is noticed. One explanation offered for this is the so-called "infection pressure" (Salisbury 1961; Baker 1986). The hypothesis is that an introduced species would have to reproduce sufficient generations locally for the population to reach a condition which allows spread from the point of introduction. Baker (1986) speculated that this could be connected to the founder effect, in that a population arising from one or a few initial seeds would require time before sufficient genotypic variation could evolve to permit colonization of variable habitats. Alternatively, the effect could be purely demographic, time being needed for populations to pass through the initial phase of exponential growth to the stage where observers would begin to notice problematic invasions. It could also be ecological, in that some change in the environment is needed before the species' requirements for spread are met. Finally, it might also be that the conditions required for spread occur episodically and infrequently.

An analysis of the information on species of *Acacia*, *Eucalyptus* and *Pinus* in Poynton (1984) together with the data on dates of introduction taken from seed-stock records and other sources, (eg Shaughnessy this volume) allows a test of the hypothesis of infection pressure. About 309 species in these genera have been introduced, mainly in the period from 1850 to 1950 (Table 2). The data indicate that about 57 of these have become invasive to a greater or lesser extent. A weighted regression analysis of the percentage of invasive species on period of introduction is highly significant (r^2 = 0,835, F = 15,1, P = 0,0301) with a negative coefficient, indicating a real increase in the incidence of invasion with time since introduction. Most of the invasive species are among the large number of plants introduced between 1850 and 1900. This provides support for the hypothesis of infection pressure and could presage more problems in future once the large numbers of species introduced since 1900 have passed through this hypothetical stage of initial adjustment. Macdonald (1984) could find only limited evidence that the extent of invasion was related to time since introduction of a plant species to South Africa, so that although the inception of an invasion by a given species may sometimes depend on the passage of time, the subsequent rate of spread may be less dependent on time.

Table 2. Summary of information on introduction of species of *Acacia, Eucalyptus* and *Pinus* to South Africa, from Poynton (1979a,b, 1984) and unpublished records of the Seed Section, South African Forestry Research Institute, Pretoria.

	Interval, year					
	<1700	1701-1800	1801-1850	1851-1900	1901-1950	1951+
Total number of species introduced	3	0	6	126	132	42
Number of invasive	2	0	5	38	12	0

Where an invasion is delayed this may involve a lag in the establishment of a required mutualistic interaction. The most obvious of these is when animals feed on fruit and disperse the seed. Richardson and Brink (1985) have inferred from the absence of early reports of *Pittosporum undulatum* invaders that the species began to spread only several decades after the initial introduction. This may have been because there was a lag before native birds switched to the new kind of fruit, after which spread became more marked (Richardson personal observation). Similar dispersal mutualisms have evolved in *Acacia cyclops*, *A saligna* and *A longifolia* now dispersed by the black korhaan, *Eupodotis afra* (Knight and Macdonald 1985) and the redwinged starling *Onychognathus morio* (Richardson personal observation) as well as in *Solanum mauritianum* (see below).

There is as yet no evidence for genetic change in invasive plants introduced to South Africa. Gill

(1978) has suggested that the hard-seededness evident in invasive populations of *Acacia cyclops* may have been strongly selected for in South Africa from an Australian stock with a very low frequency of hard-seededness (Christensen in Gill 1981). Soft-seeded strains, it is argued, would have been initially eliminated before invasion began.

Evidence that delayed invasion could be the result of infrequent occurrence of episodically suitable conditions has been adduced for *Prosopis* in the Karoo and southern Kalahari by Macdonald (1985b). Species of *Prosopis* had been introduced to the region concerned over a lengthy period from about 1910. Rapid invasion and growing concern about it were noticed in the late 1970's. Macdonald suggested that this was at least partly owing to the run of high-rainfall years beginning around 1974; for example, dense infestations became established on riparian sites on river courses which had flowed in those years but not in human memory before then.

In other cases, no initial adjustment was apparently necessary. Analyses of historical aerial photographs and the plantation history for Jonkershoek and inference of past population dynamics from the age structure of stands of *Pinus radiata* invading fynbos show that invasion began about 13 years after first planting of the species in the area. This was just sufficient time for the parent trees to begin seed production (Richardson and Brown in press).

Dispersal

Dissemination by people

Cultivation and dissemination of introduced species by humans are often implicated in the invasion process and it is frequently difficult to isolate the effect of these from that of unaided dispersal (Shaughnessy this volume). Seventeen of the 20 most important invasive aliens in the fynbos biome (Macdonald and Jarman 1984) have been actively propagated, many from as early as the mid-nineteenth century (Richardson unpublished). Even so, the majority of species that were widely planted have failed to become naturalized or invade habitats only in the immediate vicinity of plantations (Table 3). For example, *Eucalyptus globulus* was sold in greater numbers than any other species from five nurseries in the Cape between 1891 and 1902 but it has not succeeded in invading the natural vegetation in the fynbos biome, or at most invades very slowly (Taylor et al 1985). Very few *Hakea sericea* plants were supplied from the five principal forest nurseries in the biome over the period 1891-1902 (Richardson unpublished), although the species had been introduced by 1850 (Neser and Fugler 1978). The species was planted on a limited scale and at isolated places, yet *Hakea sericea* is estimated to occupy the greatest area of all woody alien plants in the fynbos biome (Macdonald and Jarman 1984; see also Shaughnessy this volume).

Table 3. List of alien tree species of which 100 000 individuals or more were sold from or supplied by the four principal forestry nurseries in the fynbos biome (Tokai, Kluitjieskraal, Concordia, George) during the period 1881 – 1902 (Richardson unpublished) and the ranking of extent of invasion of the biome (Macdonald and Jarman 1984). Large quantities of seed of *Acacia saligna* were supplied for driftsand stabilization.

Species	Plants sold 1891-1902 Rank	Plants sold 1891-1902 Number	Extent of invasion (rank according to Macdonald and Jarman 1984)
Eucalyptus globulus	1	757 569	–
Leptospermum laevigatum	2	594 648	8
Hakea suaveolens	3	571 905	14
Pinus pinaster	4	515 735	4
Acacia mearnsii	5	251 359	6
Eucalyptus diversicolor	6	238 452	14
Pinus radiata	7	198 933	7
Eucalyptus camaldulensis	8	195 348	–
Cupressus macrocarpa	9	160 541	–
Eucalyptus gummifera	10	143 216	–
Eucalyptus robusta	11	115 402	–
Acacia saligna	12	104 323	2
Eucalyptus cornuta	13	103 040	–
Quercus robur	14	101 899	14

Modes of dispersal

Successful alien plants in the different biomes are characterized by different modes of dispersal. Dispersal by animals is of cardinal importance in the forest, karoo and desert biomes and important in the savanna biome. Birds are the principal agents except in the karoo and desert where mammals rank most important. In grasslands and fynbos dispersal is mainly by wind though sometimes by animals, mainly birds (Richardson 1985a). Within the fynbos biome bird dispersal is important in Strandveld but relatively unimportant in Coastal Fynbos, Renosterveld and Mountain Fynbos (Knight 1985). Abiotic dispersal is most important in Mountain Fynbos and in Coastal Fynbos. The nutrient-deficient soils of these ecosystems preclude species with fleshy fruits (Milewski and Bond 1982).

Wind is important in mountain areas where the short stature of the grassland and fynbos allows for unimpeded dispersal of winged seed over great distances. Riverine habitats in all biomes host a suite of alien plants that are dispersed in soil and water (Richardson 1985a).

Dispersal distances

The importance of quantitative differences between species in their dispersal ability is readily illustrated by way of example. Van Wilgen and Siegfried (in press) have made a comparative analysis of invasive pines. *Pinus pinaster* and *P pinea* occupy similar habitats in Mediterranean Europe, sometimes occurring parapatrically. Abundant, apparently suitable, habitat is available to both in the south-western Cape, and both have been present since the seventeenth century (Shaughnessy this volume). Yet *P pinea* is an invader of little consequence (eg Taylor et al 1985); its seeds are large (0,6 g) and have relatively small wings, with ratio of total mass to wing surface area ("samara wing loading") of 4,6 mg mm^{-2}. Recruitment of young trees is confined to the immediate vicinity of parent trees. *Pinus pinaster* and *P radiata* share similar seed characters, and contrast with *P pinea*. Seeds have masses of 0,0485 and 0,0195 g respectively, and "samara wing loadings" of 0,25 and 0,19 g mm^{-2}. Dispersal distances for these species in excess of one kilometre are common (Richardson and Brown in press).

Studies of dispersal of seed by wind are often quoted to show how few seeds are dispersed over distances greater than a few times the height of the parent tree (Harper 1977). Whereas it is true that the majority of seed fall close to the parent plant, the few carried in turbulent air are crucial in colonization of new habitat. Such seed must initiate a two-phase invasion, involving the initial establishment of isolated individuals and the subsequent recruitment of dense daughter populations, that is so often observed in hakea and pine species (eg Richardson and Brown in press).

Despite the frequent observation of involvement of animals, especially birds, in the dispersal of introduced plants in South Africa there have been no studies undertaken to determine what distances and rates of dispersal are involved. Glyphis et al (1981) have examined patterns of dispersal by birds of *Acacia cyclops* but did not go beyond demonstrating the strong correlation of plant recruitment with perch sites so often reported for such cases.

Habitat factors

As noted previously, it is difficult to discriminate between the effects of human activity and other historical factors on the one hand and the role of ecological interactions on the other in determining the patterns and rates of invasions by introduced plants. Poynton (1984) records an average of about 51 species of introduced *Acacia*, *Eucalyptus* and *Pinus* in each of the 12 humid silvicultural zones of South Africa and 48 for the 12 subhumid zones, the difference not being statistically significant. On average about 20 species are recorded as invasive in the former case, and about 10 in the latter. The highest incidence of invasive species, 41 out of 53 introduced, was found in zone As3, ie the humid summer-rainfall zone with moderately severe frosts. This is a complex transitional forestry zone with many species of trees used in scattered forestry projects and farm woodlots, so that the high incidence of invasives may be owing to cultural rather than ecological factors. In the arid and semi-arid zones a mere handful of species are recorded as invasive, but far fewer have been introduced.

The distribution of an invasive species within the biogeographic region to which it is preadapted is limited by habitat factors other than climate. For example, an analysis of the distribution of *Hakea sericea* and congeneric invaders in the south-western Cape showed that the most important factor characterizing invaded areas is the presence of soils derived from quartzite and sandstone. Occurrence of *Hakea* species was negatively associated with shales and unconsolidated substrata (Richardson 1984).

Factors determining susceptibility of habitats to invasion have received more critical attention for *Cereus peruvianus* which is invading bushveld of the central Transvaal (Taylor and Walker 1984). Here, the invading cactus is confined to habitats with heavy soils and trees dominated by native species of *Acacia*, and does not occur on adjacent sites with leached sandy soil and broad-leaved trees. Distribution of the weed was related to availability of perch-sites for bird vectors and to conditions required for seedling establishment. Seedlings became established only in areas which have fine-textured soils and a high density of trees, which in turn related to the seasonal duration of available soil moisture in the surface horizon. This combination of factors effectively confined the invasion to acacia savanna, and excluded the cactus from the mixed savanna of oligotrophic sands. Within the acacia habitat, trees and understorey grasses interact to determine soil moisture status and hence cactus establishment.

Habitat factors may sometimes interact with biotic factors to limit distribution. *Pinus radiata*, for example, is susceptible to infection by the needle disease, *Sphaeropsis sapinea* (= *Diplodia pinea*) (Poynton 1979a). Infection of hail-damaged needles often occurs during the warm humid period following hail storms. This effectively prevents the species from invading in the summer-rainfall regions of South Africa where such storms are common.

Fire is a regular phenomenon in fynbos, grassland and savanna ecosystems of South Africa. The role of the fire regime in limiting invasions has been little studied and it is usually considered to be a disturbance factor enabling invasion (eg Macdonald 1984). Clearly fire or a particular fire regime excludes certain potential invaders. Species of *Acacia* and *Pinus* which have a youth period of four years or more and are unable in the juvenile stage to sprout after fire are excluded under normal circumstances from grasslands in which fire recurs at intervals of about two years. Fire may also directly exclude species. Of four species of *Hakea* introduced to the south-western Cape from Australia, one, *H salicifolia*, does not spread despite having being planted extensively in hedges in diverse forest, fynbos and grassland habitats. It may be excluded from grassland and fynbos because its seeds do not survive normal fire temperatures. The small follicles do not provide sufficient insulation against the heat of crown fires. The other three *Hakea* species have large follicles which provide effective protection (Richardson et al in press).

Biotic factors

Herbivores may determine the kinds of introduced species which may succeed in invading a particular habitat and hence the extent of invasion by any given species. Macdonald (1983, 1984, 1985a) has drawn attention to this in South Africa, pointing to the role of the black rhinoceros *Diceros bicornis* in eliminating plants of *Acacia mearnsii* from the river banks in the Umfolozi Game Reserve and of elephants *Loxodonta africana* doing the same for *Opuntia ficus-indica* in the Addo Elephant National Park. Effects of this kind must vary considerably with the identities, abundances, and foraging preferences of the available herbivores. There is as yet no evidence for control by native herbivores or other consumers of the distribution of introduced plant species analogous to that of introduced biological control agents on *Opuntia* species reported by Zimmerman et al (this volume).

Richardson (1985b) has quantified losses of *Hakea sericea* seed after the mass release when the parent stand is killed amounting to 99-99,9% of the 55-174 seed m^{-2} available. This loss was ascribed to rodents, but even such heavy predation has not prevented invasion, most likely because the rodents, even when abundant, do not find the last seeds.

There is tentative evidence that competition and interference between native and the alien invasive plant species may also determine the rate of invasion by some aliens, if not all. Turning once more to examples from forestry, we find that several actual or potential invaders are extremely sensitive to competition from a vigorous grass sward. *Eucalyptus* species, planted extensively in grassland ecosystems, are intolerant of competition from indigenous grasses even if the grassland has been burnt (Schonau and Stubbings 1983; Schonau 1984) and intensive site preparation is necessary to ensure successful establishment. Pits prepared in otherwise undisturbed grassland are seldom sufficient to ensure adequate survival and growth of transplants. Ploughing to a depth of at least 100 mm to kill grasses is needed to ensure successful establishment of *Acacia mearnsii* (Stubbings and Schonau 1983). Effects of competition with the grass sward have also been invoked as a factor limiting proliferation in *Cereus peruvianus* (Taylor and Walker 1984).

The role of competition at the establishment phase evidently varies between South African biomes as indicated by a survey of forestry establishment problems by Donald (1971). He found that fynbos in general requires little clearing to reduce competition for pines. In contrast, site preparation is important for establishing pines in grassland, but the intensity of preparation required varies for the different regions; the most intensive treatment is needed in the north-eastern Mountain Sourveld of the northern Transvaal (Acocks 1975). Therefore the degree of disturbance required to permit or accelerate invasions would be expected to vary among the biomes, just as the site preparation techniques required in forestry vary.

There are numerous examples where dispersal and sometimes germination is dependent on the action of native or naturalized frugivores. The invasion of species with fleshy fruit (eg *Cereus peruvianus*, *Solanum mauritianum*, *Pittosporum undulatum*) or with funicles (*Acacia cyclops*) follows a typical pattern. Establishment of second-generation individuals is generally restricted to the area around bushes or trees that bear fruit attractive to birds or around elevated perching sites (Glyphis et al 1981; Taylor and Walker 1984; Richardson and Brink 1985).

The bush clumps which arise serve as foci for further bird activity and thus for further seed deposition. The size and complexity of these clumps increase with age and with the number of native and alien plants species involved. *Solanum mauritianum* in Natal, ten years ago confined to access roads and plantation edges, now often forms a complete understorey in wattle and pine plantations (Richardson unpublished). This is due to the fact that the rameron pigeon *Columba arquatrix* is exploiting a new and expanding ecological niche (Oatley 1984), though dispersal by other bird species (I A W Macdonald personal communication) and the common duiker, *Sylvicapra grimmia*, is also implicated (T Allan-Rowlandson personal communication). Native animals may determine invasion in ways other than through dispersal of seed. The structure of native habitats may be markedly affected by activities of native ungulates and this has been invoked as a factor favouring the penetration of forest margins by *Chromolaena odorata* once browsing by nyala *Tragelaphus angasii* has opened the canopy (Macdonald and Jarman 1985).

The abundantly evident invasion of certain South African landscapes tends to belie the significance of biotic factors that act against invasion. The underlying factor permitting the invasions is some deviation from or disturbance to normal community functions. For example, changes in the fire regime in grassland may be one such disturbance, allowing invasion of moribund vegetation by pines and acacias. Another example is *Pinus patula* which is unable to invade vigorous *Themeda triandra* grasslands but does establish in forest margins, and in sparse grass swards induced by shallow soil and excessive burning and grazing (F R Smith personal communication). *Acacia* species tend to become established along streams and on other sites where erosion, shading or other factors tend to debilitate the grass sward.

Demographic factors involved in the process of invasion by plants

An alien invasive plant becomes a problem because populations spread and grow from the initial introduction to the extent that these cause some direct or indirect cost to society. This involves population changes arising from the more favourable birth and death rates in the successful invader, relative to the native species.

Seed crop sizes

Large seed crops in an invasive alien are often contrasted with low fecundity of native populations of the same species and this difference is then invoked as being the reason for its invasion (eg Neser et al this volume). However, the importance of the quantity of seed or other propagules produced over its lifespan or maintained in its seed bank by an invasive plant is not clear. Gill and Neser (1984), for example, confirm larger seed crops for *Hakea sericea* in South Africa than in Australia but found no difference for *Acacia cylops.* Taylor and Walker (1984) found for *Cereus peruvianus* that the ratio of seedlings to seed produced per parent plant was in the order of 1:10 000, about the same as for cacti in their native habitats. They concluded that a high fecundity was of lesser importance than the influence of soil moisture regime on seedling establishment.

Richardson et al (in press) have compared seed production in four species of *Hakea* in the south-western Cape. Of three species that have been declared noxious weeds in South Africa, only *H sericea* is a widespread problem. *Hakea gibbosa* and *H suaveolens* are rapid invaders only in isolated areas, despite widespread propagation following their introduction in the mid-nineteenth century. This discrepancy in invasive ability between congeners with similar, canopy-stored and wind-dispersed seeds has been attributed to differences in seed numbers produced by the three species (Richardson et al in press). For example, average seed loads on a shrub with a dry mass of 14 kg range from about 100 on *H suaveolens* to about 500 on *H gibbosa* and 2 000 on *H sericea.* Such differences are not necessarily important in the maintenance of established populations, where factors governing relative survival rate of seedlings become more important than absolute seed numbers (Harper 1977). They are important in expanding populations, for two reasons. First, the range of dispersal is affected. The relationship between seed density and dispersal distance may be expressed as a negative exponential function. Such a model reveals that with an increase in seed number of around 20%, maximum distance of disperal increases by around 80% (Harper 1977). Dispersal over a great distance is a relatively rare event but the probability of it occurring increases exponentially with an increase in seed numbers. The potential rate of spread is thus significantly affected by seed numbers. This would have been an important factor in facilitating the relatively rapid invasion by *H sericea* of its present range and no doubt also of *Chromolaena odorata* (Liggit 1983). Second, high fecundities may be assumed to be an essential part of rapid recruitment of subpopulations from colonizing individuals at the leading edge of an invasion front. High fecundity, if coupled with high recruitment, is also important for outcrossing colonizing species because rapidly expanding populations regain genetic diversity relatively quickly (Barrett and Richardson 1986). The greater the number of seeds, the greater the chance that a successful new genotype will be added to the population and, with qualifications, the greater the number of genotypes accruing to the population (Templeton and Levin 1979; Cavers 1983). These aspects have not yet been studied in the species which invade South African ecosystems.

Growth to maturity

The age at which a plant first reaches reproductive maturity has relevance principally to disturbance regime, as noted above for fire regime. Invasive aliens enjoy no special advantage in this respect. Species such as *Pinus pinaster* (Kruger 1977), *Hakea sericea* and *Chromolaena odorata* mature early, at ages between two and six years. However, native species are equally precocious for example in *Protea* (Kruger and Bigalke 1984). Nevertheless, late attainment of reproductive age will often preclude invasion by an introduced species. The low fecundity in *Hakea suaveolens* results from a long youth period. *Eucalyptus nitens*, favoured for afforestation on the eastern Transvaal highveld, seeds for the first time at about 15-16 years (Darrow 1984). Since it is grown on a rotation shorter than this it is likely to invade slowly, if at all.

Survival and mortality

The hallmark of many successful invaders lies perhaps in the fact that crowding of seedling populations causes little increase in mortality. Taylor and Walker (1984) found for *Cereus peruvianus*, for example, that population density and recruitment rate were positively correlated: "there is no negative feedback on new seedling establishment", and populations increased rapidly to very high densities, in excess of 20 000 stems ha^{-1}. The same phenomenon may be found in *Hakea sericea*, *Pinus pinaster* and *Chromolaena odorata* (Richardson unpublished). Mortality in adult populations is also often marked by weak self-thinning; instead, individuals in crowded populations are smaller than normal (K von Gadow personal communication). Such invasive species rapidly displace competitors. It may be that species with these characteristics can invade natural communities with less need for initial disturbance than do invaders with variable mortality at the seedling stage, such as *Acacia* (see Milton 1981) and *Eucalyptus* species.

Conclusions

The relatively limited degree of understanding of the processes involved in the invasion of South African ecosystems by introduced plants prevents any useful generalization at this stage. Comparative studies involving introduced congeners, one successful and one not, could provide the quickest means to making progress with the problem, as would studies involving the same species under contrasting conditions.

If the history of introduction and invasion is anything to go by then many problems lie ahead. An understanding based on much more thorough studies of the processes involved will be needed to predict and manage invaders of the future. Dispersal is crucial. The dynamic relationships between dissemination and animal vectors appear to be complex. South African biomes are distinguished from each other by invaders with different syndromes. It is not clear whether these patterns are determined by habitat or whether the interactions between species will change with time and shift the pattern, presenting new problems with weed control.

There is sufficient evidence that, even for major weeds, clearly quantifiable and hence predictable relationships between habitat factors and invasion may be obtained. Determining these relationships could provide an important tool in evaluating introduced plants in terms of potential spread.

The patterns of success and failure among introduced plants point clearly to the governing role of biotic interactions, disruption of which is often needed before invasion begins. However, there is at least tentative evidence that in some biomes and with certain introduced species the conditions for invasion exist without undue disturbance. Speculation about the relative roles of fire, competition between plants, and herbivory and predation, in permitting or preventing invasion must be addressed experimentally if management of the habitat is to be effectively adjusted to counter biological invasions.

Acknowledgements

This work was undertaken as part of the programme of conservation foresty research of the Department of Environment Affairs. We thank Shane Abbott for assistance in the preparation of data.

Acocks J P H 1975. Veld types of South Africa. Memoirs of the Botanical Survey of South Africa 40, 1-128.

Baker H G 1986. Patterns of plant invasion in North America. In: Mooney H A and Drake J A (eds) (in press). Ecology of biological invasions in North America and Hawaii. Ecological Studies No 53, Springer-Verlag, New York.

Barrett S C H and Richardson R J 1986. Genetic attributes of invading species. In: Groves R H and Burdon J J (eds) Ecology of biological invasions: an Australian perspective. Australian Academy of Science, Canberra. pp 21-33.

Bigalke R C 1980. Vertebrate damage to plantation forests in southern Africa. In: Proceedings, Joint Symposium on plantation forests as wildlife habitat and problems of damage. IUFRO Thessaloniki - Athens pp 14-38.

Cavers P B 1983. Seed demography. Canadian Journal of Botany 61, 3578-3590.

Darrow W K 1984. Provenance studies of frost-resistant eucalypts in South Africa. South African Forestry Journal 129, 31-39.

De Winter B, Vahrmeijer J and Von Breitenbach 1978. The national list of trees. 2nd edition. Van Schaik's, Pretoria.

Donald D G M 1971. Clearing operations in South African Forestry. Forestry in South Africa 12, 55-66.

Harper J L 1977. Population biology of plants. Academic Press, London.

Henderson L and Musil K J 1984. Exotic wood, plant invaders of the Transvaal. Bothalia 15, 297-313.

Gill A M 1978. Fire adaptive traits of vascular plants. In: Mooney H A, Bounicksen T M, Christensen N L, Nolan J E and Reiners W A (Technical Coordinators) Fire regimes and ecosystem properties. USDA General Technical Report WO-26, 208-230.

Gill A M 1981. Coping with fire. In: Pate J S and McComb A J (eds) The biology of Australian plants. University of Western Australia Press, Nedlands, Western Australia. pp 65-87.

Gill A M and Neser S 1984. *Acacia cyclops* and *Hakea sericea* at home and abroad. In: Dell B (ed) Proceedings of the 4th International Conference on Mediterranean Ecosystems. University of Western Australia, Nedlands. pp 57-58.

Gleadow R M 1982. Invasion by *Pittosporum undulatum* of the forests of Central Victoria. II. Dispersal, germination and establishment. Australian Journal of Botany 30, 185-198.

Glyphis J P, Milton S J and Siegfried W R 1981. Dispersal of *Acacia cyclops* by birds. Oecologia 48, 138-141.

Gordon A J 1978. St John's Wort. In: Stirton C H (ed) Plant invaders - beautiful but dangerous. The Department of Nature and Environmental Conservation of the Cape Provincial Administration, Cape Town. pp 84-87.

King N L 1938. Historical sketch of the development of forestry in South Africa. Journal of the South African Forestry Association 1, 4-16.

Knight R S 1985. Attributes of seed dispersal at different fynbos sites in relation to the presence of aliens. Abstract, 7th Annual Research Meeting of the Fynbos Biome Project, Stellenbosch. CSIR, Pretoria.

Knight R S and Macdonald I A W 1985. Seed dispersal of alien acacias by the Black Korhaan. 7th Annual Research Meeting of the Fynbos Biome Project, Stellenbosch. Unpublished presentation.

Kruger F J 1977. Invasive woody plants in the Cape fynbos with special reference to the biology and control of *Pinus pinaster*. Proceedings of the 2nd National Weeds Conference of South Africa. Balkema, Cape Town. pp 47-55.

Kruger F J and Bigalke R C 1984. Fire in fynbos. In: Booysen P de V and Tainton N M (eds) Ecological effects of fire in South African ecosystems. Ecological studies 48, Springer-Verlag, Berlin. pp 67-114.

Le Roux P J 1980. Encroaching plant species in state plantations in South Africa with special reference to their occurrence in the afforested areas of the Transvaal. Proceedings of the 3rd National Weeds Conference, 1979, Pretoria.

Liggitt B 1983. The invasive alien plant *Chromolaena odorata*, with regard to its status and control in Natal. Institute of Natural Resources, Pietermaritzburg. Unpublished report.

Macdonald I A W 1983. Alien trees, shrubs and creepers invading indigenous vegetation in the Hluhluwe-Umfolozi Game Reserves Complex in Natal. Bothalia 14, 949-959.

Macdonald I A W 1984. Is the fynbos biome especially susceptible to invasion by alien plants? – a reanalysis of available data. South African Journal of Science 80, 369-377.

Macdonald I A W 1985a. The Australian contribution to southern Africa's invasive alien flora: an ecological analysis. Proceedings of the Ecological Society of Australia 14, 225-236.

Macdonald I A W 1985b. Mesquite - a 'bomb' ready to explode. Custos 14(6), 35-39.

Macdonald I A W and Jarman M L 1984. Invasive alien organisms in the terrestrial ecosystems of the fynbos biome, South Africa. South African National Scientific Programmes Report No 85. CSIR, Pretoria.

Macdonald I A W and Jarman M L 1985. Invasive alien plants in the terrestrial ecosystems of Natal, South Africa. South African National Scientific Programmes Report No 118. CSIR, Pretoria.

Mack R N 1986. Alien plant invasion into the Intermountain West; a case history. In: Mooney H A and Drake J A (eds) (in press). Ecology of biological invasions in North America and Hawaii. Ecological Studies No 53, Springer- Verlag, New York.

Milewski A V and Bond W J 1982. Convergence of myrmecochory in mediterranean Australia and South Africa. In: Buckley R C (ed) Ant-plant interactions in Australia. W Junk, The Hague. pp 89-98.

Milton S J 1981. Litterfall of the exotic acacias in the South Western Cape. Journal of South African Botany 47, 147-155.

Neser S and Fugler S F 1978. Silky hakea. In: Stirton C H (ed) Plant invaders - beautiful but dangerous. Department of Nature and Environmental Conservation of the Cape Provincial Administration, Cape Town. pp 76-79.

Oatley T B 1984. Exploitation of a new niche by the Rameron Pigeon *Columba arquatrix* in Natal. In: Proceedings of 5th Pan-African Ornithological Congress, 323-330.

Poynton R J 1979a. Tree planting in southern Africa: Vol 1. The pines. Government Printer, Pretoria.

Poynton R J 1979b. Tree planting in southern Africa: Vol II. The eucalypts. Government Printer, Pretoria.

Poynton R J 1984. Characteristics and uses of selected trees and shrubs cultivated in South Africa. Directorate of Forestry, Pretoria.

Richardson D M 1984. A cartographic analysis of physiographic factors influencing the distribution of *Hakea* spp in the south-western Cape Province. South African Forestry Journal 128, 36-40.

Richardson D M 1985a. Major alien plants of terrestrial biomes in southern Africa: Dispersal modes and patterns of invasion. Unpublished report 85/28. Jonkershoek Forestry Research Centre, Stellenbosch.

Richardson D M 1985b. Studies on aspects of the integrated control of *Hakea sericea* in the south-western Cape Province, South Africa. MSc Thesis, University of Cape Town, Cape Town.

Richardson D M and Brink M P 1985. Notes on *Pittosporum undulatum* in the south-western Cape. Veld and Flora 71, 75-77.

Richardson D M and Brown P J (in press). Invasion of mesic mountain fynbos by *Pinus radiata*. South African Journal of Botany.

Richardson D M, van Wilgen B W and Mitchell D T (in press). Aspects of the reproductive ecology of four Australian *Hakea* species in South Africa. Oecologia.

Roughgarden J and Diamond J 1985. Overview: the role of species interactions in community ecology. In: Diamond J and Case T J (eds) Community ecology. Harper and Row, New York.

Rycroft H B and Wicht C L 1948. Field trials of geographical races of *Pinus pinaster* in South Africa. Journal of the South African Forestry Association 16, 55-66.

Salisbury E J 1961. Weeds and aliens. Collins, London.

Schonau A P G 1984. Silvicultural considerations for high productivity of *Eucalyptus grandis*. Forest Ecology and Management 9, 295-314.

Schonau A P G and Stubbings J A 1983. Silviculture of Eucalypts. Forestry Handbook 73-80. SAIF.

Scott L and Van Zinderen Bakker E M Sr 1985. Exotic pollen and long-distance wind dispersal at sub-Antarctic island. Grana 24, 45-54.

Steyn D J 1980. Bushpigs selective in rooting pines. Forestry News 2/80, 22.

Stirton C H (ed) 1978. Plant invaders. Beautiful, but dangerous. Department of Nature and Environmental Conservation of the Cape Provincial Administration, Cape Town.

Stubbings J A and Schonau A P G 1983. Silviculture of Black Wattle. Forestry Handbook 81-90. SAIF.

Tarr P W and Loutit R 1985. Invasive alien plants in the Skeleton Coast Park, western Damaraland and western Koakaland. In: Brown C J, Macdonald I A W and Brown S E (eds) Invasive alien organisms in SWA/Namibia. South African National Scientific Programmes Report No 119. CSIR, Pretoria. pp 19-23.

Taylor S E and Walker B H 1984. Autecology of an invading population of the cactus *Cereus peruvianus* (Queen of the night) in the central Transvaal. South African Journal of Botany 3, 387-396.

Taylor H C, Macdonald S A and Macdonald I A W 1985. Invasive alien woody plants in the Cape of Good Hope Nature Reserve II. Results of a second survey from 1976 to 1980. South African Journal of Botany 51, 21-29.

Templeton A R and Levin D A 1979. Evolutionary consequences of seed pools. American Naturalist 114, 232-249.

Van Wilgen B W and Siegfried W R (in press). Dispersal properties of three alien pine species in South African fynbos. South African Journal of Botany.

Chapter 12

Seed biology of invasive alien plants in South Africa and South West Africa/Namibia

S J Dean*, P M Holmes* and P W Weiss†

*University of Cape Town and †Woden College of Technical and Further Education, Australia.

Introduction

The presence in most plants of morphological, chemical and phenological traits which facilitate dispersal indicates strong selection for separation of the offspring from the parent (Balda 1980; Frost 1980; Howe and Smallwood 1982). A seed germinating remote from the parent plant may escape severe infestation by host-specific pathogens, parasites and predators (Janzen 1969) and avoid competition with the established plant for light and water. Thus far-flung progeny, if adequately pollinated, should have a higher reproductive output than those close to the parent plant. Seed dormancy could be considered a form of long-distance dispersal in that it separates parent and offspring in time. Germination cues such as exposure of the seed to heat or light, scarification, or leaching of inhibitors, permit germination after loss of the canopy or translocation of the seed.

Translocation of plants by man within continents has given rise to plant populations with low pathogen and predator loads, large crops of viable seed and consequent competitive advantages over local vegetation. The invasive behaviour of translocated *Acacia cyclops*, *A dealbata* and *Pittosporum undulatum* in Australia (Gleadow and Ashton 1981; Kloot 1985) as well as the health and fecundity of *Virgilia oroboides* and *Dodonea viscosa* outside their natural ranges in South Africa are examples of the response of some species to establishment far from parent populations.

The world-wide problem of invasion of natural vegetation by alien plants may be seen as a large-scale demonstration of the advantage certain plant species may derive from long-distance dispersal of seed or propagules. Alien plants are most successful in their host countries when growing in environments similar to those in which they evolved (Milton 1980). There is often considerable similarity in the seed biology of indigenous and alien species within a habitat. Differences in the recruitment of alien and indigenous plants must therefore be related to differences in seed availability or seedling survival.

Prolific production of long-lived seeds or vegetative disseminules characterizes the majority of the 789 alien plant species naturalized in South Africa (Wells et al this volume). These features permit seed stores of some alien species freed of their native seed predators and diseases, to build up to greater levels than those in indigenous plant communities. The species which dominate the seed bank have a competitive advantage, particularly when disturbance creates opportunities for recruitment (Bond and Slingsby 1984). In this way the species with the greatest number of viable seeds become invasive and eventually replace the original vegetation.

In this paper a number of the more aggressive alien weeds of the indigenous flora of southern Africa are grouped into broad categories defined in terms of seed biology (Table 1). Available information on crop sizes, seed store, dormancy mechanisms and dispersal agents is summarized for each category. These groupings sometimes transgress taxonomic boundaries, and inclusion of poorly documented species in one or other category is sometimes debatable as is the validity of the categories themselves. However, the classification serves both to summarize available information and to identify deficiencies in our knowledge of these plants and of the processes of invasion. The frequency of invasive alien species with the different types of reproductive biology is given for six contrasting habitats in South Africa and South West Africa/Namibia (Figure 1).

Address for correspondence: S J Dean (née Milton), Department of Botany, University of Cape Town, Private Bag, Rondebosch, 7700.

Canopy-stored seed

Canopy storage of seed (serotiny) is common in zones with winter rainfall and summer drought and

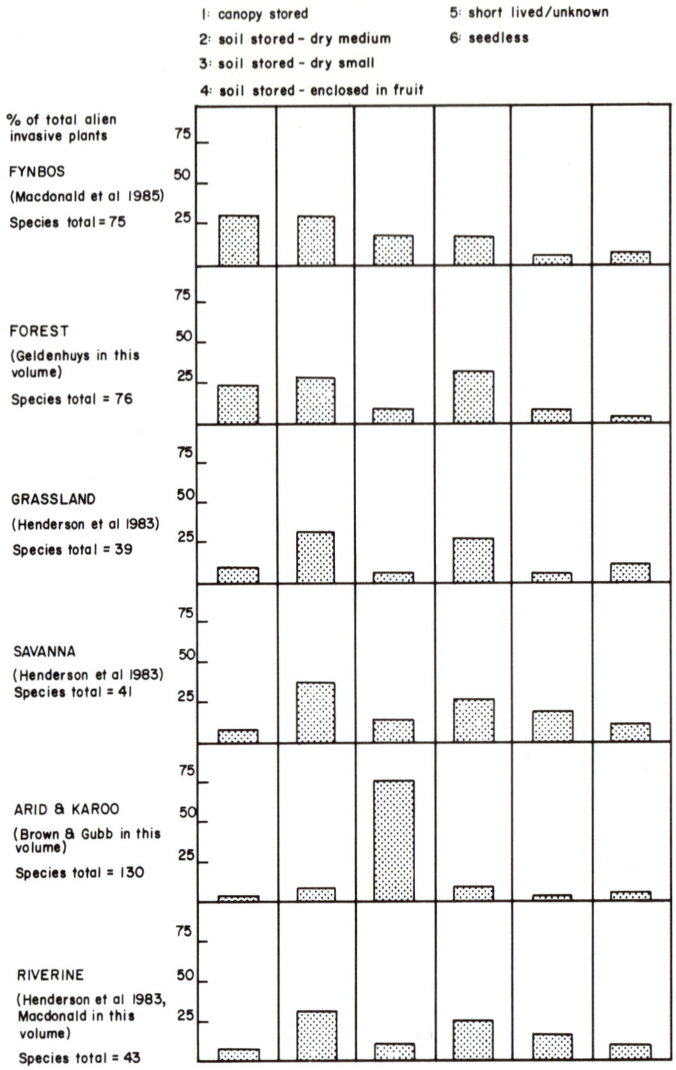

Figure 1. Distribution of five seed types and seedlessness in six vegetation types.

where hot fires occur periodically. The dormancy of seed in serotinous species is enforced as long as the seed is enclosed in the woody fruit or inflorescence. Seed accumulates on the parent plant until released by desiccation of the cone or follicle caused by weather or by death of the plant or organ. Few of the serotinous species regenerate vegetatively and their mass mortality after fire results in mass release of seed into an environment relatively free of predators and plant competition (Bond 1985).

Seeds of serotinous plants are generally wind dispersed (Table 1): biotic dispersers are scarce after fire but serotinous fruits on branches bare of foliage are well placed for wind dispersal. The winged seeds of *Pinus pinaster* can travel at least one kilometre in the wind (Kruger 1977; Kruger et al this volume) and those of *P radiata* up to 16 km (Bannister 1965).

Despite continual depletion by coevolved predators (Neser and Annecke 1973; van Staden 1978) seed stores of serotinous species may exceed the current seed crop (Bond 1985). When growing as aliens, free of coevolved defoliators and seed predators, serotinous plants store large quantities of viable seed. Seed crops of mature *Hakea sericea* trees are in the region of 7 500 seeds m^{-2} soil surface (Kluge 1983), and there are 600 trees ha^{-1} in dense mature hakea stands (Fenn 1980). In a younger *H sericea* stand, with a density of 0,5-0,9 plants m^{-2}, seed release after felling was between 50 and 170 seeds m^{-2}. Even though over 99% of this seed may be destroyed by predators in the 12 months prior to germination, the density of the regenerating stand would be no less than that of the parent stand (Richardson 1985).

Whereas *Hakea* and *Leptospermum* seeds (Johnson 1978) are poorly defended against predation and fungal attack after release, seeds of those *Pinus* and *Eucalyptus* species which are sporadically released during hot windy conditions or after dieback of branches may have a longer life expectancy in the soil (Gerber and Herbst 1982). The latter two genera invade indigenous flora outside the fynbos biome as well as within it, possibly because they are not wholly dependent on occasional hot fires for successful regeneration. Stratification is needed to break dormancy in *Pinus* species and in those *Eucalyptus* species that originated in cool, moist regions eg *E nitens* (Immelman et al 1973; Gerber and Herbst 1982).

Of the alien serotinous species only the Myrtaceae resprout after fire. The low capacity for vegetative regeneration together with the relatively long period between seed germination and seed production (two to three years for *H sericea*, four to five years for *Leptospermum*, five to ten years for *H suaveolens*, *Eucalyptus* and *Pinus* species) (Table 1) excludes all but *Eucalyptus* from fire regimes with annual or nearly annual periodicity. Vegetation types such as forest where fire is rare are equally unsuitable for serotinous species because of the low survival rate of seeds and seedlings under an established canopy.

Strongly serotinous species have successfully invaded indigenous vegetation only in the south-western and southern Cape, regions where canopy storage of seed is common among endemic species (Bond 1985). Seeds of the alien *Hakea sericea* contain greater phosphorus reserves than those of the indigenous Proteoideae tested (Mitchell and Allsopp 1984). This may provide an early competitive advantage for the alien over the indigene in the nutrient-poor soils of the south-western Cape. Further work indicates that *Hakea gibbosa*, *H suaveolens* and *H salicifolia* also have higher phosphorus levels than indigenous Proteoideae (Richardson unpublished).

Jacaranda mimosifolia has woody fruits which open within a year to release small winged seeds which germinate without pretreatment (Gerber and Herbst 1982) and are probably short lived. *Jacaranda* coppices in response to felling and fire (Macdonald 1983a) and is spreading into moist savanna in Africa and India. *Jacaranda* resembles *Eucalyptus* in that its woody fruits serve as a temporary protection against predation and unseasonal fires rather than as a mechanism for the accumulation and mass release of seed.

Soil-stored seed

Soil-stored seeds survive in a dormant condition in the soil for at least a year. The mechanisms controlling dormancy vary with plant family. Many members of the Fabaceae are hard-seeded and will germinate as soon as the water-impermeable testa is damaged by heat, abrasion or decay. Dormancy broken by chemical changes in the embryo brought about by light, temperature fluctuation, carbon dioxide levels or ageing processes, is found in the Asteraceae, Solanaceae, Poaceae and Rosaceae.

While family affiliation partly determines the dormancy behaviour of a species, selection by dispersal agents, climate and disturbance regimes modify it further. Herbaceous annuals exist solely as seed under conditions unfavourable for plant growth. Their abundant small, dormant seeds germinate in response to cues which indicate the return of conditions suitable for growth. Many annuals have polymorphic seeds, which by responding to different germination cues prevent rapid depletion of the seed bank. Perennial plants invest more energy in individual seeds, the dormancy of which is broken by factors which indicate reduction of competition. In southern Africa alien plants with soil-stored seed can be grouped as follows: species with small dry seeds, species with medium sized dry seeds and species with seeds embedded in soft fruit.

Small dry seeds

Abundant production of small long-lived seed is characteristic of short-lived plants adapted for rapid opportunistic colonization. The group includes ephemerals of desert environments as well as ruderals. The seed banks of these plants are generally larger than those found in perennial plant communities: 10 000-40 000 seeds m^{-2} for South African desert ephemerals and up to 86 000 seeds m^{-2} in the arable lands of Europe and America (van Rooyen and Grobbelaar 1982).

Table 1. Families, origin and seed characteristics of some alien plants invasive in natural vegetation of eight habitats in southern Africa.
1: Fynbos 2: Grassland 3: Savanna 4: Forest 5: Karoo and desert 6: Riverine 7: Ruderal 8: Aquatic

Species	Family	Place of origin	Habitat invaded	Seed size	Store seed m^{-2} $\times 10^{-3}$	Veg propa-gation	Dispersal agent	% Seed dormant or held	Generat time (yr)	Seed life (yr)	Dormancy breaker	Reference
CANOPY STORED SEED												
Pinus pinaster	Pinac	S Europe	1	m	1-3	x	wind,baboon	some	5-10	1	rel;strat	31
Pinus radiata	Pinac	California	1	m	3	x	wind	some	10	many	rel;strat	4
Pinus halepensis	Pinac	W Medit	1	m		x	wind	some	–	10	rel;strat	1
Leptospermum laevigatum	Myrt	SE Australia	1	s		x	wind,water		4-5	1+	rel	44
Hakea sericea	Prot	SE Australia	1,4	m	7,5	x	wind	all	2-3	1+	rel	40,29
Hakea gibbosa	Prot	SE Australia	1	m		x	wind	all	1-2	1+	rel	44
Hakea suaveolens	Prot	SW Australia	1	m		x	wind	all	6	1+	rel	44
Eucalyptus cladocalyx	Myrt	S Australia	1,4	s		c,x	wind	some	3+	10+	rel	17,1
Eucalyptus lehmanii	Myrt	SW Australia	1	s		c,s	wind	some	3+	10+	rel	17
Eucalyptus grandis	Myrt	E Australia	2,3,1	vs		c,s	wind	some	3+	10+	rel	17,1,12
SMALL DRY SEED: SOIL STORED												
Bidens bipinnata	Aster	Asia, USA	7,5,6,4	s		c	wind,mammal	10+80	1	1+	lt,sc,lch	7,2
Bidens pilosa	Aster		7,5,6,4				wind, mammal			1+	lt,sc,lch	
Chromolaena odorata	Aster	Cent America	4	s		c,s	wind			1+	lt,alt T	13,34
Tagetes minuta	Aster	S America	7,5,6	s			wind	70-100	1	1+	lt,T	46
Stipa trichotoma	Poac	Argentina	1,2	s	30-50	d	wind,water	90	1	20	lt	26,27,44,49
Hypericum perforatum	Gentian	Europe	1,7,6	vs	1-6	c	water,mammal		1	6-10	lt	30,47,44
Acanthospermum spp	Aster	S America	7,5	s		c,r	mammals		1			
Argemone mexicana	Papaver	Mexico	7,5,6	s			water		1			
Altenanthera pungens	Amaranth	S America	5,7	s		c	mammals		1			
Nicotiana glauca	Solanac	Argentina	5,6,2,3	vs		c	wind,water		1			44
Nerium oleander	Apocyan	S Mediterra	5,6	s		c	wind,water					44
Datura innoxia	Solanac	–	5,6	s			water		1	20		6,32
Cortaderia jubata (female)	Poac	S America	2	s(haploid)		d	water,wind					42
Azolla filliculoides	Salvin	S America	8	vs(spores)		d	water,birds		1			3
MEDIUM SIZED DRY SEED: SOIL STORED												
Acacia saligna	Fabac Mim	SW Australia	1,4	m*	10-13	c	water,mammal	90-100	2	10+	hear or	35,33,5
Acacia longifolia	Fabac Mim	SE Australia	1,4	m*	6-8	x	water, bird	95-100	2	5+	scarif	35,48,39

Acacia cyclops	Fabac Mim	SW Australia	1	m**	2-18	x	bird,mammal	80-100	2	2+	cracking piercing	20,35,19, 10,5
Acacia mela-noxylon	Fabac Mim	E Australia	4,6,1	m**	3-90	c	bird,water	95-100	5	50+	of testa by	18,11,15, 24
Acacia mearnsii	Fabac Mim	SE Australia	6,1,2,3,4	m*	20-30	c	water	90-100	3-6	50+	fire,sun	23,44
Acacia dealbata	Fabac Mim	SE Australia	6,1,2,3,4	m*		c	water	95-100	5+	100+	heat digestion	38
Acacia decurrens	Fabac Mim	E Australia	1,2,3,6	m		c	water	95-100			abrasion or acid	33,38
Albizia lophantha	Fabac Mim	W & SW Aust	1,6,4	m	4	x	water	100	1			13,44
Spartium junceum	Fabac Pap	S Europe	1	m		x			1			36
Caesalpinia decapetala	Fabac Caes	India	3,6,4	m		c	water,cattle					34
Cassia spp	Fabac Caes	S Asia	3,6,4	m								
Alhagi camelorum	Fabac Pap	Central Asia	5	m			cattle					
Prosopis cf *glandulosa*	Fabac Mim	SW USA	5,6,2,3	m		c	cattle					28,1,44
Sesbania punicea	Fabac Pap	Argentina	6,5,3,1	m		x	water	55-65	1			14,22,21
Ricinus communis	Euphorb	Asia, Africa	5,6,3,2	m*		c	birds?		1			
SEED EMBEDDED IN SOFT FRUIT: SOIL STORED												
Opuntia ficus-indica	Cactac	Mexico	5,2,6,4	s		cladode	mammal,bird					44
Pereskia aculeata	Cactac	Mexico	4	s		cladode	birds					
Cereus peruvianus	Cactac	Peru	5,2,3	s	1	cladode	birds					45
Solanum mauri-tianum	Solanac	trop Asia	4,6,3,2,1	m		c,s	birds		2-4	10	lt,alt T	9,34,37,43
Cestrum laevigatum	Solanac	Brazil	3,6,4	s		c	bird		1+			
Psidium gujava	Myrtac	W Indies	4,6,3	m		c,s	bird		1+			34
Lantana camara	Verbenac	S America	4,6,3,5,1	m		c,s	bird		1+			44
Pittosporum undulatum	Pittosp	S Australia	4,1	m			bird		1+			41
Rubus cuneifolius	Rosac	N America	4,3,6,2,1	m		c,s	bird		1+		scar,strat	8,34
Pyracantha spp	Rosac	S Europe	2,6	s			bird		1+		strat	17
Solanum sea-forthianum	Solanac	trop America	3	s		c	bird					
Rosa rubiginosa	Rosac	Europe	3,6,4	s		c r					strat	
Schinus molle	Anacard	Peru	5,6,2,3	m		c	bird		2+		heat	17,25
Physalis peruviana	Solanac	Peru	5,7	s		x	bird		1			
SHORT LIVED SEED												
Cedrela toona	Meliac	trop Asia	3,6,4	s		s	wind	nil			rel	17,25
Melia azedarach	Meliac	SW Asia	6,4,5,2,3	m		c,s	birds	nil			rel	1,17
Jacaranda mimosifolia	Bignon	Argentina	3,6,4	m		c	wind	nil			rel	17,34
Schinus tere-binthifolius	Anac	Brazil	4,6,4	m		c	bird					17

(continued)

Table 1. Families, origin and seed characteristics of some alien plants invasive in natural vegetation of eight habitats in southern Africa.
1: Fynbos 2: Grassland 3: Savanna 4: Forest 5: Karoo and desert 6: Riverine 7: Ruderal 8: Aquatic

Species	Family	Place of origin	Habitat invaded	Seed size	Store seed m^{-2} $\times 10^{-3}$	Veg propagation	Dispersal agent	% Seed dormant or held	Generat time (yr)	Seed life (yr)	Dormancy breaker	Reference
SEEDLESS INVADERS												
Populus alba/ canescens	Salicac	Europe	6,2,1	sterile		s,d	water					25
Salix babylonica	Salicac	S China	6,2	female		s,d	water					25
Eichhornia crassipes	Ponteder	S America	8	s(sterile SA)		d	water		1	15		44
Myriophyllum aquaticum	Halorag	S America	8	female		d	water					44
Salvinia x *molesta*	Salvin	Brazil	8	sterile		d	water					44
Elodea canadensis	Hydrochar	Canada	8	female		d	water					50
Opuntia aurantiaca	Cactac	S America	5,4	sterile		cladode	cattle					2,44

Seed size: m = over 1 000 seeds/100g, s = over 10 000 seeds/100g, vs = over 100 000 seeds/100g

Explanation of symbols used in Table 1:
*seed with small white elaiosome, **seed with large red elaiosome
Vegetative propagation: x = none, c = coppice, s = sucker, r = runner, d = division or rooting of vegetative organ
Dormancy breaker: rel = release from woody fruit, strat = prolonged exposure to damp and cold, scar = scratching or etching of seed coat by mechanical or chemical agents, lt = exposure to white light, alt T = exposure to alternating temperature, T = specific germination temperature requirement.

Number	Authors	Number	Authors	Number	Authors
1	Anonymous 1980	18	Gilbert 1959	35	Milton 1980
2	Arnold 1977	19	Gill 1985	36	Mountjoy 1979
3	Ashton 1977	20	Glyphis, Milton and Siegfried 1981	37	Oatley 1984
4	Bannister 1965	21	Graaf and van Staden 1984	38	Osbourne and Osbourne 1931
5	Holmes and I A W Macdonald unpublished	22	R N Greener unpublished	39	Pieterse 1985
6	Brown et al, 1985	23	Hendry and Van Staden 1982	40	Richardson 1985
7	Brown and Mitchell 1984	24	Howard 1974	41	Richardson and Brink 1985
8	Campbell van Staden 1983a	25	Immelman, Wicht and Ackerman 1973	42	Robinson 1984
9	Campbell and van Staden 1983b	26	Joubert 1984	43	Stam 1978
10	Dennill 1983	27	Joubert and Small 1982	44	Stirton 1978
11	D G M Donald unpublished	28	Jurriaanse 1973	45	Taylor and Walker 1984
12	Drake 1979	29	Kluge 1983	46	Van Staden and Forsyth 1983
13	Erasmus and van Staden 1985	30	Kluge and Gordon 1983	47	Van Zyl and Kluge 1983
14	Erb 1980	31	Kruger 1977	48	Weiss and Milton 1984
15	Farrell and Ashton 1978	32	Landsdell 1927	49	Wells 1977
16	Forsyth and Brown 1981	33	Larsen 1964	50	Willis 1973
17	Gerber and Herbst 1982	34	Macdonald 1983a		

In southern Africa the alien species in this group, mostly of American and European origin, are commonly found as crop and pasture weeds. However, several invaders of natural habitats, such as *Stipa trichotoma*, *Hypericum perforatum*, *Chromolaena odorata* and possibly *Nicotiana glauca*, *Portulaca oleracea* and *Nerium oleander*, are also included. In southern Africa, alien plants with this type of reproductive biology are not restricted to any one climatic zone or fire regime but their success appears dependent on vegetation disturbance and they are more common on fertile soils which permit rapid growth and prolific seed production.

Seed crops of *Stipa trichotoma* in the Cape average 2 500 seeds per plant (Joubert 1984). The small seeds are encased in long-awned glumes which facilitate dispersal by wind, water and mammals. *Hypericum perforatum* plants produce hundreds of thousands of very small sticky seeds during their long flowering season. *Chromolaena* has small wind-borne achenes and the achenes of the annuals *Bidens bipinnata* and *Tagetes minuta* are small with a bristly pappus which probably aids dispersal by mammals and water. *Nicotiana glauca* bears great numbers of very fine seed, which are dispersed by wind and water.

Ruderals and short-lived plants appear to have few host-specific parasites or pathogens (Moran et al this volume). Of the small-seeded perennial plants invasive in southern Africa, *Stipa trichotoma* and *Hypericum perforatum* are known to have a suite of natural enemies in their places of origin (Clark 1953; Wells 1977; Kluge and Gordon 1983). Host-specific seed predators and pathogens appear to be poorly known for the other small-seeded species invasive in southern Africa.

Seed polymorphism is a common occurrence among plant species in this category. The Asteraceae in particular are known for the production of seed morphs with differing germination requirements and which originate on different parts of the capitulum inflorescence (Brown and Mitchell 1984).

In the Cape up to 25 000-50 000 seeds m^{-2} of *S trichotoma* accumulate in the top 50 mm of soil. These seeds have life expectancies of up to 20 years (Joubert 1984). About 30% of this store is expected to survive any fire. *Hypericum* stands have an average of 1 000-6 000 seeds m^{-2} with a life expectancy of six to ten years (Kluge and Gordon 1983). Less than 30% of *H perforatum* seed remains dormant, the balance being capable of immediate germination (van Zyl and Kluge 1983). *Chromolaena odorata* rapidly colonizes burned and cleared land and there is evidence for a large soil seed store (Macdonald 1983a; Erasmus and van Staden 1985).

In these small seeds, dormancy is physiological. Germination may be induced by temperature fluctuations, eg 20-25° C for *Tagetes minuta* (van Staden and Forsyth 1983) and 15-30° C for *C odorata* (Erasmus and van Staden 1985). Temperature requirements for germination may be subject to a light requirement and the response is often more marked in imbibed seeds. *Chromolaena* (Erasmus and van Staden 1985), *Stipa* (Joubert and Small 1982), *Hypericum* (Kluge and Gordon 1983), *Tagetes* (van Staden and Forsyth 1983) and some forms of the polymorphic achenes of *Bidens bipinnata* (Brown and Mitchell 1984) and *B pilosa* (Forsyth and Brown 1981) all require light for germination. Chemical inhibitors in the seed coverings have been postulated for *Stipa* and *Bidens*. Germination of seeds of *Hypericum perforatum* is inhibited under the canopy of the parent plant. Dormancy occurs in over 90% of *Stipa trichotoma* seeds and is broken by damage to or removal of the lemma (Joubert and Small 1982).

Longevity of seeds in this category is not known for all aliens invading indigenous vegetation. However, small seeds can survive under natural conditions in the soil for up to a century as has been shown experimentally for species of *Malva*, *Oenothera* and *Rumex* (Kivilaan and Bandurski 1981) in North America.

Mass producers of small long-lived seed have even greater potential as weeds when they are able to grow after fire or slashing from roots or stem bases, as in the case of *Stipa trichotoma*, *Hypericum perforatum* and *Chromolaena odorata*. Two other invasive alien plants, *Nicotiana glauca* and *Nerium oleander*, may belong in this category since they produce abundant very small seeds, have long flowering seasons, resprout after damage and tend to invade bare areas, particularly watercourses. Information on their seed biology appears to be lacking in the literature.

Medium-sized dry seeds

An estimated 85% of the Fabaceae have water-impermeable seed dormancy (Rolston 1978). Of the 481 alien plant species invading natural vegetation in South Africa, 66 belong to the Fabaceae (Wells et al 1983). These include many of the woody weeds which at present have the greatest impact on the fynbos and on river systems throughout the country (Macdonald 1983b). Hard-seededness protects the seed from fungal attack and by preventing water imbibition allows it to remain dormant and viable for a century or more.

Hard-seeded species of Mediterranean climate

regions are often myrmecochorous, that is, they bear elaiosomes which attract ants to collect and bury seeds in caches 40-200 mm below the soil surface. Here seeds are beyond the reach of predators and they survive fire to germinate after a burn (Majer 1978; Milewski and Bond 1982; Bond and Slingsby 1983). In myrmecochorus acacia species the funicle is white, swollen and oily and serves as an elaiosome (Table 1). *Acacia* species of savanna and arid vegetation types, where hard-seeded species are dispersed by mammals (Lamprey 1967; Janzen 1982), have inconspicuous funicles. *Acacia cyclops*, *A melanoxylon* and *A implexa*, which occur in dune or forest communities seldom exposed to fire, have enlarged red funicles attractive to frugivorous birds (Knight this volume). The development of a showy funicle in members of a genus that lacks the genetic potential to produce succulent fruit indicates the importance of dispersal to the survival of plant species.

All of these dispersal mechanisms have been shown to allow part of the fairly large seed crops produced annually by hard-seeded species to escape the insect and mammalian predators which in their natural habitat destroy most of any seed exposed on the soil (Halevy 1974; van den Berg 1980; Bond and Slingsby 1984; Gill and Neser 1984).

Australian acacias generally produce far larger seed crops in South Africa than in their native habitats. For example, mature *A longifolia* stands in the south-western Cape yield 2 500-3 500 seeds m^{-2} annually, but similar stands in their natural habitat yield only 300-400 seeds m^{-2} (Weiss and Milton 1984). These differences can be attributed to predispersal predation of seed by Hymenopteran and Coleopteran larvae (Weiss 1983). Seed production of *A cyclops* in Australia (1 000-5 000 seeds m^{-2}) is similar to that in the Cape (1 500-3 000 seeds m^{-2}) (Milton and Hall 1981; Gill and Neser 1984) although severe predispersal destruction of seed by insects occurs patchily in the natural habitat (van den Berg 1980). Gall rust severely reduces reproductive output of *A saligna* in south-western Australia (van der Berg 1977) but in South Africa this species when mature produces 5 000-10 000 seeds m^{-2} annually (Milton and Hall 1981).

The difference in the ratio of seed produced to seed stored of acacias in Australia and South Africa may explain the invasive behaviour of alien acacias in this country. Ratios measured for *A cyclops* in Australia were 23 and 53, compared with 1,0 in South Africa (Gill and Neser 1984). In Australia five to seven viable seeds m^{-2} were found in *A longifolia* seed banks, as compared with 6 000-9 000 seeds m^{-2} under similar stands in South Africa. The ratios of seed production to storage were 60 and 0,4 respectively (Weiss 1983; Weiss and Milton 1984). *Acacia melanoxylon* seed density beneath mature trees of this species in the southern Cape reaches 90 000 seeds m^{-2} (Milton and Hall 1981), a thousand times higher than in the blackwood forests of Australia and Tasmania (Howard 1974; Farrell and Ashton 1978). Although high seed densities have been recorded for some Australian acacias in their natural habitat, up to 3 900 seeds m^{-2} in *A victoriae* (Drake 1979; Grice 1985), the stability of these seed banks is unknown. Successive studies of alien acacias in the southern and south-western Cape have consistently revealed large seed crops and seed stores (Phillips 1927; Shelton 1975; Milton and Hall 1981; Holmes and I A W Macdonald unpublished; D G M Donald unpublished).

Bruchid beetles are the major destroyers of seed of African and South American acacias (Janzen 1969; Centre and Johnson 1974), but in Australia, most of the parasites and predators of acacia seeds are weevils, moths, wasps and bugs (van den Berg 1977; Auld 1983). Because most predators of legume seed are highly specific few switch to alien hosts. Whereas about 90% of the seed of indigenous South African acacias is destroyed by insects, the seed of the alien species of Australian origin is seldom attacked (van den Berg 1977). The reverse is partly responsible for the rapid spread of the African species, *A nilotica* through Queensland, Australia (Harvey 1981). Seed stores of *Sesbania punicea* in the south-western Cape exceed 3 000 seeds m^{-2} (R N Greener unpublished) and seed viability exceeds 70%. However, in South America where the species originated, up to 90% of the fruit is parasitized by weevils and attacked by at least four other species-specific predators (Erb 1980).

The effect of predispersal predation on the dynamics of the seed banks of alien acacias in the Cape should soon become apparent as the successful introduction of *Trichilogaster acaciaelongifoliae*, a host specific parasite of *A longifolia* has reduced and even eliminated seed production in some stands of this species here (Neser et al this volume).

Hard seeds are well adapted for dispersal by herbivores. Cattle have spread *A nilotica* in Queensland (Harvey 1981) and *Prosopis* hybrids throughout the arid parts of southern Africa (Stirton 1978).

Primates and birds, both alien and indigenous, have facilitated the spread of hard-seeded alien plants in South Africa (Glyphis et al 1981; Pieterse 1985). These dispersers may negate biological control measures by transporting viable seed to predator-free sites. Man's activities, particularly horticulture and road building, account for much of the spread of hard-seeded alien plants as he both transports the seed and creates conditions suitable for its establishment by clearing and burning natural vegetation. Ants are important dispersers of acacia seed in Australia, but evidence for seed caching by ants, such as the clumped distribution of seeds or of seedlings appearing after fire (Majer 1978; Gill 1985) has not been reported for alien acacias in South Africa. Fires in alien acacia stands in the Cape result in uniformly dense, even-aged recruitment.

Water-impermeable seed may be scarified by any process which cracks or abrades the testa and allows the embryo to imbibe water. In nature dormancy is generally broken by damage to the seed coat during ingestion by vertebrate dispersers, by abrasion in the soil, or exposure to solar radiation or fire (Brown and Booysen 1969; Rolston 1978). Few hard-seeded species are shade tolerant: seedlings germinating under dense cover become etiolated and die (Phillips 1927; Milton 1982). Hard seeds can remain dormant for decades under cool moist conditions, but exposure to high temperatures or to the sun's rays breaks dormancy. In this way mass germination of hard-seeded plants follows removal of the cover by fire, felling, erosion or overgrazing. Hard-seeded plants are therefore successful colonizers wherever periodic removal of vegetation permits their establishment: fynbos, disturbed forest, riverine, grassland and arid environments.

Hard-seeded plants such as *Acacia* species which can fix atmospheric nitrogen and which have a high relative growth rate, have an early competitive advantage after fire, because they can utilize the increased supplies of phosphate from the ash (Barrow 1977). In contrast, plants with high seed phosphorus content (eg Proteaceae) do not benefit greatly from phosphorus in the ash.

Generation times of plant species are a response to the frequency of disturbance in their natural habitat. *Sesbania punicea* and *Albizia lophantha* which occur on levees scoured by periodic floods have large seeds and rapid life-cycles, flowering within two years of germination. Longer-lived species of more stable environments, eg *A melanoxylon*, take five years or more to reach reproductive maturity. Many hard-seeded species are obligate reseeders, but some are capable of vegetative regrowth (*A saligna*, *A mearnsii*) and others can multiply vegetatively by suckering (*A melanoxylon*, *Prosopis* species) or rhizome division (*Canna* species).

Longevity of the seed in hard-seeded species is variable. Reports indicate a century for *A dealbata*, at least 50 years for *A melanoxylon* and *A mearnsii* and 10-12 years for many African acacias (Gilbert 1959; Sherry 1971; Howard 1974; Nongonierma 1978). The percentage of dormant seeds ranges from 98-100% for *A mearnsii* and *A longifolia* to a low of 13% for *S punicea* (Milton and Hall 1981; Hendry and van Staden 1982; Graaff and van Staden 1983). The degree of dormancy in the seeds of a given species appears to be genetically determined since the frequency of dormant seeds in the seed crop of a species is rapidly altered in the process of domestication of field crops (Harper 1965). The percentage of seeds remaining dormant for more than one season may be reduced through mild damage by predators or dispersers which stimulates early germination thereby reducing the number of seeds entering the seed bank annually (Halevy 1974; Holmes and Macdonald unpublished).

Seed banks of *A saligna* in the Cape not only accumulate rapidly: they decrease slowly. Following the clearing of an *A saligna* thicket the seed load decreased at the rate of about 1 500 seeds $m^{-2}yr^{-1}$ falling from 11 000 seeds m^{-2} to 2 seeds m^{-2} in six years (Holmes and Macdonald unpublished). Burning can reduce the seed bank of a thicket to 10% of its original level by killing embryos or stimulating germination. Where regeneration was left to grow, the seed bank was restored to its previous size within seven years (Holmes and Macdonald unpublished).

The rate of decline in the seed bank probably decreases with time. Changes in seed distribution with soil depth over time indicate early loss of seed from the litter and surface soil and possible downward movement of seed through the soil (Milton and Hall 1981). Six years after an *A saligna* stand had been cleared, seed density in the top 40 mm of soil had decreased by 95% but at 80-120 mm by only 43%. Below 120 mm seed density had remained constant (Holmes unpublished).

Seed banks in mature monospecific thickets of *A cyclops* in the Cape range from 2 000-18 000 seeds m^{-2} (Holmes and Macdonald unpublished). The reason for the slow rate of seed accumulation at some sites is not clear, but could be the result of its utilization by a local seed-sucking bug. Seeds damaged by the bug become water-permeable so that they germinate or rot rather than entering the seed bank (Dennill 1983; Holmes and Macdonald unpublished).

Seed embedded in soft fruit

Species with soft fruits dispersed by birds or mammals are usually restricted to fertile soils. The parent plants are usually long-lived and occur in environments where fire is rare (Milewski and Bond 1982). Seeds in their fruits tend to be short-lived (Geldenhuys 1975). Most of the equivalent alien species in South Africa are exceptions in that they have dormant, long-lived seeds. Introduced from both temperate and tropical regions mainly as ornamental and hedge plants, the alien fruiting species have become important invaders of riverine, forest, forest margin, plantation, savanna and bushveld habitats in South Africa (Table 1). The majority of the invasive species with soft fruits are multiseeded, in contrast to indigenous soft-fruited plants which are mostly single-seeded (Knight this volume).

There appears to be no published information on the quantities of seed produced or stored by fruiting alien invasives in South Africa. Seventy per cent of the seeds of a *Rubus* species germinate within one year but the remainder may stay viable and dormant in the soil for many years (Campbell and van Staden 1983a). *Solanum mauritianum* seeds are known to remain viable for at least 10 years (Stam 1978) and there is some indication of dormancy in *Cereus peruvianus* seeds (Taylor and Walker 1984).

Dormancy mechanisms of seeds embedded in fleshy fruits are physiological rather than physical. They appear to have been selected by factors that indicate successful dispersal (removal of inhibitors, scarification) reduction of competition (light, temperature fluctuation) or the passing of the cold season (low temperatures).

Dormancy of many of the Rosaceae is terminated by stratification and this requirement may limit the distribution of species of *Rubus*, *Rosa*, *Pyracantha* and *Cotoneaster* to regions where winter temperatures fall below 5° C (Gerber and Herbst 1982; Campbell and van Staden 1983a). Germination of *Rubus* seed occurs only after both scarification of the hard seed coat and stratification. These prerequisites are met intermittently in an ageing seed store, giving rise to the continual germination of stored seeds. Since *Rubus* does not require light for germination, seeds may emerge from depths as great as 12 cm below soil surface (Campbell and van Staden 1983a).

Dormancy in *Solanum mauritianum* seeds is controlled by the embryo. Seeds germinate only after being subjected to fluctuating temperatures in the range 10-20° C. Germination is further stimulated by exposure of the imbibed seed to light (Campbell and van Staden 1983b) and this probably facilitates rapid re-establishment of this plant after vegetation clearing. In other species, substances in the fruit pulp inhibit germination. These inhibitors are removed in the process of digestion by bird and mammal vectors, or by decay. Passage of the seed through vertebrate digestive systems would also result in scarification of the seed coat, leaching of inhibitors and subjection of the embryo to temperatures ranging from 36-39° C in mammals to 44° C in birds (Welty 1964). The seeds of *Pittosporum undulatum* do not germinate under their own canopy: this appears to be the result of chemical inhibition rather than light requirement since *Pittosporum* germinates under the canopies of other plants as well as growing as an epiphyte in other trees (Richardson and Brink 1985).

Melia azedarach germinates without any pretreatment and it is possible that the seeds of this species, like those of the related *Azedirachta indica*, are oily and remain viable for less than a year (Anonymous 1980). *Schinus terebinthifolius*, which has started to spread beyond gardens in the south-western Cape, is reported to germinate without pretreatment (Gerber and Herbst 1982).

Vegetative regeneration, either by shedding of cladodes (*Opuntia* species, *Cereus peruvianus*), suckering (*Rubus*, *Solanum*, *Melia*, *Psidium*, *Lantana*) or resprouting (*Pittosporum*, *Cestrum*) is a feature of plants in this category.

Short-lived seed

With the possible exception of *Melia azedarach* and *Schinus terebinthifolius* mentioned above and *Jacaranda mimosifolia* and *Cedrela toona* we have been unable to place any of the more important invasive alien plants listed by Moran and Moran (1982), Macdonald (1983b) or Henderson and Musil (1983) into this category.

Seedless invaders

Seven of the 64 invasive species listed in Table 1 are incapable of sexual reproduction, proving that rapid multiplication and effective dispersal can be achieved without seeds. Disseminules which can survive and even photosynthesize under unfavourable conditions and root rapidly in suitable substrates are particularly effective colonizers.

The sterile hybrid cactus *Opuntia aurantiaca* grows from shed cladodes or even sterile fruits, which being small and prickly are transported on the hides of mammals to root where they fall and produce new plants and clones (Zimmerman in Stirton 1978).

Many invaders of streambanks and water bodies are sterile. It is possible that seed provides no competitive advantage in these habitats, where water subdivides and redistributes plants and the probability of desiccation is low. *Salix babylonica* and *Populus alba* x *canescens* are the most important riverine invaders in the Transvaal (Henderson and Musil 1983) although in South Africa the former is represented solely by female plants whereas the latter is a sterile hybrid of which only male plants have been recorded (Immelman et al 1973). Both species sucker abundantly and become established in new sites when shoots or suckers broken off during floods are deposited to root on river banks down stream.

Of the floating aquatic invaders, only *Eichhornia crassipes* and *Azolla filiculoides* reproduce sexually. Fertile *Eichornia* plants have not been recorded in South Africa and *Azolla* produces spores only when under conditions of stress. Sexual reproduction is of minor importance in a plant with a vegetative doubling time of approximately two weeks and few natural enemies (Ashton 1977; Kluge in Stirton 1978). *Salvinia molesta* is a sterile hybrid and only the female forms of *Elodea canadensis* and *Myriophyllum aquaticum* are known in South Africa (Jacot Guillarmod in Stirton 1978).

Hybridization

Of the 64 species listed in Table 1, seven are known to be hybrids. Hybridization occurred among alien species either prior to introduction to southern Africa (*Salvinia molesta*, *Lantana camara*, *Opuntia aurantiaca*) (Arnold 1977), or after introduction (*Populus*, *Prosopis*, *Cortaderia*) (Jurriaanse 1973; Stirton 1978; Robinson 1984). The *Rubus* species complex appears to be the result of hybridization between alien and indigenous species (Campbell and van Staden 1983a). A number of alien species are represented by a range of cultivars: *Nerium oleander*, *Rosa rubiginosa*, *Cestrum laevigatum*, *Opuntia ficus-indica*.

The possible implication of hybridization in our understanding of the reproductive biology of these species is that the new gene pool created may result in a wider ecological tolerance for the species (Anonymous 1980), opportunistic flowering and fruiting (Stirton 1978; Campbell and van Staden 1983a), and changes in seed dormancy (Harper 1965; Gill 1984). Not only may the reproductive biology of the plant differ from that in its place of origin, but hybrids are in some cases rejected by the host-specific parasites and predators of the parent species (Arnold 1977; Cilliers 1983).

Unsuccessful aliens

Some alien plant species, though widely planted in the Cape and having all the attributes of the plants discussed above, have failed to spread in this region. Examples are *Hakea salicifolia*, *Acacia cultriformis*, *A baileyana* and *A podalyriifolia*. All these species set viable seed in the south-western Cape and are easily propagated from seed under nursery conditions. They have no specific seed predators in the Cape. Despite this they are not considered problem species in this region. One possible explanation is that their requirements for seed set (cf acacias), germination or establishment are not well met in this region. Fire destroys all the seed of *H salicifolia*, but less than 50% of the seed of the invasive *Hakea* species (Richardson unpublished). The subject of the seed biology of other unsuccessful alien plants requires investigation.

Conclusion

A summary of information on the seed biology of alien species invading South Africa indicates that successful invaders have a number of the following characteristics:

(1) place of origin climatically and edaphically similar to place of introduction;
(2) disturbance regimes for which germination and establishment of the species are adapted;
(3) long-distance dispersal agents (including man);
(4) species-specific seed predators or pathogens in the place of origin;
(5) large annual seed crops;
(6) dormancy mechanisms;
(7) long-lived seeds;
(8) multiseeded fruits;
(9) means of vegetative regeneration.

Investigation of seed biology of invasive alien species is a recent development. This paper points to the lack of information on seed production, seed storage and on dormancy mechanisms of alien invasive plants. This is particularly so for the bird-dispersed, fruiting species, the spread of which has only recently become evident.

Acknowledgements

The CSIR provided funding for S J Dean and P M Holmes. The authors wish to thank Professor J van Staden, University of Natal, for his comments on the manuscript.

Anonymous 1980. Firewood Crops: shrub and tree species for energy production. National Academy of Sciences, Washington DC.

Arnold T J 1977. The origin and relationships of *Opuntia aurantiaca* Lindl. Proceedings of the 2nd National Weeds Conference of South Africa. Balkema, Cape Town. pp 269-286.

Ashton P J 1977. Factors affecting the growth and development of *Azolla filiculoides* Lam. Proceedings of the 2nd National Weeds Conference of South Africa. Balkema, Cape Town. pp 249-286.

Auld D 1983. Seed predation in native legumes of south eastern Australia. Australian Journal of Ecology 8, 367-376.

Balda R P 1980. Are seed caching systems co-evolved? Symposium on co-evolutionary systems in birds. Actis XVII Congressus Internationalis Ornithologici, Berlin, 1185-1191.

Bannister M H 1965. Variation in the breeding system of *Pinus radiata*. In: Baker H G and Stebbins G L (eds) The genetics of colonising species. Academic Press, New York. pp 353-372.

Barrow N J 1977. Phosphorus uptake and utilization by tree seedlings. Australian Journal of Botany 25, 571-584.

Bond W J 1985. Canopy stored seed reserves (serotiny) in Cape Proteaceae. South African Journal of Botany 51, 181-186.

Bond W J and Slingsby P 1983. Seed dispersal by ants in shrublands of the Cape Province and its evolutionary implications. South African Journal of Science 79, 231.

Bond W J and Slingsby P 1984. *Proteas*, ants and invaders: disruption of a delicate dependence. South African Journal of Science 80, 201.

Brown C J, Macdonald I A W and Brown S E (eds) 1985. Invasive alien organisms in South West Africa/Namibia. South African National Scientific Programmes Report No 119. CSIR, Pretoria.

Brown N A C and Booysen P de V 1969. Seed coat impermeability in several *Acacia* species. Agroplantae 1, 51-60.

Brown N A C and Mitchell J J 1984. Germination of the polymorphic fruits of *Bidens bipinnata*. South African Journal of Botany 3, 55-58.

Campbell P J and Van Staden J 1983a. The seed physiology of *Rubus cuneifolius* Hirsch. In: Proceedings of 5th National Weeds Conference of South Africa. Balkema, Cape Town. pp 27-34.

Campbell P J and Van Staden J 1983b. Germination of seeds of *Solanum mauritianum*. South African Journal of Botany 2, 301-304.

Centre T D and Johnson C D 1974. Coevolution of some seed beetles (Coleoptera Bruchidae) and their hosts. Ecology 55, 1096-1103.

Cilliers C J 1983. Insekte vir biologiese beheer van die onkruid *Lantana camara* L. Proceedings of the 5th National Weeds Conference of South Africa. Balkema, Cape Town. pp 181.

Clark N 1953. The biology of St John's Wort in Owens Valley Victoria with particular reference to biological control. Australian Journal of Botany 1, 95-120.

Dennill G B 1983. The effect of seed feeding hemipterans (Alydidae) on the seeds of *Acacia cyclops* in the south western Cape. Proceedings of the 5th National Weeds Conference of South Africa. Balkema, Cape Town. pp 182-187.

Drake W E 1979. Seed populations and regeneration potential of coastal *Eucalyptus* forest of south east Queensland. PhD Thesis, University of Queensland, Australia.

Erasmus D J and Van Staden J 1985. The germination of *Chromolaena odorata* achenes. Abstracts of the South African Association of Botanists Congress, Potchefstroom. pp 52.

Erb H E 1980. The natural enemies and distribution of *Sesbania punicea* in Argentina. Proceedings of the 3rd National Weeds Conference of South Africa. Balkema, Cape Town. pp 205-210.

Farrell T P and Ashton D H 1978. Population studies on *Acacia melanoxylon* I. Variation in seed and vegetative characteristics. Australian Journal of Botany 26, 356-376.

Fenn J A 1980. Control of *Hakea* in Western Cape. Proceedings of the 3rd National Weeds Conference of South Africa. Balkema, Cape Town. pp 167-173.

Forsyth C and Brown N A C 1981. Germination of the dimorphic fruits of *Bidens pilosa* L. New Phytologist 90, 151-164.

Frost P G H 1980. Fruit-frugivore interactions in a South African Coastal Dune Forest. Symposium on co-evolutionary systems in birds. Actis XV, Congressus Internationalis Ornithologici, Berlin, 1179-1184.

Geldenhuys C J 1975. Raising indigenous trees. Department of Environment Affairs, Directorate of Forestry Pamphlet 150. 32 pp.

Gerber H L and Herbst F A 1982. Tree nursery handbook. Department of Environment Affairs, Directorate of Forestry, Pamphlet 197. 100 pp.

Gilbert J M 1959. Forest succession in Florentine Valley, Tasmania. Royal Society of Tasmania 93, 129-151.

Gill A M 1985. *Acacia cyclops* G. Don in Australia: distribution and dispersal. Journal of the Royal Society of Western Australia 67, 59-65.

Gill A M and Neser S 1984. *Acacia cyclops* and *Hakea sericea* at home and abroad. Proceedings of Medecos Conference, Perth. pp 57-58.

Gleadow R M and Ashton D H 1981. The invasion by *Pittosporum undulatum* of the forests of central Victoria. I. Invasion patterns and plant morphology. Australian Journal of Botany 29, 705-720.

Glyphis J, Milton S J and Siegfried W R 1981. Dispersal of *Acacia cyclops* by birds. Oecologia (Berl) 48, 138-141.

Graaff J L and Van Staden J 1983. Germination of *Sesbania punicea*. In: van de Venter H A and Mason M (eds) Proceedings of the 4th National Weeds Conference of South Africa. Balkema, Cape Town. pp 35-44.

Graaff J L and Van Staden J 1984. The germination characteristics of two *Sesbania* species. South African Journal of Botany 3, 59-62.

Grice A C 1985. The demography of the leguminous shrubs *Acacia victoriae*, *Cassia menophila* and *C phyllodinea* in semi-arid south-eastern Australia. PhD Thesis, Macquarie University, Sydney.

Halevy G 1974. Effects of gazelles and seed beetles (Bruchidae) on germination and establishment of *Acacia* species. Israel Journal of Botany 23, 120-126.

Harper J L 1965. Establishment, aggression and cohabitation in weedy species. In: Baker H G and Stebbins G L (eds) The genetics of colonizing species. Academic Press, New York, 243-264.

Harvey G J 1981. Recovery and viability of Prickly Acacia (*Acacia nilotica)* seed ingested by sheep and cattle. Proceedings of the 6th Australian Weeds Conference 1, 197-201.

Henderson L and Musil K J 1983. Exotic woody plant invaders of the Transvaal. Proceedings of the 5th National Weeds Conference of South Africa. Balkema, Cape Town. pp 125-158.

Hendry N S and Van Staden J 1982. Seed germination and the potential for control of *Acacia mearnsii* as a weed. South African Journal of Science, 78, 206.

Howard T M 1974. *Nothofagus cunninghamii* ecotonal stages: buried viable seed in north-west Tasmania. Royal Society of Victoria Proceedings 86, 137-142.

Howe J F and Smallwood J 1982. Ecology of seed dispersal. Annual Review of Ecology and Systematics 1982, 13, 201-228.

Immelman W F E, Wicht C K and Ackerman D P 1973. Our green heritage. Tafelberg Uitgewers, Cape Town.

Janzen D H 1969. Seed eaters versus seed size, number, toxicity and dispersal. Evolution 23, 1-27.

Janzen D H 1982. Differential seed survival and passage rates in cows and horses, surrogate Pleistocene dispersal agents. Oikos 38, 150-156.

Johnson C 1978. Australian Myrtle. In: Stirton C H (ed) Plant invaders - beautiful but dangerous. The Department of Nature and Environmental Conservation of the Cape Provincial Administration, Cape Town.

Joubert D C 1984. The soil seed bank under Nasella tussock infestations at Boschberg. South African Journal of Plant and Soil 1, 1-3.

Joubert D C and Small J G C 1982. Seed germination and dormancy of *Stipa trichotoma* I. Effect of dehulling, constant temperatures, light, oxygen, activated charcoal and storage. South African Journal of Botany 1, 142-146.

Jurriaanse A 1973. Are they fodder trees? Department of Environmental Affairs, Directorate of Forestry, Pamphlet 116. 32 pp.

Kivilaan A and Bandurski R S 1981. The one hundred year period for Dr Beal's seed viability experiment. American Journal of Botany 68, 1290-1292.

Kloot P M 1985. The spread of native Australian plants as weeds in South Australia and otherMediterranean regions. Journal of the Adelaide Botanic Gardens 7, 145-157.

Kluge R L 1983. Progress in the fight against *Hakea*. Veld and Flora 69, 136-138.

Kluge R L and Gordon A J 1983. *Chrysolina quadrigemina* (Chrysomelidae: Coleoptera) and the control of *Hypericum perforatum* in a new area in South Africa. Proceedings of the 5th National Weeds Conference of South Africa. Balkema, Cape Town. pp 200-208.

Kruger F J 1977. Invasive woody plants in the Cape Fynbos with special reference to the biology and control of *Pinus pinaster*. Proceedings of the 2nd National Weeds Conference of South Africa. Balkema, Cape Town. pp 57-74.

Lamprey H F 1967. Notes on the dispersal and germination of some tree seeds through the agency of mammals and birds. East African Wildlife Journal 5, 179-180.

Landsdell K A 1927. Weeds of South Africa. Department of Agriculture, Pretoria, Bulletin No 18.

Larsen E 1964. Germination response of *Acacia* seeds to boiling. Australian Forest Research 1, 51-53.

Macdonald I A W 1983a. Alien trees, shrubs and creepers invading indigenous vegetation in the Hluhluwe-Umfolozi Game Reserve complex in Natal. Bothalia 14, 949-960.

Macdonald I A W 1983b. List of the invasive alien plants of southern Africa. Being a list accompanying the questionnaire survey of invasive alien plants in southern African protected areas. Percy FitzPatrick Institute of African Ornithology, University of Cape Town, Cape Town. 100 pp.

Macdonald I A W, Jarman M L and Beeston P (eds) 1985. Management of invasive alien plants in the fynbos biome. South African National Scientific Programmes Report No 111. CSIR, Pretoria.

Majer J D 1978. The seedy side of ants. Western Australian Institute of Technology Gazette 11, 7-9.

Milewski A V and Bond W J 1982. Convergence of myrmecocrhory in Mediterranean Australia and South Africa. In: Buckley R D (ed) Ant-plant interactions in Australia. W Junk, The Hague. pp 89-98.

Milton S J 1980. Australian acacias in the south-western Cape: pre-adaptation, predation and success. Proceedings of the 3rd National Weeds Conference of South Africa. Balkema, Cape Town. pp 69-78.

Milton S J 1982. Effects of shading on nursery grown *Acacia* seedlings. Journal of South African Botany 48, 245-272.

Milton S J and Hall A V 1981. Reproductive biology of Australian *Acacias* in the south-west Cape, South Africa. Transactions of the Royal Society of South Africa 44, 456-487.

Mitchell D T and Allsopp N 1984. Changes in the phosphorus composition of seeds of *Hakea sericea* (Proteaceae) during germination under low phosphorus conditions. New Phytologist 96, 239-247.

Moran V C and Moran P M 1982. Alien invasive vascular plants in South African natural and semi-natural environments; bibliography from 1830. South African National Scientific Programmes Report No 65. CSIR, Pretoria.

Mountjoy J H 1979. Broom - a threat to native plants. Fremontia 6, 11-15.

Neser S and Annecke D P 1973. Biological Control of Weeds in South Africa. Department of Agricultural Technical Services, Entomology Memoir 28, 27.

Nongonierma A 1978. Contribution a l'etude biosystematique du genre *Acacia* Mill en Afrique occidentale. VII. Caractres biologiques des graines: la germination. Bulletin Institut Fondam Afrique Noire 40 ser A, 480-510.

Oatley T B 1984. Exploitation of a new niche by the Rameron Pigeon *Columba arquatrix* in Natal. In: Ledger J A (ed) Proceedings of the 5th Pan African Ornithological Congress, 323-330.

Osbourne J B and Osbourne E 1931. Studies on the germination of seed of Black Wattle (*Acacia mollissima*) and Green Wattle (*Acacia decurrens*). South African Journal of Science 28, 222-237.

Phillips J F V 1927. The behaviour of *Acacia melanoxylon* RBv in the Knysna Forests: an ecological study. Transactions of the Royal Society of South Africa 16, 31-43.

Pieterse P J 1985. Verspreiding van langblaarwattelsaad. South African Weed Science Society Newsletter 19, 5.

Richardson D M 1985. Studies on aspects of the integrated control of *Hakea sericea* in the south-western Cape. MSc Thesis, University of Cape Town, Cape Town.

Richardson D M and Brink M P 1985. Notes on *Pittosporum undulatum* in the south-western Cape. Veld and Flora 71, 75-77.

Robinson E R 1984. Naturalized species of *Cortaderia* (Poaceae) in southern Africa. South African Journal of Botany 3, 343-346.

Rolston M P 1978. Water impermeable seed dormancy. The Botanical review 44, 365-396.

Shelton P 1975. Analysis of the stomach contents of the striped field mouse. Unpublished Zoology Honours project, University of Cape Town, Cape Town.

Sherry S P 1971. The Black Wattle. University of Natal Press, Pietermaritzburg. pp 385.

Stam I 1978. Seasonal influences on the chemical control of Bugweed *Solanum mauritianum*. South African Forestry 106, 41-44.

Stirton C (ed) 1978. Plant invaders - beautiful but dangerous. The Cape Department of Nature and Environmental Conservation, Cape Town.

Taylor S E and Walker B H 1984. Autecology of an invading population of the cactus *Cereus peruvianus* (Queen of the Night) in the central Transvaal. South African Journal of Botany 3, 387-396.

Van den Berg M A 1977. Natural enemies of certain *Acacias* in Australia. Proceedings of the 2nd National Weeds Conference of South Africa. Balkema, Cape Town. pp 75-82.

Van den Berg M A 1980. Natural enemies of *Acacia cyclops* and *Acacia saligna* in Western Australia. I Lepidoptera. II Coleoptera. III Hemiptera. Phytophylactica 12, 165-167, 169-171, 223-226.

Van Rooyen M W and Grobbelaar N 1982. Seed populations in the soil of the Hester Malan Nature Reserve in the Namaqualand Broken Veld. South African Journal of Botany 1, 41-50.

Van Staden J 1978. Seed viability in *Protea neriifolia*. Agroplantae 10, 65-72.

Van Staden J and Forsyth C 1983. Germination of *Tagetes minuta*. Proceedings of the 5th National Weeds Conference of South Africa. Balkema, Cape Town. pp 14-26.

Van Zyl W A and Kluge R L 1983. Aspects of the seed production of *Hypericum perforatum* in the south-western Cape. Proceedings of the 5th National Weeds Conference of South Africa. Balkema, Cape Town. pp 13.

Weiss P W 1983. Invasion of coastal *Acacia* communities by *Chrysanthemoides*. PhD Thesis, Australia National University, Canberra.

Weiss P W and Milton S J 1984. *Chrysanthemoides monilifera* and *Acacia longifolia* in Australia and South Africa. Medecos Conference, Perth. pp 2.

Wells M J 1977. Progress with research on Nasella Tussock. Proceedings of the 2nd National Weeds Conference of South Africa. Balkema, Cape Town. pp 47-56.

Wells M J, Engelbrecht V M, Balksinhas A A and Stirton C H 1983. Weed flora of South Africa. I. Major groupings. Bothalia 14, 945-948.

Welty J C 1964. The life of birds. Constable and Company, London.

Willis J C 1973. A dictionary of flowering plants and ferns. Eighth edition. Cambridge University Press, Cambridge.

Chapter 13

A comparative analysis of fleshy fruit displays in alien and indigenous plants

R S Knight
University of Cape Town.

Introduction

Plants whose seed are dispersed by vertebrates are usually distinguished by the presence of a fleshy covering around the seed eg a pericarp or sarcotesta, or a fleshy appendage attached to the seed ie an aril or pseudaril. Although birds and mammals are important dispersal agents (Ridley 1930), fishes and reptiles have also been observed to consume fleshy fruits (van der Pijl 1969).

Much information exists for frugivory by birds and mammals (Ridley 1930), but there is little data on actual dispersal of plants and information on post-dispersal establishment is almost nonexistant (Herrera 1985). In this field a considerable body of theory has been advanced with Snow (1971), McKey (1975) and Howe and Estabrook (1977) proposing that the morphology of both frugivores and fleshy fruits are interdependent and a product of mutual fine-scale evolution between the plant and its dispersal agents. Recent studies suggest that there is little evidence for the existence of fine-scale mutual evolution between plants and their dispersal agents (Howe 1980, 1981, 1983; Frost 1980; Herrera and Jordano 1981; Herrera 1982; Howe and Smallwood 1982; Jordano 1982; Moore and Willson 1982; Morden-Moore and Willson 1982; Herrera 1984). In contrast to indigenous plants the vertebrate-dispersal of introduced plant species has largely been neglected, though there is some recent work undertaken in Africa (Glyphis et al 1981; Alexandre 1982; Dean 1983; Oatley 1984; Knight 1986).

If the relationship between indigenous plants and their dispersal agents is the product of fine-scale mutual evolution (Snow 1971; McKey 1975; Howe and Estabrook 1977), one would predict that introduced plant species possessing fleshy fruits would have little chance of being dispersed by indigenous vertebrates and consequently the noninvasive flora should have a higher proportion of species bearing fleshy fruits than the invasive alien flora. This would also result in a very small vertebrate-dispersed component occurring in the invasive woody flora relative to the indigenous flora.

If a significant proportion of alien species do rely on indigenous vertebrate-dispersal agents, one would predict that the fruits of invasive alien species will be comparable in morphology to those of the indigenous species. This will promote their dispersal by indigenous frugivores presumably which are adapted to the fruit of indigenous plants.

The effects of preadaptation of fleshy fruits for consumption and possible dispersal may be enhanced if fruit displays of invasive alien plants possess certain advantages over their indigenous counterparts. One would predict that those alien plants that have successfully invaded southern Africa possess advantages with regard to presentation of their fruit displays.

The study reported here was designed to test these predictions, firstly by determining the proportion of invasive and noninvasive alien woody plant species in southern Africa that are potentially vertebrate-dispersed and to compare these with those of the indigenous flora.

The second test involved the classification of the noninvasive alien, invasive alien and indigenous woody plant species by fruit colour and size, and the last was made by comparing fruit displays of vertebrate-dispersed invasive alien plants and indigenous plants co-occurring in the south-western Cape.

Here, fleshy fruits were taken to be plant propagules that possess a fleshy covering or a conspicuous fleshy appendage. Berries, drupes, arils, pseudarils and sarcotesta were all considered to be fleshy.

Amphibians and reptiles are not important dispersers in southern Africa, so that vertebrate-

Address for correspondence: R S Knight, Percy FitzPatrick Institute of African Ornithology, University of Cape Town, Private Bag, Rondebosch, 7700.

dispersal refers to dispersal by birds and mammals. Bird names used in the text follow Maclean (1985).

Invasive alien plants were taken to be the introduced plants that are capable of establishing self-sustaining populations in areas of natural and seminatural vegetation.

Woody plants included trees, shrubs and climbers that possessed woody or semiwoody stems.

Methods

A list of alien woody plant species that have been introduced to southern Africa was compiled from data in Kruger (1981). Information in Wells et al (1986), supplemented from Powrie and Macdonald (1985), was used to identify alien species that have invaded areas of natural and seminatural vegetation in southern Africa. Any species considered to be invasive by Wells et al (1986) and Powrie and Macdonald (1985) was excluded from the list derived from Kruger (1981). A total of 1 698 alien woody plant species were considered to be noninvasive, whereas 165 species were considered to be invasive. The genera of the noninvasive alien species which are potentially vertebrate-dispersed are presented in Appendix 1. The 51 invasive, vertebrate-dispersed species are listed in Appendix 2. The fruiting characteristics of these species were classified by colour (white, red, orange, yellow, green, brown, black/blue/purple) and size (less than 15 mm length, 15-30 mm, more than 30 mm) based on information in Everett (1980-1982), Kruger (1981), Pienaar (1984) and Polunin and Everard (1985). These data were compared with data for the indigenous tree and shrub flora of southern Africa (Knight and Siegfried 1983).

To determine if individual alien plants possess advantages with regard to fruit display, a field study (Knight 1986) was undertaken on the most important invasive alien and some indigenous vertebrate-dispersed tree and shrub species co-occurring at coastal and montane sites in the south-western Cape. The six invasive alien species were *Acacia cyclops*, *Lantana camara*, *Myoporum serratum*, *Pittosporum undulatum*, *Rubus* species (*fruticosa*?) and *Solanum mauritianum*, and the eight indigenous species were *Chrysanthemoides monilifera*, *Colpoon compressum*, *Euclea racemosa*, *Maytenus acuminata*, *Myrica cordifolia*, *Myrsine africana*, *Rapanea melanophloeos* and *Sideroxylon inerme*. The fruit displays of each species were described by the following eight variables collected in the field from 1981 to 1983.

(1) SIZE - estimated size of standing fruit crop at peak season. Fruit was counted on a sample of individual branches and extrapolated for the entire plant. Since fruit size varied greatly, even on the same plant, fruit sizes were not recorded in the field. The fruits of all species analysed in this study varied between five and 12 mm in length.

(2) SEAS the length of fruiting season, determined from accumulated observations made on marked branches during the entire study period. The cut-off points for defining the season were selected to include circa 95% of yearly fruit production.

(3) SYNC - fruit synchronization: fruiting on marked plants was compared with five adjacent conspecifics and expressed as a percentage.

(4) OCUR - plant occurrence: the spatial aggregation of species was ranked from solitary to monospecific stands.

(5) P-RP - preripe colour/conspicuousness: monitored during preripening.

(6) RP - ripe colour/conspicuousness: monitored during ripening.

(7) REM fruit removal: estimated by comparing 'marked' with 'bagged' branches bearing ripe fruits and calculated on a weekly basis.

(8) ROT - fruit decay: ripe fruit were picked, placed in bags and then tied back onto the plant. The picked fruit in these bags together with the unpicked 'bagged' fruits were assessed for signs of fruit decay every day.

The range of each of the eight variables was split into four categories and each species was rated on a one to four scale for the corresponding variable. In this manner species could be ranked with regard to each other and the fruit displays of indigenous and alien species could be compared. The data and techniques will be found in Knight (1986). In this way, each species was arranged along a gradient from intensive to prolonged fruit displays. Intensive fruit displays were large, highly conspicuous, and synchronized with conspecifics, but were presented during fairly short fruiting periods, whereas prolonged fruit displays were smaller, less conspicuous and less synchronized but were presented for longer periods. Species with prolonged displays tended to have less perishable fruits and to show slower removal rates, but were more likely to be utilized by resident frugivores such as sombre bulbuls *Andropadus importunus*. Stiles (1982) predicted that species with prolonged fruit displays should possess a preripe colour phase which would allow resident frugivores to 'trapline' for ripe fruit.

Results and discussion

Vertebrate-dispersal in alien and indigenous floras

Among noninvasive alien species introduced to southern Africa circa 32% were identified as having fruits potentially suitable for vertebrate-dispersal. The proportion in invasive alien species was 31% (Table 1). Since the ratios of fleshy to nonfleshy fruits in the noninvasive and invasive alien plants were similiar, it is unlikely that a fine-scale mutual evolution between plant and dispersal-agent exists for the southern African flora. The proportions of fleshy fruits were lower than that occurring in the indigenous flora (χ^2 = 139,69; P < 0,01).

Table 1. The contribution of plant species with fleshy fruits in the indigenous, noninvasive alien and invasive alien floras of southern Africa.

Fruit type	Indigenous	Alien	
		Noninvasive	Invasive
Fleshy	698	545	51
Non-fleshy	631	1153	114

It seems that some species may require periods of adaptation in their introduced environment before they become invasive (Kruger et al this volume). It may be speculated that this process will be more pronounced in species that require biotic agents for either their pollination or seed dispersal. Under such circumstances both pollinators and frugivores may need to change foraging behaviour before exploitation of invasive alien resources can occur. Within the Cape Peninsula the bird-dispersed *Myoporum serratum* was referred to as an uncommon escapee in the 1940's (Adamson and Salter 1950). However, in 1985 it was extremely abundant at Sandvlei and Zeekoevlei (personal observation). A similar process is thought to be occurring in the bird-dispersed *Pittosporum undulatum* (Kruger et al this volume).

It may well be that vertebrate-exploitation of alien fruits is initiated by environmental disturbances which change the availability of indigenous fruits. Oatley (1984) suggested that the change of diet in the frugivorous rameron pigeon *Columba arquatrix* from indigenous plants (*Olea*, *Ocotea* and *Prunus* species) to alien plants, such as *Solanum mauritianum* and, to a lesser extent, *Morus nigra* and the plantation weed *Phytolacca octandra*, was primarily due to fruit shortages created by the extensive commercial exploitation of indigenous trees. Other avian species that have been noted to exploit *Solanum mauritianum* fruits include purple crested lourie *Tauraco porphyreolophus*, speckled mousebird *Colius striatus*, blackheaded oriole *Oriolus larvatus*, blackeyed bulbul *Pycnonotus barbatus*, Cape white-eye *Zosterops pallidus* and streakyheaded canary *Serinus gularis* (Dean 1985). It therefore appears that *Solanum mauritianum* has secured a diversified suite of frugivores since its introduction to southern Africa. *Acacia cyclops* also appears to have attracted a large number of avian frugivores. Glyphis et al (1981) recorded Cape turtle dove *Streptopelia capicola*, laughing dove *S senegalensis*, whitebacked mousebird *Colius colius*, Cape bulbul *Pycnonotus capensis*, European starling *Sturnus vulgaris*, redwinged starling *Onychognathus morio*, Cape sparrow *Passer melanurus*, Cape weaver *Ploceus capensis* and bully canary *Serinus sulphuratus* to feed on *Acacia cyclops* fruits, although not all of these species are typical avian seed dispersers.

Parallel fruit colours and sizes in alien and indigenous floras

The number of species represented by the different fruit size and colour classes for noninvasive, invasive alien and indigenous floras are summarized in Table 2. In comparing these three floras it appears that similar proportional representation in each category of colour and size occurs. This may indicate that with regard to fruit size and colour, alien plants introduced in southern Africa are preadapted for vertebrate dispersal. The most notable differences are the lower proportions of small white fruits (less than 15 mm) and large yellow fruits (more than 30 mm) in both categories of alien plants. The indigenous flora also appears to have fewer species with small (less than 15 mm) black fruits than either the invasive or noninvasive alien floras.

An alternative explanation for this similarity in fruit morphology would be that most vertebrate-dispersed woody plants have converged on a few generalized dispersal syndromes regardless of phylogenetic and geographical origin. Figure 1 compares the relative proportion of species producing fleshy ripe fruits of various colours for the three floras of southern Africa and four floras elsewhere in the world. It can be seen that proportional representations of fruit colours in the seven floras are fairly similar, especially with regard to the dominance of red and black fruits. Therefore, if any

Table 2. The percentage of total number of plant species for noninvasive, invasive alien and indigenous floras by fruit colour and size classes

Flora	Fruit length size (mm)	Fruit colour						
		White	Red	Orange	Yellow	Green	Brown	Black
Noninvasive alien N = 148	<15	6,33	20,89	5,70	3,80	1,27	0,63	36,70
	15-30	1,27	3,16	1,27	3,80	0,63	0	1,90
	>30	0	0,63	0	8,23	2,53	0,63	0,63
Invasive alien N = 40	< 15	0	22,50	10,00	5,00	0	2,50	35,00
	15-30	0	2,50	0	5,00	0	0	0
	>30	0	7,5	0	10,00	0	0	0
Indigenous N = 349	<15	1,43	20,92	12,61	8,88	2,01	4,30	23,50
	15-30	0,29	4,58	1,72	6,02	1,43	0,29	3,44
	>30	0,29	1,43	1,43	2,29	2,29	0,57	0,29

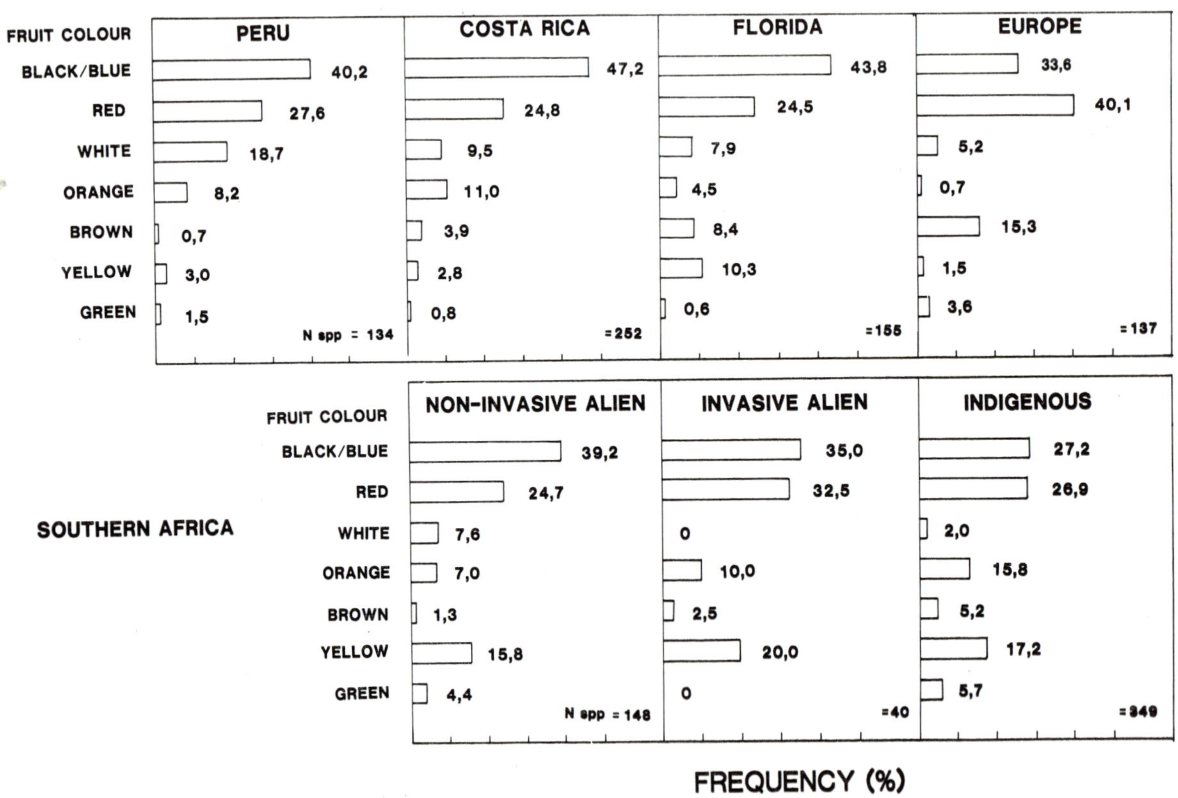

Figure 1. Schematic representation of an intensive-prolonged fruit display gradient. Relative position of species was determined by similarity distances between each species and the fruit display strategies intensive and prolonged. Attributes of each strategy are represented on a one to four ordinal scale (adapted from Knight 1985).

mutual evolution between plants and dispersers does exist it will only reflect adaptations within suites of similar species rather than at individual species levels (Howe and Smallwood 1982).

Parallel and divergent displays of alien and indigenous species

The results obtained from the fieldwork undertaken on the 14 plant species are presented in Table 3 (from Knight 1986). The alien and indigenous species are plotted in Figure 2 along a gradient of fruit-display, from intensive to prolonged. The indigenous species *Chrysanthemoides monilifera* represents the extreme of an intensive fruit display and the alien species *Acacia cyclops* is the next closest. The *Rubus* species, *Myoporum serratum* and *Lantana camara* are alien plant species which may also be considered to have intensive fruit displays. Parallel examples in the indigenous species include *Myrica cordifolia* and *Sideroxylon inerme*. The alien species *Solanum mauritianum* and *Pittosporum undulatum* show neither an intensive nor a prolonged fruit display. The rest of the indigenous plants diverge from the alien plants in possessing prolonged fruit displays.

Fruit displays and ecological perturbations

It appears that those species associated with early colonization of disturbances on sand dunes, such as *Acacia cyclops* and the indigenous *Chrysanthemoides monilifera* and *Myrica cordifolia* (Heydorn 1975), exhibit intensive fruit displays. In areas of disturbance, such as road verges and electricity way lines, monospecific stands of colonizing *Acacia cyclops* and *Chrysanthemoides monilifera* frequently occur (personal observation). Of six alien species three can be considered to exhibit intensive fruit displays (*Acacia cyclops*, *Myoporum serratum* and *Rubus* species). All three and *Pittosporum undulatum*, under certain cir-

Table 3. Summary of fruit display characteristics for each of eight indigenous and six alien species occurring in the south-western Cape, South Africa (see text).

Species	Crop size Numbers of fruit per plant × 10^3	Length of fruiting season, months	Degree of synchronization %	Occurrence	Preripe phase	Ripe phase	Removal per week %	Rotting time, weeks
Chrysanthemoides monilifera	>10	<3	>60	Solitary	Slight change	Highly conspicuous	>10	<1
Colpoon compressum	<1	>9	<20	Solitary-thickets	Conspicuous	Highly conspicuous	<1	5-12
Euclea racemosa	5-10	3-6	20-40	Thickets	Visible	Visible	1-5	1-5
Myrica cordifolia	5-10	3-6	>60	Thickets	None	Visible	>10	>12
Maytenus acuminata	<1	>9	<20	Solitary-thickets	None	Inconspicuous	<1	>12
Myrsine africana	1-5	3-6	<20	Thickets	Visible	Visible	<1	5-12
Rapanea melanophloeos	>10	3-6	40-60	Thickets	Visible	Conspicuous	<1	5-12
Sideroxylon inerme	>10	6-9	20-40	Thickets	None	Conspicuous	1-5	1-5
Acacia cyclops	>10	3-6	>60	Monospecific stands	Slight	Highly conspicuous	>10	5-12
Lantana camara	5-10	3-6	40-60	Thickets	None	Conspicuous	1-5	1-5
Myoporum serratum	5-10	<3	>60	Solitary	None	Conspicuous	6-10	1-5
Pittosporum undulatum	5-10	3-6	>60	Monospecific stands	Visible	Conspicuous	6-10	1-5
Rubus sp	5-10	>9	40-60	Thickets	Visible	Conspicuous	6-10	<1
Solanum mauritianum	1-5	3-6	40-60	Thickets	Visible	Conspicuous	6-10	1-5

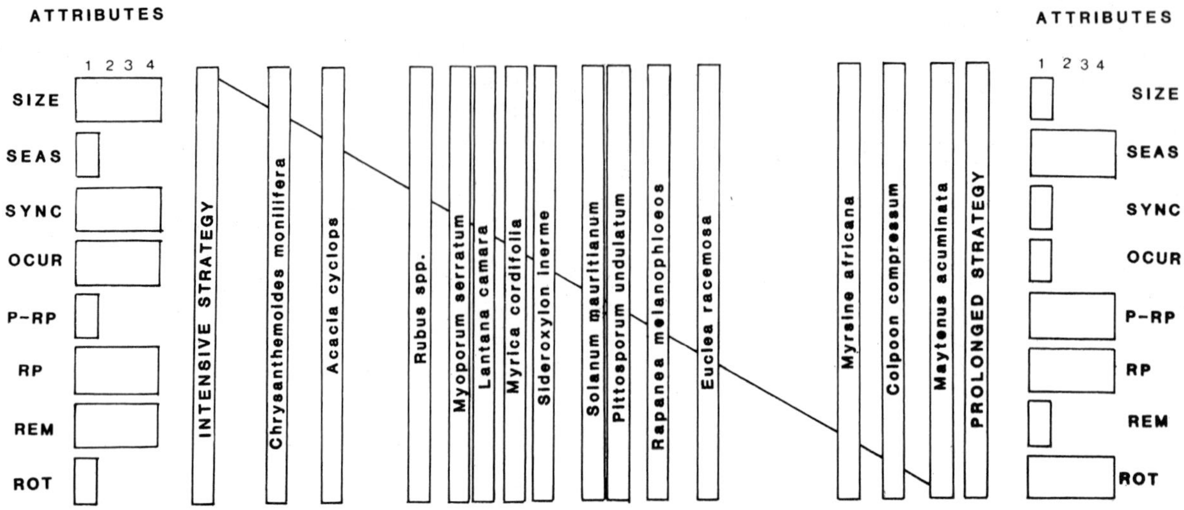

Figure 2. A comparison of vertebrate-dispersed plant species producing ripe fruits of various colours in four floras worldwide (adapted from Wheelwright and Janson 1985) with the non-invasive alien, invasive alien and indigenous floras occurring in southern Africa.

cumstances, exhibit a high potential for colonization, even in the presence of little or no disturbance. *Myoporum serratum* is invasive in wetland areas of the southern Cape Peninsula (personal observation). *Pittosporum undulatum* is invasive in tall scrub and riparian forest (Richardson and Brink 1985), even where little disturbance has occurred (personal observation; see also Geldenhuys et al this volume). This species also invades open, undisturbed *Eucalyptus* forests within its native Australia (Gleadow 1982). Finally the effective fruit displays of *Rubus* may contribute to the observed encroachment of this species into lower mountain slopes, valleys and along streambanks (Marais 1979).

Not only do alien plants exploit disturbances in the environment but they may also impose new disturbances with regard to the availability of fruit resources. Possible results of avian consumption of alien fruits are the failure of dispersal in indigenous fruits and a concomitant decrease in the recruitment and regeneration of these species. In the Cape Hangklip area of south-western Cape, fruit displays of *Acacia cyclops* attract large flocks of redwinged starlings from November to March (personal observation). After March, most *A cyclops* fruits have been consumed, whereas the indigenous fruits of *Rhus* species (*R crenata*, *R glauca*, *R lucida*, *R laevigata*, *R tomentosa*) which provide a continuous supply of fruits throughout the summer (personal observation) are largely unconsumed. During 1983-1984 the fruit crops of *A cyclops* at Hermanus were poor and the avian frugivores utilized the succession of fruiting *Rhus* during the summer. Prior to the introduction of *Acacia cyclops* to the Cape in the last century (Shaughnessy 1980), Rhus fruits were presumably heavily utilized during the summer months and were likely to have been pivotal species (sensu Howe 1984), and remain pivotal today if *A cyclops* fruit crops fail.

Conclusions

It appears that the proportion of bird-dispersed invasive alien species is lower than that of the indigenous woody flora. However, the proportion of invasive alien species that are bird-dispersed is indistinguishable from that existing in the noninvasive alien floras. It is, therefore, unlikely that a widespread but fine-scale evolution between plants and disperser exists in the southern African subcontinent.

It also appears that fruits of the invasive and noninvasive alien floras are similar to each other with respect to fruit size and colour, and to that of the indigenous flora and even other floras of the world. This is interpreted as indicating that vertebrate-dispersed plants have evolved fleshy fruits which tend to converge into generalized dispersal syndromes regardless of phylogenetic and geographical origins.

Alien species that are successful invaders of the

south-western Cape have, in general, larger and more conspicuous fruit displays than many of the indigenous species and in this respect their fruit displays diverge from the indigenous flora. However, those indigenous species that are dune pioneers appear to have fruit displays that converge with most alien plant species.

Although this study of fruit displays has indicated that the performance of an invasive alien plant species might be related to its adaptations for vertebrate-dispersal, it is apparent that at the level of floras these adaptations are not critical to invasive success. Knowledge of the vertebrate-dispersal of plants will have to be considerably improved before any but the most general predictions relating to the invasive success of alien species will be possible.

Acknowledgements

I would like to thank G J Breytenbach, R K Brooke, I A W Macdonald and M J Wells for advice and comments on earlier drafts. This study was supported financially by the South African Council for Scientific and Industrial Research through the Fynbos Biome Project.

Adamson R S and Salter T M 1950. Flora of the Cape Peninsula. Juta, Cape Town.

Alexandre D Y 1982. La dispersion de *Solanum verbascifolium* en Côte-d'Ivoire: role des cephalophes. Revue d'écologie (La terre et la vie) 36, 293-295.

Dean W R J 1985. The utilization of *Solanum mauritianum* woodland by birds in the north-eastern Transvaal. In: Bunning L J (ed) Proceedings of the Birds and Man Symposium Johannesburg 1983, Witwatersrand Bird Club, Johannesburg. pp 89-97.

Everett T H 1980-1982. The New York Botanical Garden Illustrated Encyclopedia of Horticulture. 10 volumes. Garland Publishing Incorporated, New York.

Frost P G H 1980. Fruit-frugivore interactions in a South African coastal dune forest. In: Nohring R (ed) Acta 17. Congressus Internationalis Ornithologici Band II. Verlag Deutschen Ornithologen-Gesellschaft, Berlin. pp 1179-1184.

Gleadow R M 1982. Invasion by *Pittosporum undulatum* of the forests of central Victoria. II. Dispersal, germination and establishment. Australian Journal of Botany 30, 185-198.

Glyphis J P, Milton S J and Siegfried W R 1981. Dispersal of *Acacia cyclops* by birds. Oecologia (Berlin) 48, 138-141.

Herrera C M 1982. Seasonal variation in the quality of fruits and diffuse coevolution between plants and avian dispersers. Ecology 63, 773-785.

Herrera C M 1984. A study of avian frugivores, bird-dispersed plants and their interaction in Mediterranean scrublands. Ecological Monographs 54, 1-23.

Herrera C M 1985. Determinants of plant-animal coevolution: the case of mutualistic dispersal of seeds by vertebrates. Oikos 44, 132-141.

Herrera C M and Jordano P 1981. *Prunus mahaleb* and birds: the high efficiency seed dispersal system of a temperate fruiting tree. Ecological Monographs 51, 203-218.

Heydorn A F P J 1975. Plant succession on the sea dunes of Betty's Bay. Veld and Flora 61, 26-29.

Howe H F 1980. Monkey dispersal and waste of a neotropical fruit. Ecology 61, 944-959.

Howe H F 1981. Dispersal of a neotropical nutmeg (*Virola sebifera*) by birds. Auk 98, 88-98.

Howe H F 1983. Annual variation in a neotropical seed-dispersal system. In: Sutton S L, Whitemore T and Chadwick A C (eds) The tropical rain forest: ecology and management. Blackwell, Oxford. pp 211-227.

Howe H F 1984. Implications of seed-dispersal by animals for tropical reserve management. Biological Conservation 30, 261-281.

Howe H F and Estabrook G F 1977. On intraspecific competition for avian dispersers in tropical trees. American Naturalist 111, 817-832.

Howe H F and Smallwood J 1982. Ecology of seed disperal. Annual Review of Ecological Systematics 13, 201-228.

Jordano P 1982. Migrant birds are the main seed dispersers of blackberries in southern Spain. Oikos 38, 183-193.

Knight R S and Siegfried W R 1983. Inter-relationships between type, size and colour of fruits and dispersal in southern African trees. Oecologia (Berlin) 56, 405-412.

Knight R S (1986). Fruit displays of indigenous and invasive alien plants in the south-western Cape. South African Journal of Botany 52, 249-255.

Kruger T J 1981. Trees, shrubs and climbers. Privately published, Bethal.

MacLean G L 1985. Robert's birds of southern Africa. The trustees of the John Voelcker Bird Book Fund, Cape Town.

Marais J 1979. Kirstenbosch Gardens and a "magnificent obsession". Veld and Flora 65, 26-29.

McKey D 1975. The ecology of coevolved seed dispersal systems. In: Gilbert L E and Raven P H (eds) Coevolution of animals and plants. University of Texas Press, Austin. pp 159-191.

Moore L A and Willson M F 1982. The effect of microhabitat, spatial distribution, and display size on dispersal of *Lindera benzoin* by avian frugivores. Canadian Journal of Botany 60, 1554-1560.

Morden-Moore A L and Willson M F 1982. On the ecological significance of fruit color in *Prunus serotina* and *Rubus occidentalis*: field experiments. Canadian Journal of Botany 60, 1554-1560.

Oatley T B 1984. Exploitation of a new niche by the Rameron Pigeon *Columba arquatrix* in Natal. Proceedings of the 5th Pan-African Ornithological Congress, Southern African Ornithological Society, Johannesburg. pp 323-330.

Pienaar K 1984. The South African what flower is that?Struik, Cape Town.

Polunin O and Everard B 1985. Trees and bushes of Europe. Oxford University Press, Oxford.

Powrie F J and Macdonald I A W 1985. Invasive alien plants in southern Africa. Unpublished manuscript at the FitzPatrick Institute, University of Cape Town, Cape Town.

Richardson D M and Brink M P 1985. Notes on *Pittosporum undulatum* in the south-western Cape. Veld and Flora 71, 75-77.

Ridley H N 1930. The dispersal of plants throughout the world. L Reeve and Company, Kent.

Shaughnessy G L 1980. Historical ecology of alien woody plants in the vicinity of Cape Town, South Africa. PhD Thesis, University of Cape Town, Cape Town.

Snow D W 1971. Evolutionary aspects of fruit-eating by birds. Ibis 113, 194-202.

Stiles E W 1982. Fruit flags: two hypotheses. American Naturalist 120, 500-509.

Van derPijl 1969. Evolutionary action of tropical animals on the reproduction of plants. In: Lowe-McConell R H (ed) Speciation in tropical environments. Academic Press, London. pp 85-96.

Wells M J, Balsinhas A A, Joffe T, Engelbrecht V M, Harding G and Stirton C H 1986. A catalogue of problem plants in southern Africa incorporating the national weeds list of South Africa. Memoirs of the Botanical Survey of South Africa, 53. Government Printer, Pretoria.

Wheelwright N T and Janson H C 1985. Colors of fruit displays of bird-dispersed plants in two tropical forests. American Naturalist 126, 777-799.

Appendix 1

Plant genera that include noninvasive species introduced to South Africa that are potentially suitable for vertebrate-dispersal. An asterisk indicates that species within this genus have fruits that have been recorded as being consumed by potential vertebrate-dispersal agents (Ridley 1930).

*Actinidia, Aesculus**, *Amelanchier**, *Ampelopsis**, *Andromeda**, *Arbutus**, *Arctostaphylos**, *Ardisia**, *Aristotelia**, *Aronia**, *Aucuba, Azara, Berberis**, *Callicarpa**, *Ceanothus**, *Celastrus**, *Celtis**, *Cestrum**, *Chaenomelas, Chionanthus, Cinnamomum**, *Citharexylum**, *Clerodendrum**, *Coprosma**, *Cordyline**, *Cornus**, *Corokia, Cotoneaster**, *Crataegus**, *Cyphomandra, Daphne**, *Dichroa, Diospyros**, *Dracaena**, *Drimys**, *Elaeagnus, Elaeocarpus**, *Epigaea, Eriobotrya**, *Eugenia**, *Euonymus**, *Ficus**, *Fuchsia, Gardenia**, *Gaultheria**, *Grewia**, *Griselinia**, *Hedera**, *Heteromeles, Hippophae**, *Hymenanthera, Idesia, Ilex, Iochroma, Ixora, Jasminum, Jatropha**, *Juniperus**, *Kadsura, Lantana**, *Laurus, Ledum**, *Leea, Leycesteria**, *Ligustrum**, *Lonicera, Macadamia, Macleania, Maclura**, *Mahonia**, *Malus**, *Melia**, *Meliosma, Menispermum, Mespilus**, *Morus**, *Murraya, Myoporum**, *Myrica**, *Myrtus**, *Nandina, Nyssa**, *Osmanthus**, *Osmaronia**, *Osteomelas**, *Paliurus, Panax, Parthenocissus, Passiflora**, *Pernettya**, *Persea, Persoonia, Phellodendron, Phillyrea**, *Phoenix, Photinia, Pistacia, Pittosporum**, *Podocarpus**, *Prunus**, *Psychotria**, *Pyracantha**, *Pyrus**, *Raphiolepis, Rhamnus**, *Rhodotypos, Rhus**, *Ribes, Rosa**, *Rubus**, *Ruscus**, *Sambucus**, *Sarcococca, Sassafras**, *Schizandra, Serissa, Smodingium, Solanum**, *Sorbus, Stachyurus, Stranyaesia, Symphoricarpus**, *Taxus**, *Ternstroemia, Tilia**, *Torreya, Umbellularia, Vestia, Viburnum**, *Vitex**, *Vitis**, *Yucca**, *Zanthoxylum**.

Appendix 2

Invasive alien plants that are potentially vertebrate-dispersed. A single asterisk indicates that the species possesses fleshy fruits, and that the fruits of congenerics are consumed by potential vertebrate dispersal agents (Ridley 1930). A double asterisk indicates that this particular species was noted to possess fruits that are consumed by potential vertebrate-dispersal agents (Ridley 1930). A triple asterisk indicates that the fruit of the species have been seen to be eaten by potential vertebrate-dispersal agents (personal observation).

*Acacia cyclops****, *A melanoxylon***, *Anacardium occidentale***, *Cestrum laevigatum****, *Citrus limon**, *Cotoneaster franchetii**, *Duranta repens**, *Harrisia martinii* ?, *Lantana camara****, *L trifolia**, *Ligustrum lucidum**, *Litsea glutinosa* var *brideliifolia**, *Lycopersicon pimpinellifolium**, *Mangifera indica***, *Melia azedarach***, *Morus alba***, *Myoporum serratum****, *Opuntia ficus-indica***, *Passiflora caerulea**, *P edulis**, *Pereskia aculeata* ?, *Physalis peruviana***, *Phytolacca dioica**, *Pittosporum undulatum****, *Prunus persica**, *Psidium cattleianum**, *P guajava***, *Punica grantum* ?, *Pyracantha angustifolia****, *P coccinea**, *Rosa eglanteria***, *R odorata**, *Rubus affinis**, *R cuneifolius**, *R fruticosus****, *R* hybrid*, *R idaeus***, *R niveus**, *R phoenicolasius**, *R rosifolius**, *Schinus molle***, *S terebinthifolius****, *Solanum elaeagnifolium****, *S hispidum**, *S jasminoides**, *S mauritianum****, *S nigrum****, *S rostratum**, *S seaforthianum****, *S sisymbrifolium**, *Syzygium cuminii**, *Yucca aliofolia***.

Chapter 14

Photosynthesis and carbon allocation in invasive plants of South African biomes

B H Whiting*, G C Bate* and D J Erasmus†
University of Port Elizabeth and †Plant Protection Research Institute.

Introduction

Of the wide variety of exotic plants introduced to southern Africa, only certain species are successful invaders. Of these some appear to be more aggressive than others (Macdonald 1983). Plant success in a given environment and in relation to its competitors may be linked to its total carbon gain (Schulze 1982) which in turn is dependent on plant physiological, morphological and life history characteristics. This chapter reviews available information on the processes of photosynthesis and carbon allocation in aliens of southern African biomes and addresses the hypothesis that the invasive success of aliens can be accounted for in terms of these processes alone.

Photosynthesis and carbon allocation

Fundamental to plant performance is the maintenance of a positive carbon balance (Schulze 1983). This is only possible if the rate of carbon gain exceeds that of carbon loss (mainly dark respiration and photorespiration). Carbon balance under any specific set of environmental conditions strongly influences plant productivity as does the way in which assimilated carbon is allocated within plants.

Photosynthesis

Photosynthesis in terrestrial plants is represented at the biochemical level by three major pathways, namely, the C_3, C_4 and Crassulacean Acid Metabolism (CAM) pathways (Osmond et al 1982). Recently, however, a number of plants showing characteristics intermediate between C_3 and C_4 as well as a range of CAM-type variants have been identified (Holaday and Chollet 1984; Monson et al 1984; Sipes and Ting 1985). The C_4 and CAM pathways are thought to be evolutionary modifications of the more primitive C_3 pathway (Kluge and Ting 1978; Pearcy and Ehleringer 1984).

The biochemical differences between the C_3, C_4 and CAM pathways are reflected at the gas exchange level as differences in the patterns and magnitudes of net CO_2 exchange. However, as Pearcy and Ehleringer (1984) point out, 'typical' comparative data frequently cited for C_3 and C_4 leaves under optimum conditions are of limited value to understanding plant performance in the field.

Salisbury and Ross (1985) provide a general overview of the fundamental processes involved in the different photosynthetic pathways. More detailed accounts are available in the general literature (Hatch 1975; Kluge and Ting 1978; Holaday and Chollet 1984; Sharkey 1985).

According to Long (1983), C_3 plants occur in almost every terrestrial habitat while C_4 plants are most common in warmer tropical and subtropical environments, uncommon in cool or shaded environments and absent from polar climates. This distribution may be due to the relatively recent evolution of the C_4 pathway rather than any intrinsic disadvantages of the pathway as such (Pearcy and Ehleringer 1984). Plants with the CAM pathway are most commonly found in semi-arid areas with seasonal rainfall, coastal fog deserts and epiphytic habitats in tropical rain forests (Osmond et al 1982).

Carbon allocation

Fundamental to plant growth, development, reproduction and performance is the allocation of fixed carbon to basic functions and life events (Mooney 1972; Mooney and Parsons 1974; Osmond et al 1982). Carbon allocation begins at the biochemical level and continues to the whole plant level (Gifford et al 1984). Movement of carbon in plants takes place from sites of carbon supply (sources) to sites of utilization and/or transformation (sinks). General rules governing patterns of translocate movement have been presented by Canny (1984). At the level of the whole plant carbon

Address for correspondence: B H Whiting, Department of Botany, University of Port Elizabeth, P O Box 1600, Port Elizabeth, 6000.

allocation occurs in a number of characteristic ways correlated with plant life forms and Schulze (1982) has proposed that these patterns reflect optimum carbon return. At the community level, allocation strategies can profoundly influence performance through the amount and timing of carbon committed to such functions as leaf development, root growth and storage.

Readers are referred to texts such as Trlica and Singh (1979), Moorby (1981), Osmond et al (1982), Schulze (1982), Canny (1984) and Gifford et al (1984) for a more comprehensive coverage of the creation, distribution and utilization of reserves by plants as well as the implications of allocation strategies for plant performance and competition.

Photosynthetic pathways of alien and indigenous vegetation occurring in southern African boimes

In agronomic situations the possession of C_4 as opposed to C_3 photosynthesis appears important in the attainment of competitive advantage (Black et al 1969; Hatch 1975). Whether this generalization holds for alien invasion of indigenous communities is examined. There are, however, a great many gaps in our information regarding the distribution of C_3, C_4 and CAM pathways amongst indigenous and alien vegetation within southern African biomes.

Photosynthetic pathways of aliens

Macdonald (1983) listed 418 invasive alien plant species in southern African protected areas. Of the terrestrial aliens, approximately 42% were considered to be aggressive invasives. According to a recent survey (Richardson 1985), only 13% (23 species) of the aggressive invasives (or 6% of the total terrestrial alien flora) listed by Macdonald (1983) have become major alien plant invaders in southern African biomes. Table 1 shows the distribution of photosynthetic pathways in aggressive and major terrestrial invaders.

Of the 174 aggressive aliens belonging to some 39 families, 128 have the C_3 pathway, six have the C_4 pathway and a further 6 have CAM. The pathway of photosynthesis in the remaining 34 species is not known for certain. No data are available for those plants which belong to families known to contain either C_3 and C_4 taxa or C_3, C_4 and CAM taxa (Osmond et al 1982). None of the major aliens identified by Richardson (1985) have the C_4 pathway. Twenty are C_3 plants and three are CAM. Of the C_3 plants 16 are trees, three are shrubs and one is a grass. All, however, are perennials. From this we conclude that C_4 and CAM biochemistry per se does not explain invasive success in southern African biomes.

Photosynthetic pathways of indigenous vegetation

The distribution of photosynthetic pathways within the indigenous flora of southern Africa has not been fully examined. Other than the comprehensive investigation into the $\delta\ ^{13}C$ values of southern African grasses by Vogel et al (1978), most studies have been relatively localized (Mooney et al 1977; Werger and Ellis 1981). For this reason some indigenous species are classified on the basis of information drawn from similar biomes elsewhere.

Grasses

Vogel et al (1978) showed that amongst the Poaceae the C_4 pathway was dominant in most southern African grassland, savanna and karoo biomes. Only on the summits of the Drakensberg and some other mountain ranges in the eastern Cape does the C_3 pathway assume numerical significance.

In the western succulent karoo which receives its rain mainly in winter, the relative abundance of C_3 and C_4 species depends upon geographical locality, with C_3 species increasing in importance in the cooler coastal zones. In the fynbos grasses are unimportant (Huntley 1984).

Trees

All forest trees examined appear to utilize the C_3 pathway (Jarvis and Leverenz 1983; Medina and Klinge 1983). There is no evidence to suggest that a similar situation does not occur in southern Africa. Woodland and savanna tree species are almost certainly also C_3.

Table 1. Distribution of photosynthetic pathways amongst aggressive[1] and major[2] terrestrial aliens of southern African biomes.

Photosynthetic pathway	Number of aggressive alien species	Number of major alien species
C_3	128	20
C_4	6	0
CAM	6	3
Uncertain	34	–
Total	174	23

[1]Data from Macdonald (1983)
[2]Data from Richardson (1985)

Shrubs and arborescent species

The distribution of C_3, C_4 and CAM amongst woody shrubs, succulent and semi- succulent shrubs and other arborescent species is essentially unknown. Mooney et al (1977) showed that the more massive succulent forms of the Eurphorbiaceae are strictly CAM while smaller leafy succulents appear to be largely faculative-CAM plants. The succulent karoo flora is almost exclusively CAM although some C_3 species have been noted (von Willert et al 1985). In general terms, it seems likely that woody shrubs will be C_3 species although the possibility of C_4 species cannot be excluded. In the fynbos which is characterized by evergreen sclerophyllous heath plants, fine leaved shrubs and leafless tufted grasslike plants (Huntley 1984), C_3 is almost certainly the dominant mode of CO_2 assimilation (Ehleringer and Mooney 1983).

Photosynthesis in specific alien invaders

Few specific studies of photosynthesis in alien invaders have been undertaken in South Africa, and only in jointed cactus and nassella tussock has any attention been given to the environmental control of carbon assimilation. In addition, photosynthesis in *Agrostis stolonifera*, an alien invasive grass on Marion Island has been studied in relation to that in the dominant indigenous grass, *Agrostis magellanica*. The special advantages which may accrue to each of these species are discussed here.

Jointed cactus *Opuntia aurantiaca*

The introduction and spread of jointed cactus in South Africa has been reviewed by several authors (Moran and Annecke 1979; Zimmermann 1981). The densest infestations exist in the semi-arid regions of the eastern Cape where some 800 000 ha are infested (Malan 1985). Several factors appear to contribute to the success of the plant in South Africa (Whiting and Campbell 1983). Among these are the possession of CAM and the ability to develop an underground tuber.

Jointed cactus is an obligate-CAM (O-CAM) plant, a characteristic shared with all other species of Cactaceae investigated to date. Under controlled conditions it has been shown to exhibit features which are typical of O-CAM plants in general (Whiting 1980; Whiting and Campbell 1983).

Unstressed jointed cactus plants held under controlled conditions exhibit a typical full-CAM pattern of gas exchange in which all four phases of CAM-linked CO_2 exchange are evident. Net CO_2 uptake via open stomata during the night (phase 1) contributes up to 82% of the total carbon fixed per 24 hours and results in a high efficiency of water use in relation to carbon gain (Table 2).

Table 2. Transpiration/photosynthesis ratios (H_2O loss/CO_2 gain) of jointed cactus and published (Kluge 1976) ranges for CAM, C_3 and C_4 species (Whiting and Campbell 1983).

Types of species	Day	Night
Jointed cactus	135	13
CAM	150 – 600	15 – 150
C_3	450 – 600	
C_4	250 – 350	

A reliable indicator of the extent to which CAM plants engage in light and/or dark CO_2 fixation under natural conditions is the carbon isotope discrimination ratio expressed as a $\delta^{13}C$ value (Bender 1971). The $\delta\ ^{13}C$ values (– 10,9‰ to – 14‰) of naturalized populations of jointed cactus over most of the distribution range in the eastern Cape fall near the upper limit for C_4 photosynthesis (Figure 1). Similar results have been found for other species of *Opuntia* in arid regions (Sutton et al 1976) and indicate an almost exclusive reliance on nocturnal CO_2 fixation for the maintenance of a positive carbon balance.

The adaptive value of CAM in relation to growth forms, life cycles and habitats of CAM plants has been questioned by von Willert et al (1985). In the case of jointed cactus the value of CAM as a survival mechanism assumes special significance because the sole propagules of the plant are its cladodes. Following abscission or dislodgement, significant nocturnal carbon dioxide uptake may be supported by endogenous water reserves. The time during which this can occur for cladodes of various turgidities and environmental conditions at the time of dislodgement remains to be determined.

Survival for extended periods of moisture stress by whole jointed cactus plants and propagules is achieved via the phenomenon of 'CAM idling'. In this state both carbon and water are conserved through tight stomatal closure and a low level of metabolic activity reflected in reduced diurnal fluctuations of malic acid relative to those in unstressed plants (Ferreria 1978) and insignificant CO_2 exchange (Whiting and Campbell 1983). By retaining functional photosynthetic tissues during periods of moisture stress,

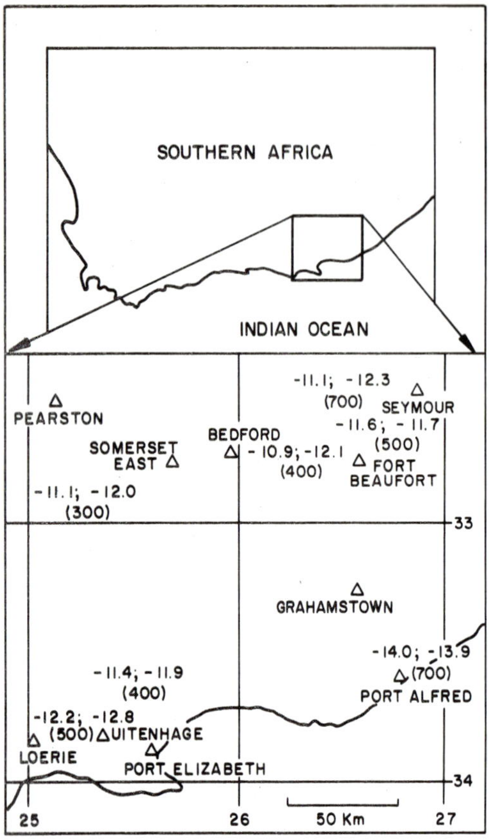

Figure 1. $\delta^{13}C$ (−‰ values of terminal (first value) and basal (second value) cladodes from naturalized populations of *Opuntia aurantiaca* in the eastern Cape. Mean annual rainfall (mm) for areas sampled is shown in brackets.

jointed cactus is able to respond rapidly to the absorption of water. The significance of this characteristic for invasive success requires evaluation in terms of the adaptations possessed by native species. With respect to the environmental control of nocturnal carbon assimilation in jointed cactus, optimal nighttime temperatures lie between 10° C and 20° C with substantial inhibition of CO_2 uptake above 20° C (Whiting unpublished). Daytime chlorenchyma temperatures ranging from 20° C to 40° C have relatively little influence on nocturnal acidification in jointed cactus. Optimum temperature during the day (photosynthetic photon flux density 430 $\mu mol^{-2}s^{-1}$) is 30° C when nighttime temperature is 15° C. However, the acclimation potential of this plant is unknown and requires investigation since it could have important implications for seasonal performance. Unlike certain plants with CAM capability, photoperiod does not appear to significantly influence nocturnal acidification although a trend towards slightly lower nocturnal acid accumulation has been noted with decreasing nightlength (Whiting unpublished).

The CAM mode of carbon assimilation in jointed cactus appears to provide an effective mechanism for conserving water and carbon and maintaining functional photosynthetic tissues during periods of moisture stress. This is of particular relevance to the successful propagation of the plant and may therefore be seen as an important contributory factor to its success as an invader. However, CAM alone does not guarantee success and must be complemented by other anatomical, morphological and physiological characteristics. Amongst these are the low stomatal frequency of 24 mm^{-2} and 40 mm^{-2} in mature and juvenile cladodes respectively (Malan 1985), well developed intercladode abscission zones that facilitate propagule release in response to minor physical disturbance and spines which act not only as a feeding deterrent but also as an effective attachment mechanism to various dispersal agencies. This is made possible by their barbed structure (Malan 1985).

Nassella tussock *Stipa trichotoma*

Wells (1978) has documented the history of the introduction of nassella tussock to South Africa as well as the localities and extent of subsequent infestation. Particularly successful invasions occur along the tops of a number of mountain ranges in the eastern Cape where nassella is a strong invader of *Themeda/Merxmuellera* grassveld in cool, moist situations (Bate 1983). Nassella is both unpalatable and a prolific producer of seed (up to 100 000 per plant per annum). The seeds remain viable for as long as 20 years (Joubert 1983). These features are seen as major contributory factors to the success of this invader.

The photosynthetic characteristics of nassella as well as a number of co-occuring species from a heavily infested site on the Boschberg range adjacent to Somerset East have been investigated by Bate (1983) and D C Joubert (personal communication). Joubert found the daily pattern of leaf photosynthesis to be typical of many C_3 species. Data relating to the light saturation characteristics of nassella (Figure 2) are incomplete having been assessed only up to 200 μmol m^{-2} s^{-1} (approximately 10% of full sunlight).

The light compensation point of nassella lies between 40 and 50 μmol m^{-2} s^{-1} which falls in the C_3 range of values as does the optimum temperature (10-15° C) for net CO_2 assimilation. At high temperatures (30° C) net photosynthesis is markedly depressed (up to 76%) over that at 15° C. An investigation (Bate 1983) of leaf photosynthetic temperature responses of nassella and eight co-occuring grasses from the Boschberg infestation (Table 3) confirmed the well documented observation that C_3 grasses, on average, have lower photosynthetic rates at optimum temperature than C_4 grasses. Furthermore the data of Bate (1983) shows that the temperature optima of the indigenous C_3 and C_4 species are within the same range. This is in agreement with the general observation of Pearcy and Ehleringer (1984) that within a habitat C_3 and C_4 species exhibit similar photosynthetic temperature responses. However, two potentially significant observations emerged from Bate's study. Firstly, the alien C_3 species exhibited notably higher (about 150%) maximum photosynthetic rates than the indigenous C_3 species. Secondly, the T opt for photosynthesis was between four and nine degrees lower in the aliens than in the indigenous species. Could this not reflect better adaptation to the invaded habitat by the aliens than by the indigenous species? The comparatively high T opt of the indigenous C_3 grasses are not expected to be of ecological significance with respect to resisting invasion because their photosynthetic rates are very much lower than those of the aliens. The observation that *Merxmuellera stricta* does in fact invade *Themeda* veld is thus more likely related to some other factor, such as unpalatibility than to its carbon acquiring capacity (Bate 1983).

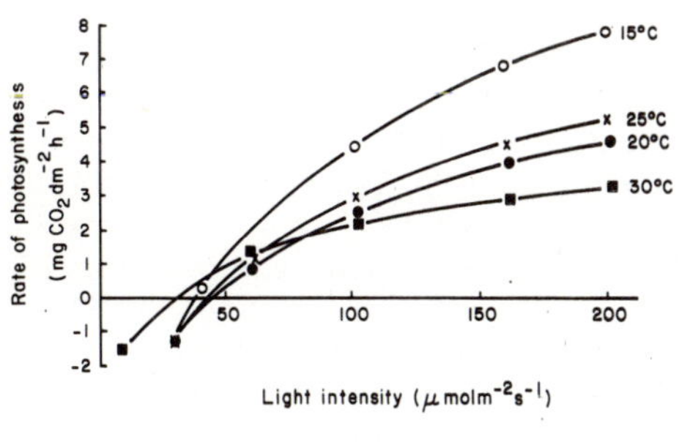

Figure 2. Photosynthetic rates of *Stipa trichotoma* as a function of irradiance at four temperatures. (D C Joubert personal communication)

Table 3. Photosynthetic pathway, optimum temperatures for photosynthesis, rates of photosynthesis at optimum temperature and rates of photosynthesis at low and high temperature for two alien (*) and six indigenous grasses of the Boschberg near Somerset East (Bate 1983).

Species	P[1]	T(opt)[2]	P(M)[3]	P(L)[4]	P(H)[5]
Stipa trichotoma(*)	C3	10 – 15	11,2	8,6(6)	3,5(29)
Bromus unioloides(*)	C3	14	11,9	10,9(6)	4,5(32)
Tristachya leucothrix	C3	19	4,9	1,6(5)	4,5(29)
Merxmeullera stricta	C3	29	4,2	2,8(7)	4,2(29)
Themeda triandra	C4	18	12,8	3,2(6)	2,8(29)
Sporobolus africanus	C4	19	24,8	0,1(5)	1,4(29)
Cymbopogon validus	C4	28	20,6	6,4(5)	20,6(28)
Eragrostis sp.	C4	28	40,1	23,7(6)	40,1(28)

[1]P = Photosynthetic pathway
[2]T (opt) = Temperature at which P_{max} was reached
[3]P (M) = Maximum photosynthetic rate (mg CO_2 dm^2 h^{-1}) at 390 mol $m^{-2}s^{-1}$
[4]P(L) = Photosynthetic rate at lowest temperature (in brackets) measured
[5]P(H) = Photosynthetic rate at highest temperature (in brackets) measured

Cool temperatures associated with the high-altitude habitats invaded by nassella tussock are not conducive to strong competitive performance by most C_4 species. The high leaf photosynthetic rates of the *Eragrostis* species at low temperature might suggest that this plant could be an exception. However, it does not appear to be a strong competitor, the reasons for which are not clear. Possible explanations may reside in nutritional requirements (Bate 1983), the adverse influence of low nighttime temperatures on the performance of most C_4 plants (Osmond et al 1982) or factors relating to seed production, germination and survival.

It is tempting to speculate that the low temperature optimum for photosynthesis in nassella tussock, together with numerous persistent seeds (Wells 1978; Joubert 1983), contribute towards the competitive advantage over *Themeda triandra* and other co-dominants. The unpalatability of nassella to livestock undoubtedly reinforces the advantage, because neighbouring native grasses are usually heavily grazed.

However, as Osmond et al (1982) and Pearcy and Ehleringer (1984) point out photosynthetic performance alone cannot always guarantee competitive success. Other factors such as the area and display of assimilatory tissues, period and duration of growth, acclimatization potential and responses to moisture will all require evaluation to clarify the invasive success of this plant and the management strategies best suited to limit it.

Agrostis stolonifera

Pammenter et al (1986) have compared photosynthesis in the alien *Agrostis stolonifera*, which is an invader of sheltered drainage lines, and the indigenous *Agrostis magellanica*, which is the dominant grass of the coastal mires of Marion Island. Although no marked differences in photosynthetic capacity or rates of CO_2 assimilation (expressed per unit of leaf area) were noted for these C_3 grasses, potentially important differences in the shape of the photosynthesis/irradiance curves were found at low irrandiances. Photosynthesis in *A stolonifera* responded more rapidly to low irradiances than in the indigene, leading Pammenter et al (1986) to propose that the alien is better adapted to the characteristically low irradiance prevailing under a high incidence of cloud cover on this subantarctic island. Furthermore, leaf anatomical detail revealed a greater investment of carbon in support tissue in the indigene suggesting that potentially more is available for allocation to assimilatory tissue in the alien. Pammenter et al (1986) argue that the lack of support tissue in the leaves of the alien could account for its restriction to the comparatively sheltered drainage lines and conclude that *A stolonifera* is unlikely to successfully invade the gale-swept open mire grasslands dominated by *A magellanica*.

Carbon allocation in specific alien invaders

Carbon allocation in alien invasive species occurring in southern Africa has received only scant attention. The major emphasis of the few studies conducted has been aimed at predicting the movement of foliar-applied systemic herbicides, since these are thought to be transported with assimilates in the phloem (Crafts and Crisp 1971).

Jointed cactus

Jointed cactus is a taxon of hybrid origin which reproduces exclusively via vegetative propagation (Arnold 1977; van de Venter et al 1984). In comparison to most other plants capable of vegetative propagation, jointed cactus is unusual in that virtually all assimilated carbon (up to 93%) may be allocated to the only propagules of the plant, namely the cladodes and sterile fruit. Furthermore the propagules are capable of rapid rooting (Figure 3) to replenish depleted water and carbon reserves when favourable conditions prevail.

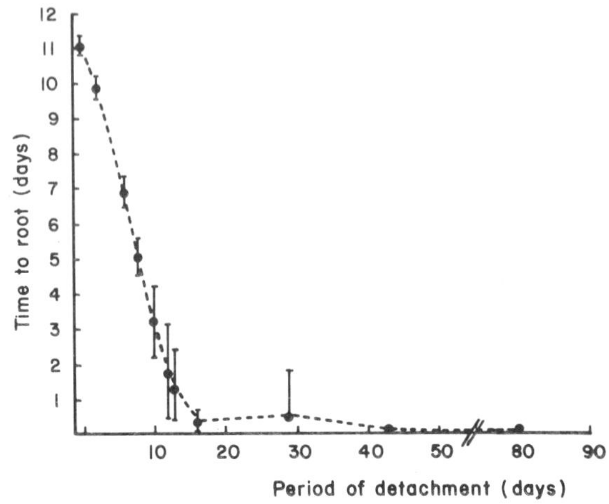

Figure 3. Time taken by dislodged cladodes of *Opuntia aurantiaca* to root as a function of period of detachment up to 80 days. Day/night photoperiod 12/12; day/night temperature 30/15° C, photosynthetic photon flux density 300 μmol m^{-2} s^{-1} (Bar = ± SD; n = 10)

Carbon allocated to new propagules is strongly correlated with plant phenology and occurs mainly from September to April. Jointed cactus possesses an enormous capacity for vegetative propagule production. A single 60 mm cladode planted in a root observation chamber and maintained outdoors under a favourable moisture regime for the duration of the cladode production phenophase, produced 720 cladodes in 186 days. This capacity for vegetative propagule production is much higher than that of some other invasive cacti in southern Africa. Under controlled conditions jointed cactus plants produced 310% and 266% more propagules than *Opuntia rosea* and *O imbricata* respectively over a period of 100 days (Figure 4).

Under natural conditions moisture stress depresses propagule production. The maximum number of propagules produced by insect-free plants under natural conditions near Uitenhage varied from 17-145 over a 192-day growing period during 1980/81. Biotic pressure from introduced insects frequently complements the influence of moisture stress in limiting the number of propagules available for dispersal (H G Zimmermann personal communication).

Relatively little is known about the dynamics of carbon allocation in relation to source/sink activity in succulents as a whole let alone in jointed cactus. E E Campbell and Whiting (unpublished) studied acropetal and basipetal translocation in jointed cactus during late autumn/early winter, a time when the plants are dormant with respect to cladode initiation. Within 16 days of ^{14}C-sucrose application to apical cladodes 50% of the total radioactivity in the plant was located in the tuber. Negligible acropetal translocation was noted when tracer was applied to basal cladodes. These observations are in accordance with the general rules of source/sink relationships reported by Canny (1984). Plants utilized in this experiment were small, comprising a tuber, basal cladode and an apical cladode. Investigations (Whiting unpublished) of plants raised from buried cladodes under a variety of controlled conditions suggest that, provided moisture is available to support exogenous CO_2 assimilation, tuber growth will occur under both long days and high temperatures as well as short days and low temperatures. However under these conditions total carbon allocation as indicated by dry matter distribution is markedly different. Long days and high temperatures favour the allocation of carbon to developing cladodes and fine root formation while short days and low temperatures favour allocation to spines, coarse roots and tubers. Questions relating to the distribution of sources involved in tuber maintenance and growth in large plants, as well as the periodicity of tuber sink strength remain to be answered, as do those relating to the influence of temperture and moisture stress on carbon allocation to tubers and the relevance of tuberization to plant performance in the field.

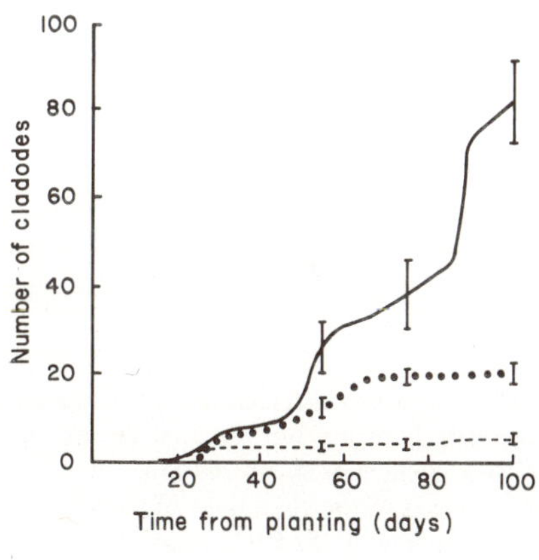

Figure 4. Production of vegetative propagules by three species of invasive Opuntia under controlled conditions. *O aurantiaca* ____; *O rosea* .....; *O imbricata* -----. Day/night photoperiod 12/12; day/night temp. 25/15° C photosynthetic photon flux density 410 μmol m^{-2} s^{-1} (Bar = ± SD; n = 4)

American bramble *Rubus cuneifolius*

Rubus cuneifolius is a prolific and troublesome invader in Natal and the eastern Transvaal. Detailed descriptions of the history of introduction, locality of infestations, factors contributing to its proliferaton, as well as habitat and climatic preferences have been provided by various authors (Campbell and van Staden 1983; Erasmus and van Staden 1983). The root system of bramble is perennial with widely spreading, vigorous, horizontal roots along which vegetative buds are located. The shoot system is characterized by a biennial growth cycle.

Erasmus and van Staden (1983) showed that the rules governing source/sink relationships and patterns of assimilate translocation in plants in general (Canny 1983) were applicable to bramble in Natal. The cane apical region acts as a sink for assimilates imported from its nearest sources, namely the uppermost leaves. Assimilates from the lowermost leaves are exported to the root system, their nearest sink. Leaves in between export assimilates both acro- and basipetally. Thus the source/sink relationships of *Rubus cuneifolius* are similar to those reported for a wide variety of plant species (Wardlaw 1968). However, in *Eupatorium adenophorum* (= *Ageratina adenophora*), an alien invader of the mist-belt region of Natal, acropetal translocation predominates in actively growing plants regardless of the position of the leaf (P H Bennet personal communication).

With respect to the seasonality of carbon allocation in *R cuneifolius*, tracer studies (Erasmus and van Staden 1983) have shown that during early- and mid-summer, translocation in sterile, first year stems (primocanes) is predominantly acropetal, while in autumn basipetal movement to the roots occurs. Assimilate movement in second year fertile stems (floricanes) is basipetal to the developing replacement canes, which act as strong sinks for assimilates during early- and mid-summer. Sink strength of the root system increases as extension growth of the replacement canes declines during late summer and early autumn. Concurrent analyses of root and rootcrown material indicates a decrease in total nonstructural carbohydrates during early summer and an increase during mid- and late summer (Erasmus 1982). While these data give little insight into the role of carbon allocation in invasive success, they have provided information on the dynamics of assimilate movement, which, it must be remembered, was the original goal of the research.

General

The occurrence of an alien or any other plant in an ecosystem is related to migration dynamics. Survival there, however, implies biological compatability with environmental conditions and successful competition with co-occuring species. Persistence, on the other hand, is related to successful reproduction. Proliferation relative to co-occuring species suggests ecological imbalance, the possibilities for which are numerous. Among these are factors such as intrinsically more successful physiological and morphological characteristics for more efficient resource acquisition and utilization. Altered resource availability in favour of the invasive species through man-made or natural disturbance (Sousa 1984), may enable the alien to prosper as will new biological pressures on the indigenous plants.

Assessments of the roles of isolated plant processes such as carbon assimilation and allocation in plant performance are not easily made, especially when they rely on an incomplete data base. Osmond et al (1982) and Pearcy and Ehleringer (1985) have pointed out that processes such as photosynthesis are just single components of a much larger suite of characters comprising the plant genome that influence plant performance in a habitat. The dangers of drawing invalid conclusions from broad generalizations concerning photosynthetic pathways and plant performance have been emphasized by Woolhouse (1983).

Conclusions and needs for future research

Of significance to plant biological invasions in southern African biomes is the virtual absence of invasive alien species with the C_4 pathway of photosynthesis. The vast majority of species possess the C_3 pathway and of the major invaders most are arborescent. This suggests that life form (and hence carbon allocation) in conjunction with other features (possibly reproductive capacity and specific adaptations to invaded habitats) is of more value to performance than mode of carbon assimilation. This is unlike the situation in agronomy where many of the world's worst weeds possess the C_4 photosynthetic pathway (Black et al 1969; Hatch 1975). Hence, there is a need to compare invasive life form to invaded life forms and establish the contribution of photosynthesis and carbon allocation to performance at different life history stages. However, when invasive life form is similar to the dominant life form comprising the invaded community eg grassland invaded by an alien grass, then leaf photosynthetic characteristics appear to be of importance in so far as they reflect, to a degree, compatability with the irradiance and thermal regimes of the invaded habitat. Finally the possibility that the attainment of competitive advantage by aliens in an invasive situation is realized through lower carbon drains to predators must also be considered.

While it is premature to assign any outstanding significance to the photosynthetic and carbon allocation processes in nassella tussock and American bramble as far as the attainment of elevated performance is concerned, the picture is somewhat clearer in the case of *Agrostis stolonifera* on Marion Island. Likewise in the case of jointed cactus, a clearer picture emerges. Here, the CAM photosynthetic pathway (which in itself can be regarded as an ecological adaptation to aridity) and the specific allocation of carbon to spiny, water-storing photosynthetic propagules can be viewed as significant contributory factors to a life history unique amongst the terrestrial flora of the semi-arid regions of southern Africa. However, a number of questions regarding ecophysiological tolerances to temperature and desiccation as well as propagule survival and establishment still remain to be addressed. It is evident from the information presented in this chapter that before any definite conclusions regarding the roles of photosynthesis and carbon allocation in invasive success can be arrived at, a considerable expansion of the data base on these processes in aliens and neighbouring species will have to be made.

Acknowledgements

The Department of Agriculture is thanked for financial support. We are grateful to all who participated by returning questionnaires. Finally we would like to thank Mrs A C Whiting and Mrs J Henegan for typing the manuscript.

Arnold T H 1977. The origin and relation of *Opuntia aurantiaca* Lindley. Proceedings of the 2nd National Weeds Conference of South Africa, Stellenbosch. pp 269-286.

Bate G C 1983. Aspects of the physiology of nassella grass *Stipa trichotoma*. Proceedings of the 5th National Weeds Conference of South Africa, Stellenbosch. pp 89-95.

Black C C, Chen T M and Brown E H 1969. Biochemical basis for plant competition. Weed Science 17, 338-344.

Bender M M 1971. Variations in the $^{13}C/^{12}C$ ratios of plants in relation to photosynthetic carbon dioxide fixation. Phytochemistry 10, 1239-1244.

Campbell P L and van Staden J 1983. The seed physiology of *Rubus cuneifolius* Pursh. Proceedings of the 5th National Weeds Conference of South Africa, Stellenbosch. pp 27-34.

Canny M J 1984. Translocation of nutrients and hormones. In: Wilkins M B (ed) Advanced Plant Physiology. Pitman Publishers Ltd, London and Marshfield, Massachusetts. pp 277-296.

Crafts A S and Crisp C E 1971. Phloem transport in plants. Freeman and Company, San Francisco.

Ehleringer J and Mooney H A 1983. Productivity of desert and Mediterranean climate plants. In: Lange O L, Nobel P S, Osmond C B and Ziegler H (eds). Ecosystem processes: Mineral cycling, productivity and man's influence. Encyclopaedia of Plant Physiology (New series) Vol 12D. Springer-Verlag, Berlin. pp 205-232.

Erasmus D J 1982. Translocation studies in the weed, *Rubus cuneifolius* Pursch. MSc Thesis, University of Pietermaritzburg, Pietermaritzburg. 109 pp.

Erasmus D J and van Staden J 1983. Translocation studies in the weed *Rubus cuneifolius* (Bramble). Proceedings of the 5th National Weeds Conference of South Africa, Stellenbosch. pp 79-88.

Ferreira B 1978. The effect of water stress on detached joints of *Opuntia aurantiaca* Lindley. Honours project, University of Port Elizabeth, Port Elizabeth. pp 20.

Gifford R M, Thorn J H, Hitz W D and Giaquinta R T 1984. Crop productivity and photoassimilate partioning. Science 225, 801-808.

Hatch M D 1975. The C_4 pathway of photosynthesis. In: Burris R H and Black C C (eds) CO_2 metabolism and plant productivity. University Park Press, Baltimore, London and Tokyo. pp 59-81.

Holaday S A and Chollet R 1984. Photosynthesis/photorespiratory characteristics of C_3-C_4 intermediate species. Photosynthesis Research 5, 307-323.

Huntley B J 1984. Characteristics of South African biomes. In: Booysen P de V and Tainton N M (eds) Ecological effects of fire in South African ecosystems. Ecological Studies 48. Springer-Verlag, Berlin. pp 2-17.

Jarvis P G and Leverenz J W 1983. Productivity of Temperate, Deciduous and Evergreen forests. In: Lange O L, Nobel P S, Osmond C B and Ziegler H (eds) Ecosystem processes: Mineral cycling, productivity and man's influence. Encyclopaedia of Plant Physiology (New Series) Vol 12D. Springer-Verlag, Berlin. pp 233-280.

Joubert D C 1983. The soil seed bank under nassella tussock infestations at Boschberg. South African Journal of Plant and Soil 1, 1-3.

Kluge M 1976. Crassulacean acid metabolism (CAM): CO_2 and water economy. In: Lange O L, Kappen L and Schultz E-D (eds) Water and plant life: Problems and modern approaches. Springer-Verlag, Berlin. pp 313-322.

Kluge M and Ting I P 1978. Crassulacean acid metabolism. Analysis of an ecological adaptation. Springer-Verlag, Berlin.

Long S P 1983. C_4 photosynthesis at low temperatures. Plant Cell and Environment 6, 345-363.

Macdonald I A W 1983. Invasive alien plants in southern Africa. List accompanying questionnaire survey of invasive alien plants in southern African protected areas. FitzPatrick Institute, University of Cape Town, Cape Town.

Malan H 1985. A characterisation of the morphological boundary layers of jointed cactus (*Opuntia aurantiaca* Lindley) with special reference to the uptake of xenobiotics. MSc Thesis, University of Port Elizabeth, Port Elizabeth. pp 228.

Medina E and Klinge H 1983. Productivity of Tropical Forests and Tropical Woodlands. In: Lange O L, Nobel P S, Osmond C B and Ziegler H (eds) Ecosystem Processes: Mineral cycling, productivity and man's influence. Encyclopaedia of Plant Physiology (New Series) Vol 12D. Springer-Verlag, Berlin. pp 281-304.

Monson R K, Edwards G E and Ku M S B 1984. C_3-C_4 intermediate photosynthesis in plants. Bioscience 34, 563-574.

Mooney H A 1972. The carbon balance of plants. Annual Review of Ecology and Systematics 3, 315-346.

Mooney H A and Parsons D J 1974. Plant development in Mediterranean climates. In: Leith H (ed) Phenology and Seasonality modelling. Springer-Verlag, Berlin. pp 255-267.

Mooney H A, Troughton J H and Berry J A 1977. Carbon isotope ratio measurements of succulent plants in southern Africa. Oecologia 30, 295-305.

Moorby J 1981. Transport systems in Plants. Longman, London and New York.

Moran V C and Annecke D P 1979. Critical reviews of biological pest control in South Africa III. The jointed cactus, *Opuntia aurantiaca* Lindley. Journal of the Entomological Society of South Africa 42, 299-329.

Osmond C B, Winter K and Ziegler H 1982. Functional significance of different pathways of CO_2 fixation in photosynthesis. In: Lange O L, Nobel P S, Osmond C B and Ziegler H (eds) Water. relations and carbon assimilation. Encyclopaedia of Plant Physiology (New Series) Vol 12B. Springer-Verlag, Berlin. pp 281-304.

Pammenter N W, Drennan P M and Smith V R 1986. Physiological and anatomical aspects of photosynthesis of two *Agrostis* species at a sub-Antarctic island. New Phytologist 102, 143-160.

Pearcy R W and Ehleringer J 1984. Comparative ecophysiology of C_3 and C_4 plants. Plant Cell and Environment 7, 1-13.

Richardson D M 1985. Major alien plants of terrestrial biomes southern Africa. Dispersal modes and patterns of invasion. Unpublished report 85/28. Jonkershoek Forestry Research Centre, Stellenbosch.

Salisbury F B and Ross C W 1985. Plant physiology (3rd edition) Wadsworth, Belmont, California.

Schulze E-D 1982. Plant life forms and their carbon, water and nutrient relations. In: Lange O L, Nobel P S, Osmond C B and Ziegler H (eds) Water relations and carbon assimilation. Encyclopaedia of Plant Physiology (New Series) Vol 12B. Springer-Verlag, Berlin. pp 616-676.

Schulze E-D 1983. Rock-shoot interactions and plant life forms. Netherlands Journal of Agricultural Science 4, 291-303.

Sharkey T D 1985. O_2 insensitive photosynthesis in C_3 plants. Its occurrence and possible explanation. Plant physiology 78, 71-75.

Sipes D L and Ting I P 1985. Crassulacean acid metabolism modifications in *Peperomia camptotricha*. Plant Physiology 77, 59-63.

Sousa W P 1984. The role of disturbance in natural communities. Annual Review of Ecology and Systematics 15, 353-391.

Sutton B G, Ting I and Troughton J H 1976. Seasonal effects on isotope composition of cactus in a desert environment. Nature 261, 42-43.

Trlica M J and Singh J S 1979. Translocation of assimilates and creation, distribution and utilization of reserves. In: Goodall D W, Perry R A and Howes K M W (eds) Arid-land ecosystems: structure, functioning and management. Vol 1. Cambridge University Press, Cambridge, London, New York and Melbourne. pp 537-571.

Van de Venter H A, Hosten L, Lubke R A and Palmer A R 1984. Morphology of *Opuntia aurantiaca* (jointed cactus) biotypes and its close relatives *O discolor* and *O salmiana* (Cactaceae). South African Journal of Botany 3, 331-339.

Vogel J C, Fuls A and Ellis R P 1978. The geographical distribution of Kranz grasses in South Africa. South African Journal of Science 74, 209-215.

Von Willert D J, Brinkman E, Scheitler B and Eller B M 1985. Availability of water controls crassulacean acid metabolism in succulents of the Richtersveld (Namib desert, South Africa). Planta 164, 44-55.

Wardlaw I F 1968. Control and pattern of movement of carbohydrates in plants. Botanical Review 34, 79-105.

Wells M J 1978. *Stipa trichotoma* Nees. In: Stirton C H (ed) Plant invaders - beautiful but dangerous. Department of Nature and Environmental Conservation of the Cape Provincial Administration, Cape Town. pp 140-143.

Werger M J A and Ellis R P 1981. Photosynthetic pathways in the arid regions of South Africa. Flora 171, 64-75.

Whiting B H 1980. Crassulacean acid metabolism in jointed cactus (*Opuntia aurantiaca*, Lindley) with special reference to the effects of Hexaflurate. MSc Thesis, University of Port Elizabeth, Port Elizabeth. pp 104.

Whiting B H and Campbell E E 1983. Some adaptations of jointed cactus (*Opuntia aurantiaca* Lindley) to semiarid ecosystems with special reference to CO_2 and H_2O vapour exchange. Proceedings of the 5th National Weeds Conference of South Africa, Stellenbosch. pp 104-112.

Woolhouse H W 1983. The effects of stress on photosynthesis. In: Marcelle R, Clijsters H and van Pouke M (eds) Effects of st.ess on photosynthesis. W Junk, The Hague. pp 1-28.

Zimmermann H G 1981. The ecology and control of *Opuntia aurantiaca* in South Africa in relation to the cochineal insect *Dactylopius austrinus*. PhD Thesis, Rhodes University, Grahamstown. 154 pp.

Chapter 15

Standing crops, growth rates and resource use efficiency in alien plant invaded ecosystems

M C Rutherford, F M Pressinger and C F Musil
Botanical Research Institute.

Introduction

The rapid spread of alien vegetation in many areas of southern Africa (Hall and Boucher 1977; Macdonald 1984) has been attributed to various factors, including a high reproductive potential and growth rate in the plants concerned. The former is often considered more critical for successful invasion. Little information has previously been collated on standing crops, growth rates and resource use efficiency of alien plants in southern Africa. This paper reviews these aspects and the influence of some important environmental factors. The nature of many published results on growth rates in southern Africa very often precludes valid statistical tests of specific hypotheses due to incomplete data sets. Rather than address untestable hypotheses in this review, a comparative approach has been followed. We compare growth rates of invasive alien plant species in different categories and assess these rates relative to those of indigenous species subject to availability of data. Where possible, comparisons are limited to those indigenous plants found in ecosystems invaded by alien plants. Relevant data on the growth rates of alien plants are nevertheless included even where suitable comparative data for indigenous plants are lacking.

There is a general lack of uniformity in expression of growth rate data in the literature which often leads to a confusion of concepts. The growth rate of a plant or plant part can be expressed in absolute or relative terms. Absolute growth rate (AGR) is the increase in plant matter, usually dry mass, per unit time, whereas relative growth rate (RGR) is the increase in plant matter per unit plant matter per unit time (Hunt 1978). The former is independant of initial mass, whereas the latter takes into account the available growth capital and is, therefore, a measure of the efficiency of plant growth. Blackman (1919) termed RGR a physiologically important efficiency index of dry weight production. RGR is calculated as an instantaneous rate. However, in practice, experimental data are available over time intervals and mean growth rates apply. Mean relative growth rates, derived from published data, were calculated as follows (Radford 1967; Hunt 1978):

$$RGR = (\ln W_2 - \ln W_1)/(t_2 - t_1)$$

where: W_1 is plant mass at time t_1 , and W_2 is plant mass at time t_2

Relative growth rates calculated or derived from some published data could not be compared where: W_1 equalled zero or values close to zero; W_2 did not reflect the total amount of plant material produced in time interval t_2 - t_1 due to losses, particularly in deciduous plants (Rutherford 1984); the interval between t_2 and t_1 was large; different generations of plants occurred in the interval t_2 - t_1; there existed different stages in plant development, age, density and life form and where different units of expression (eg mass, basal area cover, height) were used. These constraints determined the categories within which comparisons could be made.

Water-use efficiency of productivity is normally given by dry matter production per unit water consumed (Larcher 1983) but when direct measures of water consumption are unavailable, units of available water have also been employed (Noy-Meir 1973). Efficiency of dry matter production per unit nutrient utilized is difficult to express because it incorporates nutrient uptake efficiency with the often unavailable measures of efficiency of nutrient retention within the plant and efficiency of the use of nutrients in the metabolic processes. More readily expressed are nutrient uptake efficiencies.

Address for correspondence: M C Rutherford, Botanical Research Institute, Experimental Ecology Division, Private Bag X16, Rondebosch, 7700.

Comparative growth rates

Invasive aliens versus indigenous plants

Comparisons are made firstly on the basis of relative growth rate, followed by absolute growth rate, standing crop (often approximated by basal stem cross-sectional area in woody plants) and root/shoot standing crop ratio.

Relative growth rates (in terms of height) of three-year-old seedlings of *Pinus pinaster* (0,592 m $m^{-1}yr^{-1}$), *Acacia longifolia* (0,587 m $m^{-1}yr^{-1}$), and *Hakea sericea* (0,531 m $m^{-1}yr^{-1}$) are significantly higher than that of indigenous *Protea repens* (0,339 m $m^{-1}yr^{-1}$) in the south-western Cape (Richardson 1985). Similarly, the RGR (in terms of basal stem area) of *Eucalyptus grandis* (0,093 $m^2m^{-2}yr^{-1}$) (Poynton 1979b) is much greater than that found for southern Cape forest trees (0,013 $m^2m^{-2}yr^{-1}$) (Rutherford 1978) of comparable size. In some savanna woody plants in the northern Transvaal, RGR varies from 0,022-0,047 $m^2m^{-2}yr^{-1}$ (Rutherford 1978) which is often less than half the corresponding rates in alien plants. Among terrestrial alien plants, eucalypts appear to have the highest productivity (Schonau and Boden 1982). Relative growth rates of 0,14 $m^2m^{-2}yr^{-1}$, for example, have been reported for three- to five-year-old *Eucalyptus grandis* in the eastern Transvaal (Herbert et al 1982). Among the aquatics, *Azolla filiculoides* appears to have the highest RGR under natural conditions. Ashton (1982) recorded a rate as high as 0,19 g $g^{-1}day^{-1}$ in the eastern Cape during summer, but decreasing almost to zero during winter.

In terms of above-ground absolute growth rates, terrestrial alien production usually varies from about six to about 26 ton $ha^{-1}yr^{-1}$ for a variety of plantations of different ages (Schonau and Pennefather 1975; Milton 1980; Van Laar 1983; Darrow 1984). When the mean annual above-ground production for large areas of South African biomes (Rutherford and Westfall in press) is considered, only forest and the most productive areas of savanna fall within the range of production of these alien plants. Fully comparable AGR data for southern African aquatics are not available. However, other data indicate that the AGR of alien floating macrophytes, for example, 15-44 ton $ha^{-1}yr^{-1}$ for *Eichhornia crassipes* (Westlake cited in Mitchell 1974), substantially exceeds that of indigenous floating Lemnaceae (Sutton and Ornes 1975, 1977).

Meaningful comparisons can also be made using widely available rates of radial stem increment. In the southern Cape forests, a typical annual radial growth rate is 0,5-2,0 mm yr^{-1} (Rutherford 1978). The growth rate of certain indigenous species, such as *Buxus macowanii* which averages only 0,4 mm yr^{-1} over 30 years, contrasts strongly with the growth rate of the alien *Pinus radiata* with a rate of 7,2 mm yr^{-1} for the same period (King 1951). Similarly, the average growth rate of species in the Xumeni forest of Natal is 0,8 mm yr^{-1}, and includes 0,5 mm yr^{-1} for *Podocarpus henkelii* and 1,1 mm yr^{-1} for *Kiggelaria africana* (Moll and Woods 1971) which is lower than the 5,9 mm yr^{-1} for a 53 year old *Pinus patula* population south of Xumeni (Poynton 1979a). The annual radial increment of the indigenous *Trema orientalis* in the Westfalia estate area of the north-eastern Transvaal escarpment is approximately 8,4 mm (Rutherford 1978 calculated from data of Keet 1962). This is lower than the 14,5 mm yr^{-1} for the exotic *Eucalyptus grandis* in the Rooikoppies area of the eastern Transvaal (Poynton 1979b).

The above-ground standing crop of mature mountain fynbos vegetation varies from about 15-36,5 tons ha^{-1} (Kruger 1977; van Wilgen 1982) which is much lower than the 104 tons ha^{-1} for a mixed stand of *Acacia saligna* and *Acacia cyclops* in the fynbos biome (Milton 1980) and the 265 tons ha^{-1} for a *Pinus radiata* stand within an area of Mountain Fynbos (Van Laar 1983).

The basal stem cross-sectional area of 40-65 year old *Pinus* and *Eucalyptus* plantations range from, for example, 38,5 m^2ha^{-1} for *Pinus patula* in Natal (Marsh 1972), 49,5 and 65,2 m^2ha^{-1} for *Pinus pinaster* at Elgin, south-western Cape and Belfast, eastern Transvaal respectively (Poynton 1979a) to 143,9 m^2ha^{-1} for *Eucalyptus grandis* at Entabeni, north-eastern Transvaal (Poynton 1979b). Similarly, basal area of indigenous forests in the southern Cape range from 39,6-49,7 m^2ha^{-1} (C J Geldenhuys personal communication) and values as high as 144,4 m^2ha^{-1} have been reported for forests in watercourses at Richards Bay, Natal (Venter 1972). In contrast, the basal area of alien plantation stands is much higher than that of savanna woodland (eg Figure 1) where the latter seldom exceeds 16 m^2ha^{-1} (Rutherford 1982). It is clear that of the three southern African biomes with vegetation containing the largest woody plants (forest, savanna and fynbos) the standing crop of woody aliens is usually matched only by that of forest.

Many common woody alien genera are evergreen or have a relatively long (approaching 12 months) leaf duration, for example, *Acacia*, *Eucalyptus*, *Hakea*, *Pinus* and *Opuntia*. The duration of active shoot

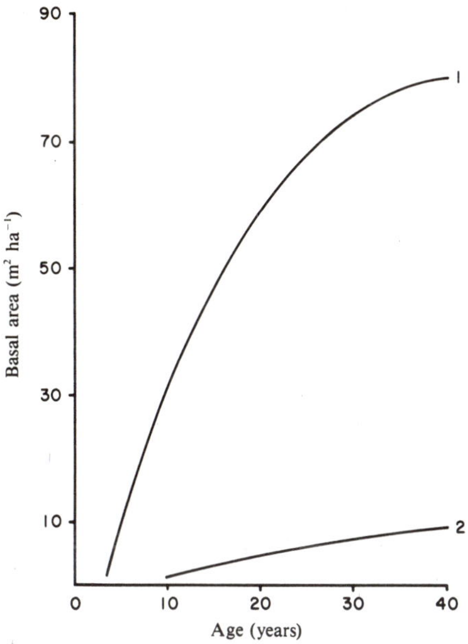

Figure 1. Stem basal area increment in 1) an unthinned stand of *Pinus radiata* in the south-western Cape (Van Laar 1982), and 2) *Brachystegia spiciformis* woodland of equivalent mean annual rainfall on the Zimbabwean highveld (Strang 1974).

growth in many of these plants occurs over a greater part of the year than many indigenous species. For example, in the fynbos biome, shoots of *Hakea sericea* (Fugler 1979) and *Acacia longifolia* (Milton and Moll 1982) grow for at least nine months of the year compared to the two to seven months for most of the indigenous species tabulated by Pierce (1984) for this biome. The period of growth of some indigenous species, for example, various thicket species, can be similar to that of alien plants (Sommerville 1981; Pierce 1984). Plant evergreenness may be linked to a higher efficiency of nutrient retention within the plant than in deciduous plants (Monk 1966).

Opportunistic shoot growth of indigenous plants of the fynbos biome is of limited occurrence (Pierce 1984). In contrast, various invasive alien species are capable of effective opportunistic shoot growth behaviour through specific shoot adaptation. Indefinite shoots with naked buds in species of, for example, *Eucalyptus* (Poynton 1979b) allow continuous use of growth capital whenever favourable conditions occur. This contrasts with many indigenous plants with definite buds and discontinuous shoot growth in which there are periodic delays in the utilization of growth capital.

All the above comparisons of standing crop are for the above-ground components. The proportion of above- and below-ground components (root/shoot standing crop ratios) can vary greatly for different plant species and can be important in qualifying comparisons of above-ground standing crops or growth rates. In most alien woody species, these are often low. For example, 0,12 in one-year-old *Hakea sericea* (Fugler 1979 - in the south-western Cape), slightly higher than 0,23 in a mixture including three-year old *Eucalyptus regans* and *Acacia dealbata* (Adams and Attiwill 1984a - in Australia), 0,06 in 35-year-old *Pinus pinea* (Rapp and Cabanettes 1981 - in southern Europe) and 0,18 in mature *Pinus radiata* (Van Laar 1983 - in the south-western Cape). Root/shoot ratios in indigenous woody plants can also vary greatly from high values in excess of 0,60, for example, in several savanna (Rutherford 1983) and resprouting fynbos plants (K B Higgins personal communication) to low ratios below 0,30, for example, in obligate seeding fynbos plants (K B Higgins personal communication). Clearly, the use of only above-ground growth rates and standing crops to derive growth rates and standing crops for whole plants may overestimate the differences between many alien species and those indigenous species with high root/shoot ratios.

Life Form

The relative differences in growth rates between various life form categories, based on available data, are presented diagramatically in Figure 2. Data for each pair of life forms are compared in turn. Data of certain indigenous plant species with a weedy nature have been incorporated to complete the comparisons.

It appears that the AGR of some herbaceous alien species is lower than that of several woody invasive alien species. For example, in the first ten months *Acacia saligna* has a growth rate of 1,9 g per month (Milton 1982) compared with 0,3 g per month for *Stipa trichotoma* (Harding 1982).

The mean annual production of submerged macrophytes is often lower than that of emergent aquatics (Gaudet 1974). For example, the annual dry matter production (above- and below-ground components) of the submerged macrophytes, *Ruppia spiralis*, a mixture of *Chara globularis* and *Lamprothamnium papulosum*, and *Potamageton pectinatus* in the southern Cape lakes is 0,8, 2,3 and 4,2 ton $ha^{-1}yr^{-1}$ respectively (Howard-Williams 1980),

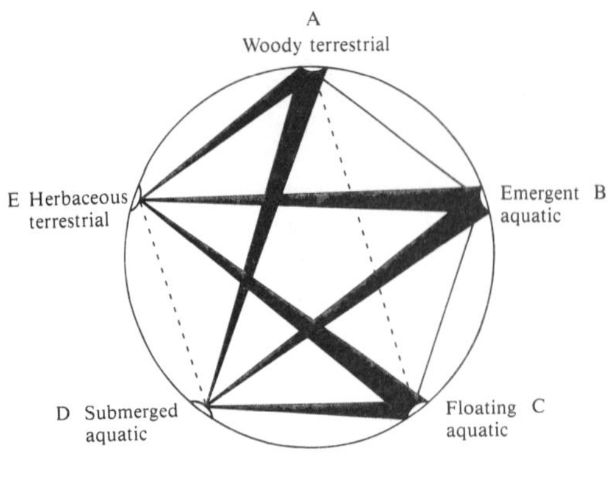

Figure 2. A diagramatic representation of relative differences in growth rates between five life-form categories, based on available data. A solid line denotes approximate equality eg B = C; a tapered band denotes inequality eg B > E; a broken line denotes uncertainty.

whereas that of the above-ground components only of the emergent macrophytes *Phragmites australis*, *Scirpus littoralis* and *Typha latifolia* is 9,0, 10,0 and 12,5 ton $ha^{-1}yr^{-1}$ respectively (Howard-Williams 1980). In the southern Orange Free State, the apparent annual production of *Phragmites australis* is 30 ton $ha^{-1}yr^{-1}$ (Wiltshire 1984). When the below-ground components of emergent aquatics is included, the difference between the submerged and emergent groups is much enhanced. Boyd (1970) derived a maximum production of 192,0 ton $ha^{-1}yr^{-1}$ for *Typha latifolia*. This, however, assumed unrealistic year-round growing conditions and can only, at best, be viewed as potential maximum production, since maximum seasonal growth rates were used to calculate annual production.

The annual production of floating macrophytes, for example, 15-44 ton $ha^{-1}yr^{-1}$ estimated elsewhere for *Eichhornia crassipes* (Westlake, cited in Mitchell 1974) is higher than that of submerged aquatics. However, it is similar to that of many southern African emergent macrophytes and some faster growing woody terrestrials, for example, 26,0 (Herbert et al 1982) to 26,7 ton $ha^{-1}yr^{-1}$ (Schonau and Pennefather 1975) for *Eucalyptus grandis* though higher than that of others, for example, 6,6 ton $ha^{-1}yr^{-1}$ for *Pinus radiata* (van Laar 1983) and 9,0 ton $ha^{-1}yr^{-1}$ for *Acacia saligna* and *A cyclops* thickets (Milton 1980).

Relative growth rates of alien floating macrophytes, for example, 0,0305-0,1227 g g^{-1} day^{-1} for *Eichhornia crassipes* in a lake along the Natal coast (Musil 1982), 0,0485-0,0861 g $g^{-1}day^{-1}$ for *Salvinia molesta* in Lake Kariba (Mitchell and Tur 1975) and circa 0,0050-0,1900 g $g^{-1}day^{-1}$ for *Azolla filiculoides* at a riverine station and farm dam in the eastern Cape (Ashton 1982), appear considerably higher than those available for woody alien terrestrials, for example 0,096 $m^2m^{-2}yr^{-1}$ (Marsh 1972 - pooled South African data) for 10-year-old and 0,018 (Burgers 1972 - eastern Transvaal) to 0,040 $m^2m^{-2}yr^{-1}$ (Marsh 1972) for 40-year old *Pinus patula*. The apparent extreme difference in RGR between the two life forms is probably due to large differences in mean plant age of individuals within populations of each group (see section on age and density).

The AGR of aquatic emergents and that of alien woody terrestrials are approximately in the same range of up to about 30 ton $ha^{-1}yr^{-1}$ (above-ground). Not unexpectedly, the above-ground standing crop of emergent macrophytes, for example, nine to 30 ton ha^{-1} (Howard-Williams 1980; Wiltshire 1984) is exceeded by an order of magnitude only by that of alien woody terrestrials, for example, 265 ton ha^{-1} for *Pinus radiata* (Van Laar 1983). The lack of successful alien emergent aquatic weeds in South Africa may be linked with the high productivity of indigenous aquatic emergents.

The AGR of submerged aquatics is lower than that of woody terrestrials. No comparisons could be made between submerged aquatics and terrestrial herbaceous forms due to inadequate data.

Age and density

RGR usually decreases with plant age, for example, in *Pinus patula* and *P elliottii* (Figure 3). This is because the proportion of growth capital decreases with an increase in plant age, most noticeably in tree forms, as necessary structural material is accumulated. In contrast, the high RGR (in excess of one per cent per day) that is maintained by many aquatic plant populations is due to their constant rejuvenation by a high rate of turnover of individuals.

Many invasive aliens can endure crowding during early stages of development. For example, 14-year-old *Pinus pinaster* in the south-western Cape tolerates densities of 150 000 stems ha^{-1} (Poynton 1979a) before further self-thinning occurs. However, high population densities can affect the growth rate of a plant, its components and its morphology. For example, the growth rate of individual *Pinus radiata* plants is inversely related to their density (Figure 4). Increasing the density of *Acacia saligna* seedlings from 722 to 1 264 plants m^{-2} decreases the average

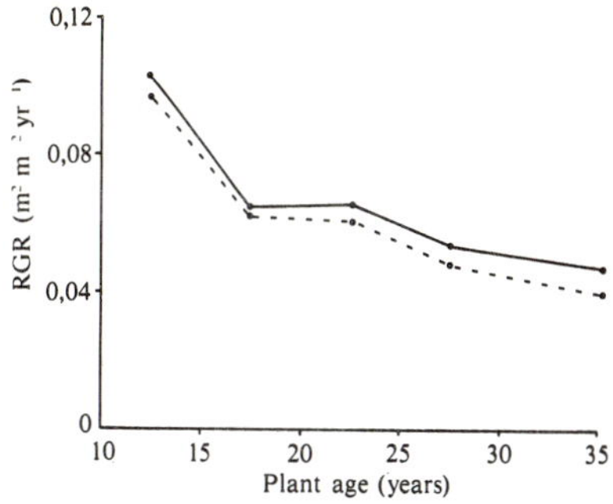

Figure 3. The relationship between plant age and relative growth rate of *Pinus patula* (broken line) and *Pinus elliottii* (solid line) on average forestry sites in South Africa (based on Marsh 1972).

plant mass from 78 to 55 g per plant over a six week period under nursery conditions (Pressinger unpublished). Similarly, the average plant biomass of young *Eucalyptus grandis* in the eastern Transvaal is reduced by 65,4% (15 056 to 5 204,1 g per plant) with a fourfold increase in density (0,25 to 1,00 plant m^{-2}) (Darrow 1984). An increase in population density also affects the proportion of leaves, bark and stemwood production in this species with the greatest growth of stemwood occurring at the highest densities (Darrow 1984).

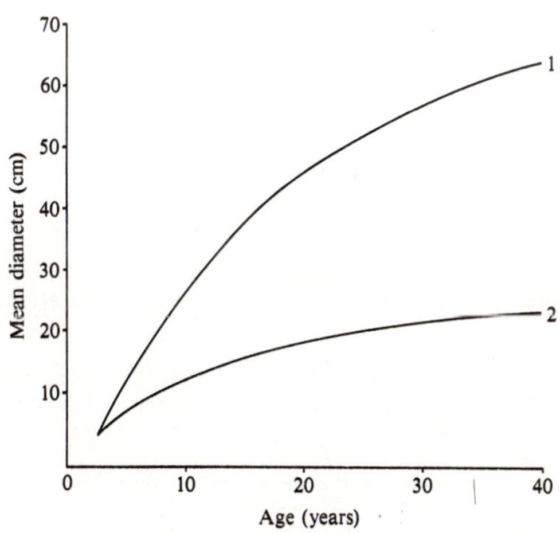

Figure 4. The relationship between plant age and mean stem diameter of *Pinus radiata* at densities of 1) 124 plants ha^{-1} and 2) 2 965 plants ha^{-1} in the south-western Cape (Van Laar 1982).

Alien aquatic macrophytes show similar responses in growth rate and morphology to changes in population density. For example in Zimbabwe, the RGR of *Salvinia molesta* plants of the "open-water-colonizing form" found in loosely crowded populations is 0,0802 leaf $leaf^{-1}day^{-1}$ compared with 0,0597 leaf $leaf^{-1}day^{-1}$ for plants of the "mat form" found in densely crowded populations (Mitchell and Tur 1975). Similarly, the average RGR of *Eichhornia crassipes* plants of the "marginal form" found in loosely crowded populations is 0,1271 g $g^{-1}day^{-1}$ compared with 0,0281 g $g^{-1}day^{-1}$ for plants of the "central form" found in densely crowded populations in Natal (Musil 1982; Musil and Breen 1985c). The change from a "marginal" to a "central" form with an increase in population density in *E crassipes* is usually reflected in a decrease in the laminar area ratio (Center and Spencer 1981), ie a reduction in the ratio of photosynthetic area to plant mass. This may be partly responsible for the reduced growth rate of densely crowded plants.

Although an increase in population density decreases the growth rate of individual plants, the greater number of plants per unit ground area compensates for the lowered individual production, so production per unit ground area tends to remain within relatively narrow limits (Yoda et al 1963). Debusk et al (1981), for example, have shown that the productivity of *Eichhornia crassipes*, the product of RGR and density, defines a bell-shaped curve with maximum productivities being achieved at intermediate densities.

Environment

Light

The high growth rates of invasive alien plant species usually occur at high light levels. A 75% reduction in daylight, through shading, reduces whole plant biomass of three to ten-month-old Acacia seedlings by 37% in *A saligna*, 38% in *A longifolia*, 54% in *A cyclops* and 56% in *A melanoxylon* (Milton 1982). Similarly, a 70% reduction in daylight reduces growth of *Stipa trichotoma* by 83% (Harding 1982), while the growth rate (annual volume increment) of juvenile *Pinus pinaster* plants in fynbos is reduced by 59% through shading by the natural vegetation (Poynton 1979a). In contrast to the intolerance to shading shown by for example, most eucalypts in southern Africa (Poynton 1979b), some aliens are shade tolerant and have high seedling survival rates, for example, *Pinus pinaster* (Poynton 1979a), which may contribute to their invasive success. Reduced light intensities also modify the relative proportion of above- and below-ground components. A 75% reduction in daylight, for example, decreases the root/shoot ratio of 10 week old *Acacia* from 0,43 to 0,25 in *A cyclops*, 0,56 to 0,35 in *A longifolia*, 0,79 to 0,37 in *A saligna* and 0,96 to 0,40 in *A melanoxylon*. These reduced ratios are due to a greater decrease in root than shoot mass (Milton 1982). Under low light conditions, the relatively smaller reduction in shoot mass in these *Acacia* species may allow compensatory reduced loss in photosynthetic area.

Changes in water turbidity, with which are associated changes in light intensity and composition, have important effects on the growth of submerged aquatic populations (Sculthorpe 1967). Musil et al (1976), for example, observed a direct relationship between the cover of *Potamogeton crispus* in a Pongola River pan and the water turbidity, plotted as percentage transmitted light. The high water turbidities found in many South African inland waters may partly explain the relatively low abundance of both indigenous and alien submerged aquatics.

Temperature

Shoot growth of *Acacia saligna* and *A cyclops* is significantly correlated with temperature in areas where water supply is not limiting for growth (Figure 5) (Milton and Moll 1982). In contrast, Taylor (1974) found no relationship between altitude (0-1280 m), normally inversely related to temperature, and annual growth of *Eucalyptus grandis*.

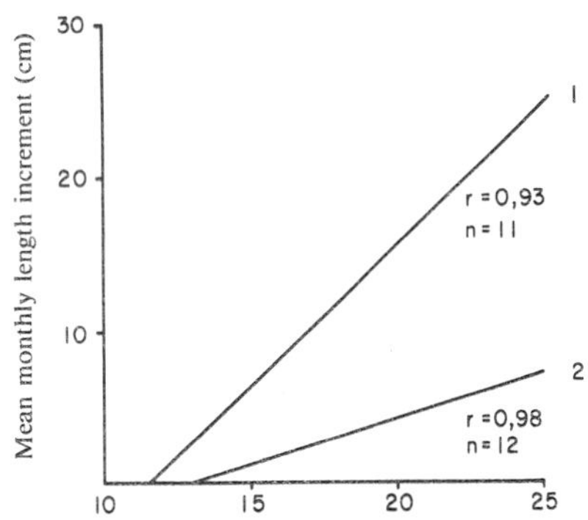

Figure 5 Relationship between shoot length increment and temperature at a site in the south-western Cape where the water supply is not limited throughout the year. 1) *Acacia saligna*; 2) *Acacia cyclops* (Milton and Moll 1982).

The growth rate of floating alien aquatics is highly correlated with temperature. For example, the RGR of *Azolla filiculoides* (Ashton 1982) and *Eichhornia crassipes* (Musil 1982) closely follows seasonal fluctuations in air and water temperature. Arrhenius plots of relative growth rates against temperatures (°K) for *Eichhornia crassipes* and *Salvinia molesta* (Table 1) give high activation energies and Q_{10} values. These indicate that the plants are extremely sensitive to changes in temperature. For example, the Q_{10} value of 2,14 in the range 15-25°C, determined for *E crassipes* (Musil and Breen 1985b) indicates that the growth rate approximately doubles with a 10°C rise in temperature.

Water

Most invasive alien plant species occur in relatively mesic habitats in southern Africa. A few notable, yet relatively restricted, exceptions include succulents, for example, *Opuntia* species. In many arid areas, alien plants are largely linked to watercourses and drainage areas. In addition, *Prosopis glandulosa* has been reported to be associated with high water tables (Vinjevold et al 1985; G B Harding personal communication). These findings indicate a higher water requirement than that of the indigenous plants of the drier surrounds. Florence (cited in Sinclair 1980) suggests that eucalypts do not close their stomata rapidly when water stressed and maintain transpiration and photosynthetic rates by relying more on access to water supply than on restriction of water loss. The sustained high growth rates of many invasive aliens is usually associated with high levels of water loss, an unavoidable consequence of common apertures for CO_2-uptake and H_2O loss. Many abundantly occurring alien invaders in southern Africa are trees. This growth form is frequently associated with a high water uptake (Box 1981) which suggests a greater water use than in other forms. Phillips (1928) reported that the major difference between *Acacia melanoxylon* and several indigenous forest species in the southern Cape is its greater water use. Similarly, Johnson (1978) indicated that water uptake by *Leptospermum laevigatum* is much more effective than that of adjacent indigenous plants. Water deficits seriously affect the growth of many alien invasives. For example, the high evapotranspiration found in species such as *Acacia pycnantha* in South Australia results in a cessation of shoot growth when plants are subjected to drought (Maconochie 1975). The basal area increment of *Pinus pinaster* in the south-western Cape is more than halved from 2,49 to 0,82

Table 1. The relationship between relative growth rate and temperature (°K) for 1) *Salvinia molesta* and 2) *Eichhornia crassipes* grown in sewage treatment ponds.

Arrhenius equation	Activation energy ($kcal\ mol^{-1}$)	Q_{10}	r	n	Reference
1 $RGR = 1{,}39 \times 10^{13}\ e^{-9476/T}$	18,84	3,01	0,78	58	Toerien et al 1983
2 $RGR = 5{,}26 \times 10^{8}\ e^{-6540/T}$	12,98	2,12	0,81	28	Musil and Breen 1985a

$m^2ha^{-1}yr^{-1}$ with a decrease from 937 to 610 mm mean annual rainfall. Many individuals of this species occurring on fine-textured soils in the summer rainfall zone of South Africa died from drought during dry winter months, even in some areas with "moderately good annual rainfall" (Poynton 1979a). The acetylene reduction activity and, possibly, the associated nitrogen fixation in *Acacia saligna* seedlings is also significantly reduced by water stress (Nakos 1977). *Acacia saligna* seedlings are most susceptable to water stress in the first four days following germination during which high mortalities can occur (Pressinger unpublished). The incomplete evidence presented may suggest that most of the important woody alien plant species in southern Africa are not efficient users of water. Among aquatic aliens, such as *Eichhornia crassipes*, research elsewhere has shown that minimum plant water-use efficiency is attained not only at low but also high extremes of plant water stress (Idso et al 1984).

Nutrients and substrate

The RGR of *Acacia cyclops* on calcareous substrates varies from 0,006 to 0,014 mm $mm^{-1}day^{-1}$ depending upon the degree of exposure to wind, but is reduced to 0,002 mm $mm^{-1}day^{-1}$ on a shallow, rocky, non-calcareous soil in the southwestern Cape (Sommerville 1981). In Australia, Hopmans et al (1983) found an increase in salinity decreased acetylene reduction activity in *Acacia dealbata* which might lower productivity.

Nutrient levels in the growth medium (substrate or water) have important effects on the growth rate of many alien plant species. An increase in soil phosphorus levels most consistently raises the productivity of many tree species in South Africa, for example, *Pinus radiata*, *Acacia mearnsii* and *Eucalyptus grandis* (Schonau 1983). Reaction to potassium and nitrogen additions is more variable. Schonau (1983) reported a positive response in growth to nitrogen additions in *Acacia dealbata*, but no response in *Acacia mearnsii*. *Eucalyptus grandis* shows a positive response in growth to potassium additions (Schonau 1983), whereas in *Pinus pinaster* the response is sometimes negative (Donald and Glen 1974). An increase in growth with the addition of phosphorus in some alien species may be associated with the coincident development of mycorrhizas, found for example, in *Pinus radiata* in Australia (Barrow 1977). A consistent response to phosphorus additions is not suprising since phosphorus levels are low in many soils of areas invaded by woody alien species in South Africa. Such additions often lead to an increase in foliar phosphorus concentrations in, for example, *Eucalyptus grandis* (Schonau and Herbert 1983). Combined nutrient applications (NPK) usually result in increased production. This is well illustrated in *Eucalyptus grandis* in the eastern Transvaal where such an application increased the mass of this plant by 29,8% and its AGR to 31,4 ton $ha^{-1}yr^{-1}$ in the first 8,5 years of growth (Herbert et al 1982). In many combined applications, however, it is still phosphorus which has the dominant effect on growth.

Aquatic alien plants show similar responses in growth to increases in nutrient levels. For example, the average seasonal RGR of *Eichhornia crassipes* in a sewage maturation pond (with nitrogen of 20746 $\mu g\ l^{-1}$ and phosphorus of 6569 $\mu g\ l^{-1}$) in the Durban district of Natal is 0,1144 g $g^{-1}day^{-1}$ compared with 0,0705 g $g^{-1}day^{-1}$ in a lake (with nitrogen of 10206 $\mu g\ l^{-1}$ and phosphorus of 150 $\mu g\ l^{-1}$) in the same area (Musil 1982). This species is exceptionally effective in the absorption of nutrients at relatively high concentrations. Potential quantities of nutrients (kg $ha^{-1}yr^{-1}$) that could, under continuous optimal growing conditions, be taken up by various floating and submerged species are given by Ashton (1979). *Eichhornia crassipes* absorbs the largest quantities of nutrients, for example, 2 112, 344, 3 400, 840 and 264 kg $ha^{-1}yr^{-1}$ for nitrogen, phosphorus, potassium, magnesium and sulphur respectively

(Ashton 1979). In many aquatic weeds, limitation of growth rate by inorganic nitrogen appears to be the most widely manifest (Mitchell 1970), though phosphorus can also be limiting, for example in Natal (Musil and Breen 1985c). Generalization on nutrient relations in aquatic weeds, however, is complicated by their relative dependencies on nutrients in the water versus those in the sediment.

Several important terrestrial alien plant species, for example, *Acacia mearnsii*, *A melanoxylon* and *A longifolia*, and some aquatic aliens, for example, *Azolla filiculoides* (Ashton and Walmsley 1976) are able to fix atmospheric nitrogen which allows them to maintain a high growth rate on nutrient poor sites. Amounts of nitrogen fixed per annum have been estimated at 0,746 kg $ha^{-1}yr^{-1}$ for *A mearnsii* at a density of 13 972 plants ha^{-1} in Australia (Laurie 1981). The rate of fixation is, however, dependent on temperature and moisture. For example, in Australia, the high rate of fixation found in *A dealbata* is dependent on sustained levels of adequate moisture supply (Adams and Attiwill 1984b).

Efficiencies of dry matter production per available nutrient by terrestrial alien plants is best evaluated under nutrient-poor conditions. An example is *Pinus pinaster* which thrives better than any other timber tree on the infertile, sandy soils of the western and southern Cape Province (Poynton 1979a). The high productivity of *P pinaster* on low-nutrient soils, together with its much lower response to the addition of fertilizers than any other introduced pine (Schutz 1976), indicates that this plant maintains optimum growth at low nutrient levels which in turn suggests efficient nutrient uptake. At certain higher nutrient levels, the uptake of some nutrients becomes less efficient. For example, on nutrient-poor substrates in the south-western Cape, the volume of *Pinus pinaster* increases by 5,4% with the first application of 19 kg phosphorus ha^{-1}, but only by 3,2% with double the amount applied (Donald and Glen 1974). *Eucalyptus grandis* responds similarly. In soils derived from Table Mountain Sandstone (generally deficient in phosphate) in the Natal midlands, the mass of *Eucalyptus grandis* increases by 30,5% with the addition of 224,5 kg phosphorus ha^{-1} (superphosphate), but only by a further 11,5% with double the amount of fertilizer added (Pennefather and MacGillivray 1971).

Efficiencies of nutrient utilization by aquatic alien plants are indicated by their kinetics of nitrogen and phosphorus limited growth. Half-saturation coefficients derived for *Eichhornia crassipes* (976 μg N l^{-1}; 94,1 μg P l^{-1}) (Musil and Breen 1985a) and *Salvinia molesta* (16,5 μg N l^{-1}; 1,0 μg P l^{-1}) (Toerien et al 1983) are low. This indicates that these plants will maintain high growth rates at low levels of nitrogen and phosphorus (given no other limiting or inhibiting condition) which suggests efficient dry matter production per available nutrient.

Schutte (1960) postulates that some invasive alien plant species in the fynbos biome may absorb trace elements more efficiently than indigenous species. Although generalizations are difficult to make for such a heterogenous group as the invasive alien plant species in South Africa, the evidence suggests that many are efficient utilizers of nutrients in terms of their productivity, though luxury consumption may apply at higher levels of nutrient availability. The capacity to utilize nutrients efficiently is potentially important in the colonization of nutrient poor habitats by such plants.

Certain alien plants contain high nutrient levels in the seed, for example, *Hakea sericea* seeds contain more phosphorus than some indigenous Proteaceae (Mitchell and Allsopp 1984). Such concentrations may boost the growth rate of young seedlings and afford a competitive advantage over other plants with lower concentrations.

Concluding remarks

Annual production of invasive alien terrestrial species usually exceeds that of indigenous vegetation. *Eucalyptus* species and *Azolla filiculoides* appear to have the highest growth rates in terrestrial and aquatic environments respectively. Irrespective of the alien or indigenous status of plants in southern Africa, productivity of emergent aquatic and woody terrestrial forms is much greater than that of submerged aquatic and herbaceous terrestrial forms. The effects of age and density on the growth rate of individual alien plants conforms to the theoretical expectations derived from a wide range of plants.

Various reasons may be advanced to explain the high growth rate of many invasive alien plant species in South Africa. Introduced species were initially selected for their rapid growth which was often associated with structural, protective and soil binding properties and specific product requirements in South Africa (Stirton 1978). The low root/shoot ratio could have been an important criterion in selecting aliens to maximize accessible and utilizable above-ground material. Many common alien species possess several of the following properties:

(1) are well adapted to high moisture conditions

and tend towards avoidance of water limitation;

(2) exhibit efficient nutrient uptake at low nutrient levels;

(3) are evergreen or have a relatively long (approaching 12 months) leaf duration;

(4) are capable of opportunistic shoot growth behaviour;

(5) contain high nutrient levels in the seed;

(6) have high nitrogen fixation capacities under favourable environmental conditions; and

(7) promote high growth rates by regular rejuvenation of growth capital through rapid population turnover rates, for example, in most invasive aquatic plant species.

Alien plant invaders of southern African ecosystems generally have a high growth rate but it is unlikely to be the sole cause of their success. Often more critical factors are those contributing to initial establishment, for example, their high reproductive capacities (see Kruger et al; Neser and Kluge this volume).

Acknowledgements

We gratefully acknowledge those who provided unpublished data, Messrs J de W Bosenberg and A P Flynn for assistance in the production of this document and the reviewers and referees for their comments.

Adams M A and Attiwill P M 1984a. Role of *Acacia* spp in nutrient balance and cycling in regenerating *Eucalyptus regnans* F. Muell. forests. I Temporal changes in biomass and nutrient content. Australian Journal of Botany 32, 205-215.

Adams M A and Attiwill P M 1984b. Role of *Acacia* spp in nutrient balance and cycling in regenerating *Eucalyptus regnans* F. Muell. forests. II Field studies of acetylene reduction. Australian Journal of Botany 32, 217-223.

Ashton P J 1979. The production and potential uses of macrophytes grown in wastewater. Unpublished report, Council for Scientific and Industrial Research. Reprint RW No 794. Pretoria.

Ashton P J 1982. An autecological investigation of *Azolla filiculoides* Lamarck, with special reference to its occurrence in the catchment area of the Hendrick Verwoerd Dam. PhD Thesis. Rhodes University, Grahamstown. pp 345.

Ashton P J and Walmsley R D 1976. The aquatic fern *Azolla* and its *Anabaena* symbiont. Endeavour 35, 39-43.

Barrow N J 1977. Phosphorus uptake and utilization by tree seedlings. Australian Journal of Botany 25, 571-584.

Blackman V H 1919. The compound interest law and plant growth. Annals of Botany 33, 353-360.

Boyd C E 1970. Vascular aquatic plants for mineral nutrient removal from polluted waters. Economic Botany 24, 95-103.

Box E 0 1981. Macroclimate and plant forms: an introduction to predictive modeling in phytogeography. W Junk, The Hague.

Burgers T F 1972. Growth prediction for sawlog, pulpwood and veneer-log regimes for *Pinus patula*. Forestry in South Africa 13, 31-44.

Center T D and Spencer N R 1981. The phenology and growth of water hyacinth *Eichhornia crassipes* (Mart) Solms in a eutrophic north-central Florida lake. Aquatic Botany 10, 1-32.

Darrow Wm K 1984. Biomass production of *Eucalyptus grandis* in South Africa planted at various close espacements: two-year results. South African Forestry Journal 131, 34-39.

Debusk T A, Ryther J H, Hanisak M D and Williams L D 1981. Effects of seasonality and plant density on the productivity of some freshwater macrophytes. Aquatic Botany 10, 133-142.

Donald D G M and Glen L M 1974. The response of *Pinus radiata* and *Pinus pinaster* to N, P and K fertilizers applied at planting. South African Forestry Journal 91, 19-28.

Fugler S R 1979. Some aspects of the autecology of three *Hakea* species in the Cape province, South Africa. MSc Thesis, University of Cape Town, Cape Town. pp 295.

Gaudet J J 1974. The normal role of vegetation in water. In: Mitchell D S (ed) Aquatic vegetation and its use and control. Unesco, Paris. pp 24-37.

Hall A V and Boucher C 1977. The threat posed by alien weeds to the Cape flora. In: Proceedings of the 2nd National Weeds Conference of South Africa, Balkema, Cape Town. pp 35-45.

Harding G B 1982. The autecology of *Stipa trichotoma* Nees. (Nasella tussock) on Rhodes Estate, in the south western Cape. MSc Thesis, University of Cape Town, Cape Town. 226 pp.

Herbert M A, Jackson D A and Verloren van Themaat R 1982. Final findings on the response of *Eucalyptus grandis* to fertilizing at De Kaap, eastern Transvaal. South African Forestry Journal 122, 53-58.

Hopmans P, Douglas L A and Chalk P M 1983. Nitrogen fixation associated with *Acacia dealbata* Link. seedlings as estimated by the acetylene reduction assay. Australian Journal of Botany 31, 331-339.

Howard-Williams C 1980. Aquatic macrophyte communities of the Wilderness lakes: community structure and associated environmental conditions. Journal of the Limnological Society of southern Africa 6, 85-92.

Hunt R 1978. Plant growth analysis. Studies in biology no 96. Edward Arnold, London.

Idso S B, Pinter P J (Jr), Reginato R J and Clawson K L 1984. Stomatal conductance and photosynthesis in water hyacinth: effects of removing water from roots as quantified by a foliage-temperature-based plant water stress index. Agricultural and Forest Meteorology 32, 249-256.

Johnson C 1978. Australian myrtle. In: Stirton C H (ed) Plant invaders - beautiful but dangerous. The Department of Nature and Environmental Conservation of the Cape Provincial Administration, Cape Town. pp 92-95.

Keet J D M 1962. The Trema plantations of Westfalia Estate. In: South African Forestry Journal 41, 15-27.

King N L 1951. Tree-planting in South Africa. Journal of South African Forestry Association 21, 1-102.

Kruger F J 1977. A preliminary account of aerial plant biomass in fynbos communities of the Mediterranean-type climate zone of the Cape Province. Bothalia 12, 301-307.

Larcher W 1983. Physiological plant ecology. Second corrected edition. Springer-Verlag, Berlin.

Lawrie A C 1981. Nitrogen fixation by native Australian legumes. Australian Journal of Botany 29, 143-157.

Macdonald I A W 1984. Is the fynbos biome especially susceptible to invasion by alien plants? A re-analysis of available data. South African Journal of Science 80, 369-377.

Maconochie J R 1975. Shoot and foliage production of five shrub species of *Acacia* and *Hakea* in a dry sclerophyll forest. Transactions of the Royal Society of South Australia 99, 177-181.

Marsh E K 1972. Some observations on the growth and characteristics of *Pinus kesiya*, *Pinus oocarpa* and *Pinus merkusii* in South Africa. Forestry in South Africa 13, 45-50.

Milton S J 1980. Studies on Australian acacias in the south-western Cape, South Africa. MSc Thesis, University of Cape Town, Cape Town. 259 pp.

Milton S J 1982. Effects of shading on nursery grown *Acacia* seedlings. Journal of South African Botany 48, 245-272.

Milton S J and Moll E J 1982. Phenology of Australian acacias in the SW Cape, South Africa, and its implications for management. Botanical Journal of the Linnean Society 84, 295-327.

Mitchell D S 1970. Autecological studies of *Salvinia auriculata*. PhD Thesis, University of London, London.

Mitchell D S 1974. The development of excessive populations of aquatic plants. In: Mitchell D S (ed) Aquatic vegetation and its use and control. Unesco, Paris. pp 38-49.

Mitchell D S and Tur N M 1975. The rate of growth of *Salvinia molesta* (*S auriculata* Auct.) in laboratory and natural conditions. Journal of Applied Ecology 12, 213-225.

Mitchell D T and Allsopp N 1984. Changes in the phosphorus composition of seeds of *Hakea sericea* (Proteaceae) during germination under low phosphorus conditions. The New Phytologist 96, 239-247.

Moll E J and Woods D B 1971. The rate of forest tree growth and a forest ordination at Xumeni, Natal. Bothalia 10, 451-460.

Monk C D 1966. An ecological significance of evergreenness. Ecology 47, 504-505.

Musil C F 1982. The use of growth kinetics in the development of a predictive model for the growth of *Eichhornia crassipes* (Mart.) Solms in the field. PhD Thesis, University of Natal, Pietermaritzburg.

Musil C F, Bornman C H and Grunow J 0 1976. Some observed interrelationships between the cover of aquatic vegetation and various physical properties of the water medium. Journal of South African Botany 42, 157-169.

Musil C F and Breen C M 1985a. The development from kinetic coefficients of a predictive model for the growth of *Eichhornia crassipes* in the field. I. Generating kinetic coefficients for the model in greenhouse culture. Bothalia 15, 689-703.

Musil C F and Breen C M 1985b. The development from kinetic coefficients of a predictive model for the growth of *Eichhornia crassipes* in the field. II. Testing and refining the model under field conditions. Bothalia 15, 705-724.

Musil C F and Breen C M 1985c. The development from kinetic coefficients of a predictive model for the growth of *Eichhornia crassipes* in the field. IV. Application of the model to the Vernon Hooper Dam - a eutrophied South African impoundment. Bothalia 15, 733-748.

Nakos G 1977. Acetylene reduction (N_2-fixation) by nodules of *Acacia cyanophylla*. Soil Biology and Biochemistry 9, 131-133.

Noy-Meir I 1973. Desert ecosystems: environment and producers. Annual Review of Ecology and Systematics 4, 25-51.

Pennefather M and MacGillivray F 1971. Fertilising of *Eucalyptus grandis* (*Saligna*) on old wattle soils. South African Forestry Journal 76, 1-5.

Phillips J F V 1928. The behaviour of *Acacia melanoxylon* R.Br. ('Tasmanian Blackwood') in the Knysna forests: an ecological study. Transactions of the Royal Society of South Africa 16, 31-43.

Pierce S M 1984. A synthesis of plant phenology in the fynbos biome. South African National Scientific Programmes Report No 88. CSIR, Pretoria.

Poynton R J 1979a. Tree planting in southern Africa. 1. The pines. Department of Forestry, Republic of South Africa, Pretoria.

Poynton R J 1979b. Tree planting in southern Africa. II. The eucalypts. Department of Forestry, Republic of South Africa, Pretoria.

Radford P J 1967. Growth analysis formulae - their use and abuse. Crop Science 7, 171-175.

Rapp M and Cabanettes A 1981. Biomass and productivity of a *Pinus pinea* L. stand. In: Margaris N S and Mooney H A (eds) Components of productivity of Mediterranean-climate regions. Basic and applied aspects. W Junk, The Hague. pp 131-134.

Richardson D M 1985. Height growth of seedlings of *Protea repens*, *Hakea sericea*, *Pinus pinaster* and *Acacia longifolia*. Unpublished report. Report 85/26 Jonkershoek Forestry Research Centre, Stellenbosch.

Rutherford M C 1978. Primary production ecology in southern Africa. In: Werger M J A (ed) Biogeography and ecology of southern Africa. W Junk, The Hague. pp 621-659.

Rutherford M C 1982. Woody plant biomass distribution in *Burkea africana* savannas. In: Huntley B J and Walker B H (eds) Ecology of tropical savannas. Ecological studies, Vol 42. Springer-Verlag, Berlin. pp 120-141.

Rutherford M C 1983. Growth rates, biomass and distribution of selected woody plant roots in *Burkea africana - Ochna pulchra* savanna. Vegetatio 52, 45-63.

Rutherford M C 1984. Relative allocation and seasonal phasing of growth of woody plant components in a South African savanna. Progress in Biometeorology 3, 200-221.

Rutherford M C and Westfall R H (in press). Biomes of southern Africa: an objective categorization. Memoirs of the Botanical Survey of South Africa.

Schonau A P G 1983. Fertilisation in South African forestry. South African Forestry Journal 125, 1-19.

Schonau A P G and Boden D I 1982. Preliminary biomass studies in young eucalypts. South African Forestry Journal 120, 24-28.

Schonau A P G and Herbert M A 1983. Relationship between growth rate, fertilizing and foliar nutrient concentrations for *Eucalyptus grandis*; preliminary investigations. Fertilizer Research 4, 369-380.

Schonau A P G and Pennefather M 1975. A first account of profits at harvesting as a result of fertilizing *Eucalyptus grandis* at time of planting in southern Africa. South African Forestry Journal 94, 29-35.

Schutte K H 1960. Trace element deficiencies in Cape vegetation. Journal of South African Botany 26, 45-50.

Schutz C J 1976. Fertilization of fast-growing pines and eucalypts in South Africa. South African Forestry Journal 98, 44-47.

Sculthorpe C D 1967. The biology of aquatic vascular plants. Edward Arnold, London.

Sinclair R 1980. Water potential and stomatal conductance of three *Eucalyptus* species in the Mount Lofty Ranges, South Australia: responses to summer drought. Australian Journal of Botany 28, 499-510.

Sommerville J E M 1981. A comparison of the seasonal growth of indigenous and non-indigenous species in the south-western Cape. Journal of South African Botany 47, 797-805.

Stirton C H 1978. Where did plant invaders come from? In: Stirton C H (ed) Plant invaders - beautiful but dangerous. The Department of Nature and Environmental Conservation of the Cape Provincial Administration, Cape Town. pp 30-35.

Strang R M 1974. Some man-made changes in successional trends on the Rhodesian Highveld. Journal of Applied Ecology 11, 249-263.

Sutton D L and Ornes W H 1975. Phosphorus removal from static sewage effluent using duckweed. Journal of Environmental Quality 4, 367-370.

Sutton D L and Ornes W H 1977. Growth of *Spirodela polyrhiza* in static sewage effluent. Aquatic Botany 3, 231-237.

Taylor F W 1974. Differences in the wood of *Eucalyptus grandis* grown in different parts of South Africa. South African Forestry Journal 91, 14-18.

Toerien D F, Cary P R, Finlayson C M, Mitchell D S and Weerts P G J 1983. Growth models for *Salvinia molesta*. Aquatic Botany 16, 173-179.

Van Laar A 1982. The response of *Pinus radiata* to initial spacing. South African Forestry Journal 121, 52-63.

Van Laar A 1983. Sampling for above-ground biomass for *Pinus radiata* in the Bosboukloof Catchment at Jonkershoek. South African Forestry Journal 125, 75-79.

Van Wilgen B W 1982. Some effects of post-fire age on the above-ground biomass of fynbos (macchia) vegetation in South Africa. Journal of Ecology 70, 217-225.

Venter H J T 1972. Die plantekologie van Richardsbaai, Natal. DSc Thesis, University of Pretoria, Pretoria.

Vinjevold R D, Bridgeford P and Yeaton D 1985. Invasive alien plants in the Namib-Naukluft Park. In: Brown C J, Macdonald I A W and Brown S E (eds) Invasive alien organisms in South West Africa/Namibia. South African National Scientific Programmes Report No 119. CSIR, Pretoria. pp 24-27.

Wiltshire G H 1984. Growth of *Phragmites* and *Typha*. Unpublished paper. Short course on aquatic weeds, identification, use and control, 17-19 July 1984. University of the Orange Free State, Bloemfontein.

Yoda K, Kira T, Ogawa H and Hozumi K 1963. Self-thinning in overcrowded pure stands under cultivated and natural conditions (Intraspecific competition among higher plants XI). Journal of Biology 14, 107-129.

Chapter 16

Life history styles of invasive fishes in southern Africa

M N Bruton
JLB Smith Institute of Ichthyology.

Introduction

The manager of an aquatic ecosystem is often involved with the introduction of fish species and must know whether a species to be introduced will become invasive, ie be able to establish self-sustaining populations in the wild. There has been a tendency in the past to address this question by reference to the taxonomic relationship of the fish, ie if one species in a family has survived, then another species is likely to do so as well. However, it is more important to consider the ability of a particular species to exploit a new environment when evaluating its likely success as an invader. Invasive ability is considered here to be a combination of the overall life history style of the species (sensu Balon 1985), as well as the more traditionally considered characteristics of environmental tolerance and trophic plasticity. The ideas of Balon (1975, 1981, 1985 and references therein), which developed mainly from studies on Northern Hemisphere fishes, are tested here on freshwater fish species from southern Africa. In this paper invasive fishes may be alien or translocated indigenous species (Bruton and Merron 1985).

Life history characteristics of fishes

Fishes are the most diverse group of living vertebrates (ie they have a greatly varied genotype) and it is becoming increasingly apparent that they may, in addition, modify their phenotype ontogenetically in order to enhance survival in a given environment (Balon 1985). Although the Linnaean system adequately classifies fish according to their phylogenetic relationships, no classification was evolved until recently to show the extent to which fish are adapted to their environment. Consequently, Balon (1975, 1985) proposed an eco-ethological classification of fishes based on reproductive guilds (Table 1). This distinguishes classes of species each similar in their use of the ecosystem and with convergent behavioural, morphological and physiological adaptations, irrespective of phyletic origin. The eco-ethological classification bears no resemblance to the Linnaean classification since some phylogenetically primitive fishes are found in advanced guilds and vice versa (Balon 1985).

Table 1. Summary of the reproductive guilds of fishes proposed by Balon (1975, 1981).

Non-guarders
- Spawners on open substrata (egg scatterers): seven guilds of pelagic, gravel, plant and sand spawners which produce buoyant or adhesive eggs and do not hide or guard the young
- Brood hiders: five guilds of beach, gravel, cave and other specialized spawners which hide the young but do not guard them

Guarders
- Substratum choosers: four guilds of pelagic, above water, rock and plant spawners whose young are tended by the parent/s
- Nest spawners: eight guilds of froth, gravel, plant, sand and other specialized spawners which make protective cavities in which they guard the young

Bearers
- External bearers: five guilds of mouth, gillchamber, pouch and other spawners which carry the young in a cavity opening to the exterior in order to protect them
- Internal bearers: four guilds of spawners which carry the young inside the body cavity where they may receive some nutrition from the parent

Reproductive guilds may be represented as alternate states from generalized to specialized types (Figure 1). The most general group is the non-guarders which spawn on open substrata and exercise no parental care. Non-guarders have a high fecundity with ova poorly endowed in nutrients. They invest a relatively small amount of energy in each of

Address for correspondence: M N Bruton, JLB Smith Institute of Ichthyology, Private Bag 1015, Grahamstown, 6140.

a large number of young. The more specialized guilds are the guarders and bearers which spawn in specially prepared nests or carry the young internally or externally but usually in the mouth in southern African freshwater fishes. They exercise intensive parental care, have a low fecundity but large-yolked ova, and invest a large amount of energy in each of a small number of young (Balon 1975, 1985).

Superimposed on this genotypic state is a phenotypic one (Figure 2) whereby a fish can alter its life history style within a generation in response to the environment (Balon 1979; Noakes and Balon

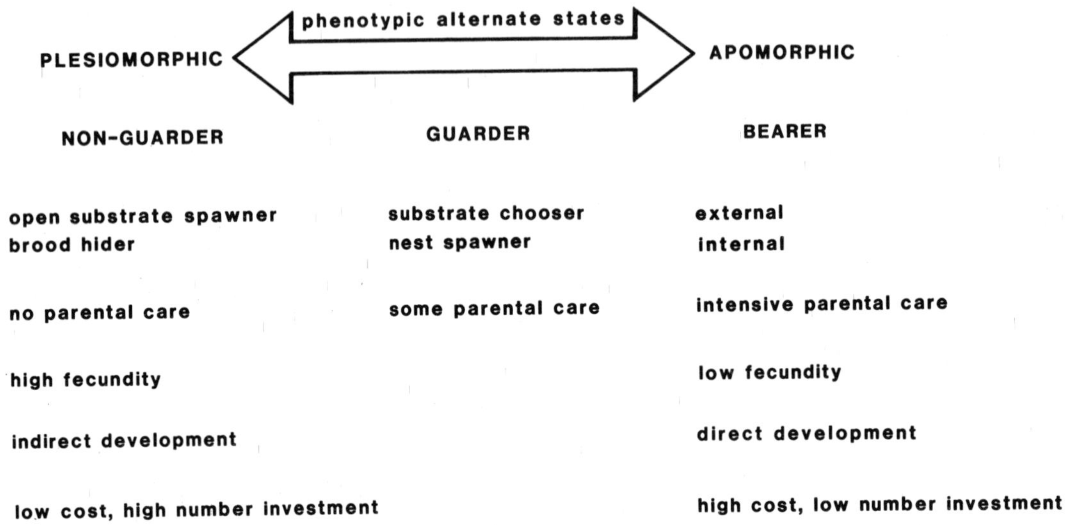

Figure 1. The genotypic alternate states represented by the different reproductive guilds of Balon (1975, 1981), showing some characteristics of different guild groups.

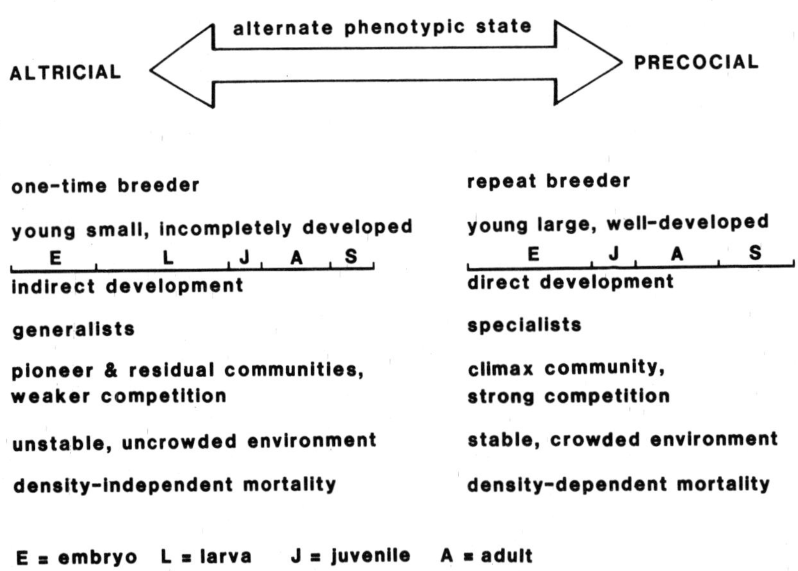

Figure 2. The alternate phenotypic states from altricial to precocial life styles in fishes following Balon (1985).

1982). Altricial fishes produce small, incompletely developed young and are generalists capable of surviving in an unstable, uncrowded environment in which they are mainly subjected to density-independent mortality. Precocial forms produce large, well-developed young and are specialists adapted to survive in a stable, crowded environment which is subject to density-dependent mortality. The interaction between the genotypic and phenotypic states allows the fish to respond successfully to different environments. The population of *Oreochromis mossambicus* in Lake Sibaya, Zululand, is regarded by Noakes and Balon (1982) as a 'well-documented example of a shift in the life history of a species as an adaptive response to .... environmental conditions'. In Lake Sibaya juvenile and adult *O mossambicus* feed on detritus, but the quality of the food differs according to water depth (Bowen 1979). Juveniles obtain a diet adequate in protein for growth, but as they become larger and feed in deeper waters, they progressively encounter protein deficiency and as a result adults in this lake grow slowly and are severely emaciated (Bruton and Allanson 1974; Bowen 1979). With water levels at or above normal, only larger adult males are reproductively active (Bruton and Boltt 1975). With higher than normal water levels, successively smaller (younger) males are recruited into the breeding stock, indicative of a shift towards a more altricial mode (ie earlier age of first reproduction). The earlier age of first reproduction (thus a shorter generation time) has a marked effect on net reproductive effort and greatly accelerates population growth (Bruton 1973; Bruton and Boltt 1975) as virtually all available resources are channelled into reproduction. This is a typical altricial response to harsh environmental conditions (ie food-restricted, fluctuating lake levels). In contrast, *O mossambicus* in the nearby Pongolo floodplain in Zululand have abundant food supplies, have a late onset of maturity and reach a large size (Bowen 1979; Bruton 1979).

Noakes and Balon (1982) reviewed the life history characteristics of 19 tilapia species in Africa and the Middle East and found that the more altricial state was characterized not only by an earlier onset of breeding but also by higher fecundity, smaller eggs, smaller egg yolks, a smaller size at first feeding, higher juvenile mortality and a shorter life span. The mechanism for changing from an altricial to a precocial state (or vice versa) is simply an alteration in the relative timing of key developmental events and intervals eg first exogenous feeding, onset of breeding (Noakes and Balon 1982; Balon 1985). The trend in evolution will be from altricial to precocial, based on the assumption that, all other things being equal, communities will tend towards a more mature, competitive (K-selected) condition through ecological succession, favouring a more precocial life style in fishes

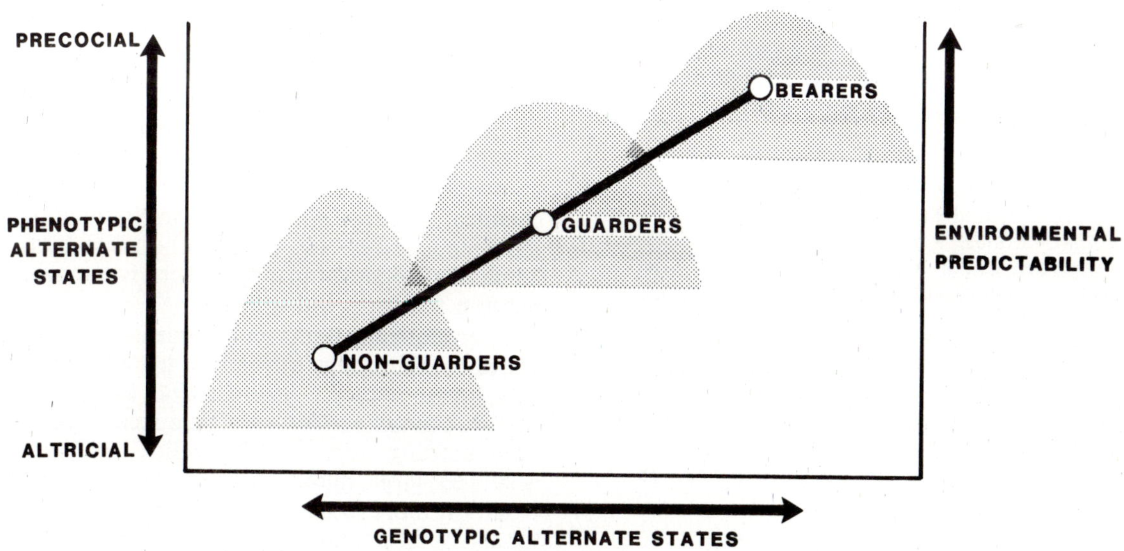

Figure 3. The proposed relationship between genotypic and phenotypic states in fishes and environmental predictability. The shaded area represents the degree of phenotypic plasticity.

(Balon 1980; Noakes 1981). Juvenilization remains as the process available for 'retracing' phylogeny, and prolonging early ontogeny, with its concomitant flexibility and adaptability (Balon 1979).

Certain fishes have a greater ability to shift from one phenotypic state to another than others and it is predicted in Figure 3 that the more generalized breeding guilds will have a more flexible phenotype than the more specialized ones. Evidence from African lakes, rivers and floodplains (Bruton and Merron unpublished) suggests that fishes from the more generalized guilds ie spawners on open substrata, are better adapted to environments characterized by abiotic harshness eg with wide and less predictable variations in water temperature, oxygen, pH, and light levels, such as in the rivers and floodplains, whereas those in more specialized guilds are better adapted to conditions of biotic harshness eg with strong competition for food and space and high rates of predation, as found in the relatively stable rift lakes in east Africa.

Life history styles of invasive fishes

The reproductive guilds of invasive fishes which occur in southern Africa, and of invasive fishes which are internationally regarded as pests or nuisance animals (Welcomme 1984), are given in Table 2. There is more or less equal representation in the two groups of guilds. Eleven invasive species which are widespread in North America, Oceania, Australia and New Zealand and also occur in southern Africa, are listed in Table 3. The first five species are non-guarders and the remaining six are guarders and bearers. It is also interesting that the same five fish families which have contributed 95% of the alien fish species in southern Africa (Table 4) also contributed 69% of the alien species which have most frequently been translocated to various countries in the world (Welcomme 1984). In the case of the southern African forms, the species in these families included 58% guarders and bearers and those in the international group, 52%.

Table 2. The reproductive guilds (sensu Balon 1975, 1981) of invasive fishes. *according to Welcomme (1984); **including South Africa, Transkei, Ciskei, Bophuthatswana, Venda, Lesotho, Swaziland, Botswana and South West Africa/Namibia.

	Non-guarders	Guarders and bearers	Total
Internationally regarded as pests*	11	16	27
Internationally regarded as nuisance animals*	10	10	20
Invasives in southern Africa**	39	25	64

Table 4. Numbers of alien fish species, and their reproductive guilds, in different families in southern Africa and internationally (data from Bruton and Merron 1985; Welcomme 1984) (NG = non-guarders, G = guarders, B = bearers).

Fish family	No of alien species: Southern Africa	No of alien species: International	Reproductive guilds
Salmonidae	4	18	NG
Cyprinidae	4	36	NG
Poeciliidae	3	10	B
Centrarchidae	5	12	G
Cichlidae	3	36	B
Other families	1	51	
Total	20	163	

The equal proportions of the two groups of reproductive guilds would suggest that invasive fishes do not have one particular style of life history. On closer examination, however, it is apparent that two kinds of invasions have taken place in southern Africa (and elsewhere), ie on the one hand into abiotically harsh environments subject to physico-chemical extremes which are sparsely inhabited by indigenous fishes, and on the other, into biotically harsh environments in which there is intense competition for food and space.

The abiotically harsh environments have been invaded by fishes such as rainbow trout *Salmo gairdneri*, common carp *Cyprinus carpio* and sharptooth catfish *Clarias gariepinus*. These species belong to non-guarding reproductive guilds and produce relatively large numbers of young. The rainbow trout, the more specialized of the three (Balon 1980), inhabits cold streams at high altitudes where no large indigenous fish predator (except *Sandelia bainsii* in the eastern Cape) occurs, whereas the carp and sharptooth catfish thrive in unstable and often turbid man-

Table 3. Invasive fish species in southern Africa which have successfully invaded several other countries or continents (data from Courtenay and Stauffer 1984; Bruton and Merron 1985).
*numbers refer to the number of states in which the species has been recorded (after Courtenay et al 1984).
(P = internationally regarded as pests or N = nuisance animals, Welcomme 1984, I = indigenous).

		Place invaded							
Fish species	Common name	Continental USA*	Canada	Mexico	Hawaii	Oceania	Puerto Rico	Australia	New Zealand
Non guarders									
Egg scatterers									
Cyprinidae									
Carassius auratus	goldfish (N)	43	X	X	X	X	X	X	X
Cyprinus carpio	common carp (N)	48	X	X	X	X	X	X	X
Brood hiders									
Salmonidae									
Salmo gairdneri	rainbow trout (N)	I	I	X	X		X	X	X
S trutta	brown trout (N)	40	X		X		X	X	X
Salvelinus fontinalis	brook trout (N)	I	I	X	X				
Guarders									
Centrarchidae									
Lepomis macrochirus	bluegill sunfish	I	I		X		X		
Micropterus dolomieu	smallmouth bass	I	I	X	X	X			
M salmoides	largemouth bass (N)	I	I		X	X	X		
Bearers									
Poeciliidae									
Gambusia affinis	mosquitofish (N)	I	X	X	X	X	X	X	X
Poecilia reticulata	guppy (P)	7	X	X	X	X	X	X	X
Xiphophorus helleri	swordtail (P)	3	X	X	X	X	X	X	
Cichlidae									
Oreochromis mossambicus	Mozambique tilapia (P)	5		X	X	X	X	X	

made lakes and manipulated rivers (Bruton and de Moor in press). Likewise, the freshwater sardine *Limnothrissa miodon*, a non-guarding egg scatterer, has successfully invaded the species-poor (contrary to Moyle 1985) pelagic habitat in man-made Lake Kariba and also invaded the open water sections of the Middle Zambezi River (CIFA 1985; Bruton et al 1982).

Biotically harsh, highly competitive environments have been invaded by bluegill *Lepomis macrochirus*, largemouth bass *Micropterus salmoides*, guppy *Poecilia reticulata*, mosquitofish *Gambusia affinis* and Mozambique tilapia *Oreochromis mossambicus*. These species belong to guarding and bearing reproductive guilds and produce a few, well-protected young. They are capable of outcompeting indigenous fishes and establishing viable populations within existing fish communities. *Gambusia affinis* (a live bearer) is notorious for its ability to outcompete indigenous fish populations through predation on the young (Myers 1965; Hurlbert et al 1972; Arthington et al 1983), and viable populations have been established in several localities in southern Africa (Bruton and de Moor in press). *Oreochromis mossambicus*, a mouthbrooder which is indigenous to southern Africa, is one of the most successful and vagile invaders internationally, having established populations in North and South America, Australia,

Hawaii and elsewhere (Courtenay and Stauffer 1984).

An organism's tolerance of a wide range of environments is dependent on several factors, including its physiological adaptability (Calow 1978) and its ability to adopt different life history options (Balon 1985). *Oreochromis mossambicus* is both tolerant of wide ranges of water temperatures, salinities, depths and oxygen saturations (Philippart and Ruwet 1982) as well as able to adopt altricial or precocial life history styles in accordance with the nature of the environment (Bruton and Boltt 1975; Noakes and Balon 1982). The plasticity of the phenotype, as reflected in the physiology and life history style, appears to be a characteristic of successful invasive fishes. It is also likely that euryphagic fishes will be successful invaders (eg the sharptooth catfish *Clarias gariepinus*, smallmouth bass *Micropterus dolomieu*). There are, however, also several examples of successful invasive fishes which are stenophagic but which feed on a food resource which is usually abundant (eg *Oreochromis mossambicus* which mainly feeds on detrital and epiphytic diatoms and bacteria, and the Chinese carps which feed on aquatic macrophytes).

Altricial fishes also appear to be well-adapted to take advantage of underutilized food resources in invaded environments, even if there is a well established fish community there. A good example is the Nile perch *Lates niloticus* which has successfully invaded the multispecific community of Lake Victoria and now preys heavily on the endemic cichlid fishes there (Barel et al 1985; Coulter et al 1986). There are no large indigenous fish predators in Lake Victoria (other than crocodiles and piscivorous birds) with the result that the Nile perch is preying on 'naive' prey with no previous experience of a large fish predator.

Thus we can predict that altricial fishes will be most competitive in abiotically harsh environments and precocial fishes in biotically harsh environments. Some altricial fishes, with their wide phenotypic plasticity, higher fecundity and greater adaptability are also likely to succeed in environments in which a major food resource is underutilized. Thus, although Moyle (1985) has suggested that 'It would seem that successful introductions are theoretically possible to virtually any environment that can support fish', a pattern does seem to be emerging. This suggests agreement with Kohler and Stanley (1984) and Moyle (1985), that 'reproductive strategy per se is of little use in predicting whether or not an introduced species is likely to be successful in a new environment'. However it appears as if the ability of a species to shift from one phenotypic state to another in order to 'match' its environment does have some significance.

Invasive dormancy

Although some very competitive species, eg *Gambusia affinis*, *Oreochromis mossambicus*, may exert their dominance in stable climax communities, other invasive species appear to lie dormant as a relatively insignificant element of a host community eg tench *Tinca tinca* and perch *Perca fluviatilis* in southern Africa (Bruton and de Moor in press) and in Australia (Tilzey 1980). Other species are dormant under some circumstances but may thrive and outcompete the indigenous community if the environment is stressed (Welcomme 1984) by overfishing, pollution, the downstream effects of impoundments, or increased turbidity. The common carp is an example of an invasive species which thrives in a stressed environment in southern Africa eg in rivers or dams with increased turbidity (Bruton 1985). The species which benefit from the environmental perturbations are likely to be non-guarders or the altricial forms of guarders and bearers, which are more tolerant of abiotic harshness.

Invasive vigour

It appears that certain invasive species (eg *Oreochromis mossambicus*) attain enhanced vigour and colonize habitats which they do not normally occupy in their natural range when they are introduced into new ranges (Philippart and Ruwet 1982). *Oreochromis mossambicus* has been widely translocated to islands of the Pacific Ocean and breeding populations have been recorded on coral reefs in the sea (Philippart and Ruwet 1982; Welcomme 1984), whereas this species is confined to fresh and brackish waters in its natural range in Africa. The sharptooth catfish *Clarias gariepinus* has invaded rivers in the eastern Cape through the Orange-Fish tunnel and now inhabits environments with a different temperature regime from those in its natural range (Laurenson and Hocutt 1986). We can expect that environmental manipulations by man will in future provide many opportunities for invasion by adaptable fishes. Whereas invasive vigour may be explained by the absence of natural controls eg predators, parasites and disease, or as a result of selection in the alien environment, it may also be a result of their ability to adopt alternate life styles in response to a new environment.

Phenotypic plasticity has long been recognized as a major adaptation and it has been regarded as a typical quality of colonizing or early successional plants (Bradshaw 1965; Hickman 1975). However, plasticity covers a variety of complex phenomena and hard evidence in support of it being adaptive in animals, ie having arisen by natural selection, is sparse (Stearns 1982). Invasive animals would appear to provide excellent subjects for study in this field, as would pioneering communities of endorheic lakes, such as Lakes Ngami and Chilwa in southern Africa, which undergo periodic desiccation and refilling.

Conclusions

These findings have relevance to the management of invasive fishes. A knowledge of the reproductive style of a species and of its ability to shift from one phenotypic state to another may allow predictions of the likelihood of an introduced species being able to establish self-sustaining populations in different environments. It also appears that species with wide environmental tolerances within either their native or invaded ranges are most likely to be successful invaders. The latter is an easy rule of thumb for administrators to use, but needs to be further validated before being put into general use.

The successful management of southern African aquatic systems impacted by invasives requires a clear understanding of the status, behaviour and ecological potential of the species concerned. It is clear that there is not one optimal strategy for an invasive fish, but several, both between and within species. Although some extrapolations can be made from studies in northern temperate environments, an understanding of the ecology of invasions in southern Africa will depend on our knowledge of local interactions. This will be a long process as we are only now beginning to comprehend their complexity. In the mean time it is essential that we do not exacerbate matters by unwisely introducing and translocating further invasives.

Acknowledgements

I am grateful for the assistance of J Cooper, A A Ferrar, P B N Jackson, R Kluge, F J Kruger, A J Ribbink, P H Skelton and H Tomlinson in compiling this manuscript. I have also benefited greatly from discussions with Professor E K Balon of the University of Guelph, Canada, on the life history styles of fishes. This research was funded by the Invasive Biota Working Group of the National Programme for Ecosystem Research of the CSIR.

Arthington A H Milton D A and McKay R J 1983. Effects of urban development and habitat alterations on the distribution and abundance of native and exotic freshwater fish in the Brisbane region, Queensland. Australian Journal of Ecology 8, 87-101.

Balon E K 1975. Reproductive guilds of fishes: a proposal and definition. Journal of the Fisheries Research Board of Canada 32, 821-864.

Balon E K 1979. The juvenilization process in phylogeny and the altricial to precocial forms in the ontogeny of fishes. Environmental Biology of Fishes 4, 193-198.

Balon E K 1980. Charrs: Salmonid fishes of the genus *Salvelinus*. Perspectives in vertebrate science. 1. W Junk, The Hague.

Balon E K 1981. Additions and amendments to the classification of reproductive styles in fishes. Environmental Biology of Fishes 6, 377-389.

Balon E K (ed) 1985. Early life histories of fishes. New developmental, ecological and evolutionary perspectives. W Junk, The Hague.

Barel C D B, Dorit R, Greenwood P H, Fryer G, Hughes N, Jackson P B N, Kawanabe H, Lowe-McConnell R H, Nagoshi M, Ribbink A J, Trewavas E, Witte F and Yamoaka K 1985. Destruction of fisheries in Africa's lakes. Nature 315, 19-20.

Bowen S H 1979. A nutritional constraint in detritivory by fishes: the stunted population of *Sarotherodon mossambicus* in Lake Sibaya, South Africa. Ecological Monographs 49, 17-31.

Bradshaw A D 1965. Evolutionary significance of phenotypic plasticity in plants. Advances in Genetics 13, 115-155.

Bruton M N 1973. A contribution to the biology of *Tilapia mossambica* Peters in Lake Sibaya, South Africa. MSc Thesis, Rhodes University, Grahamstown.

Bruton M N 1979. The fishes of Lake Sibaya. In: Allanson B R (ed). Lake Sibaya. Monographiae Biologicae 36. W Junk, The Hague.

Bruton M N 1985. The effects of suspensoids on fishes. Hydrobiologia 125, 221-241.

Bruton M N and Allanson B R 1974. The growth of *Tilapia mossambica* Peters (Pisces: Cichlidae) in Lake Sibaya, South Africa. Journal of Fish Biology 6, 701-715.

Bruton M N and Boltt R E 1975. Aspects of the biology of *Tilapia mossambica* Peters (Pisces: Cichlidae) in a natural freshwater lake (Lake Sibaya, South Africa). Journal of Fish Biology 7, 423-446.

Bruton M N and De Moor I (in press). Invasive aquatic animals in southern Africa: an atlas and detailed assessment of the problem. South African National Scientific Programmes Report.

Bruton M N, Jackson P B N and Skelton P H 1982. Pocket guide to the freshwater fishes of southern Africa. Centaur, Cape Town.

Bruton M N and Merron S V 1985. Alien and translocated aquatic animals in southern Africa: a general introduction, checklist and bibliography. South African National Scientific Programmes Report No 113. CSIR, Pretoria.

Calow P 1978. Life cycles. An evolutionary approach to the physiology of reproduction, development and ageing. Chapman and Hall, London.

CIFA 1985. Introduction of species and conservation of genetic resources. Committee for Inland Fisheries of Africa Report 85/13. Food and Agriculture Organization. 18 pp.

Coulter G H, Allanson B R, Bruton M N, Greenwood P H, Hart R C, Jackson P B N and Ribbink A J (in press). Special features of the African Great Lakes and their susceptibility to the effects of human activities. Environmental Biology of Fishes.

Courtenay W R, Hensley D A, Taylor J N and McCann J A 1984. Distribution of exotic fishes in the continental United States. In: Courtenay W R and Stauffer J R (eds) Distribution, biology and management of exotic fishes. Johns Hopkins University Press, Baltimore. pp 41-77.

Courtenay W R and Stauffer J R (eds) 1984. Distribution, biology and management of exotic fishes. Johns Hopkins University Press, Baltimore.

Hickman J 1975. Environmental unpredictability and plastic energy allocation strategies in the annual *Polygonum cascadense* (Polygonaceae). Journal of Ecology 63, 689-701.

Hurlbert S H, Zedler J and Fairbanks D 1972. Ecosystem alteration by mosquitofish (*Gambusia affinis*) predation. Science 175, 639-641.

Kohler C C and Stanley J G 1984. A suggested protocol for evaluating proposed exotic fish introductions in the United States. In: Courtenay W R and Stauffer J R (eds) Distribution, biology and management of exotic fishes. Johns Hopkins University Press, Baltimore.

Laurenson L J B and Hocutt C H 1986. Colonization theory and invasive biota: the Great Fish River, South Africa, a case history. Environmental Monitoring and Assessment 6, 71-90.

Moyle P B 1985. Exotic fishes and vacant niches. Environmental Biology of Fishes 13, 315-317.

Myers G S 1965. *Gambusia*, the fish destroyer. Tropical Fish Hobbyist No 16, 31-32, 53-54.

Noakes D L G 1981. Comparative aspects of behavioural ontogeny: a philosophy from fishes. In: Immelmann K, Barlow G W, Petrinovich L and Man M (eds) The interdisciplinary study of early development. Cambridge University Press, Cambridge. pp 491-508.

Noakes D L G and Balon E K 1982. Life histories of tilapias: an evolutionary perspective. In: Pullen R S V and Lowe-McConnell R H (eds) Biology and culture of tilapias. ICLARM, Manila. pp 61-82.

Philippart J-Cl and Ruwet J-Cl 1982. Ecology and distribution of tilapias. In: Pullen R S V and Lowe-McConnell R H (eds) Biology and culture of tilapias. ICLARM, Manila. pp 15-59.

Stearns S C 1982. The role of development in the evolution of life histories. In: Bonner J T (ed) Evolution and development. Springer-Verlag, Berlin. pp 237-258.

Tilzey R D J 1980. Introduced fish. In: Williams W D (ed) An ecological basis for water resource management. ANU Press, Canberra.

Welcomme R L 1984. International transfers of inland fish species. In: Courtenay W R and Stauffer J R (eds) Distribution, biology and management of exotic fishes. Johns Hopkins University Press, Baltimore. pp 22-40.

Chapter 17

The differential invasion of southern Africa's biomes and ecosystems by alien plants and animals

I A W Macdonald, F J Powrie and W R Siegfried
University of Cape Town.

Introduction

It is commonly observed that natural ecosystems differ in the extent to which they have been invaded by alien organisms (Elton 1958). For example, islands and mediterranean climate regions often show a high incidence of alien invasions (Ferrar and Kruger 1983).

Within southern Africa, variations in the extent of alien plant invasions between biomes and between ecosystem types within biomes have been noted by several authors.

Differential invasion of southern African biomes by plants

Using as their measure of the extent of invasions the number of species of invasive alien plants recorded from different types of ecosystem within South Africa, Wells et al (1983) concluded that aquatic ecosystems were the most severely affected with streambank ecosystems second and the terrestrial habitats of the winter rainfall region third. Without data several other authorities have contended that the fynbos biome is especially susceptible to invasion by alien trees and shrubs (Hall and Boucher 1977; Taylor 1977; Campbell et al 1979). In attempting to test this hypothesis, Macdonald (1984) concluded that there were strong indications that the fynbos biome had more extensive infestations of alien trees and shrubs than the other South African biomes. However, the quality of the available data did not allow statistical tests of the hypothesis. In addition, any such differences in the extent of invasions could not be taken as being proof of differences in the inherent susceptibility to invasion of the biomes, because the patterns are confounded by differences in the extent of human disturbance and the history and extent of alien plant introductions to the various biomes (Macdonald 1984).

Address for correspondence: I A W Macdonald, Percy FitzPatrick Institute of African Ornithology, University of Cape Town, Rondebosch, 7700.

Differential invasion of southern African ecosystems by plants

There have been several southern African studies which allow comparisons to be made of the relative extent of alien plant invasions of different types of ecosystem within a biome. However, most of these studies have been carried out for individual species and it is difficult to extract an overall index of the extent of alien plant invasions of the different ecosystems involved (eg Henderson and Musil 1984; Richardson 1984; Taylor et al 1985). In a study in the Hluhluwe-Umfolozi Game Reserve a crude measure of the relative frequency of occurrence of 16 species of tree, shrub and woody creeper indicated that disturbed areas around human habitations were the most invaded, with riverine ecosystems being second (Macdonald 1983). In an intensive survey of the central Transvaal highveld, Wells et al (1979) found the average number of alien woody plant species and mean abundance rating to be highest in bushveld ecosystems, intermediate in bankenveld (a transitional grassland type with scattered trees) and lowest in true grassland ecosystems. However, working as they did in an extensively transformed and otherwise man-modified area, they found alien plant infestations to be generally highly correlated with intentional plantings. Thus their data on the relative extent of infestations might not reflect inherent differences in susceptibility to invasion. Although the techniques used by Wells et al (1979) do not allow for a direct comparison between riverine and upland ecosystems it is apparent from their account that they, too, found the riverine ecosystems to be the most severely invaded. They classified the riverine ecosystems sampled into those along rivers, along perennial streams, and along temporary streams. Alien woody species were present in the riverine fringing vegetation in 100%, 69% and 16% of the samples in these three types of riverine ecosystems, respectively, with more than four species being present in 34,8%, 34,9% and 2,9% of the samples (Wells et al 1979). Alien plant species in the arid regions of southern Africa are restricted almost entirely to riverine ecosystems, as has been recorded in the Kalahari Gemsbok National Park (Macdonald in preparation) and in the Namib Desert (Tarr and Loutit 1985; Vinjevold et al 1985; Macdonald and Nott in press).

The only quantitative study in southern Africa which has not shown riverine ecosystems to be more densely invaded than adjacent upland ecosystems is the aerial photographic assessment of Boucher (1981), working in the south-western Cape lowlands. Here he found 30% of the remaining areas of Coastal Renosterveld to be invaded by alien woody plants, 27% having dense stands. Seventeen per cent of the length of watercourses were invaded, with only seven per cent densely invaded. Strandveld had 16% of its remaining area invaded and Coastal Fynbos, eight per cent. The relatively low level of invasion of riverine ecosystems in this study is at variance with the consensus opinion of a panel of fynbos researchers and managers (Macdonald and Jarman 1984). It is possible that alien trees were undercounted in the aerial photographic analysis of riverine ecosystems, since this habitat has indigenous tree species present which would be difficult to distinguish from alien species.

Faunal invasions

The extent of faunal invasions of different biomes or ecosystems within southern Africa has not been analysed. However, Van Bruggen (1964) provides data on the number of alien terrestrial Mollusca recorded from different localities. It is possibly significant that the three port cities in the fynbos biome, Cape Town, Port Elizabeth and East London all showed high numbers of species (17, 9 and 5 respectively) compared to the two subtropical ports, Durban (4) and Maputo (1). The cities of Johannesburg and Bloemfontein, in the grassland biome, both had only a single species (Van Bruggen 1964).

Background to the present study

Alien rodent invasions in southern Africa

In this study, we have attempted to analyse the results of two naturally occurring experiments as a preliminary test of the hypothesis that South African biomes and ecosystems vary in their susceptibility to invasion. The first of these is the differential invasion of biomes and ecosystems by two introduced rodents the house rat *Rattus rattus* and the house mouse *Mus musculus*, as reflected in the incidence of their remains in owl pellets.

These rodents are commensal with man and have been unintentionally carried to localities scattered throughout the globe. Individuals have evidently been introduced into all but the most remote areas as they are accidentally translocated by road and rail transport and have been present in the sub-continent for centuries (cf Smithers 1975, 1983; Plug et al 1979). Both species are capable of surviving under a much wider range of climatic conditions than is present in the subcontinent (Hanney 1975). In a variety of insular ecosystems and in certain continental ecosystems both species have been recorded invading natural and seminatural ecosystems (Fox 1983; Smithers 1983; Atkinson 1985; Cooper and Brooke this volume). Although *Mus musculus* is stated to be "closely confined to dwelling houses, outbuildings, ... especially where foodstuffs are stored" within southern Africa (Smithers 1983), it has been recorded up to 1,5 km from human habitation and is known to live in burrows in these outlying areas (Shortridge 1934). *Mus musculus* has established a feral population in a small fynbos nature reserve completely surrounded by a suburban area (Middlemiss 1975). This is possibly analogous to the hillside habitat within Francistown, Botswana, where the species was also collected (Smithers 1971). The species is considered to have established a "feral population" at Sandvis Harbour in the Namib Desert (Stuart 1975a,b; Griffin and Panagis 1985), centered on a deserted building (C T Stuart personal communication). *Rattus rattus* has been found in several localities within natural and seminatural vegetation within the subcontinent: Davis (1974) states "wild-living populations not uncommon, but main habitat attachment is to houses, outbuildings and huts". *Rattus rattus* has been recorded in areas of natural and seminatural vegetation in the forests of the Transkei (Shortridge 1934), around Durban (Shortridge 1934) and at Oribi Gorge Nature Reserve in Natal, "two kilometres from nearest habitation" (Bourquin and Mathias 1984), on the Springbok Flats in Transvaal savanna "one kilometre from the nearest deserted farmhouse" (J Mendelsohn 1982 and personal communication), in the fynbos biome at the Cape of Good Hope (D L Clark personal communication) and Assegaaibos Nature Reserves (Stuart 1976), and at Sandvis Harbour on the Namib Desert coast (Griffin and Panagis 1985). Since the degree of invasion by *Mus musculus* and *Rattus rattus* is variable, an improved understanding might be achieved of factors limiting alien invasions by analysing the characteristics of localities where either or both of these rodents have invaded the small mammal community and contrasting these with those of localities where they are not known to be present.

The use of barn owl *Tyto alba* pellets for surveying for the occurrence of alien rodents

Rodents make up >75% of vertebrate prey items in the pellets of the barn owl *Tyto alba* in all regions of southern Africa (Vernon 1972a). The species composition of the diet varies between regions, within a region between sites, and at a site between years and seasons (Kolbe 1946; Vernon 1972a; Dean 1973, 1978; Macdonald and Dean 1984). *Tyto alba* is a generalist predator feeding on whatever prey items it can secure but with a preference for small mammals where these are available (Vernon 1972a; Macdonald and Dean 1984).

This low selectivity in prey choice is understandable given the owl's ability to locate its prey using sound cues alone (Payne 1971). However, the occurrence of prey species in the diet does not always reflect their abundance in the environment as determined by trapping (Hanney 1962; Coetzee 1963). This might be due to behavioural differences between prey species (Glue 1970; Perrin 1982) and habitat selection by the owls when hunting (Morton and Martin 1979). But the biases in the trapped sample are often marked (see references in Macdonald and Dean 1984) and there are no published data to show greater bias in the diet of the barn owl.

In southern Africa both *Mus musculus* and *Rattus rattus* are rarely recorded in *T alba* pellet analyses (Vernon 1972a; Dean 1978; Macdonald and Dean 1984). Experience elsewhere has shown that these species are not behaviourally immune to predation by barn owls (Glue 1974; Herrera and Jaksic 1980). In some localities where these rodents have been introduced they have become the dominant prey items (Abs et al 1965; Morton and Martin 1979; Valente 1981).

The use of the number of alien species recorded from nature reserves as an index of their susceptibility to invasion

The second naturally occurring experiment considered here is the invasion of nature reserves by alien species. In these areas the general management objective is the maintenance of indigenous communities of species. Even in the most intensively managed reserves in the subcontinent an invasive alien plant species, once present, is very seldom completely eradicated from the reserve (Macdonald unpublished). Comparisons of the numbers of alien species recorded in reserves might provide some indication of the susceptibility of a reserve to invasion. This is more particularly the case when one considers only those species invading the reserve which are not known to have been intentionally introduced to the reserve at some time in its history. Although there are uncontrolled variables present, such as the number of species that have been introduced into the reserve since its inception, there is little chance of controlling for these variables. Given the wide range of successfully invasive species present in the subcontinent (Wells et al 1986), and the extremely efficient dispersal mechanisms shown by many of these species, one may assume that the number of species established within a particular reserve is a measure of its susceptibility to invasion rather than simply reflecting its exposure to invasion. In order to test this assumption various measures such as the distance of the reserve from towns of different sizes, the presence of influent rivers and the annual number of visitors, were incorporated in the analysis.

Methods

The alien rodent/owl pellet data set

Data on the prey in pellets of the barn owl *Tyto alba* were collected for sites scattered throughout southern Africa (Figure 1). These data were assembled from the published literature (excluding those cited in previous sections these were Davis 1958, 1959; Bateman 1960; De Graaf 1960; Malherbe 1963; Pocock 1963; Nel and Nolte 1965; Winterbottom 1966; Niethammer 1968, 1975; Wilson 1970; Vernon 1971, 1972b, 1980a,b,c, 1981, 1983; Grindley et al 1973; Dean 1975; Avery 1977; Van der Merwe 1980; Brain 1981; Tilson and Le Roux 1983; Steyn 1984), from one of the authors' own records (Macdonald), and from unpublished records of ornithologists. These prey analyses were paired with data on the biome, climate, ecosystem type, extent of human disturbance and the number of species of predators of rodents likely to be present at the collection locality. The data were analysed to test for association between the presence or absence of alien rodents (*Rattus rattus* and *Mus musculus*) at each collection locality and such categorical information as biome type and ecosystem type. In addition, different derived variables describing the prey collection itself were analysed for differences between collections with and without alien rodents.

The collections made in different years at a site were analysed separately as, in some cases, these collections had been made by different workers and the exact collection locality might have differed. Also, invasions by alien rodents could have occurred in the

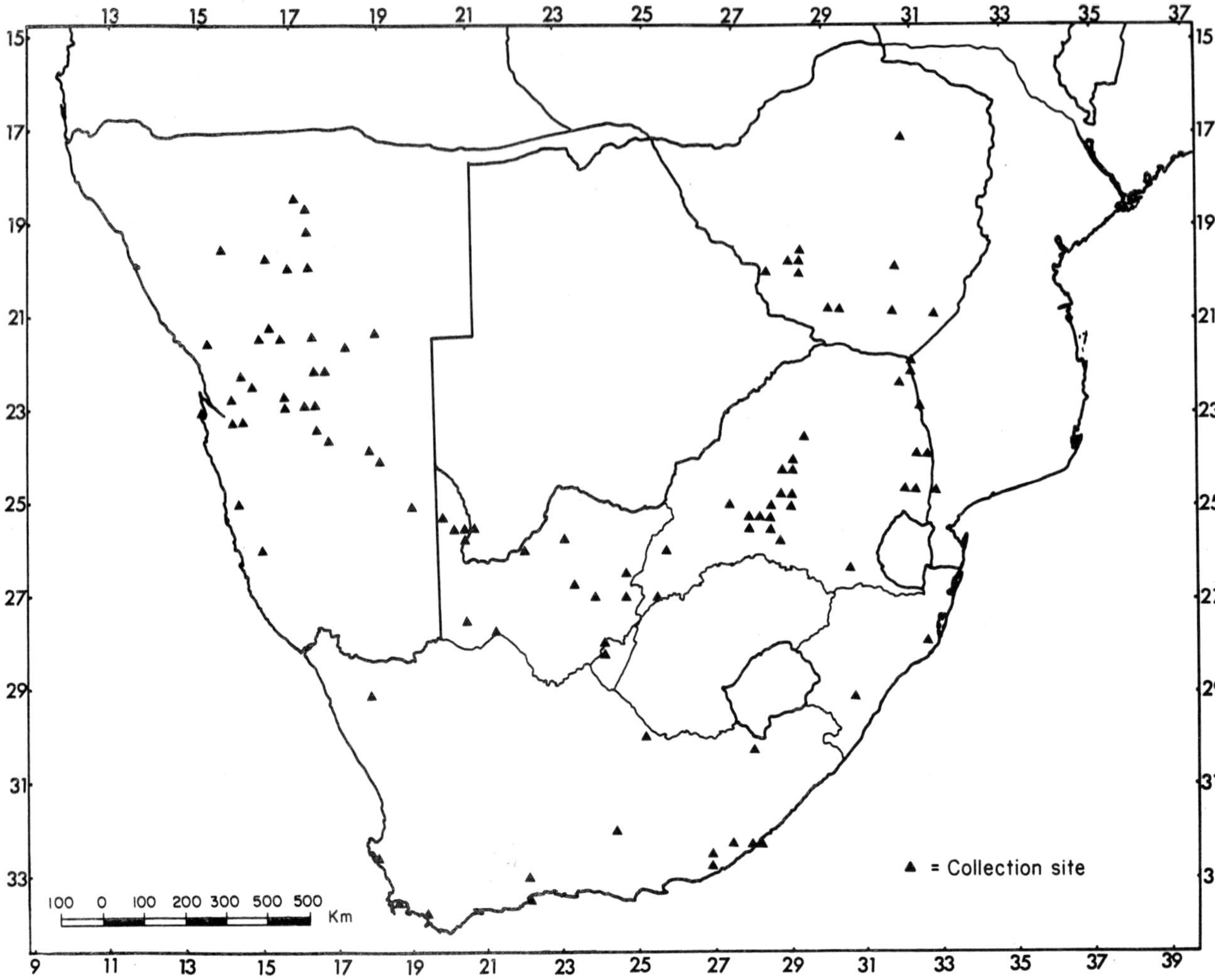

Figure 1. Map of the distribution of *Tyto alba* pellet collection sites by quarter degree grids.

interval between collections which sometimes exceeded a decade. Finally, the prey composition of the sample often differed markedly between years (cf Macdonald and Dean 1984).

Statistical tests employed in the alien rodent analysis

Statistical tests were applied to the data with the purpose of generating useful hypotheses. The analysis should not be viewed as definitive as there are problems with the randomness of the data sets, the cross-correlations between some of the classifications and the accuracy with which individual samples can be allocated to certain classes (eg those relating to extent of habitat modification). In the latter connection the sizes of *Tyto alba* hunting ranges in southern Africa have not yet been measured but a European estimate is that these have a mean radius of 2,5 km (Taberlet 1983). We classified collections on the basis of the location of the roost or nest where the collections were made; in some cases this might be different to the classification of the hunting range.

In all cases the frequencies of owl pellet collections with or without alien rodents were compared for two or more classes of these collections using the likelihood ratio chi-square G^2 test. Computations were carried out using the BMDP analysis P4F (Dixon 1985). Where the sample sizes for some of the classes were small, and where the classes could not be meaningfully combined, eg in the inter-biome comparisons, the requirement for a valid contingency test, that the minimum expected value should not be less than one (Everitt 1977), could not be met. In these cases the results of the test are not presented, the data simply being tabulated for discussion.

The nature reserves data set

The second data set comprised the total number of indigenous and invasive alien vascular plants and vertebrate species recorded for a sample of nature reserves within southern Africa. The sample was selected to represent reserves in each biome, preference being given to those reserves which were known to have been reasonably well studied and those reserves which had been proclaimed for some time (ie those which can be considered reasonably "natural" rather than profoundly modified). In several cases it was found that data were not available for one of the selected reserves, and in most cases data were not available for every taxonomic group within a reserve. Some members of the initial sample were therefore rejected and, in a few cases, reserves with good species lists which were not previously in the sample were added. All were selected independent of their known invasion status. Most of the species lists were assembled from the unpublished records of the conservation agency responsible for the reserve, with a minority being obtained from published reports. Data on alien plant invasions were supplemented with the results of a questionnaire survey (Macdonald unpublished). Each reserve's species richness data were paired with data on the biome and climate, as well as various data relating to the reserve itself. In many cases these data were obtained from Greyling and Huntley (1984).

Statistical tests employed in the nature reserves analysis

The 57 reserves in the sample were classified into two or more groupings according to their environmental or reserve characteristics (see Tables 3 and 7 for these characteristics).

The numbers of invasive alien species in a taxonomic group recorded for reserves were compared (either as absolute numbers or as percentages of the number of indigenous species in the reserves) using the Kruskal-Wallis nonparametric analysis of variance. Computations were carried out using the P3S programme of BMDP (Dixon 1985). In the case of the comparison of the fynbos biome reserves with all other reserves this test is identical to the Mann-Whitney U-test. Even here a two-tailed test of significance was appropriate as the mean number of invasive alien species in fynbos reserves was not always greater than that for the other reserves, even though the a priori hypothesis being tested was that fynbos reserves are more heavily invaded than other reserves.

The results of these analyses are presented simply as the significance of the overall analysis of variance. No detailed analyses using least significant differences between different classes of reserves have been attempted. The use of the number of species rather than a more quantitative measure of the extent of the invasions makes such analyses almost spurious. The intention of the analysis of this data set is the same as that of the alien rodent data set, ie the identification of potentially useful hypotheses

Table 1. A summary of the *Tyto alba* prey analyses used in the present study. All collections made from a site within a single calendar year were summed and called one collection. Collections made in different years were kept separate. Published sources of data are presented in the reference list.

Biome	No of sites	No of collections	No of vertebrate prey items	Decades of collections
Forest	0	0		
Fynbos	4	4	604	1950 – 60
Moist savanna	15	19	3 557	1960 – 70
Arid savanna	99	113	24 635	1950 – 70
Grassland	23	29	7 332	1930 – 80
Karoo/semidesert	14	16	2 722	1960 – 70
Desert	9	12	2 460	1960 – 70
Total	164	193	41 310	1930 – 80

Table 2. A summary of the nature reserves used in the analysis of indigenous and alien species richness.

Biome	No of reserves in sample	No of reserves for which species lists available (and mean number of invasive alien species per reserve)					
		Vascular plants			Vertebrate Classes		
		Angio-sperms	Spermato-phyta	All vascular + angiosperms	Reptilia	Aves	Mammalia
Forest	5	1(15)	1(32)	2(23,5)	3(0)	5(1,2)	2(0)
Fynbos	17	3(16,7)	6(40)	14(29,3)	7(0,1)	12(2,8)	11(2,7)
Savanna (arid or moist)	17	4(26,2)	4(68,7)	11(40,9)	15(0,1)	16(1,5)	12(1,2)
Grassland	8	0	5(32,2)	7(30,2)	5(0)	8(0,9)	6(0,5)
Karoo/semidesert	8	3(22,3)	1(15)	5(17,8)	4(0,2)	8(1)	5(1,2)
Desert	2	0	2(12,5)	2(12,5)	1(0)	2(1)	1(1)
Total	57	11	19	41	35	51	37

Table 3. Collection site or nature reserve data used in the analyses.

Parameter	Used in *Tyto alba* Study	Used in Nature Reserve Study	Main Sources
Biome	+	+	Huntley (1984); Brown et al (1985); Wild and Barbosa (1967)
Mean annual rainfall (mm)	+	+	Greyling and Huntley (1984); Weather Bureau (undated); Department of Met Services (1977)
Rainfall in 6 wettest months as % of annual rainfall	+	+	Department of Met Services (1977); Weather Bureau (undated)
Koppen climate classification	+		Schulze and McGee (1978, Figure 10)
Mean annual incidence of days having frost		+	Schulze (1965, Figure 85)
Reserve size (ha) and number of visitors per year		+	Greyling and Huntley (1984); NAKOR (1985)
Presence of influent rivers and distance from urban area		+	Trig Survey Maps, RSA and SWA
Number of species of rodent predators likely to be present in community:			
Birds	+		Tarboton (1968); Steyn (1982); Boshoff et al (1983); Tarboton and Allan (1984); Kemp et al (1985)
Mammals	+		Stuart (1975a, 1981); Smithers and Wilson (1979); Smithers (1983)
Reptiles	+		Rose (1962); FitzSimons (1970)

relating to the extent of invasions.

Summaries of the two data sets used

The full data sets are available in the archives of the Percy FitzPatrick Institute of African Ornithology. Summaries of the barn owl prey analysis data and reserve species richness data are presented in Tables 1 and 2. The parameters used in the paired data sets, together with their main source references, are listed in Table 3.

Patterns of incidence of alien rodents in owl pellets

Alien rodents were present at 19 (11,6%) of the 164 sites for which *Tyto alba* prey analyses were available (Figures 2 and 3). At three sites alien rodents occurred in the collections for more than one year, giving a total of 22 (11,3%) of the 195 collections with alien rodents.

The probabilities of the likelihood ratio chi-square tests of the frequency of occurrence of alien rodents in samples of these sites having different habitat or biotic characteristics are presented in Tables 4 and 5. Since alien rodents generally only occurred in very low numbers in any collection (15 of the 19 sites had fewer than five individuals present) it appeared likely that collections with few prey would be inaccurate predictors of the presence or absence of an alien ro-

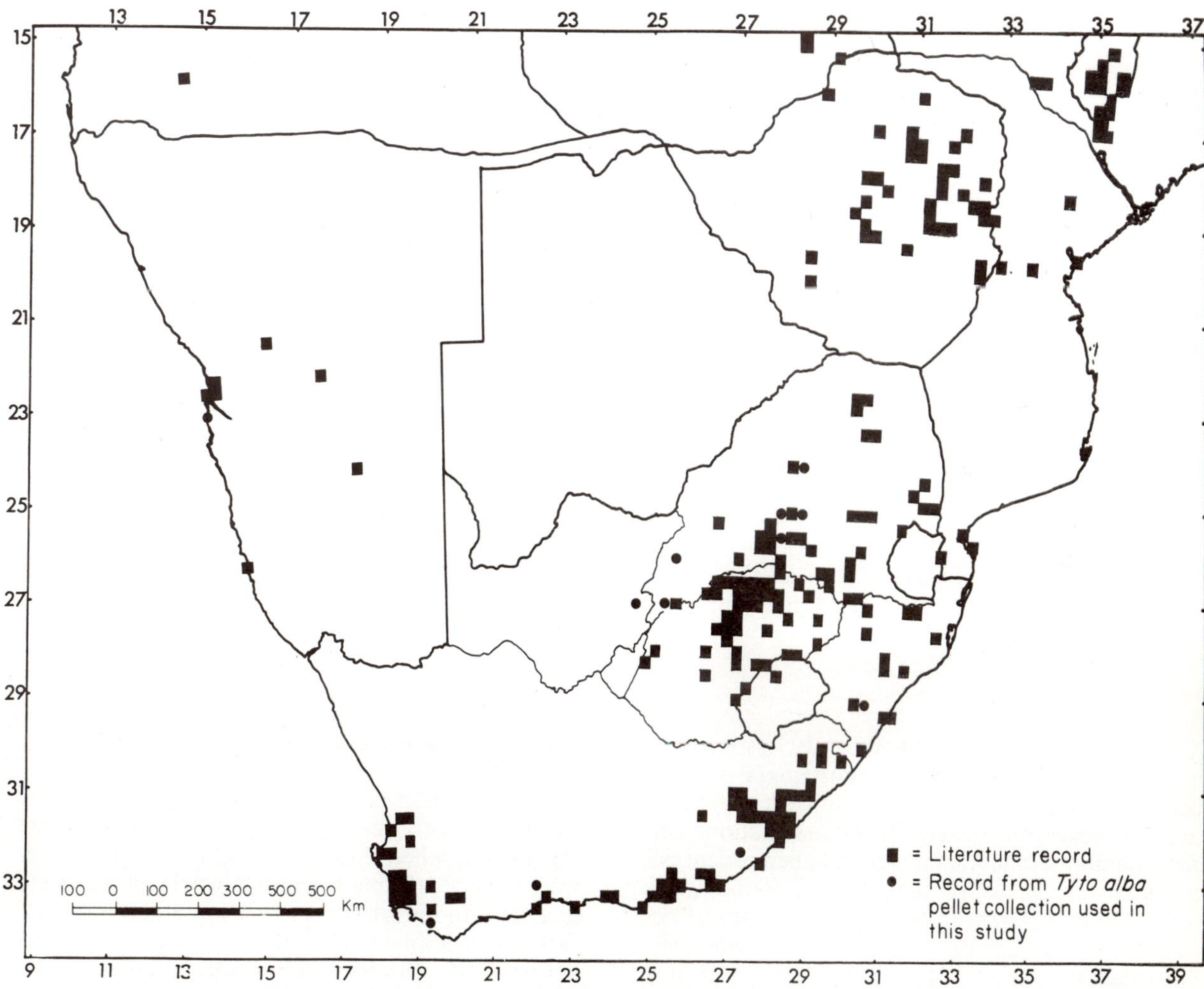

Figure 2. Map of the known distribution of *Rattus rattus* according to the quarter degree grid system (literature records from Davis 1974; Smithers and Tello 1976; Smithers and Wilson 1979; Bourquin and Mathias 1984; Brown et al 1985).

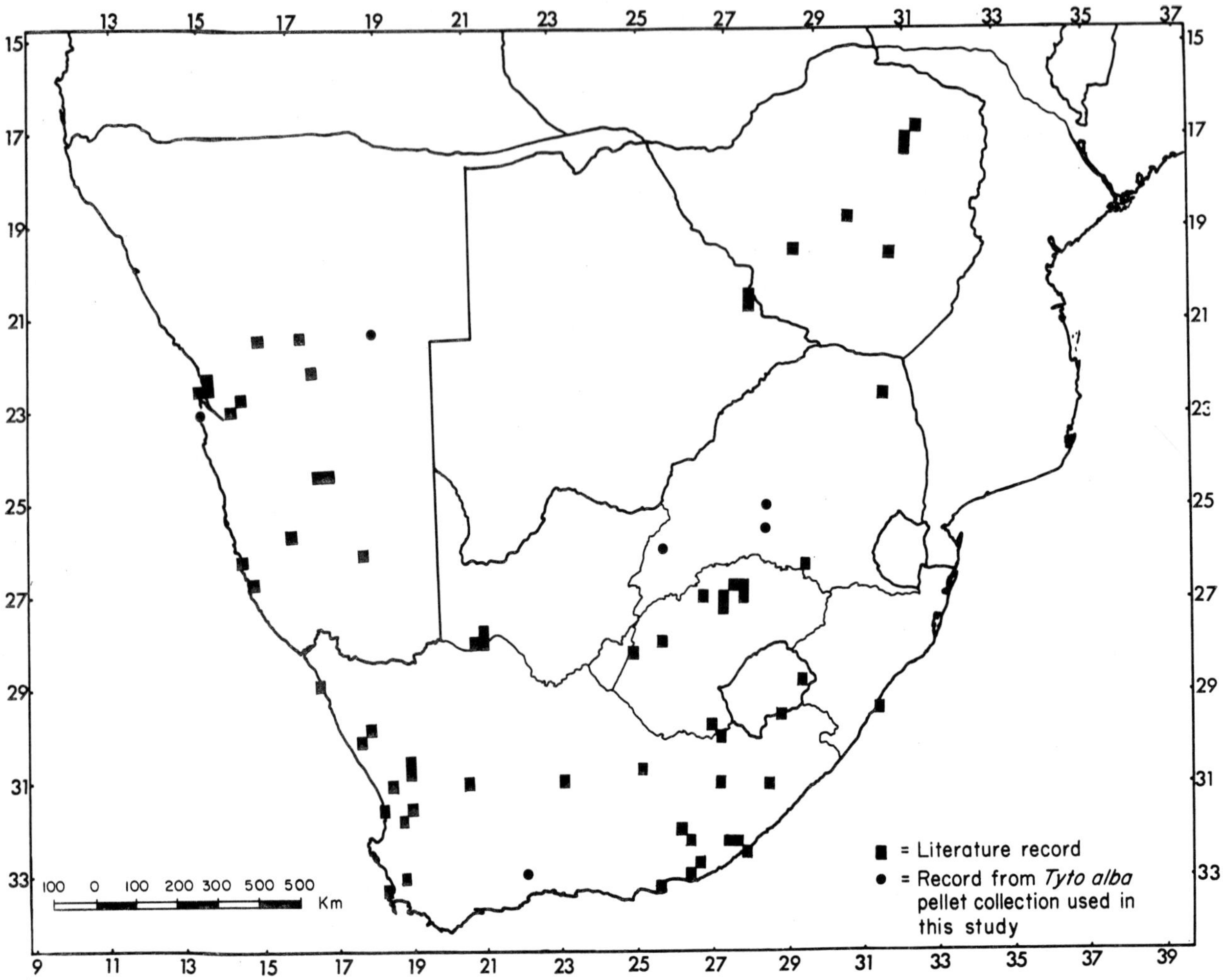

Figure 3. Map of the known distribution of *Mus musculus* according to the quarter degree grid system (literature records from Smithers 1971; Davis 1974; Smithers and Wilson 1979; Brown et al 1985).

dent. Alien rodents were significantly less frequent in collections with less than 75 prey items ($P < 0,01$), being present in only two per cent of such collections ($n = 59$). Several of the significant associations between habitat and biotic factors and alien rodent occurrence for the entire sample were found to be significantly different in terms of frequency of occurrence of sites having different prey sample sizes. The smallest category of prey sample size (<75 prey items) was therefore eliminated from the final analyses, although the results of the tests using all samples are also included (Tables 4 and 5).

Differences between biomes

The ranking of those biomes showing relatively high occurrences of aliens does not appear to be significant. This ranking changed markedly between the analysis when sites with small prey sample sizes were excluded (Table 4) and that where all sites were included. In the latter, savanna was still the lowest, five per cent ($n = 114$), but desert was the second lowest, 22% ($n = 9$), fynbos third, 25% ($n = 4$), grassland fourth, 26% ($n = 23$) and karoo the highest, 29% ($n = 14$). In neither of these analyses could the results be tested statistically as minimum expected values were much less than one for some cells in the contingency table (Table 4).

If the savanna sites are compared with those from all other biomes the statistical analysis is valid and the frequency of occurrence of aliens is significantly lower in the savanna ($P<0,01$). The only reason that the desert biome shows such a high frequency of occurrence is the presence of aliens in both collections from Sandvis Harbour. The susceptibility to invasion of mesic coastal sites surrounded by desert is sug-

Table 4. Probability values obtained in the likelihood ratio chi-square tests of the frequency of samples with or without alien rodents where samples were classified by habitat factors. Asterisks indicate cases where valid tests could not be made.

Factor	P: All samples	P: Samples >75 items	Frequency of occurrence (per cent) of alien rodents including only collections with more than 75 prey items (number of collections)
Biome	*	*	Savanna – 7 (70), Fynbos – 25 (4), Grassland – 33 (18), Desert – 40 (5), Karoo – 50 (8)
Ecosystem types	*	*	Hillsides – 0 (15), Fynbos – 0 (1), Agricultural fields – 9(11), Open woodland – 12 (33), Riverine fringe – 19 (31), Grassland – 56 (9), Coastal dunes/estuaries – 67 (3)
Extent of modification	0,289	0,685	Natural – 12 (24), Seminatural – 17 (47), Transformed – 21 (33)
Pellet collection site	0,017	0,012	Cliff/cave roost – 5 (39), building roost – 20 (30), Tree roost – 30 (33)
Koppen climatic classification	*	*	BWh, Cfb, Dwa and Csa – 0 (7,6,1,1) Bsh – 15 (61), BWk – 18 (11), CWh – 20 (10), Csb – 50 (2), BSk – 67 (6)
Mean annual rainfall (mm)	0,951	0,632	400 to 600 – 14 (49), >600 – 16 (25), <400 – 23 (31)
% Rainfall in 6 main months	0,197	0,293	>85 – 10 (39), <80 – 20 (38), 80 to 85 – 23 (26)

Table 5. Probability values obtained in the likelihood ratio chi-square tests of the frequency of samples with or without alien rodents where samples were classified by biotic factors.

Factor	P: All samples	P: Samples >75 items	Frequency of occurrence (per cent) of alien rodents including only collections with more than 75 prey items (number of collections)
Number of spp avian predators	0,001	0,006	>23 – 5(21), <15 to 23 – 9(47), 15 – 35(37)
Number of spp mammalian predators	0,174	0,386	>13 – 13(29), <11 to 13 – 15(26), 11 – 22(50)
Number of spp reptilian predators	0,009	0,060	<6 – 10(50), 6 to 8 – 24(55)
Total number of spp of predators	0,004	0,067	>40 – 7(29), <35 to 40 – 12(25), 35 – 25(51)
Total number of prey items from site	0,0016	0,200	75 to 300 – 13(61), 300 to 2150 – 23(44)
Total number of indigenous mammalian prey species from site	0,009	0,023	8 to 15 – 10(62), <8 – 21(28), 16 to 21 – 40(15)

gested by other results from the Namib Desert. In a limited sample of jackal *Canis mesomelas* scats from the small lagoon at Luderitz Bay only *Rattus rattus* and *Mus musculus* were present (Niethammer 1968). Similarly, pellets of an unidentified predator from Walvis Bay sewage settling ponds were composed entirely of *R rattus* (Niethammer 1975). Completely xeric sites and riverine ecosystems in the desert show no alien penetration (Niethammer 1968, 1975; Vernon 1972a; Tilson and Le Roux 1983). Mesic coastal isolates in the Namib Desert are possibly analogous to oceanic island ecosystems which have proven to be highly susceptible to alien rodent invasions (Atkinson 1985).

Differences between ecosystems

The individual ecosystem categories all had too few sites to allow for accurate statistical analysis (Table 4). The absence of aliens in hillside samples is possibly because most of these sites were from natural areas, eg the Kruger and Matopos National Parks. That hillside ecosystems are susceptible to invasion is exemplified by the occurrence of *Mus musculus* on a hill in Francistown, Botswana (Smithers 1971). The high frequency of occurrence in grassland and estuarine ecosystems (two of the latter were from different collections at Sandvis Harbour, one containing *R rattus* and the other *M musculus*) is in accordance with Smithers's (1983) observation that *R rattus* is dependent on dense low ground cover in natural and seminatural areas.

Extent of ecosystem modification

That this analysis failed to show significant differences in the frequency of occurrence of alien rodents in response to gross differences in the extent of ecosystem modification is puzzling (Table 4), given that the species are known to be human commensals. Where all sites are included in the analysis the frequencies of occurrence of alien rodents are: natural - six per cent (n = 48), seminatural - 12% (n = 64) and transformed - 16% (n = 50). These results suggest higher levels of incidence in modified ecosystems (cf Hanney 1962; Macdonald and Dean 1984), but are not statistically significant.

Pellet collection site

The highly significant difference between the three categories of pellet collection sites (Table 4) is difficult to explain. The very low frequency of occurrence at sites on cliffs or in caves appears to be linked both to the high occurrence of such sites in natural areas and to the low occurrence of alien rodents in samples from hillside habitats.

Climatic factors

The Koppen climate zones each had too few sites to allow for valid statistical testing. However, there was a high occurrence of alien rodents in the Bsk climate zone, because of the presence of alien rodents at several sites in the Little Karoo around Oudtshoorn. Davis (1964) notes that this is one of the few areas in southern Africa that has not been penetrated by the bubonic plague bacillus *Yersinia pestis*. This disease, which was introduced to several South African ports in the early 1900's by infected *Rattus* species, has now become "endemic" throughout the region and causes periodic mass mortalities in indigenous rodents. It may also limit the penetration of indigenous rodent communities by the aliens.

Both mean annual rainfall and the seasonality of this rainfall (as measured by the percentage falling in the six consecutive heaviest rainfall months) had no significant effect on the incidence of alien rodents (Table 4).

Characteristics of the predator community

The highest level of statistical significance (P = 0,006) was obtained for the association between the incidence of alien rodents and the number of species of avian predators of rodents likely to be present at a site (Table 5). It appears that sites having alien rodents were distinguished by a depauperate avian predator community. This trend was weaker and not significant for mammalian predators, possibly because the distribution data for mammalian predators was less exact than for avian predators. The relationship of reptilian predator species to alien rodent occurrence, which was significant (P = 0,009) when all the sites were included in the analysis, was found to be highly confounded by differences in the total prey numbers of sites in the two categories of reptile species richness. The distribution data used for reptile species and the definition of which species preyed upon rodents were much less precise than was the case for both birds and mammals. The total number of species that prey on rodents that were likely to be present at a site was also linked to alien rodent occurrence when all samples were included in the analysis (P = 0,004).

These results can be considered either to allow one to refute the hypothesis that ecosystems characterized by having different numbers of species which prey on rodents are equally susceptible to invasions by alien rodents, or to suggest the hypothesis that predation pressure exerted by a diverse predator community is an important factor limiting alien rodent penetration of an ecosystem. These results definitely do not "prove" this latter hypothesis as there is no indication of causality in the observed correlation. That there is in fact causality in the relationship has been suggested as an explanation for the general lack of success of alien terrestrial vertebrates in southern Africa (Ferrar and Kruger 1983; Brooke et al this volume). Although it is widely recognized that predators do not eliminate established populations of these alien rodents it has been suggested that predation pressure exerted by domestic cats *Felis catus* can prevent the colonization of new areas by these species (De Graaf 1981).

Characteristics of the small mammal community

The number of species of indigenous small mammals present in the prey community (as reflected in the *Tyto alba* pellet analysis) was another biotic factor significantly associated with the frequency of occurrence of alien rodents (Table 5). In the analysis of all sites, regardless of the number of prey items, highest frequencies of alien rodents were found in the most species-rich prey communities. However, richness was strongly linked to the number of prey items in a sample. Moreover the results of this analysis were strongly influenced by the choice of cut-points between the different categories of species richness. If these were changed to six and eight there was no statistical significance to the differences in frequencies of occurrence of alien rodents in the analyses of all samples and samples having more than 75 prey items. The frequencies of occurrence for the latter analysis were: 8-21 species - 16% (n=68), 6-8 species - 18% (n=17), and <6 species - 20% (n=20).

In order to test for a relationship between the occurrence of alien rodents and the diversity of the indigenous small-mammal community which would not be so heavily influenced by the variations in total prey numbers, the Shannon-Wiener index of diversity was calculated for each collection. This index was calculated twice for each collection, using both the total small-mammal prey community excluding bats, Chiroptera, and the total rodent prey community. The values of this index for sites with and without alien rodents and for sites having either one or two species of alien rodents present were compared using the Kruskal-Wallis non-parametric test. The comparisons of sites without alien rodents, with one species and with two species of alien rodents were not statistically significant for either the entire small-mammal community (P=0,69) or for the rodent community (P=0,90). The same comparisons between sites free of aliens and all sites with alien species were also not significant (P=0,91 and P=0,94).

The hypothesis that the species richness or diversity of the indigenous small-mammal community is irrelevant to the invasion of the community by alien species can thus not be convincingly refuted on the basis of this sample. This result is in accordance with the observation that *R rattus* is capable of invading southern African areas having highly diverse indigenous small mammal faunas (Shortridge 1934; Mendelsohn 1982; Bourquin and Mathias 1984). The species has, however, never been found to be common in such areas. Smithers (1983) suggests that the invasion of continental ecosystems in southern Africa by *Mus musculus* is prevented by the "high level of competition from indigenous species".

Results of the nature reserve analyses

The statistical significance of the Kruskal-Wallis analyses of variance of the number and proportion of invasive introduced species in the reserves are presented in Tables 6 and 7. All significance values with P=0,1 or less are shown.

Differences between biomes

Two tests were carried out for each taxonomic group; the first comparing each biome independently and the second comparing the fynbos biome reserves with all other reserves regardless of biome. This latter test was carried out to test the a priori hypothesis that the fynbos biome is more susceptible to alien invasions than the other southern African biomes.

The number of accidentally introduced invasive species of Spermatophyta was significantly higher in arid savanna reserves with forest second, grassland third, fynbos fourth, karoo fifth and desert last. This effect was not manifest in any other grouping of plant species. When reserves were analysed for the number of invasive alien species of vascular plants (or of angiosperms where total vascular plant lists were unavailable) which had originally been intentionally introduced to the reserve, a total of 41 reserves could be compared. The biomes were shown to be significantly different (P=0,05) in their mean rankings of this parameter: grassland reserves had the highest number, with arid savanna second and fynbos third.

When the fynbos biome reserves were compared against all the other reserves the only variable which was significantly different was the number of accidentally introduced invasive species expressed as a percentage of the total indigenous flora. This was significantly lower for fynbos where reserves with total lists of vascular plant species were compared

Table 6. Significance of the Kruskal-Wallis tests for differences in the number or proportion of alien plant and vertebrate species in the sample of nature reserves when classified by biomes and climatic parameters.

Taxonomic Group	Category of alien species	Probability of significance (P) Biome	Fynbos vs all others	Mean annual r/fall	% Rainfall in 6 main months	No days frost per year	No of reserves in sample
Plants							
Angiosperms	all invasive spp	NS	NS	0,08	NS		11
	accidental spp	NS	NS	0,07	NS		
Spermatophyta	all invasive spp	NS	NS	NS	NS		19
	accidental spp	0,1	NS	NS	NS		19
All vascular	all invasive spp	NS	NS	0,08	NS		30
	accidental spp	NS	NS	0,1	NS		
	accidental spp as % of indigen spp	NS	0,03	NS	NS		
All vascular or angiosperm species	all invasive spp	NS	NS	0,04	NS	NS	41
	intentional spp	0,05	NS	NS	NS	NS	
	accidental spp	NS	NS	0,05	NS	NS	
	accidental spp as % of indigen spp	NS	0,05	0,03	NS	NS	
Reptiles	all invasive spp	NS	NS	NS	NS		35
	intentional spp	NS	NS	NS	NS		
	accidental spp	NS	NS	NS	NS		
Birds	all invasive spp	0,03	0,001	NS	0,05		51
	intentional spp	NS	NS	NS	NS		
	accidental spp	0,01	0,0002	NS	0,03		
Mammals	all invasive spp	NS	0,07	NS	NS		37
	intentional spp	NS	NS	NS	NS		
	accidental spp	0,07	0,07	NS	NS		

(P = 0,03) and where reserves with either vascular plant lists or angiosperm plant lists were compared (P = 0,05). This reflects the rich floras of fynbos reserves relative to reserves in other southern African biomes and is not related to variations in the number of alien species.

The results of these analyses do not allow us to reject the hypothesis that southern African biomes are equally susceptible to alien plant invasions.

Amongst the vertebrate fauna the fynbos biome reserves had significantly higher numbers of accidentally introduced invasive alien species (see also mean numbers of all invasive species in Table 2). This resulted from the inclusion of the southern African helmeted guineafowl *Numida meleagris* as an accidentally introduced alien species in most fynbos reserve lists (see Brooke et al this volume) as well as the presence of the European starling *Sturnus vulgaris* and the house sparrow *Passer domesticus* in most of these reserves. In most other biomes only *P domesticus* was present. Similarly, in the analyses for mammals, fynbos reserves ranked higher than all other reserves. This was due to the frequent occurrence of *Rattus* species (both *R rattus* and the brown rat *R norvegicus*) and *M musculus* in fynbos reserves, as well as localized occurrences of the Himalayan tahr *Hemitragus jemlahicus*, the fallow deer *Dama dama*, the feral pig *Sus scrofa* and the grey squirrel *Sciurus carolinensis* in these reserves. In some fynbos reserves there were also southern African ungulates which did not formerly occur in the area which had been intentionally introduced. In other biomes virtually the only accidentally introduced mammals were the alien rodents *R rattus* and *M musculus* and even these were not invariably recorded as being present. The fynbos biome appears to be more heavily invaded by birds and mammals than are the other terrestrial biomes. Whether this is a reflection of a heightened inherent susceptibility to invasion rather than of introduction pressure and the depression of indigenous predator populations in this biome is uncertain. It is possibly of note that the only feral population of *H jemlahicus* in the biome is located in the only mountain chain from which all the larger carnivores have been exterminated. In the case of *S scrofa*, predation of piglets by indigenous carnivores has been identified as one of the factors limiting the successful establishment of some of the populations within the biome (Botha 1985).

Differences associated with rainfall regimes

The different groupings of plant species all showed the same trend, with the total number of invasive alien plant species and number of accidentally introduced species positively associated with annual rainfall. This is in accord with the observation within biomes that mesic habitats tend to have more alien species than do xeric habitats (Wells et al 1979; Macdonald 1983; Taylor et al 1985).

None of the vertebrate groups showed trends associated with variations in mean annual rainfall.

Surprisingly, the seasonality of the rainfall, as reflected in the extent of concentration of annual rainfall in the six main rainfall months, appeared to have no effect on the number of alien plant species invading a reserve. Alien bird species were recorded as being present in greater numbers in reserves with aseasonal rainfall patterns (less than 70% of annual rainfall in six main rainfall months) than in those experiencing moderately (70-85%) and highly seasonal (more than 85%) rainfall regimes. This is possibly related to the autecological requirements of one of the major alien bird species in the region, *Sturnus vulgaris*. This species feeds by probing into soft ground with its beak (Feare 1984) and, as a result, it tends to be excluded from natural ecosystems in areas where surface soils harden during the prolonged dry seasons.

Differences due to variations in frost incidence

This parameter was only tested for lists of all vascular or angiosperm species, where available (Table 6). Contrary to expectations, there were no significant differences in the number of introduced plant species in reserves having less than five days of frost per year and reserves having more than sixty days of frosts.

The effect of reserve features

Although reserves varied in size from approximately 20 ha to 2 000 000 ha the number of alien plant species was not significantly associated with reserve size. Similarly, the presence or absence of influent rivers, often considered to be a factor of major importance in the introduction of alien species into

Table 7. Significance of the Kruskal-Wallis tests for differences in the number or proportion of invasive species of vascular or angiosperm plants in a reserve classified by reserve characteristics.

	Probability of significance (P)					
			Distance from			
Category of alien species	Reserve Size	Presence of influent River	Village	Town	City	Number of visitors per year
Total number of invasive alien spp	NS	NS	NS	NS	NS	0,1
Accidentally introduced alien spp	NS	NS	NS	NS	NS	0,08
Intentionally introduced alien spp	NS	NS	NS	NS	NS	NS
Accidental as % of total indigenous spp	NS	NS	NS	0,06*	NS	NS

*This is considered a spurious effect, showing no logical trend with increasing distance.

reserves (eg Macdonald 1983; Tarr and Loutit 1985; Vinjevold et al 1985), had no significant effect. The distances of the reserves from the nearest small town (= village), medium town (= town), and metropolitan area (= city), as determined from maps were shown also to have no statistically significant effect. What was significant was the relationship between the annual number of visitors to a reserve and both the total number of invasive alien plant species and the number of accidentally introduced species. This relationship might be direct (visitors or their vehicles acting as the major vector for the introduction of alien plant species) or might be indirect, the number of visitors in the reserve possibly being correlated with factors such as the extent of disturbance of natural ecosystems within the reserve (eg through rest camp or roadbuilding activities).

Conclusions

The two studies carried out here indicate that faunal and plant invasions are possibly controlled by different suites of factors. The savanna biome appears to be highly resistant to faunal invasions, while possibly being susceptible to alien plant invasions. Similarly, whereas a high mean annual rainfall appears to indicate a high susceptibility to alien plant invasions, no such relationship apparently holds for faunal invasions. There are indications from both these studies that faunal invasions might be limited by indigenous predation pressure.

The alleged heightened susceptibility of the fynbos biome to alien plant invasions was not borne out by this analysis, based as it was on the number of species invading. Quantitative analyses, based on the extent of alien plant infestations, have still to be carried out (cf Macdonald 1984). The analysis of the number of species of invasive alien animals present in nature reserves indicated that the fynbos biome has been more severely invaded.

Neither of the analyses served to indicate the relative susceptibility to invasion of 'pristine' ecosystems. The extent of human modification of southern African ecosystems appears to be positively correlated to their susceptibility to alien plant invasions (Macdonald 1984). Even ecosystems in the subcontinent's nature reserves cannot now be considered to be pristine and some have been extensively modified.

In attempting to analyse natural experiments such as these, considerable caution must always be exercised in interpretation of the observed relationships. Significant relationships should not be taken to be a proof of causation. The validity of such relationships can only be determined through carefully planned experimentation.

Acknowledgements

Guy Jenkins, Greg Espitalier-Noel and Ian Newton are thanked for technical assistance. The following are thanked for making available unpublished data: R K Brooke; P M Brooks (Natal Parks Board); E H W Baard, A de Villiers, P Lloyd (Cape Department of Nature and Environmental Conservation); N Jacobsen (Transvaal Provincial Administration), E Ashton and M Hawthorn (Cape Town Municipality); D L Clark (Cape Divisional Council); A Lamb and T Newby (Department of Environment Affairs); C T Stuart; A B Low; B Loutit, P Tarr and A J Williams (SWA Division of Nature Conservation); N Hanekom, H Braak and A B Lawson (National Parks Board); J Mendelsohn (Durban Museum); D A Ewbank; D Snyman (National Botanic Gardens of South Africa) and A J Phelan. C J Vernon and R Prys-Jones are thanked for assistance with references.

Susan Macdonald is thanked for diverse assistance at all hours of the day and night.

Abs M, Curio E, Kramer P and Niethammer J 1965. Zur ernahrungsweise der eulen auf Galapagos. Journal fur Ornithologie 106, 49-57.

Avery D M 1977. Past and present distribution of some rodent and insectivore species in southern Cape Province, South Africa: new information. Annals of the South African Museum 74, 201-209.

Atkinson I A E 1985. The spread of commensal species of *Rattus* to oceanic islands and their effects on island avifaunas. ICBP Technical Publication 3, 35-81.

Bateman J A 1960. Owl habits from pellets. Bulletin of the Zoological Society of South Africa 1, 20-21.

Boshoff A F, Vernon C J and Brooke R K 1983. Historical atlas of the diurnal raptors of the Cape Province. (Aves: Falconiformes). Annals of the Cape Provincial Museum (Natural History) 14, 173-297.

Botha S A 1985. The feral pig: an unsuccessful alien in Africa. Poster Paper presented at National Synthesis Symposium on the Ecology of Biological Invasions, Stellenbosch.

Boucher C 1981. Western Cape Province lowland alien vegetation. In: Moll E J (ed) Proceedings of the symposium on coastal lowlands of the Western Cape. University of the Western Cape, Bellville. p 67.

Bourquin O and Mathias I 1984. The vertebrates of Oribi Gorge Nature Reserve: 1. Lammergeyer 33, 35-44.

Brain C K 1981. The hunters or the hunted?An introduction to African cave taphonomy. Chicago Press, Chicago.

Brown C J, Macdonald I A W and Brown S E (eds) 1985. Invasive alien organisms in South West Africa/Namibia. South African National Scientific Programmes Report No 119. CSIR, Pretoria.

Campbell B M, McKenzie B and Moll E J 1979. Should there be more tree vegetation in the Mediterranean climatic region of South Africa. Journal of South African Botany 45, 453-457.

Coetzee C G 1963. The prey of owls in the Kruger National Park as indicated by owl pellets collected during 1960-61. Koedoe 6, 115-125.

Davis D H S 1958. Notes on the small mammals in the Kalahari Gemsbok National Park with special reference to those preyed upon by barn owls. Koedoe 1, 184-188.

Davis D H S 1959. The barn owl's contribution to ecology and palaeoecology. Ostrich Supplement 3, 144-153.

Davis D H S 1964. Ecology of wild rodent plague. In: Davis D H S (ed) Ecological studies in southern Africa. W Junk, The Hague. pp 301-314.

Davis D H S 1974. The distribution of some small southern African mammals (Mammalia. Insectivora, Rodentia). Annals of the Transvaal Museum 29, 136-184.

Dean W R J 1973. Analysis of a collection of barn owl *Tyto alba* pellets from Warmbaths, Transvaal. Zoologica Africana 8, 75-81.

Dean W R J 1975. *Tyto alba* prey in South West Africa and the northern Cape. Zoologica Africana 10, 217-219.

Dean W R J 1978. The ecology of owls at Barberspan, Transvaal. Proceedings of the Symposium on African Predatory Birds, 25-45.

De Graaf G 1960. 'n Ontleding van uilklonte van Nonnetjiesuil. Ostrich 31, 1-5.

De Graaf G 1981. The rodents of southern Africa. Butterworths, Durban.

Department of Meteorological Services 1977. Mean rainfall in Rhodesia. Government Printer, Salisbury.

Dixon W J (ed) 1985. BMDP statistical software. University of California Press, Los Angeles.

Elton C S 1958. The ecology of invasions by plants and animals. John Wiley, New York.

Everitt B S 1977. The analysis of contingency tables. Chapman and Hall, London.

Feare C 1984. The Starling. Oxford University Press, Oxford.

Ferrar A A and Kruger F J 1983. South African programme for the SCOPE project on the ecology of biological invasions. South African National Scientific Programmes Report No 72. CSIR, Pretoria.

FitzSimons V F M 1970. Snakes of southern Africa: a field guide. Collins, London.

Fox B J 1983. Mammal species diversity in Australian heathlands - the importance of pyric succession and habitat diversity. In: Kruger F J, Mitchell D T and Jarvis J U M (eds) Mediterranean-type ecosystems: the role of nutrients. Springer-Verlag, Berlin. pp 473-489.

Glue D E 1970. Avian predator pellet analysis and the mammalogist. Mammal Review 1, 53-62.

Glue D E 1974. Food of the barn owl in Britain and Ireland. Bird Study 21, 200-210.

Greyling T and Huntley B J (eds) 1984. Directory of southern African conservation areas. South African National Scientific Programmes Report No 98. CSIR, Pretoria.

Griffin M and Panagis K 1985. Invasive alien mammals, reptiles and amphibians in South West Africa/Namibia. In: Brown C J, Macdonald I A W and Brown S E (eds) Invasive alien organisms in South West Africa/Namibia. South African National Scientific Programmes Report No 119. CSIR, Pretoria. pp 44-47.

Grindley J, Siegfried W R and Vernon C J 1973. Diet of the barn owl in the Cape Province. Ostrich 44, 266-267.

Hall A V and Boucher C 1977. The threat posed by alien weeds to the Cape Flora. In: Proceedings of the 2nd National Weeds Conference. Balkema, Cape Town. pp 35-45.

Hanney P 1962. Observations upon the food of the barn owl *Tyto alba* in southern Nyasaland. Annals and Magazine of Natural History 13, 305-313.

Hanney P W 1975. Rodents: their lives and habits. Davis and Charles, London.

Henderson L and Musil K J 1984. Exotic woody plant invaders of the Transvaal. Bothalia 15, 297-313.

Herrera C M and Jaksic F M 1980. Feeding ecology of the barn owl in Central Chile and Southern Spain: a comparative study. Auk 97, 760-767.

Huntley B J 1984. Characteristics of South African biomes. In: Booysen P de V and Tainton N M (eds) Ecological effects of fire in South African ecosystems. Springer-Verlag, Berlin. pp 1-17.

Kemp M I, Kemp A C and Tarboton W R 1985. A catalogue of the birds of the Transvaal. Cyclostyled manuscript, Transvaal Museum and Transvaal Nature Conservation Division, Pretoria.

Kolbe F F 1946. The case for the barn owl. African Wildlife 1, 69-73.

Macdonald I A W 1983. Alien trees, shrubs and creepers invading indigenous vegetation in the Hluhluwe-Umfolozi Game Reserve complex in Natal. Bothalia 14, 949-959.

Macdonald I A W 1984. Is the fynbos biome especially susceptible to invasion by alien plants - a reanalysis of available data. South African Journal of Science 80, 369-377.

Macdonald I A W and Dean W R J 1984. Vertebrate prey of the barn owl *Tyto alba* in the Cape Province north of the Orange River. Proceedings of the 5th Pan-African Ornithological Congress. pp 517-541.

Macdonald I A W and Jarman M L (eds) 1984. Invasive alien organisms in the terrestrial ecosystems of the Fynbos Biome, South Africa. South African National Scientific Programmes Report No 85. CSIR, Pretoria.

Macdonald I A W and Nott T B (in press). Invasive alien organisms in central South West Africa/Namibia: results of a reconnaisance survey conducted in November 1984. Madoqua.

Malherbe A P 1963. Notes of birds of prey and some others at Boshoek, north of Rustenburg, during a rodent plague. Ostrich 34, 95-96.

Mendelsohn J M 1982. Notes on small mammals on the Springbok Flats, Transvaal. South African Journal of Zoology 17, 197-201.

Middlemiss E 1975. The Rondevlei Bird Sanctuary 1952-1974: a record of an environment. The Divisional Council of the Cape, Cape Town.

Morton S R and Martin A A 1979. Feeding ecology of the barn owl *Tyto alba* in arid Southern Australia. Australian Wildlife Research 6, 191-204.

Nakor 1985. Register of conserved areas February 1985. Department of Environment Affairs, Pretoria.
Nel J A J and Nolte H 1965. Notes on the prey of owls in the Kalahari Gemsbok National Park, with special reference to the small mammals. Koedoe 8, 75-81.
Niethammer J 1968. Gewölluntersuchungen aus Südwestafrika. Journal of South West African Scientific Society 22, 5-40.
Niethammer J 1975. Neue gewölle aus Südwestafrika. Journal of South West African Scientific Society 29, 133-170.
Payne R S 1971. Acoustic location of prey by barn owls *Tyto alba*. Journal of Experimental Biology 54, 535-573.
Perrin M R 1982. Prey specificity of the barn owl *Tyto alba*, in the Great Fish River valley of the Eastern Cape Province. South African Journal of Wildlife Research 12, 14-25.
Pocock T N 1963. Food of the barn owl in the southern Cape. Manuscript, Percy FitzPatrick Institute of African Ornithology, University of Cape Town, Cape Town.
Plug I, Dippenaar N J and Hanisch E O M 1979. Evidence of *Rattus rattus* (house rat) from Pont Drift, an iron age site in the northern Transvaal. South African Journal of Science 75, 82.
Richardson D M 1984. A cartographic analysis of physiographic factors influencing the distribution of *Hakea* spp in the south western Cape. South African Forestry Journal 128, 36-40.
Rose W 1962. The reptiles and amphibians of southern Africa. The Standard Press, Cape Town.
Schulze B R 1965. Climate of South Africa Part 8. General survey. The Government Printer and Weather Bureau, Pretoria.
Schulze R E and McGee O S 1978. Climatic indices and classifications in relation to the biogeography of southern Africa. In: Werger M J A (ed) Biogeography and ecology of southern Africa. W Junk, The Hague. pp 19-52.
Shortridge G C 1934. The mammals of South West Africa, Vol 1. Heineman, London.
Smithers R H N 1971. The mammals of Botswana. Trustees of the National Museums and Monuments of Rhodesia, Salisbury, Museum Memoirs 4, 1-340.
Smithers R H N 1975. Guide to the rats and mice of Rhodesia. Trustees of the National Museums and Monuments of Rhodesia, Salisbury.
Smithers R H N 1983. The mammals of the southern African subregion. University of Pretoria, Pretoria.
Smithers R H N and Tello J L P L 1976. Checklist and atlas of the mammals of Mocambique. Trustees of the National Museums and Monuments of Rhodesia, Salisbury, Museum Memoirs 8, 1-184.
Smithers R H N and Wilson V J 1979. Checklist and atlas of the mammals of Zimbabwe Rhodesia. Trustees of the National Museums and Monuments of Rhodesia, Salisbury, Museum Memoirs 9, 1-193.
Steyn P 1982. Birds of prey of southern Africa. David Phillip, Cape Town.
Steyn P 1984. A delight of owls. David Phillip, Cape Town.
Stuart C T 1975a. Preliminary notes on the mammals of the Namib Desert Park. Madoqua Series II, 4, 5-68.
Stuart C T 1975b. A short note on the diet of *Tyto alba* at Sandwich Harbour, Namib Desert Park, South West Africa. Madoqua Series II, 4, 103.
Stuart C T 1976. Contributions to the biology and ecology of the rodents and insectivores of the Andries Venter Research Station, Jonkershoek Valley. Cape Department of Nature and Environmental Conservation, Jonkershoek.
Stuart C T 1981. Notes on the mammalian carnivores of the Cape Province South Africa. Bontebok 1, 1-58.
Taberlet P 1983. Evaluation du rayon d'action moyen de la chouette effaie, *Tyto alba* (Scopoli, 1769), a partir de ses pelotes de rejection. Revue d'ecologie la terre et la vie 38, 171-177.
Tarboton W R 1968. Checklist of birds of the south central Transvaal. Witwatersrand Bird Club, Johannesburg.
Tarboton W R and Allan D G 1984. The status and conservation of birds of prey in the Transvaal. Transvaal Museum Monograph 3, 1-115.
Tarr P W and Loutit R 1985. Invasive alien plants in the Skeleton Coast Park, Western Damaraland and Western Kaokoland. In: Brown C J, Macdonald I A W and Brown S E (eds) Invasive alien organisms in South West Africa/Namibia. South African National Scientific Programmes Report No 119. CSIR, Pretoria. pp 19-23.
Taylor H C 1977. The Cape Floral Kingdom, an ecological view. In: Proceedings of the 2nd National Weeds Conference. Balkema, Cape Town. pp 19-33.
Taylor H C, Macdonald S A and Macdonald I A W 1985. Invasive alien woody plants in the Cape of Good Hope Nature Reserve 2. Results of a second survey in 1976-1980. South African Journal of Botany 51, 21-29.
Tilson R L and Le Roux P 1983. Resource partitioning in coexisting Namib Desert owls, *Bubo africanus* and *Tyto alba*. Madoqua 13, 221-227.
Valente A 1981. Vertebrate remains in pellets of the barn owl *Tyto alba* from Planet Downs Station, South-Western Queensland. Australian Wildlife Research 8, 181-185.
Van Bruggen A C 1964. The distribution of introduced mollusc species in southern Africa. Beaufortia 11, 161-169.
Van der Merwe M 1980. Importance of *Miniopterus schreibersi natalensis* in the diet of barn owls. South African Journal of Wildlife Research 10, 15-17.
Vernon C J 1971. Owl foods and other notes from a trip to South West Africa. Ostrich 42, 153-154.
Vernon C J 1972a. An analysis of owl pellets collected in southern Africa. Ostrich 43, 109-124.
Vernon C J 1972b. Ornithology Report. Rhodesian Schools Exploration Society, Tshibidzini Expedition Report 22, 29-39.
Vernon C J 1980a. Prey of the barn owl from the Lundi River. Honeyguide 101, 10-13.
Vernon C J 1980b. Prey of six species of owl at the Zimbabwe Ruins - 1970 to 1975. Honeyguide 101, 26-28.
Vernon C J 1980c. Prey of barn owl at Port Alfred. Diaz Diary 87, 8.
Vernon C J 1981. Owl pellets. Diaz Diary 98, 5.
Vernon C J 1983. Prey of barn owl around East London. The Bee-Eater Supplement 10, 9.
Vinjevold R D, Bridgeford P and Yeaton D 1985. Invasive alien plants in the Namib-Naukluft Park. In: Brown C J, Macdonald I A W and Brown S E (eds) Invasive alien organisms in South West Africa/Namibia. South African National Scientific Programmes Report No 119. CSIR, Pretoria. pp 24-27.
Weather Bureau (undated). Climate of South Africa Part 2. Rain-

fall statistics. Government Printer, Pretoria.
Wells M J, Duggan K and Henderson L 1979. Woody plant invaders of the central Transvaal. In: Proceedings of the 3rd National Weeds Conference. Balkema, Cape Town. pp 11-24.
Wells M J, Engelbrecht V M, Balsinhas A A and Stirton C H 1983. Weed flora of South Africa 3: more power shifts in the veld. Bothalia 14, 967-970.
Wells M J, Balsinhas A A, Joffe T, Engelbrecht V M, Harding G and Stirton C H 1986. A catalogue of problem plants in southern Africa. Memoirs of the Botanical Survey of South Africa 53, Government Printer, Pretoria.
Wild H and Barbosa L A G 1967. Vegetation map of the Flora Zambesiaca area. Collins, Salisbury.
Wilson V J 1970. Notes on the breeding and feeding habits of a pair of barn owls, *Tyto alba* (Scopoli) in Rhodesia. Arnoldia 4, 8.
Winterbottom J M 1966. Notes on the food of *Tyto alba* in the Etosha Pan. Ostrich 37, 139-140.

Impacts of invasions

It is striking to note how little has been done to evaluate the ecological and economic impacts of southern Africa's alien organisms, although the concern about these invasions and the motivation for their control stems almost entirely from their presumed deleterious impacts. The three chapters presented in this section attempt to draw together what little is known on the subject.

The study of impacts on communities and ecosystems is extremely difficult. In order to quantify an impact there have to be suitable 'control' areas, preferably areas showing a range of the likely densities of the invasive organism and all of these suitably replicated. Field study sites with all these characteristics are seldom found and even where they exist the analysis is usually confounded by historical factors and by heterogeneity in the environment. Few researchers have been able to design controlled experiments and authors have had to turn often to the long-term experiments in hydrology and silviculture established by foresters, to find evidence of the impacts of plant invasions.

In the aquatic systems covered by **Ashton and co-authors** these difficulties are to some extent overcome by comparisons of the same waterbody before and after invasion. As almost complete eradication of alien aquatic plants is often achieved over relatively short periods, and the re-establishment of the pre-invasion community apparently being fairly rapid, such comparisons appear to be valid in these systems. By contrast, in the highly complex terrestrial ecosystems being studied by **Breytenbach** and **Versfeld and van Wilgen** such comparisons are generally not possible, although some hypothetical impacts may be indicated by reference to analogous, controlled experiments, such as those involving afforestation. Much of what we think we know about the impacts of aliens on terrestrial ecosystems still requires critical testing. The recent series of investigations of the impacts of the ant *Iridomyrmex humilis* on seed dispersal provides a useful model for researchers to follow.

One of the perspectives that is not covered in any of the chapters in this section is that the impacts of an alien organism might vary according to the age of the invasion. In so far as most of our invasions are still relatively recent, almost all being less than two hundred years old and mostly less than one hundred, this might become a perspective of increasing importance. For instance, is the impact of a two-hundred-year-old stand of an alien tree species the same as that of a fifty-year-old stand? Another time-related consideration is that the impacts of a species might lessen with time, possibly as a result of the alien's vigour being reduced through the build-up of an adapted community of predators and pathogens. Well-known southern African examples of this are the post-1960 reduction in *Hakea sericea* density by the apparently indigenous fungus *Colletotrichum gloeosporioides* (Morris 1982) and the post-1970 reduction in seed accumulation rate of *Acacia cyclops* due to predation by indigenous hemipterans (Dennill 1983). The possibility exists that these 'natural' controls will serve to reduce the densities and thus the impacts of these species in the long term. This historical perspective on impacts is particularly relevant where the control measures currently being applied have undesirable impacts on the native biota, as related by **Breytenbach** for *Hakea sericea*.

The importance of long-term monitoring of the impacts of both uncontrolled and controlled stands of alien plants is stressed by considerations such as those above. Without an improved understanding of their impacts, the management of alien organisms cannot be based on rational and objective criteria.

Dennill G B 1983. The effect of seed feeding hemipterans (Alydidae) on the seeds of *Acacia cyclops* in the south western Cape. Proceedings of the 5th National Weeds Conference of South Africa. Balkema, Cape Town. pp 188-187.

Morris M J 1982. Gummosis and die-back of *Hakea sericea* in South Africa. Proceedings of the 4th National Weeds Conference of South Africa. Balkema, Cape Town. pp 51-54.

Chapter 18

Impacts of alien organisms on terrestrial communities with emphasis on communities of the south-western Cape

G J Breytenbach
Saasveld Forestry Reserch Centre.

Introduction

Communities may be seen as consisting of a number of organisms of different species, and in abstract as the interactions between organisms, and between organisms and the environment. Any extraneous or introduced factor such as an invasive alien may influence the community by changing its structure, especially the proportional abundance of different species, guilds or growth forms in the community. It may also change functional relationships such as that between pollinator and plant. Whether the change is considered to be of benefit or detriment to the system is not important.

The influences of invasive alien biota on ecosystems and communities are frequently assumed to be of sufficient magnitude to justify expenditure on control (eg Kruger 1977, 1981; Fugler 1982). In this chapter I intend to evaluate such effects by analysis of exemplary cases. I will also discuss evidence which shows that methods employed in the control of aliens have an effect on communities and that this should be taken into consideration when control techniques are evaluated. An effective control strategy should not merely get rid of the alien but should also have minimal effect on the native community.

Impact on biological diversity

In the assessment of the effects of alien invasives on communities the most commonly recorded phenomenon is the change in species diversity. Changes can occur at different levels of diversity eg at alpha, beta, or gamma diversity levels. Information content of diversity data, however, is largely based on taxonomic criteria and is expressed as a community statistic such as richness. The functional role of the organism in the community is ignored. Nevertheless, these summary statistics provide indices of the impact of aliens which are relevant to the major aims of conservation ie the maintenance of species diversity (IUCN 1980). I will therefore first discuss the effect of alien organisms on diversity and then discuss how changes in composition and structure may influence community processes and the functioning of ecosystems.

Address for correspondence: G J Breytenbach, Saasveld Forestry Research Centre, Private Bag X6515, George, 6530.

Plant communities

If alien plants have a dense canopy and are taller than plants from indigenous communities, native light-demanding plant species could be lost. No data are available on the light requirements of the native flora. Nevertheless it is evident that most species that do disappear, do so after canopy closure of alien species occurs (Cowling 1976; J Vlok and C De Ronde personal communication).

Some data are available on the effect of invasive pine species on fynbos plant communities (Cowling 1976) but these are not readily interpretable. I will therefore use data on *Pinus radiata* from a first-rotation plantation environment, and assume that it reflects patterns in areas invaded by *P pinaster* (eg Cowling 1976; Smit 1983; Smit and de Kock 1984). The mean number of indigenous plant species recorded in 0,1 m^2 plots (n=200) under a *Pinus radiata* stand established in 1948, declined significantly from 8,5 in 1945 to 2,1 in 1984. Of the 190 species recorded in the 1945 survey, 63,8% were absent in 1984 (Richardson and Van Wilgen 1986). No information on the light requirements of lost species is available. In conjunction with understorey fires, cover of shade tolerant species like *Ehrharta calycina* increased considerably under pine plantations in the southern Cape (J Vlok and C de Ronde personal communication). It is, however, not only the light regime that changes. Microclimate changes and pine litter which is in excess of 10 ton ha^{-1} in autumn (de Ronde 1983, 1984) has a large effect on the shade-tolerant species (J Vlok and C de Ronde personal communication). The influence of the various parameters needs to be quantified before definitive statements can be made.

Diversity also decreased on sites in the southern Cape invaded by *Hakea sericea*. Here the fynbos communities which were lightly and heavily invaded by *Hakea sericea* supported fewer species than vigorous fynbos in an adjoining firebreak without *Hakea*, viz 4,5 as opposed to 8,2 species per 4 m^2 plot (G J Breytenbach and J Vlok personal observation). Fynbos plant communities managed on a 12-20 year burning rotation should support some 15,4-16,7 species per square metre (Naveh and Wittaker 1980; Bond 1983). It seems highly unlikely that these changes can be wholly ascribed to changes in light regimes, especially in the lightly infested sites. The decrease in diversity in pine and hakea invaded sites reflects substantial changes at alpha diversity level. Gamma diversity will also be influenced if dense stands are allowed to occupy entire landscapes.

Alien herbivores could also have an effect on plant community diversity. Grazing in the Baviaanskloof and Kouga mountains by feral stock, which are here considered to be invasive organisms, reduces plant species diversity slightly at alpha diversity level (by some two species per 4 m^2 plot). If within-habitat variation is included (species summed over several plots within the same habitat), richness is higher in grazed sites (Breytenbach et al 1985). Species turnover between grazed and ungrazed sites was, however, considerable, ie several species only occurred in either grazed or in ungrazed sites. Within-habitat, or pattern diversity (Whittaker 1977), therefore increased overall.

Mammal communities

Alien mammals are seldom invasive in South Africa. Of the small mammals it is only *Rattus rattus*, which is still largely commensal with man, that has succeeded in invading natural communities to some degree (eg Smithers 1983; Brooke et al and Macdonald et al this volume). The Himalayan tahr *Hemitragus jemlahicus* and feral cattle and donkeys are invasive in limited areas. Donkeys are known to interbreed with Burchell's zebra *Equus burchelli* (Smithers 1983). Feral donkeys co-occur with small subpopulations of the endangered Cape mountain zebra *Equus zebra zebra* (Skinner et al 1977) in the Kammanassie mountains and hybridization could therefore jeopardize the future of the zebra. The African wild cat *Felis lybica* also interbreeds with the introduced domestic cat *Felis catus*. The problem is especially acute in the vicinity of human settlements, but concern has been expressed that pure strains of *F lybica* are likely to disappear (Smithers 1983). Other than these often hypothetical genetic effects, it is not known to what extent alien mammals influence native mammal communities in South Africa.

Invasive alien plants, however, also have an effect on mammal communities. Mature fynbos plant communities in the southern Cape for example support five to seven small mammal species (Bond et al 1980; Nel et al 1980; Breytenbach 1982). Alpha diversity of small-mammal communities in fynbos planted to pines decreases to less than three species within eight years of establishment (unpublished). By contrast, fynbos sites in the southern Cape invaded by *Hakea sericea* support up to five small mammal species which is similar to that recorded in uninvaded fynbos (Breytenbach et al 1984). The observed differences can probably be attributed to differences in the way ground cover changes under pines and hakea.

Bird communities

Alien birds seldom, if ever, invade native plant communities (Brooke et al this volume), and therefore probably do not have a large influence on indigenous bird communities. Invasion of native plant communities in the south-western Cape by alien plants, however, has a marked impact on bird communities.

The European starling *Sturnus vulgaris* has invaded the alien acacia communities of the Cape Flats and coastal zones, where they probably compete with indigenous birds for food resources (Winterbottom 1970, 1972, 1978). In the various veld types occurring in the south-western Cape between 105 and 171 bird species have been recorded whilst only 86 species were recorded in *Acacia cyclops* thickets (Winterbottom 1970).

Bird faunas of fynbos communities invaded by *Hakea sericea* have been compared with those of uninvaded fynbos. Alpha diversity (total number of species observed during visits over a two year period, in different seasons) in the 90% infested site (four species) was lower than in an uninfested site (13 species) or in the moderately infested site (12 species) (unpublished).

Arthropod communities

Alien arthropods frequently occur as pests on imported crops (Samways 1981) but little data are available to show how many alien arthropods have invaded natural communities. The Argentine ant *Iridomyrmex humilis* invades fynbos plant communities in the south-western Cape, forms nearly monospecific assemblages, and displaces most of the indigenous ant fauna. Of six dominant ant species found in a fynbos site without *I humilis*, none was found in a site infested by *I humilis* (Donnelly 1983). The dominant species at the invaded site (*I humilis* and *Monomorium schultzei*) are normally absent from communities free of aliens. Of five native species occurring in a fynbos habitat in the Kogelberg only two remained once the area was invaded by *I humilis*, which constituted more than 90% of the catch in pitfall traps (Bond and Slingsby 1984).

When the Argentine ant displaces ants such as *Anoplolepis custodiens*, *Pheidole capensis* and *Meranoplus peringueyi*, the dispersal and survival of seed of myrmecochorous, ie ant dispersed, plants are seriously curtailed (Bond and Slingsby 1984; Bond and Breytenbach 1985). Native ant dispersers quickly recover seed exposed on the soil, burying most in their nests. The Argentine ant moves seed for shorter distances from the mother plant and does not bury them. This leads to high seed predation and less successful establishment of myrmecochorous shrubs (Bond and Slingsby 1984; Bond and Breytenbach 1985).

Alien invasive plants also influence diversity of arthropod communities. The presence of *H sericea* is associated with an increase in the diversity of some groups (eg Diptera), but a decrease in others (eg Formicidae). Some (eg Apionidae), were completely absent from beat net samples collected in dense hakea sites. Similarly, ant diversity (counts from pitfall traps) decreased with increasing levels of *H sericea* invasion, whilst fly diversity increased (Breytenbach et al 1984; Breytenbach unpublished).

These exploratory results are confounded by the sampling problems which arise from the great variability in fynbos arthropod communities. They do indicate that invasion of native communities by introduced plants may have profound effects on arthropod faunas. Indigenous phytophagous insects do not feed extensively on introduced plants. In some cases introduced species (eg *Opuntia* species) are not used by any native phytophagous insect species in spite of having been present in southern Africa for more than 250 years (Moran 1980). Furthermore, I would predict from the comparatively high degree of monophagy in phytophagous arthropods, that if aliens change the species composition of the flora, large changes in the arthropod fauna could be expected.

Impact on community structure and function

Community structure is often described as a function of its constituent species (eg Begon and Mortimer 1981). It can, however, also be described by the functional attributes of the species (Gillison 1981), or by certain attributes related to dynamic characteristics (Noble 1981). Communities of certain biomes may be distinguished from those of others in terms of essential functional features. For example, nearly 16% of fynbos plants are dispersed by ants, whereas this dispersal syndrome is absent or virtually so from other South African biomes (Milewski and Bond 1982).

The way in which communities function will change significantly if the species that are lost or gained all have similar functional traits. For example, if soil arthropods such as termites or ants are lost from a system, litter breakdown and nutrient cycling must be expected to change (Buckley 1982). It is therefore important to assess observed changes in species composition in terms of possible influences on community functioning.

Plant communities

Loss of or turnover in species does not necessarily imply change in the physiognomy of plant communities. If alien species are taller than the communities they invade, or have a completely different structure, they bring about the greatest changes. This is frequently the case in South Africa where alien invasives in grasslands and shrublands are trees.

In *Pinus radiata* plantations no change in the general height of the indigenous plants occurred over a 40-year period. The number of plants per unit area, however, decreased from 260-78 plants per square metre (Richardson and Van Wilgen 1985). Reduction in density and cover greatly changed the foliage density of the remaining natural vegetation. The pines of course contributed the greatest change, since when mature they are three to twenty times taller than the dominants of the communities they invade. Marked changes in the canopy can also occur where the invader plants have a similar stature to those already occurring within the community. Foliage profiles changed substantially in sites invaded by *H sericea* in the southern Cape (Figure 1). It is likely that all invasive plants occurring at high densities will change foliage profiles significantly.

Feral stock and Himalayan tahr in mountain catchments of the south-western Cape, have as great an impact on the horizontal distribution of foliage. In areas frequented by stock in the Kouga and Baviaanskloof mountains, paths cover one third of the surface area of all slopes and foliage profiles are changed substantially (Breytenbach et al 1985). On Table Mountain tahrs almost denuded some areas (Smithers 1983). Changes in the horizontal distribution of

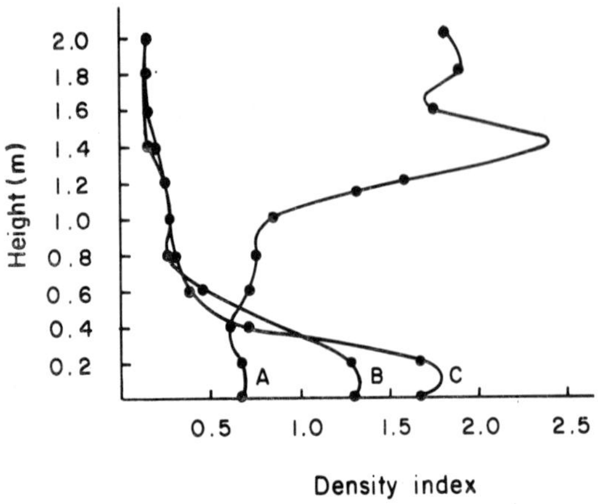

Figure 1. Foliage density profiles for plant canopies sampled along a *Hakea sericea* gradient of abundance in the southern Cape. The horizontal axis represents a horizontal foliage density index, with A = 95%, B = 35% and C = 0% hakea cover. It was possible to differentiate between profiles using discriminant function analysis.

foliage without a concomitant change in species composition may seem of little or no consequence, but in fact may have a large impact on faunal communities. Birds respond to changes in the vertical and horizontal distribution of foliage (Cody 1975, 1983) as do small mammals (Rosenzweig and Winakur 1969; Bond et al 1980; Breytenbach 1982), and arthropods (Moran 1980; Moran and Southwood 1982).

The physiognomy of the native vegetation also changes since the representation of major growth forms within a community is altered. For example, chamaephytes (low shrubs) decreased while hemicryptophytes (eg grasses, sedges and restios) increased in *P radiata* plantations (Richardson and Van Wilgen 1985). Fynbos plant communities of the Kouga and Baviaanskloof mountains undergo considerable change when grazed by feral stock. Species that sprout after fire were more commonly found in grazed plots whilst plants killed by fire which are dependent on seed for establishment decreased. The occurrence of various dispersal and pollination syndromes also changed, eg the number of myrmecochorous and bird-pollinated species decreased, whilst insect-pollinated species increased in grazed sites (Table 1) (Breytenbach et al 1985). The effects of intensity and season of grazing are not known. Recovery would be slow, since many fynbos plants (eg myrmecochorous species) have poor dispersal mechanisms and recruitment normally only occurs after fire.

Invasive species may therefore change species composition, physiognomy and structure of plant communities. These changes not only have implications for monophagous herbivores and pollinators, but also for generalist consumer organisms (Moran 1980; Strong et al 1984). Impact on consumer organisms, and monophagous arthropods in particular, will be greater in some ecosystems. For example, any disruption of small populations by aliens could quite easily result in species extinctions and loss of mutualistic linkages. This is especially true in fynbos where localized endemic plants (Hall et al 1980; Bond and Goldblatt 1984) and a certain degree of monophagy (Raubenheimer and Picker 1985), occur simultaneously.

Small mammals

As in the case of plants, replacement of fynbos by pine stands causes marked changes in the structure of the small-mammal community. Herbivores and specialist granivores disappear. *Praomys verreauxii*

Table 1. Functional attributes of plant communities that are influenced by grazing (feral stock) in fynbos plant communities of the Kouga and Baviaanskloof mountain ranges in the southern Cape. (Adapted from Breytenbach et al 1985). (UGR = ungrazed; GRA = grazed). (Mean = mean number of plant species, per 4 m² plot, n = 5; s = standard deviation).

	Fire response				Seed dispersal				Pollination			
	Sprouters		Seeders		Wind		Ants		Insect		Bird	
	UGR	GRA	UGR	GRA	UGR	GRA	UGR	GRA	UGR	GRA	UGR	GRA
Mean	5,8	9,0	13,0	7,8	9,6	10,2	6,6	5,4	6,2	7,2	10,8	8,6
s	1,9	1,2	3,2	1,5	2,3	1,3	1,1	1,1	1,5	0,8	2,9	1,1
t	6,36		6,52		1,02		3,33		2,87		3,2	
P	0,01		0,01		n.s.		0,05		0,05		0,05	

a species feeding predominantly on protea seed, *Acomys subspinosus* which feeds on seeds of Restionaceae and *Otomys irroratus* a grazer and browser (Breytenbach 1982) disappear within five to eight years of the establishment of *P pinaster* plantations in fynbos habitats (unpublished). Of the two to four insectivores that are normally found in fynbos stands, only *Myosorex varius* remained. In communities invaded by *H sericea*, however, community structure changes only slightly, with insectivores and specialist granivores surviving. However, the herbivores, which included one to three species of *Otomys*, disappear (Breytenbach et al 1984).

Bird communities

Bird communities are disrupted by alien invasive species in two ways. Firstly, some bird species disappear and this could result in the disruption of processes such as pollination and seed dispersal. Secondly, new species may appear. This is especially the case where arboreal habitats develop in vegetation types that do not support such habitats under normal circumstances, eg trees invading grasslands.

In sites invaded by hakea in the southern Cape, density of some of the insectivores (eg the spotted prinia *Prinia maculosa*) increases with increasing hakea density. This could be due to an increase in small flying insects (see above). Nectarivores (eg the Cape sugarbird *Promerops cafer*) disappear completely from dense hakea stands (unpublished). The impact on seed set of the loss of avian pollinators is largely unknown. In exclosure experiments protea seed set decreased only slightly in *Protea repens* in the absence of bird pollinators (Coetzee 1984), as it does also in *P punctata*, *P eximia* and *P lorifolia* (unpublished). Insects were the most important pollinators for protea species. The effects of species loss cannot be assumed. As is demonstrated here, experimentation is needed to show whether pollination or dispersal will not be carried out effectively by other organisms not affected by the invasive species.

Bird community structure, based on the frequency of occurrence of species in bird club lists, in the alien acacia thickets of the south-western Cape is different from those in indigenous communities (Winterbottom 1970). Nectarivore, frugivore and insectivore densities decrease in sites invaded by *Acacia cyclops* (Fraser et al 1985).

Certain species are favoured by alien invasive plants. The hadedah ibis *Bostrychia hagedash* and pied barbet *Lybius leucomelas* have undergone considerable range expansions since the turn of the century. This has been attributed, in part, to the increase in nesting and roosting sites provided by alien woody plants (Macdonald et al 1985; Macdonald in press). In the Transvaal grasslands, nesting sites have become available to birds of prey in eucalyptus plantations (Steyn 1977). The 'new' tree habitat provides roosting and nesting sites for birds normally not present in the grasslands. These 'additions' to the avifauna contribute towards a change in avifaunal community structure.

Impact on processes

Ecological processes that are dependent on the interaction between organisms eg pollination, herbivory, competition and predation, are especially prone to disruption by invasion. Because the degree of specificity involved in these interactions varies from case to case, the effects of invasive species cannot readily be predicted. Herbivory and pollination could continue in some form or other, eg another pollinator could take over the role of the missing species. One can therefore not assume that if structure changes, functions and processes will be disrupted, as we have seen from the case of pollination by sugarbirds.

Effects of *Hakea* on canopy growth and reproduction in native Proteaceae

Native plants are not always replaced by alien species. They frequently survive in or under the canopy of alien plants, especially in those cases where the alien canopy has not closed. The effects of aliens present at low densities have not been investigated in detail. This section is therefore largely speculative and I discuss the potential influence of aliens on growth, reproduction and dispersal.

Leaf duration in *Protea lorifolia* decreases significantly from 11 to six to less than five years, along a gradient of density in *Hakea sericea* with 0%, 35% and 95% hakea cover respectively. A decrease was also observed in *Leucadendron salignum* (in both male and female plants) although leaf duration in this species seldom exceeds four years (Breytenbach et al 1984). No explanation for this phenomenon has been offered. It may, however, indicate changes in soil moisture regime since leaf abscission is frequently a function of drought stress (Fitter and Hay 1981). Since leaf duration is shorter, nutrient cycling patterns (both internally and externally), are likely to change. The change in leaf retention may also have an impact on consumer organisms (herbivores and detritivores), but no data are available to substantiate this.

As part of normal control programmes hakea is

cut and allowed to dry for more than a year before being burnt. After hakea was cut in the southern Cape, native plants remaining in the area produced more leaves (Table 2), whilst *Leucadendron salignum* also produced larger leaves than those growing in adjoining uncut control blocks (Table 3). Higher insect predation on leaves of *L salignum* in unslashed hakea sites may reflect the plants reduced production of secondary compounds in the foliage (unpublished). In spite of no differences in leaf predation rates in *P lorifolia* occurring along the hakea density gradient that was sampled, leaf area was still smaller since leave retention times were shorter. Loss in leaf area may influence the carbon balance of a plant. Flower production in *P lorifolia* in areas with 0%, 35% and 95% hakea cover, for example, was 0,58, 0,43 and 0,48 flowers per annual cohort of shoots per stem respectively (Breytenbach et al 1984), indicating a decrease in reproductive output.

Table 2. Number of leaves produced annually per node per single stem in *Protea lorifolia* in two adjoining sites, one with *Hakea sericea* left standing and the other with *H sericea* felled. *H sericea* constituted 35% cover in both cases before felling. All means differ significantly with P less than 0,01.

	H sericea felled		*H sericea* left standing	
	Year after cut	Year of cut	Year after cut	Year of cut
Mean	10,0	6,6	9,3	6,2
s	3,2	2,2	2,2	2,6
N	25	25	25	25

Herbivory

Plants allocate and use nutrients for various purposes, eg photosynthesis and assimilation, support, reproduction and defence against herbivores and predators. Some plants expend a large proportion of their resources on defence against herbivores (Harborne 1982). If more resources are allocated towards defence due to an increase in predation (eg if alien herbivores are not affected by normal plant defence mechanisms), native species may have their reproductive output curtailed. Changes in reproductive output ultimately result in changes in community structure and function. It is therefore essential that the effects of aliens on herbivory and the consequences of such changes are understood.

Alien invasives change the spatial distribution of plants. Protea density in areas invaded by *H sericea* in the southern Cape can be as low as 50 plants per hectare. The inter-plant distances for conspecifics therefore change and could potentially influence the efficacy with which pollinators and herbivores find host plants. Phytophagous insects, (both mono- and poly-phagous) do not have a uniform response to changes in host plant density. Highest insect herbivore density for some species is reached at maximum host plant density (eg Ralph 1977; Lemen 1981), for others on isolated plants or patches (eg Jones 1977; Courtney and Courtney 1982), and yet

Table 3. Leaf size and levels of foliar predation for *Leucadendron salignum* in slashed and standing *H sericea* sites sampled in the southern Cape. Means and standard deviation (in parentheses) are given. Predation is expressed as measured length of leaf missing.

Sex	Leaf length	Reduction in leaf length	Percentage reduction in leaf area	N
Female				
Slashed	32,6(7,5)	3,88(5,1)	14,3(29,9)	51
Standing	37,0(7,6)	9,66(6,6)	26,7(18,8)	42
t	18,9	31,4	16,1	
P	<0,001	<0,001	<0,001	
Male				
Slashed	36,7(11,5)	3,2(3,3)	8,9(8,4)	126
Standing	28,4(9,6)	4,6(4,3)	16,5(13,4)	183
t	8,39	39,4	72,7	
P	<0,001	<0,001	<0,001	

others show no link to host plant density, but increase if the biomass or density of non-host plants are high (Bach 1980).

Levels of herbivory vary considerably, both in space and time (see examples in Strong et al 1984) and it is therefore difficult to differentiate between normal patterns and the effect of alien invasives. No significant differences in degree of herbivory on *Protea lorifolia* could be detected along a hakea density gradient. Herbivory on *L salignum* decreased during the first season after aliens were removed (Table 3; see above). The impact of arthropod herbivores has not been assessed for other species. The effect of aliens on normal plant herbivore relationships can, therefore, not be assessed adequately.

Introduced herbivores may also influence the way in which leaf material is removed from plants. Large herbivores (eg cattle and Himalayan tahr) change patterns of herbivory substantially. Insects, which normally form the major component of the fynbos herbivore biomass, feed on parts of leaves. The large herbivores consume whole leaves or whole plants. Fynbos plants, especially members of the Proteaceae, retain their leaves for long periods. Plants also retain leaves that are damaged by insects. How the change in partial loss of a leaf to total leaf loss will affect native plants is not clear at this stage, but we know that standing biomass is reduced substantially (Lloyd in Smithers 1983; Breytenbach et al 1985).

Predation on seed as a specialized form of herbivory also changes (Table 4, adapted from Breytenbach et al 1984). Under normal conditions as many as 95% of protea infructescences could be damaged by the end of their first year. The rate of seed loss however, is higher in sites invaded by hakea since more infructescences are damaged early in the season. This only applies to serotinous (canopy stored) protea seed reserves. Soil-stored seed reserves (as determined from protea seeds buried in the soil to half their length), disappeared more rapidly from noninvaded sites. Ultimately, however, seed and seedling predation accounted for complete loss of all individuals from all areas (Table 4, adapted from Breytenbach et al 1984).

Pollination and seed set

Seed production per flower in *Protea lorifolia* at sites where *Hakea sericea* has 90% cover is lower than in those from low density or alien free sites (unpublished). This can not be directly ascribed to absence of pollinators, since growth is also impeded (see above). Lower seed set may therefore be a function of poor pollination and/or due to competition for limited resources. Observations on extremely isolated patches of proteas (eg in road reserves and wheat fields), where no seed set occurred (unpublished), indicate that pollination could be completely disrupted under certain conditions. Studies of the impact of aliens on pollination are urgently required.

Table 4. Damage to infructescences and seed loss, ascribed to insects and rodents. Damage and loss is expressed as the percentage of infructescenes damaged and seeds removed from outside exclosures.

	Percentage *Hakea sericea* cover		
	0%	35%	95%
Damage to infructescences by:-			
Insects	3	18	4
Rodents	2	19	19
Total	5	37	23
Seed loss outside exclosure after:—			
four days	82	69	2
four months	100	100	100

Impacts due to control measures

Invasive alien populations are controlled by several means eg through the use of biological control agents, by means of chemical agents or by physical means such as cutting and burning. It is obvious that biological control measures are being sought because such measures are ultimately cheap whilst probably having the least impact on the native community (Samways 1981). We must, however, not assume that any technique that offers a way of getting rid of an alien is acceptable. The control measures we apply may have a greater impact than the alien itself.

Measures to control introduced species also effect indigenous biotic communities. *Hakea sericea* in fynbos plant communities is a case where control measures have had a severe impact. *Hakea sericea* is controlled by cutting adult plants, leaving them for some 12-18 months to allow serotinous seeds to be released and germinate. Once seedlings emerge, the area is burned. This is an extremely effective way of getting rid of *H sericea* (Kruger 1977; Fugler 1982; Kluge et al this volume). The slashed hakea, however, substantially increases the combustible fuel close to ground level and leads to increased fire intensities.

The effect of an intense hakea fire on the plant community is shown in Table 5 (J Vlok personal communication). Plant species which regenerate by vegetative means after fire are completely eliminated from dense hakea sites. Soil seed stores are depleted to some extent by fire. The total number of seedlings present (as determined in the nursery from germination of seed from a square metre soil core sample over a two year period) in a non-hakea site decreased by 35%, and by 22% and 39% in a medium-dense and dense hakea site respectively. These data reflect the impact of fire. Field studies subsequent to fire showed an increase in richness in the fynbos site after fire, but a decrease of 45%, 56% and 44% in the firebreak, medium dense and dense hakea sites respectively (J Vlok personal communication).

Table 5. Number of species recorded in plant communities, (plots were 20 m^2) and which germinated from soil stored seed banks (1 m^2 × 25 cm deep soil samples) taken along a *Hakea sericea* density gradient in the southern Cape.

	Site			
	Fynbos	Fire-break	Medium hakea	Dense hakea
Number of species in:-				
Seedbank before fire	20	18	23	18
Seedbank after fire	13	–	17	11
In the field before fire	18	33	34	18
Reseeding spp in the field after fire	18	12	7	7
Sprouting spp in the field after fire	3	6	8	0

Increased fire intensities also influenced faunal communities. The survival of termite mounds decreased considerably (Table 6). The impact of these faunal changes on soil formation, aeration, soil movement in the profile and nutrient cycling must be considerable, but has not been measured (cf Buckley 1982). Exposure of the soil to erosion by wind and rain increases with reductions in the plant canopy. Soil loss on an alluvial plain, estimated from root exposure, after slashing and burning of dense hakea in the southern Cape amounted to between 540 and 720 tons ha^{-1} over a three-year period (personal observation).

Table 6. Effect of fire on termite (*Amitermes* species) nest survival at two sites (fynbos and fynbos heavily infested with *Hakea sericea*) in the southern Cape. *Hakea* was felled some 10 months before the fire. Nests were classified as being dead if termites were killed, but the nest was not burned by the fire (unpublished).

	Number of nests			Percentage of nests		
Habitat	Alive	Dead	Burnt	Alive	Dead	Burnt
Fynbos	80	1	1	98	1	1
Fynbos with felled and burnt hakea	9	18	73	9	18	73

Discussion and conclusions

The effects of invasive aliens on native organisms, and ecosystems have not been investigated adequately. Conclusions are tentative, nevertheless it is evident that alien invasive species are not simply effective invaders because of inherent traits of the alien itself eg they may be preadapted (eg Schutte 1960), occupy a vacant niche (Campbell et al 1979), be good dispersers (eg Middlemiss 1963; Kruger 1977), or suffer low predation levels (eg Hall 1979). Aliens are also effective invaders because they disrupt ongoing ecological processes, especially those involving biotic interactions. In several of the studies reported on here, evidence points to the disruption of pollination, herbivory, seed dispersal and seed predation. The methods used in the control of aliens also have effects on the environment.

In the development of control strategies and techniques, it must be remembered that it is virtually impossible to manage ecological processes. The participants in a process can, however, be managed fairly effectively and the unit of management should therefore be the population. Species cannot be treated in isolation and evaluations should be done within a community context:

(1) Impact should be assessed at the level of the community by determining which species are influenced by the alien or the control methods being used ie which species appear or disappear.
(2) Impact should also be assessed in terms of the affects on community attributes such as guilds, functional groups and dynamic characteristics, eg are pollinators and dispersers being affected?
(3) It is equally important to determine how and why organisms are affected. In doing so, it is

necessary to select organisms that are representative of a guild, rather than a unique species, the premise being that if impacts on one such organism are understood, it is more likely that the data could be extrapolated to other species with similar traits.

(4) Indicators of impacts on ecosystem processes, eg changes in soil erosion rate, are also important in this context.

It is unlikely that alien invasives will all influence the same processes and species in a similar fashion. Managers and researchers should therefore also attempt to develop an understanding of alien 'traits'. Are there, for example, certain species that always influence certain processes and can we identify them from a set of biological attributes?

Acknowledgements

This paper is published with the permission of the Forestry Branch of the Department of Environment Affairs. Elmé Breytenbach, Jan Vlok, Steve Viljoen, Jeremy Midgley, Roland America and other colleagues at the Saasveld Forestry Research Centre have contributed to this paper by sharing their collective expertise and field knowledge with me. E Breytenbach, D Versveld, M Fraser and R Knight commented on earlier drafts.

Bach C E 1980. Effect of plant density and diversity on the population dynamics of a specialist herbivore, the striped cucumber beetle, *Acalymma vittata* (Fab). Ecology 61, 1515-1530.

Begon M and Mortimer M 1981. Population ecology: A unified study of plants and animals. Blackwell Scientific Publications, London.

Bond W J 1983. On alpha diversity and the richness of the Cape flora: A study in southern Cape fynbos. In: Kruger F J, Mitchell D T and Jarvis J U M (eds) Ecological Studies 43. Mediterranean-type ecosystems: the role of nutrients. Springer-Verlag, Berlin. pp 337-356.

Bond W J and Breytenbach G J 1985. Ants, rodents and seed predation in Proteaceae. South African Journal of Zoology 20, 150-154.

Bond W J, Ferguson M and Forsyth G 1980. Small mammals and habitat structure along altitudinal gradients in the southern Cape mountains. South African Journal of Zoology 15, 34-43.

Bond P and Goldblatt P 1984. Plants of the Cape Flora: A descriptive catalogue. Trustees of the National Botanic Gardens of South Africa, Kirstenbosch.

Bond W J and Slingsby P 1984. Collapse of an ant-plant mutualism: The Argentine ant (*Iridomyrmex humilis*) and myrmecochorous Proteaceae. Ecology 65, 1031-1037.

Breytenbach G J 1982. Small mammal responses to environmental gradients in the Groot Swartberg of the southern Cape. MSc Thesis, University of Pretoria, Pretoria. 226 pp.

Breytenbach G J, Breytenbach W and Vlok J 1984. Ecological disruption in Fynbos. In: Dell B (ed) MEDECOS IV: Proceedings of the 4th International Conference on Mediterranean Ecosystems. Botany Department University of Western Australia, Nedlands. pp 16-17.

Breytenbach G J, Vlok J and Viljoen S 1985. The impact of grazing on catchment conditions in the mountains of the south-eastern Cape. South African Forestry Research Institute. Saasveld Forestry Research Centre Report 85/42. pp 35.

Buckley R C 1982. Ant-plant interactions in Australia. W Junk, The Hague.

Campbell B M, MacKenzie B and Moll E J 1979. Should there be more tree vegetation in the Mediterranean climatic region of South Africa. Journal of South African Botany 45, 453-457.

Cody M L 1975. Towards a theory of continental species diversities: Bird distribution over Mediterranean habitat gradients. In: Cody M L and Diamond J M Ecology and evolution of communities. Belknap Press Harvard University, Massachusetts. pp 214-257.

Cody M L 1983. Continental diversity patterns and convergent evolution in bird communities. In: Kruger F J, Mitchell D T and Jarvis J U M (eds) Ecological Studies 43. Mediterranean-type ecosystems: the role of nutrients. Springer-Verlag, Berlin. pp 357-402.

Coetzee J H 1984. Insekte in assosiasie met *Protea repens* (L.) L. MSc Thesis, University of Stellenbosch, Stellenbosch.

Courtney S P and Courtney S 1982. The 'edge effect' in butterfly ovi-position: Causality in *Anthocharis cardamines* and related species. Ecological Entomology 7, 131-137.

Cowling R 1976. The ecological status of the understorey communities of pine forests on Table Mountain. In: Moll E J and Campbell B M (eds) The ecological status of Table Mountain. Department of Botany, University of Cape Town, Cape Town. pp 1-29.

Donnelly D 1983. Comparison of community structure of the epigaeic invertebrate fauna, with particular reference to ants and beetles, between fynbos sites under different management practices. MSc Thesis, University of Stellenbosch, Stellenbosch.

De Ronde C 1983. Controlled burning in pine stands in the Cape: The influence of crown scorch on tree growth and litterfall. South African Forestry Journal 127, 39-41.

De Ronde C 1984. Litter accumulation problems identified in *P pinaster* stands of the Cape Province. South African Forestry Journal 131, 48-52.

Fitter A H and Hay R K M 1981. Environmental physiology of plants. Academic Press, London.

Fraser M W, Prys-Jones R P and Clark D L 1985. Effects of burning and alien plants on bird communities in mountain fynbos of the Cape of Good Hope Nature Reserve. Unpublished abstract from Research for Resource Management, 7th Annual Research Meeting of the Fynbos Biome Project. CSIR, Cape Town. pp 32-33.

Fugler S R 1982. Infestations of three Australian *Hakea* species in South Africa and their control. South African Forestry Journal 120, 63-68.

Gillison A N 1981. Towards a functional vegetation classification. In: Gillison A N and Anderson D J Vegetation classification in Australia. CSIRO Canberra. pp 22-29.

Hall A V 1979. Invasive weeds. In: Day J, Siegfried W R, Louw G N and Jarman M L Fynbos ecology: A preliminary synthesis. South African National Scientific Programmes Report No 40. CSIR, Pretoria.

Hall A V, De Winter M, De Winter B and Oosterhout S A M 1980. Threatened plants of South Africa. South African National Scientific Programmes Report No 45. CSIR, Pretoria.

Harborne J B 1982. Introduction to ecological biochemistry. Academic Press, New York.

IUCN 1980. World conservation strategy. International Union for the Conservation of Nature and Natural Resources, Gland.

Jones R E 1977. Movement patterns and egg distribution in cabbage butterflies. Journal of Animal Ecology 46, 195-212.

Kruger F J 1977. Invasive woody plants in the Cape Fynbos, with special reference to the biology and control of *Pinus pinaster*. In: Proceedings of the 2nd National Weeds Conference of South Africa. Balkema, Cape Town.

Kruger F J 1981. Conservation: South African heathlands. In: Specht R L (ed) Ecosystems of the World 9B. Heathlands and related shrublands: Analytical studies. Elsevier, New York. pp 231-234.

Lemen C 1981. Elm trees and elm leaf beetles: patterns of herbivory. Oikos 36, 65-67.

Macdonald I A W (in press). Range expansion in the pied barbet and the spread of alien tree species in South Africa. Ostrich.

Macdonald I A W, Powrie F J and Richardson D M 1985. Avifaunal responses to the spread of alien trees in the fynbos biome. Unpublished abstract from: Research for Resource Management, 7th Annual Research Meeting of the Fynbos Biome Project CSIR, Cape Town. pp 37.

Middlemiss E 1963. The distribution of *Acacia cyclops* in the Cape Peninsula by birds and other animals. South African Journal of Science 59, 419-420.

Milewski A V and Bond W J 1982. Convergence of myrmecochory in Mediterranean Australia and South Africa. In: Buckley R C (ed) Ant-plant interactions in Australia. W Junk, The Hague. pp 89-98.

Moran V C 1980. Interactions between phytophagous insects and their *Opuntia* hosts. Ecological Entomology 5, 153-164.

Moran V C and Southwood T R E 1982. The guild composition of arthropod communities in trees. Journal of Animal Ecology 51, 289-306.

Naveh Z and Whittaker R H 1980. Structural and floristic diversity of shrublands and woodlands in northern Israel and other Mediterranean areas. Vegetatio 41, 171-190.

Nel J A J, Rautenbach I L and Breytenbach G J 1980. Mammals of the Kamanassie mountains, southern Cape Province. South African Journal of Zoology 15, 255-261.

Noble I 1981. The use of dynamic. characteristics in vegetation classification. In: Gillison A N and Anderson D J. Vegetation classification in Australia. CSIRO, Canberra. pp 22-29.

Ralph C P 1977. Effect of host plant density on populations of a specialized seed sucking bug, *Oncopeltus fasciatus*. Ecology 58, 799-809.

Raubenheimer D and Picker M 1985. Determinants of fynbos insect community structure and distribution. Unpublished abstract from: Research for Resource Management, 7th Annual Research Meeting of the Fynbos Biome Project. CSIR, Cape Town.

Richardson D M and Van Wilgen B W 1986. Changes in the composition of fynbos vegetation following thirty years of afforestation with pines at Jonkershoek, Stellenbosch. South African Forestry Research Institute, Jonkershoek Forestry Research Centre Report 85/10. pp 24.

Rosenzweig M L and Winakur J 1969. Population ecology of desert rodent communities: Habitats and environmental complexity. Ecology 50, 558-572.

Samways M J 1981. Biological control of pests and weeds. Edward Arnold, London. pp 58.

Schutte K H 1960. Trace element deficiencies in Cape vegetation. Journal of South African Botany 26, 46-49.

Skinner J D, Fairall N and Bothma J du P 1977. South African Red Data Book - Large mammals. South African National Scientific Programmes Report No 18. CSIR, Pretoria.

Smit H A P 1983. Die indringing van uitheemse plante teen die hange van Stellenboschberg. BSc Honours Project Report. Department of Geography, University of Stellenbosch, Stellenbosch. 19 pp.

Smit H A P and De Kock G L 1984. Ruimtelike diffusie van die indringerplant *Pinus pinaster* teen die hange van Stellenboschberg. South African Forestry Journal 129, 10-16.

Smithers R H N 1983. The mammals of the southern African subregion. University of Pretoria, Pretoria. 736 pp.

Steyn D J 1977. Occupation and use of the eucalyptus plantations in the Tzaneen area by indigenous birds. South African Forestry Journal 100, 56-60.

Strong D R, Lawton J H and Southwood R 1984. Insects on plants: Community patterns and mechanisms. Blackwell Scientific Publications, Oxford.

Whittaker R H 1977. Evolution of species diversity in land communities. Evolutionary Biology 10, 1-67.

Winterbottom J M 1970. The birds of the alien *Acacia* thickets of the south-western Cape. Zoologica Africana 5, 49-57.

Winterbottom J M 1972. The ecological distribution of birds in southern Africa. Monographs of the Percy FitzPatrick Institute of African Ornithology 1. pp 81.

Winterbottom J M 1978. Birds. In: Werger M J A. The ecology and biogeography of southern Africa. W Junk, The Hague.

Chapter 19

Impact of woody aliens on ecosystem properties

D B Versfeld and B W van Wilgen
South African Forestry Research Institute.

Introduction

In humid and subhumid regions of southern Africa, invasions by introduced woody plants cause, in the extreme, a marked change in the structure of vegetation canopies. Grasslands and shrublands are converted to forest, or sparse shrubland to dense tall shrubland. Changes of this magnitude must cause substantial changes to ecosystem processes such as erosion and the hydrological and nutrient cycles.

In this chapter we attempt to document the changes in vegetation canopies resulting from invasions and to evaluate these by examining the effects on the microclimate and water balance. Experimental and observational data are adduced to quantify the likely magnitude of the canopy changes and of the consequent changes to environmental regimes. As few quantitative data are available for invaded areas we draw heavily on data collected from plantations of species that are known to be invasive in southern Africa and which form self-sown stands that approximate plantations.

Vegetation structure

Above-ground biomass

For most ecosystems that have been studied, large increases in above-ground biomass accompany invasion. These changes will have impacts on fire hazard, nutrient cycling and the hydrology of ecosystems. Changes in biomass following invasion are depicted in Figure 1. Changes are best documented in the fynbos biome. Most other data are from studies in forest plantations. Figure 1 shows that invasion of fynbos by *Hakea* or *Acacia* species can bring about a threefold increase in biomass. There are no comparative data from sites invaded by pines, but afforested stands have a biomass of three to more than 10 times that of fynbos. Areas invaded by pines may be even more dense than carefully managed plantations. In grassland ecosystems, the biomass of afforested sites can be 70-100 times the original biomass. It should not be assumed that all invasion will necessarily increase biomass. Alien trees along streams for example, may have much the same biomass as the native vegetation, but data are lacking.

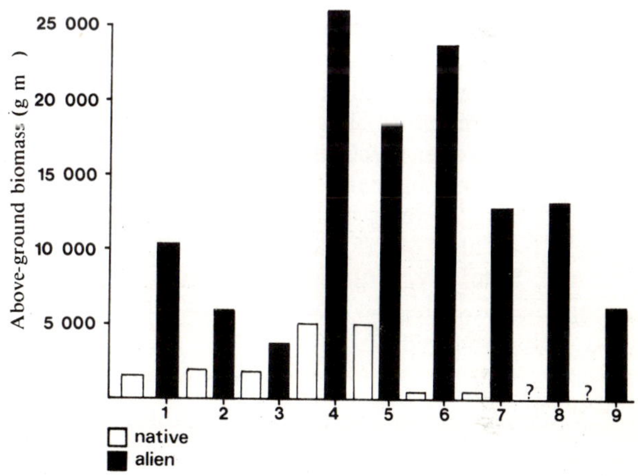

Figure 1. Increases in above-ground plant biomass through invasion or afforestation of natural vegetation. Details of the sites and sources of information are as follows: (1) fynbos invaded by *Acacia cyclops* and *A saligna* (Milton and Siegfried 1981); (2) fynbos invaded by *A saligna* (van Wilgen and Richardson 1985); (3) fynbos invaded by *Hakea sericea* (van Wilgen and Richardson 1985); (4) fynbos afforested with *Pinus radiata* (35 yr) (van Wilgen 1982 and van Laar 1984); (5) fynbos afforested with *P radiata* (30 yr) (van Wilgen 1982; van Laar and van Lill 1978); (6) grassland in Zimbabwe afforested with *P patula* (20 yr) (quoted in Morris 1984); (7) grassland in the Natal Drakensberg afforested with *P patula* (Bosch 1985), (8) Natal midlands grassland afforested with *Acacia mearnsii* (10 yr) (Schonau 1978); (9) Transvaal highveld grassland afforested with *Eucalyptus* species (5 yr) (Schonau and Boden 1982).

Address for correspondence: D B Versfeld, Jonkershoek Forestry Research Centre, Private Bag 5011, Stellenbosch, 7600.

Table 1. Changes in the height and density of dominant shrubs following the invasion of fynbos.

Natural vegetation	Height (m)	Density (stems ha^{-1})	Invader species	Height (m)	Density (stems ha^{-1})	Source
Mountain Fynbos	2,0	c.1680	*Hakea sericea*	2,6	8 900	Van Wilgen and Richardson 1985
Mountain Fynbos	2,0	c.1680	*Acacia saligna*	3,9	9 800	Van Wilgen and Richardson 1985
Coastal Fynbos and Strandveld	1 – 2	not given	*A saligna* and *A cyclops*	5 – 10	5 200	Milton and Siegfried 1981

Plant density and vegetation height

The major impact of invasion in fynbos is the replacement of indigenous plants by alien species of greater size and density. Some examples of differences in vegetation height and stand density on equivalent sites are listed in Table 1. Invasion causes considerable changes in fynbos community structure (van Wilgen and Richardson 1985), where the number and size of dominant shrubs increases at the expense of understorey herbs and shrubs. Richardson and van Wilgen (1986) record pronounced reductions in the cover and density of native plants in fynbos following 35 years of afforestation with *Pinus radiata* (see also Macdonald and Richardson this volume). Similarly, fynbos cover under a *P radiata* canopy was less than five per cent of values for undisturbed vegetation (Milton 1976).

Leaf area indices

Changes in the foliage profile brought about by invasion (Breytenbach this volume) can have important effects on microclimate. Such changes should be reflected by changes to the leaf area index. Table 2 summarizes available data on leaf area indices (LAI) for natural and afforested sites. Stands of coniferous species often have a LAI at least twice that of broad-leaved species (Whittaker and Marks 1975). *Pinus radiata* (Table 2) in the western Cape illustrates this tendency. Invasives such as *Pinus* or *Hakea* species with needle-like leaves can therefore be expected to increase the LAI. This will have marked influences on the energy balance, rainfall interception and transpiration (see below).

Litterfall and decomposition rates

Higher growth rates (Rutherford et al this volume), and greater standing crop biomass are often accompanied by greater accumulation of litter than in the native vegetation replaced. Any lengthening of the burning cycle through invasion also leads to an accumulation of litter which would otherwise be recycled by frequent burning. This is particularly apparent in grassland.

Table 2. Estimates of leaf area indices for indigenous and alien vegetation in South African ecosystems. The individual shrub estimates for the two *Protea* species were reduced by 70% to allow for incomplete cover by these species in most stands.

Species	Natural vegetation	Leaf area index (m^2 per m^2)	Source
Protea repens	fynbos	3,4	Van Wilgen unpublished
Protea neriifolia	fynbos	1,6	Van Wilgen unpublished
Eucalyptus spp	grassland	4 – 6	Versfeld and Brink unpublished
Populus canescens	riparian forest	3,1	Van Laar 1983
Pinus radiata	fynbos	8,8	Van Laar 1984

Litterfall rates and litter mass for invaded and uninvaded sites in southern Africa are summarized in Table 3. The impacts of invasion on litterfall and decomposition rates seem to vary. Slow decomposition rates allow marked accumulation of litter under *P pinaster* and *P patula*. Litter build-up is relatively low under *P radiata*. Possible causes of heavy litter accumulations include polyphenol inhibitors, or microclimatic changes which suppress decomposers. Changes in litter biomass of the order indicated in

Table 3 will have major consequences in terms of nutrient pools and cycling, and the severity of fires.

Fire hazard

Fire hazard is defined as a measure of that part of fire danger contributed by the fuels available for burning (Chandler et al 1983). Invasion of the habitat by introduced plants affects fire hazard by changing the structure of the fuel bed. Changes in biomass, size and distribution of plant parts, and plant moisture and energy contents have been determined at two fynbos sites invaded by *Hakea sericea* and *Acacia saligna* (van Wilgen and Richardson 1985), to test the hypothesis that invasion increases fire hazard through increasing fuel loads. Fuel was defined as dead and live plant parts with a diameter less than six millimetres. Here, invasion by *H sericea* resulted in a 60% increase in fuel loads and lowered the moisture content of live foliage from 155 to 110%, thus increasing fire hazard. Invasion by *A saligna* resulted in a 50% increase in fuel loads, but because the moisture content of the foliage of this shrub was high, about 270%, the fuel bed was effectively confined to the litter layer below the trees. This would cause fires to spread slowly and burn at

Table 3. Litterfall rates and litter mass for native vegetation and sites invaded by woody plants.

Region	Species or vegetation type	Litterfall rate (kg ha^{-1} yr^{-1})	Litter mass below canopy (kg ha^{-1})	Source
Western Cape	Fynbos 6 – 16 yrs postfire	not given	1 000 – 4 950	Kruger 1977
Western Cape	Fynbos 20 yrs postfire	not given	4 750	Van Wilgen et al 1985
Western Cape	Fynbos 21 yrs postfire	not given	14 300	Van Wilgen 1982
Western Cape	Fynbos 21 yrs postfire	2 170	not given	Kruger and Lamb (unpublished)
Western Cape	Fynbos 9 yrs postfire	780	6 600	Mitchell et al (in press)
Western Cape	Riparian forest (unburnt)	6 700	not given	Versfeld (unpublished)
Western Cape	*Acacia saligna* 20 yrs postfire	not given	15 020	Van Wilgen and Richardson 1985
Western Cape	*Acacia saligna*† *A cyclops*† *A longifolia*† *A melanoxylon*†	7 040	21 120	Milton 1980
Western Cape	*Hakea sericea* 9 yrs postfire	not given	6 890	Van Wilgen and Richardson 1985
Western Cape	*Pinus radiata** 37 yrs	4 000	6 500	Versfeld 1978a
Western Cape	*Pinus pinaster** 30 – 50 yrs	6 500	30 000 – 70 000	De Ronde 1984
Eastern Transvaal	*Pinus patula** 45 yrs	not given	320 000	Schutz 1982
Swaziland	*Pinus patula** 12 – 14 yrs	not given	43 600	Morris 1984

†No age given
*Plantations of given age

low intensities under moderate weather conditions. Under extreme weather conditions, however, fires in such invaded stands are known to burn live foliage in the shrub crowns. The increased energy released by such fires makes them self-sustaining, difficult to control and potentially more damaging to ecosystems than fires in natural vegetation. Under such conditions, fire hazard is increased by invasion.

Intense fires in stands of *Hakea sericea* felled as a control measure depressed the regeneration of natural vegetation (Richardson and van Wilgen in press), although it is not clear how much of this effect was due to fire and how much was due to the prior effects of the dense *H sericea* stands.

Macdonald (1983) has observed that invasions of *Chromolaena odorata* in Natal are dangerous in view of their high flammability. The species forms thickets in forest and riverine forest ecotones, which are normally fire-excluding. Fires burning in adjacent grasslands cross the ecotones invaded by *C odorata* and burn into the canopies of forest trees.

It is less clear how changes in the fuel properties of invaded stands will affect the frequency of fires, if at all. Invasion of an ecosystem prone to frequent fires (such as grassland) by forest species that are relatively fire resistant can be expected to both decrease fire frequency and have important impacts on nutrient cycling. Changes in fire frequency, and the real reasons for them, have not been documented, however.

Nutrient cycling

Increases in both the litter and live biomass on invaded sites substantially alter the size and disposition of nutrient pools. Where grassland was converted to *Pinus patula* in Swaziland, the soil levels of most nutrient elements declined (Table 4). There were some increases in litter and the nutrients contained in litter. Second rotation sites showed an 18% decrease in soil organic carbon over grassland and a 40% decrease (measured by $C0_2$ evolved) in biological activity. Changes in soil nutrient levels were ascribed to differences in biomass, with more nutrients contained in the above-ground biomass. Phosphorus however, was reduced even when the combined levels in soil and biomass were considered (Morris 1984).

Beard and Darby (1951) found soils under black wattle *Acacia mearnsii* to have little more humus than grassland soils. The pH was 4,4 under *A mearnsii* compared to consistently higher values around 5,3 under grassland. Incipient podzolization of the surface soil may be evident below dense pine litter layers with the release of organic acids the possible cause (IUFRO 1984).

The nitrogen and phosphorus content of the leaves and bark of Australian *Acacia* species growing in the south-western Cape was two to four times greater than in fynbos plants (Milton 1980). Acacias are nitrogen-fixing and they form mycorrhizal associations which may aid in phosphorus uptake. Hoffman and Mitchell (1986) attribute the success of *Acacia saligna* in part to its extensive root system, abundant root nodules and mycorrhizal associations. Milton (1980) suggests that *Acacia* species make the environment less suitable for fynbos species (adapted to nutrient-poor soils) by enriching fynbos soils.

Fire in *Pinus patula* planted in the Natal Drakensberg increased sediment yields markedly, compared with yields from adjacent grasslands burnt in the same fire, but did not cause a significant increase in total dissolved nutrients in streamflow. There was, for example, no increase in the level of

Table 4. Levels of five nutrient elements in soil covered by grassland and grassland converted to *Pinus patula* forest in Swaziland. Data are from Morris (1984).

Vegetation	Total N (%)	Organic P (ppm)	Exchangeable K (me %)	Exchangeable Ca (me %)	Exchangeable Mg (me %)
Grassland	0,152	153	0,244	0,350	0,304
Grassland replaced by *Pinus patula* (first rotation)	0,144	134	0,186	0,134	0,198
Grassland replaced by *Pinus patula* (second rotation)	0,130	120	0,166	0,126	0,130
Significance level of F-test	N.S.	1%	0,1%	0,1%	1%

phosphates. Nitrogen was the exception. Concentrations from the stream in burnt grassland were 0,015 mg ℓ^{-1} whereas those from the burnt forest were 1,9 mg ℓ^{-1} (van Wyk 1985). The impact of invasion, followed by fire, on the nutrient cycle generally appears small, but atmospheric losses have not been accounted for. Because greater amounts of nutrients accumulate in the standing crop and litter in stands of invasive plants, excessive amounts may be lost when invaded sites are burnt. A proper comparative analysis of nutrient balances would be required to quantify this, however, because the net effects of such changes are rate-dependant.

Energy balance

Impacts of invasion on the energy balance of ecosystems can be viewed only speculatively in the absence of comparative data on the major processes of photosynthesis, evaporation and transpiration. Fynbos communities have unusually low reflectivities (Morris 1981) but in a high energy environment any changes to albedo through invasion will make little difference. Light extinction through the canopy, a function of leaf area index (LAI) and canopy structure could possibly explain the loss of understorey species on invaded sites (Macdonald and Richardson this volume). Radiation attenuation may be considered in terms of a simplification of Beer's law where $I = I_0 e^{-L}$ for a uniform horizontal canopy, or $I = I_0 e^{-kL}$ for other patterns of leaf distribution, where I = radiation below canopy, I_0 = incident radiation, L = LAI, and k = extinction coefficient. The coefficient k will be small (perhaps 0,3) for very irregular clumped distributions typical in fynbos, and for vertically orientated leaves (grassland). Uniform invader canopies will increase k to a value closer to 1, thus reducing I with or without an increase in the leaf area index. Radiation beneath a canopy with LAI = 5 and k = 1 will be reduced to about two per cent of above-canopy values, effectively eliminating most understorey development (see also Ross 1981; Jones 1983).

The Penman-Monteith combination equation for transpiration and evaporation from plant communities (Monteith 1965) may be given as

$$\lambda E = \frac{sA + \rho.c_p.D.g_a}{s + \gamma\,[1 + g_a/g_c]}$$

where
λ = latent heat of vaporization of water
E = evaporation rate
A = available energy (approximately equal to net radiation)
s = slope of saturation vapour pressure curve
c_p = specific heat of air
ρ = density of air
γ = psychrometric constant
D = vapour pressure deficit
g_a = aerodynamic conductance of the canopy
g_c = canopy conductance

For tall and aerodynamically rough canopies, especially conifers, the right-hand advective (or imposed) term dominates the radiation term. To evaluate the relative importance of the term for different canopies McNaughton and Jarvis (1983) have restructured the Penman-Monteith equation, introducing a dimensionless factor omega (Ω) with

$$\lambda E = \Omega\left(\frac{s}{s + \gamma}\right) A + (1 - \Omega)\,\frac{\rho.c_p.D.g_c}{\gamma},$$

where

$$\Omega = \left[1 + \left(\frac{\gamma}{s + \gamma}\right)(g_a/g_c)\right]^{-1}$$

Omega is a measure of coupling between plant and atmosphere and depends on the ratio of aerodynamic to canopy conductances. Short, smooth crops are poorly coupled ie Ω approaches 1 and evaporation approaches the equilibrium rate (where the surface is isolated from the bulk of the airmass) with a predominant dependence on net radiation. Tall, fine-leaved canopies are well coupled ($\Omega \leqslant 0.2$): this keeps the saturation deficit at the leaf surface close to that in air and evaporation is driven by the vapour pressure deficit of the air mass (see also Jarvis and McNaughton 1986).

Invasions in relatively smooth environments, as in the case of pines or acacias invading grassland, will increase coupling with the atmosphere, conditioning the sensitivity of transpiration to stomatal conductance. Large-leaved species such as the invasive *Solanum mauritianum* will be relatively poorly coupled and even with a large stomatal conductance transpiration rates may be low, especially in windless conditions. Control of transpiration is also of course exercised by stomatal closure and the response of species to high ambient radiation, vapour pressure deficit and moisture stress may thus limit transpiration losses. These relationships underlie the effects of plant invasions on the hydrological cycle.

Hydrology

Rainfall interception

Invaded areas, through changes in both the canopy density and the radiant and sensible component of the energy balance, may be expected to have higher amounts of rainfall intercepted by and evaporated from the canopies than the native vegetation. Rainfall interception in a stand of *Protea neriifolia* (Versfeld 1985) was between five and nine per cent, compared to 20% in a nearby 25-year-old stand of *P radiata*. Interception by *A mearnsii* was 19% in the Natal mistbelt against 13% by grassland (Beard 1962). In grassland, throughfall accounted for 42-50% of gross rainfall and stemflow 35-45%. Stemflow was five per cent in *Pinus radiata* (Versfeld 1978b) while the corresponding value for *Protea neriifolia* was 10% (Versfeld 1985). Clearly invasion causes changes both to net rainfall and its subsequent fate.

Transpiration

There are few comparative data on transpiration rates. Stomatal conductances to water vapour (Richardson and Versfeld unpublished) measured in *Acacia* species, *Pittosporum undulatum* and *Leptospermum laevigatum* at Jonkershoek did not differ from those in native species at the same location. P J Dye (personal communication) measured very high conductance rates in *Eucalyptus grandis* in winter in the eastern Transvaal. Dye estimated a daily water use by four-year-old *E grandis* at 30-110 litres or 2-7 mm day^{-1}. Over 10 weeks at 4 mm day^{-1} this amounted to 280 mm. This compares with actively growing herbaceous crops but occurred at a time of year when the natural grasses would usually be dormant or nearly so. Transpiration rates were lower in the rainy summer because humidities were high and conductance was negatively related to humidity. The major impact of invaders on dry season water yield, where the native vegetation is seasonally dormant, will result from the evergreen habit of the invading species.

If these patterns are general in plantations of evergreen trees then the work by Dye would explain the results of a study contrasting the differences between grassland soils and soils under *Acacia mearnsii* (Beard 1963). Soils under grassland dried progressively from the end of the rainy season until rains of the next spring, but the soils did not become "seriously desiccated". Soils supporting *A mearnsii* crops were found to dry out rapidly after rain, and were desiccated to wilting point by the middle of winter. Soils under grassland were continuously saturated for 32 weeks in summer and the wattle soil only for two separate periods of 12 and eight weeks. Indigenous forest, when compared to grassland, had the same effect as wattle on soil moisture.

Infiltration and erosion

There are no data to show that invasives reduce infiltration rates although water repellancy has been noted under some acacias (Milton 1980). No overland flow could be detected on plots at Jonkershoek, whether covered with fynbos or *Pinus radiata* (Versfeld 1981). Observations following a fire in *Pinus radiata* at Jonkershoek suggest that water repellency develops in the soil due to secondary compounds, originating from pine litter, accumulating in the soil after fire (D F Scott personal communication). Overland flow on plots in natural grassland was six per cent, compared to 7,7% in *A mearnsii* (Beard 1956). This increase was observed mainly during spring and was ascribed to resistance of the wattle litter bed to wetting. There were no appreciable differences in the water-holding capacity, hygroscopic moisture, or moisture equivalent of the two soils, although crumb structure was less well developed in soils under wattle (Beard and Darby 1951). A problem with plot studies is that they do not necessarily relate to basin-wide stormflow percentages or erosion rates, and this must be borne in mind when interpreting the results of such studies.

Native grassland and shrubland communities recover rapidly after fire and effectively prevent erosion because of the predominance of sprouting species which are not killed by fire. The loss of an herbaceous understorey may seriously enhance soil erosion rates when the invader canopy is destroyed by fire. A fire in a *P patula* plantation in the Natal Drakensberg killed the trees and removed the litter layer. Sediment loss after fire in the adjacent grassland was 0,4 ton ha^{-1}yr^{-1} compared to about 25 ton ha^{-1}yr^{-1} from the afforested catchments for the two years after the fire (Van Wyk 1985 and personal communication). This has implications for both alien invasions and the invasion of grassland by indigenous woody species, which occurs when such areas are protected from fire. Accelerated erosion may also be observed in riverbeds invaded by *Acacia mearnsii* and other Australian acacias (Grindley 1984). *Acacia mearnsii*, being shallow rooted, is prone to collapse easily from bank-erosion or the force of floods. This exposes loose soil to further water action, while fallen stems divert the flow from channel to bank.

Water yield as streamflow

A reduction in streamflow due to invasion can be inferred from forest hydrology studies. Afforestation of fynbos with pines in the western Cape Province and of grassland with both pine and eucalypt species in Natal and Tranvaal has decreased streamflow markedly. Decreases are dependent on the rainfall regime, the degree of afforestation and the nature of the vegetation displaced.

Decreases of more than 50% (from 630-313 mm) of natural runoff have been observed on afforested fynbos catchments (van Wyk in press), 52% (from 750-360 mm) where grassland is compared with 29-year-old *P patula* (Bosch 1979) and 100% (from 250-0 mm) five years after planting *E grandis* on grassland (van Lill et al 1980). Maximum changes in water yield of more than 400 mm have been ascribed to afforestation on a number of sites. Hewlett and Bosch (1984) detected small decreases in stormflow and peakflows after afforestation of *Themeda* grassland. This is of little practical importance, however, as decreases amount to only one per cent of the mean storm rainfall of 40 mm.

Because managed eucalypt and pine forests are clearfelled and regenerated on a regular cycle, the average reduction in water yield from catchments is less than that observed for mature stands. Invasives, however, have the potential to become a permanent feature of the landscape and the effects of dense unmanaged stands can well be compared with those of mature forest stands.

Afforestation within riparian zones is strictly controlled by legislation as water use by trees will be higher in a continually moist medium than on seasonally dry sites upslope (Bosch and Versfeld 1983). Weed infestations within these zones have largely negated the effect of this legislation. Many South African streams have become afforested through invasion instead of through active planting.

Conclusions

This review has attempted to summarize the scant data on the impacts of invasion by woody species. Biomass and structure are obviously changed, with easily observable impacts on fire behaviour and on species composition. The effects of invasion on streamflow have been demonstrated in a few cases, but the exact causes of these changes in many cases remain obscure, preventing confident extrapolation of results. The effects of invasion on nutrient cycles and the energy balance are poorly understood. It is likely that fires, particularly in the grassland biome, will be reduced in frequency, but increased in intensity and impact. Nutrients in plant and litter loads may be lost on release by fire while the high fire intensities associated with dense stands of invasives will adversely affect the regeneration of natural vegetation. The results of studies carried out to date need to be incorporated into predictive models to facilitate extrapolation. More work on certain aspects, for example comparisons of the root systems of invasive and native vegetation, and the site requirements of invasive species will help to strengthen such models.

Acknowledgements

This work arises from the conservation forestry research programme of the South African Forestry Research Institute and is published with permission.

Beard J S 1956. Results of the Mountain Home rainfall interception and infiltration project in Black Wattle, 1953/54. Journal of the South African Forestry Association 27, 72-86.

Beard J S 1962. Rainfall interception by grass. South African Forestry Journal 42, 12-15.

Beard J S 1963. Soil moisture studies using the Berkeley fibreglass instrument. South African Forestry Journal 44, 24-28.

Beard J S and Darby G D 1951. An experiment on burning in wattle silviculture. Journal of the South African Forestry Association 20, 53-77.

Bosch J M 1979. Treatment effects on annual and dry period streamflow at Cathedral Peak. South African Forestry Journal 108, 29-38.

Bosch J M 1985. Above-ground biomass of a *Pinus patula* stand in the Natal Drakensberg and its hydrological significance. Unpublished report 85-05, Jonkershoek Forestry Research Centre, Stellenbosch.

Bosch J M and Versfeld D B 1983. Forestry and the management of riparian zones. Unpublished report, Jonkershoek Forestry Research Centre, Stellenbosch.

Chandler C, Cheney P, Thomas P, Trabaud L and Williams D 1983. Fire in forestry. Vol II. Forest fire management and organization. Wiley, New York.

De Ronde C 1984. Litter accumulation problems identified in *Pinus pinaster* stands of the Cape Province. South African Forestry Journal 131, 48-52.

Grindley S A 1984. A study of land use patterns in the Hout Bay river catchment with reference to the degradation of the river channel. Research report 56, Department of Environment Studies, University of Cape Town, Cape Town.

Hewlett J D and Bosch J M 1984. The dependence of storm flows on rainfall intensity and vegetal cover in South Africa. Journal of Hydrology 75, 365-381.

Hoffmann M T and Mitchell D T 1986. The root morhology of some legume spp in the south-western Cape and the relationship of vesicular-arbuscular mycorrhizas with dry mass and phosphorus content of *Acacia saligna* seedlings. South African Journal of Botany. 52, 316-320.

IUFRO 1984. Proceedings of the symposium on site and productivity of fast growing plantations. Pretoria and Pietermaritzburg. 13 April and 11 May. South African Forestry Research Institute, Pretoria.

Jarvis P G and McNaughton K G 1986. Stomatal control of transpiration: scaling up from leaf to region. Advances in Ecological Research Vol 15. Academic Press, London. pp 1-49.

Jones H G 1983. Plants and microclimate. Cambridge University Press, Cambridge.

Kruger F J 1977. A preliminary account of the aerial plant biomass in fynbos communities of the mediterranean-type climate zone of the Cape Province. Bothalia 12, 301-307.

Macdonald I A W 1983. Alien trees, shrubs and creepers invading indigenous vegetation in the Hluhluwe-Umfolozi Game Reserve Complex in Natal. Bothalia 14, 949-959.

McNaughton K G and Jarvis P G 1983. Predicting effects of vegetation changes on transpiration and evaporation. In: Kozlowski T T (ed) Water deficits and plant growth, Vol VII. Academic Press, New York. pp 1-47.

Mitchell D T, Coley P G F, Webb S and Allsopp N (in press). Litter fall and decomposition processes in the coastal fynbos vegetation, south-western Cape, South Africa. Journal of Ecology.

Monteith J L 1965. Evaporation and environment. In: Fogg G E (ed) The State and movement of water in living organisms. Symposia of the Society for Experimental Biology. Vol 19. pp 205-234.

Morris A R 1984. A comparison of soil nutrient levels under grassland and two rotations of *Pinus patula* in the Usutu forest - Swaziland. Proceedings of the IUFRO symposium on site and productivity of fast growing plantations. Pretoria and Pietermaritzburg. South African Forestry Research Institute, Pretoria.

Morris I M 1981. The variation of components in the radiation balance over different fynbos vegetation types. MSc Thesis, University of Cape Town, Cape Town.

Milton S J 1976. Natural vegetation under a *Pinus radiata* plantation at Bosboukloof, Jonkershoek. BSc Honours Project, Forestry Faculty, University of Stellenbosch, Stellenbosch.

Milton S J 1980. Studies on Australian *Acacias* in the south-western Cape, South Africa. MSc Thesis, University of Cape Town, Cape Town.

Milton S J and Siegfried W R 1981. Above-ground biomass of Australian *Acacias* in the southern Cape, South Africa. Journal of South African Botany 47, 701-716.

Richardson D M and Van Wilgen B W 1986. Effects of 35 years of afforestation with *Pinus radiata* on the composition of mesic mountain fynbos near Stellenbosch. South African Journal of Botany 52, 309-315.

Richardson D M and Van Wilgen B W (in press). The effects of fire in felled *Hakea sericea* and natural fynbos and the implications for weed control in mountain catchments. South African Forestry Journal.

Ross J 1981. The radiation regime and architecture of plant stands. W Junk, The Hague.

Schonau A P G 1978. Provisional biomass estimates of black wattle. Unpublished report, Wattle Research Institute, Pietermaritzburg.

Schonau A P G and Boden D I 1982. Preliminary biomass studies in young eucalypts. South African Forestry Journal 120, 24-28.

Schutz C J 1982. Monitoring the long term effects of management practices on site productivity in South African forestry. South African Forestry Journal 120, 3-6.

Van Laar A 1983. A case study in *Populus canescens* to estimate leaf area index. South African Forestry Journal 125, 80-84.

Van Laar A 1984. The estimation of the leaf area index for a mature *Pinus radiata* stand. South African Forestry Journal 128, 8-11.

Van Laar A and Van Lill W S 1978. A biomass study in *Pinus radiata* D. Don. South African Forestry Journal 107, 71-76.

Van Lill W S, Kruger F J and Van Wyk D B 1980. The effect of afforestation with *Eucalyptus grandis* and *Pinus patula* on streamflow from experimental catchments at Mokobulaan, Transvaal. Journal of Hydrology 48, 107-118.

Van Wilgen B W 1982. Some effects of post-fire age on the above-ground biomass of fynbos (macchia) vegetation in South Africa. Journal of Ecology 70, 217-225.

Van Wilgen B W, Le Maitre D C and Kruger F J 1985. Fire behaviour in South African fynbos (macchia) vegetation and predictions from Rothermel's fire model. Journal of Applied Ecology 22, 207-216.

Van Wilgen B W and Richardson D M 1985. The effects of alien shrub invasions on vegetation structure and fire behaviour in South African fynbos shrublands: a simulation study. Journal of Applied Ecology 22, 955-966.

Van Wyk D B 1985. The influence of catchment management on sediment and nutrient exports in the Natal Drakensberg. In: Schulze R E (ed) Proceedings of the 2nd South African National Hydrology Symposium, Pietermaritzburg. pp 266-274.

Van Wyk D B (in press). Some effects of afforestation on streamflow in the western Cape Province, South Africa (Part 1). Water SA.

Versfeld D B 1978a. Litterfall and decomposition in stands of mature *Pinus radiata*. South African Forestry Journal 116, 40-50.

Versfeld D B 1978b. Rainfall interception in stands of *Pinus radiata* at the Jonkershoek Forestry Research Station. Unpublished report 78-08, Jonkershoek Forestry Research Centre, Stellenbosch.

Versfeld D B 1981. Overland flow on small plots at the Jonkershoek Forestry Research Station. South African Forestry Journal 119, 35-40.

Versfeld D B 1985. Rainfall interception of *Protea neriifolia* - measurement and prediction. MSc Thesis, University of Stellenbosch, Stellenbosch.

Whittaker R H and Marks P L 1975. Methods of assessing terrestrial productivity. In: Leith H and Whittaker R H (eds) Primary productivity of the biosphere. Springer-Verlag, Berlin.

Chapter 20

Ecological impacts and economic consequences of alien invasive organisms in southern African aquatic ecosystems

P J Ashton*, C C Appleton† and P B N Jackson‡

*National Institute for Water Research, †University of Natal; and ‡JLB Smith Institute of Ichthyology.

Introduction

Southern African rainfall patterns are highly seasonal; most areas receive their highest rainfalls from individual storms during the austral summer (Schulze and McGee 1978). As a result, most rivers have seasonal flow patterns and those in the more arid western and central regions are dry for a large part of each year. In addition, vast areas of low topographic relief combined with a virtual absence of significant Pleistocene glaciation (Du Toit 1954) have ensured that relatively few permanent natural lakes are found.

The general aridity of climate, combined with the paucity and seasonality of natural lakes and reliable water resources, has hindered economic development of the region. Escalating rates of human population growth and industrial development and the attendant demands for water for household, industrial and agricultural purposes have worsened the water supply situation. The growing needs for water have largely been met by the construction of numerous reservoirs and inter-catchment transfer schemes adjacent to developing areas in the wetter, eastern and southern regions of the subcontinent (Figure 1A). These have regulated and stabilized the natural hydrological regimes of most river systems in southern Africa. However, the increasing use of waterborne sewage and industrial effluent disposal systems has caused a progressive deterioration in the quality in most of these rivers (Toerien et al 1975). This, in turn, has aggravated the problems associated with the provision of adequate supplies of potable water. The gravity of the situation has been emphasized by the Science Committee (1983) which indicated that water is probably the most limiting natural resource in South Africa.

Address for correspondence: P J Ashton, National Institute for Water Research, CSIR, P O Box 395, Pretoria, 0001.

The indigenous aquatic biota of southern Africa are adapted to the natural hydrological fluctuations that occur prior to river regulation (Mitchell 1974). Relatively stable open water habitats created by reservoir construction represent an entirely new environment for colonization. The aquatic fauna are, in general, sufficiently resilient in their habitat tolerances to be able to exploit the range of new habitats (Appleton 1983; Bruton and Jackson 1983). In contrast, most indigenous aquatic plant species are unable to occupy large areas of relatively deep, open water habitat and are confined to shallower marginal zones where they form a continuum between terrestrial and aquatic habitats (Twinch and Ashton 1983).

Indigenous aquatic communities are composed of a diverse array of species and life forms with various functional attributes, regulated by complex series of interactions between the organisms and their environment. Invasive alien species impact indigenous communities at all levels of organization, often completely transforming habitats and eliminating original communities. Open water habitats that are underexploited by indigenous species are particularly prone to invasion by alien species (Mitchell 1974). The colonization of reservoirs and regulated rivers by alien organisms that prevent or inhibit optimal utilization of the scarce water resources has immense financial and ecological implications for the subcontinent.

In this review we assess the major economic and ecological impacts (where these are known) of alien plants, invertebrates and vertebrates in southern African aquatic ecosystems and emphasize the urgent need for investigations into the effects of those invasive species whose impacts are not understood, and for comprehensive, multidisciplinary management strategies.

The impacts of alien species in marginal regions, such as river banks and lake shores, have been omitted from this review. In the assessment of economic impacts, both the original costs and their 1985 Rand values adjusted for inflation using the Consumer Price Index have been given to facilitate comparisons.

Plants

Aquatic plants provide some of the most spectacular examples of successful invasions when introduced into habitats with which they are not in ecological equilibrium (Mitchell 1974). The competitive ability of alien species is manifest in rapid population growth leading ultimately to complete dominance of the available habitat at the expense of indigenous species. In southern Africa, this sequence of events is particularly evident in those cases where free-floating alien species have been introduced into man-made reservoirs (Ashton et al 1979).

Four species of alien aquatic plants, *Eichhornia crassipes*, *Myriophyllum aquaticum*, *Salvinia molesta* and *Azolla filiculoides*, are of particular concern in southern African aquatic ecosystems. Of these four species, only *M aquaticum* is rooted in soil either on river banks or in shallow water; the other three are free-floating. *Eichhornia crassipes*, *M aquaticum* and *S molesta* were apparently imported deliberately by man during the first half of this century as ornamental plants for fish ponds and aquaria (Jacot Guillarmod 1979). The exact mode of entry of the fourth species, *A filiculoides*, into the Cape Province is as yet unresolved (Ashton and Walmsley 1984). The main foci of infestations of these species are the south-western, southern and eastern portions of the Cape Province, the coastal areas of Natal, the Vaal and Crocodile River systems in Transvaal and the Kwando, Linyanti, Chobe and Zambezi Rivers in the north of the region (Figures 1B, C and D). In almost every instance, man has been implicated in the spread of these species through southern Africa. This is particularly evident in the recent records of *M aquaticum* from northern Ovamboland in SWA/Namibia (Brown et al 1985); natural dispersal could not account for this huge extension in range. The occurrence of *A filiculoides* in the Transvaal is also due to man; in 1973 this fern was transferred from Colesberg in the Cape Province to the Florence Bloom Bird Sanctuary in Johannesburg as a 'duck weed' and has since been dispersed by flood waters (N Bloom personal communication).

When introduced into suitable habitats in southern Africa, population explosions of these four species parallel their invasions in other parts of the world (Mitchell 1974). The high degree of success attained by these species is due to their ability to modify their morphology to suit environmental conditions, regenerate from small pieces of vegetative material, and sustain very high rates of vegetative reproduction at low nutrient concentrations (Mitchell and Tur 1975; Ashton and Walmsley 1976; Musil and Breen 1977). Indeed, vegetative reproduction is the only means of propagation in *S molesta*, a sterile species (Mitchell 1974), and in *M aquaticum*, of which only the female form of this dioecious plant is present in southern Africa (Jacot Guillarmod 1979). In addition to vegetative reproduction, both *E crassipes* and *A filiculoides* reproduce sexually by means of seeds and spores, respectively. The seeds of *E crassipes* can remain viable for at least 15 years (Matthews 1967) and provide an important means of recolonization. The lack of sexual reproduction in *S molesta* and *M aquaticum* suggests that these species may have spread from a single clone and, because of their reduced gene flow and the resultant lack of genetic variability, might be particularly susceptible to an introduced biological control agent (Mitchell 1974).

The rapid production of large areas of photosynthetic tissue by all four species enables them to shade out competing species. Free-floating species possess a significant additional advantage in an almost complete independence of substrate conditions and water-level fluctuations (Mitchell 1974). This, and their mobility due to wind and water movements, allows rapid occupation of the available water surface. When several species of floating plants compete for the same habitat, the largest and most vigorous species eventually dominates. This was shown very clearly by Bond and Roberts (1978) in the Cahora Bassa reservoir, Mozambique, where *E crassipes*, *S molesta*, *Pistia stratiotes* and *Azolla nilotica* were in competition. The smaller species *A nilotica* and then *S molesta* were eliminated first and the largest species *E crassipes* eventually dominated the flora.

The primary ecological impacts of successful invasions by aquatic plants are brought about by reduced water movements, decreased oxygenation from the atmosphere and an almost complete loss of light penetration into the water. The production and accumulation of large quantities of detritus and organic matter also promote bacterial activity with the consequent transformation of food webs to a detritus base. This transformation is accelerated with increasing levels of eutrophication. Secondary effects include stabilization and compaction of plant mats and their colonization by other aquatic and terrestrial species, leading to further mat compaction and the development of sudd or floating islands of secondary vegetation (Mitchell 1974; Ashton et al 1979).

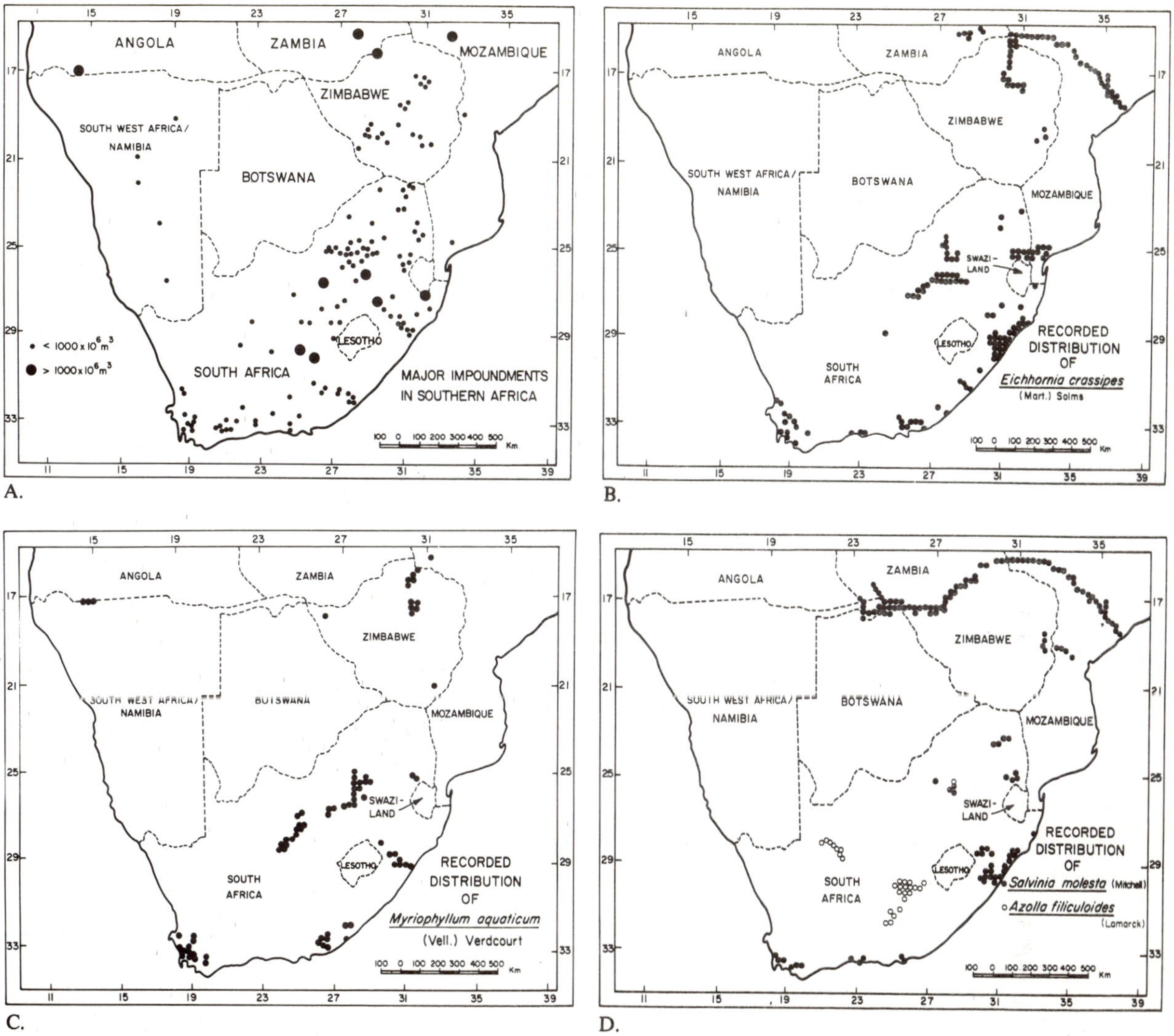

Figure 1. A. The locations of major impoundments in southern Africa. The recorded distribution of: B. *Eichhornia crassipes*; C. *Myriophyllum aquaticum*; D. *Salvinia molesta* and *Azolla filiculoides*.
(Each symbol on a distribution map indicates the presence of the particular plant within a quarter-degree grid square).

The major economic impacts of invasive aquatic plants arise from river blockages and interrupted water flows, evapotranspiration losses, difficulties in the treatment of water to attain potable standards, development of increased breeding sites for vectors and intermediate hosts of human diseases such as malaria and bilharzia, and inhibition of recreational uses of water bodies (Mitchell 1974). To these can be added the costs of eradicating or controlling the infestation, plus the costs of actions taken to ameliorate the ecological effects of the infestation and of control measures. In most cases, inadequate data preclude effective assessment of the financial implications. This is particularly true of attempts to quantify the aesthetic value of uninfested systems and is evident in the uncertainties surrounding estimates of the degree to which recreational activities have been inhibited (Viljoen and Haynes 1985). Three arbitrary levels of 25, 50 and 100% inhibition of activity have therefore been used in this analysis to illustrate the financial implications of reduced recreational activity. In addition, because the cost of water to a community depends on its availability, an average value of 10c per cubic metre has been used to assess the 1985 value of water lost by evapotranspiration. In southern Africa, evapotranspirative water loss from *E crassipes* plants usually varies between 1,2 and 2,4 times the evaporative loss from an open water surface (Ashton unpublished; C A Bruwer personal communication). In the wetter southern and eastern regions of the subcontinent where *E crassipes* occurs, the average evaporation rate is 1,5 m per year (Noble and Hemens 1978). Thus, assuming the average evapotranspirative loss by *E crassipes* to be double that from an open water surface, *E crassipes* could lose 1,5 m of water per year, over and above the loss due to evaporation.

Hofmeyer (1976) calculated that 695 000 visitor-days per year (Table 1) and a sum of R26 million were spent during 1975 (equivalent to R81,4 million in 1985) on recreation-related activities at Hartbeespoort Dam, Transvaal, prior to the development of the *E crassipes* infestation. This 1 200 ha infestation (60% of the lake's surface area) greatly reduced angling, yachting and power boating and even a 50% loss of recreation-related revenue would therefore have amounted to R40,7 million at 1985 Rand values. Evapotranspirative losses could have amounted to 18 million m^3yr^{-1} valued at R1,8 million in 1985. At Hartbeespoort Dam water quality improved at first because *E crassipes* prevented the development of excessive algal blooms. However, this had a negligible effect on costs because of subsequent increases in taste and odour problems due to high concentrations of detritus in the water (Ashton et al 1979). Mechanical and biological control techniques were considered inappropriate for the crisis condition that prevailed and the infestation was ultimately eliminated by herbicide application during 1977/78 at a cost of R250 000. This would be equivalent to R633 000 in 1985. The submergence and decay of sprayed plant material caused severe deoxygenation of the lake water but a large flood during January 1978 (280 million m^3; 1,4 times the lake volume) alleviated the problem (Ashton et al 1979). A follow-up programme is presently under way to prevent recolonization by germinated seedlings.

A similar chemical control programme conducted in 1976 at nearby Bon Accord Dam destroyed the 100% *E crassipes* cover of the 200-ha lake at a cost of R25 000 (equivalent to R71 000 in 1985). Rapid deoxygenation killed most of the fish in the lake and fish stocks were replenished slowly by those which had sought refuge in the inflowing Apies River (Ashton personal observation). Prior to chemical control, the local hotel lost an estimated 95% of its trade from recreational visitors, some R40 000 (equivalent to R125 000 in 1985). The lake's water quality deteriorated significantly but no remedial measures were necessary since its water was used solely for irrigation.

Table 1. Recreational use of Hartbeespoort Dam — 1976 (from Hofmeyer 1976).

	Visitor-days per year
Angling	220 000
Yachting	15 000
Power boating	30 000
Water skiing	30 000
Pleasure drivers	400 000
Total	695 000

Another system with severe aquatic plant problems and which supports a large community of recreational visitors is the Vaal River, Transvaal. At different times since 1951, the Rand Water Board has undertaken mechanical, manual and chemical control of *E crassipes*, *M aquaticum* and *S molesta* infestations on the 62 km of navigable river between Vaal Dam and the downstream barrage. Cost estimates are not available for the earlier operations, though chemical control of *S molesta* was undertaken successfully at a cost of R30 000 in 1962 (F C Viljoen personal communication); this would be equivalent to R186 000 in 1985. A permanent labour force has been established for manual and mechanical removal of water plants at an annual cost of R55 000 (1985 values). The extent of the aquatic plant infestations controlled by the Rand Water Board for the period 1974-1984 are shown in Table 2. *Myriophyllum aquaticum* accounted for approximately 93% of the area cleared, the remainder being *E crassipes* (1,6%) and indigenous species of *Potamogeton* and *Phragmites* (F C Viljoen personal communication). At present, manual and mechanical methods for the control of alien invasive plants are nearly four times as expensive as chemical control methods. However, chemical control programmes have far greater ecological consequences and should only be used as a last resort. Viljoen and Haynes (1985) calculated that R13,48 million was spent annually on recreation-related activities on the Vaal

Barrage (Table 3). Some 62% of this income is derived from power-boating, an activity that is impossible to continue when infestations of aquatic plants are present.

Table 2. Extent (hectares) of aquatic plants on Vaal Barrage controlled by Rand Water Board (data provided by F C Viljoen).

Year	Species		
	Eichhornia	*Myriophyllum*	Other
1974	0,6	1,3	0,6
1975	0,9	1,3	–
1976	0,8	10,0	–
1977	–	12,0	–
1978	–	6,5	–
1979	–	11,0	–
1980	–	97,0	–
1981	0,4	4,6	1,2
1982	0,1	3,4	1,7
1983	0,4	18,0	5,2
1984	–	24,2	2,6

Table 3. Impact (million Rands) of a reduction in recreational usage of the Vaal Barrage (from Viljoen and Haynes 1985).

Activity	Reduction		
	25%	50%	100%
Boating	R2,10	4,20	8,40
Recreation	R0,75	1,50	3,00
Fishing	R0,52	1,04	2,08
Total	R3,37	6,74	13,48

Below the Vaal Barrage, *E crassipes* has spread downstream to Bloemhof Dam. At several places along the river, *E crassipes* infestations form a complete cover and prevent all forms of water use. No data are available to assess the financial impact caused by the loss of recreation though it must be considerable. Similarly, the extent of the *E crassipes* infestation and its effects on water losses to downstream users have not been assessed. Evapotranspirative losses could amount to 15 000 $m^3ha^{-1}yr^{-1}$, with a value of R1500 ha^{-1} in 1985. During the last five years (1981-1985) the Department of Water Affairs has spent approximately R1,5 million on the chemical control of *E crassipes* in the Vaal River (C A Bruwer personal communication); however, this programme has had limited success to date.

The Department of Water Affairs also monitors a biological control programme against *S molesta* on the Letaba River, Transvaal. Periodic checks, estimated to cost R5 000 per annum in 1985, are made of the degree of damage and rate of spread of the coleopteran control agent *Cyrtobagous salviniae* (C A Bruwer personal communication). *Cyrtobagous salviniae* has been used successfully to control *S molesta* in parts of the Lake Liambezi-Linyanti River area of Caprivi (Schlettwein 1985). No cost estimates for this operation are available, however (C Schlettwein personal communication). Previous attempts to control *S molesta* in the Caprivi with herbicide sprays were not successful, due mainly to inaccessibility of the many plants sheltered by wide expanses of *Phragmites* reed-beds.

Invertebrates

The most obvious invasive aquatic invertebrates are pulmonate snails. Two species, *Physa acuta* and *Lymnaea columella*, were introduced from North America and are widely distributed within southern Africa (Van Eeden and Brown 1966; Hamilton-Attwell et al 1970). *Physa acuta* was introduced some time after 1939 and now dominates the gastropod fauna of several rivers and at least one small lake in Natal, but is not known to act as an intermediate host for any trematodes in southern Africa.

Lymnaea columella, first recorded in 1942 (Brown 1980), now occurs in every major river system in the country and is a known intermediate host for the liver flukes *Fasciola hepatica* and *F gigantica* which infect both domestic stock and wild ungulates. Although no published data exist on whether or not the spread of *L columella* has exacerbated the fascioliasis problem in South Africa, this does seem to be likely (Horak in Boray 1974; I G Horak personal communication). The increasing prevalence of the disease in both South Africa and Zimbabwe can be at least partly attributed to *L columella* since, (1) it occurs commonly in artificial water sources such as drinking troughs where it could facilitate transmission to stock having no contact with natural water sources; (2) it has colonized areas of South Africa in which neither of the indigenous snail hosts of fascioliasis, *L natalensis* and *L truncatula*, occur and thus has the potential for extending the transmission of both *Fasciola* species to new areas; (3) *L columella* is semiamphibious and its habits would promote the efficiency of transmission because *Fasciola* cercariae emerging from infected snails would be more likely to encyst on marginal vegetation than from *L natalensis*, a purely aquatic snail.

Several other gastropods have been introduced but their distribution is restricted. A localized population of a large, unidentified species of *Biomphalaria* occurs in central Durban. This snail, which may reach a diameter of 27 mm, was first collected in 1969 (J D Agnew personal communication) but does not seem to have spread far. The North American planorbid *Helisoma duryi*, a popular aquarium snail, is known from artificial water bodies in several South African towns and cities but so far from only one natural watercourse - the Liesbeeck River in Cape Town (Appleton 1977). Although this species is of no medical or veterinary importance, it is favoured for use as a competitor against the bilharzia snails in biological control programmes. No such trials have been done in South Africa but this approach was contemplated in Swaziland in 1977/78 using an alien competitor (J Bailey personal communication). Outside its natural range, *H duryi* has not been recorded as an intermediate host for any trematodes in its adopted countries.

Another potential invasive freshwater gastropod is the tetraploid *Bulinus truncatus*. This snail is widely distributed in Africa north of the equator (Brown 1980) but has been reported increasingly further south culminating in a record from southern Malawi (Brown and Rollinson 1982). Similar though diploid snails from Harare (Mandahl-Barth et al 1976) and Nelspruit (Lo et al 1970), are susceptible to *Schistosoma haematobium* (human urinary bilharzia). *Bulinus truncatus* is a particularly efficient transmitter of *S haematobium* when infected with certain North African strains of the parasite (Frandsen 1979). These strains are associated with a more severe clinical condition and greater morbidity than is generally seen in South Africa (Mayet and Cooppan 1983).

Bruton and Merron (1985) and Bruton and Van As (this volume) list a number of alien fish parasites that have been introduced into southern Africa. Two economically important species are *Argulus japonicus*, an opportunistic, ectoparasitic crustacean (Kruger et al 1983) and a tapeworm, *Bothriocephalus acheilognathi* (Van As et al 1981). *Argulus japonicus* now infects a broad spectrum of indigenous fish while *B acheilognathi* seems to be limited to the family Cyprinidae. Both species have a significant impact on fish condition in artificial impoundments and fish production ponds, and Brandt et al (1981) regard the appearance of bothriocephalosis in South African fish as a threat to the still small fish-farming industry. They consider this to be sufficiently important for steps to be taken for the control of the tapeworm and cestocides have been used in some ponds. It appears that *B acheilognathi* was introduced either with the Aischgrund variety of *Cyprinus carpio* from West Germany in 1952 or with the grass carp *Ctenopharyngdon idellus* imported from Israel for aquatic weed control purposes in 1975 (Brandt et al 1981). This highlights the urgent need for stricter measures to regulate the importation of alien fish.

Turning to human parasites, probably the most dramatic and economically important example of an invasive invertebrate organism entering South African river systems is the cholera bacterium, *Vibrio cholerae*. First introduced from Mozambique and Swaziland in 1974, the disease was successfully contained until 1980 when it reappeared in the south-eastern Transvaal lowveld. From here it spread quickly to the interior of the country and via the eastern lowlands to the Eastern Cape (Küstner et al 1981; Küstner 1984). Küstner (1984) considered that the epidemiological status of the disease was changing from epidemic between 1980 and 1983 to endemic in 1984. The economic implications of this invasion, an inevitable consequence of the seventh pandemic which began in Indonesia in 1961, have not been assessed but the enormity of the cost involved can be gauged from the fact that during the outbreak in KwaZulu spanning the 1982/83 summer, 51 240 adults were treated in hospitals during a 16-month period (Seedat 1983). This is highly significant when one realizes that one of the characteristics of the El Tor biotype of *V cholerae* (the strain involved in South Africa) is its huge carrier rate; 75% remain asymptomatic carriers, able to pass the disease on. An estimated 342 people have died from cholera in South Africa since 1980, representing a mean annual mortality rate, based on bacteriologically proven cases, of 1,4% (Küstner 1984).

Many textbooks on human parasitology (eg Fripp 1979) cite records of *Diphyllobothrium latum,* a tape worm indigenous to northern Europe and causing intestinal disorders in man, from Lake Ngami, Botswana. It was first reported by Dr Livingstone who visited the lake in 1849/50, and it was found sporadically until the 1930's (P J Fripp personal communication). Lake Ngami supports a thriving fishery and *D latum* infections are acquired by eating raw or undercooked fish.

Undesirable potential immigrants are freshwater crayfish species from North America and Australia. The North American *Procambarus clarkii* is popular with aquarists and is currently available in pet shops. The hazards of this state of affairs have been stressed by Appleton (1983), Hart (1983) and Van Eeden et al (1983). Preliminary experiments conducted by Coetzee (1985) on the Australian marron *Cherax tenuimanus* suggest that if released into natural waters around Jonkershoek, Cape, it could survive and possibly oust the indigenous crab, *Potamonautes perlatus*. Coetzee (1985) also noted that *C tenuimanus* could invade the upper reaches of estuaries and lagoons since it tolerated salinities up to 50% that of sea water. *Cherax tenuimanus* is well suited to commercial farming ventures but extreme care is needed to ensure that it remains securely in captivity and well away from natural watercourses. Both *P clarkii* and *C tenuimanus* are omnivores but feed extensively on detritus and macrophytes. Their introduction into South African freshwaters might constitute a repetition of the lesson that should have been learned from the release of *P clarkii* into Lake Naivasha, Kenya (Hart 1983). The Cape Department of Nature and Environmental Conservation has refused a number of applications for permits to import *C tenuimanus*. Nevertheless, two commercial marron farms have been started, one near George in the Cape Province and the other at Pirie in the Ciskei (M N Bruton personal communication).

Vertebrates

In contrast to aquatic invertebrates, where the introduction of alien species was often accidental through eggs and other inconspicuous life-cycle stages being unwittingly imported, the introduction of alien vertebrates to southern Africa has almost invariably been intentional. Apart perhaps from some aquarium ornamentals which have not escaped into the wild, and the possibility that other salmonid species or subspecies were introduced with rainbow trout (*Salmo gairdneri*) eggs, no fish or aquatic birds have been mistakenly imported through confusion with another species. The original introductions of alien aquatic vertebrates (twenty fish, five birds, one reptile and one mammal) (Bruton and Merron 1985) were all deliberate, in the expectation of some economic benefit. The alien species which have established feral breeding populations are listed by Bruton and Merron (1985). This review concentrates on ecological and economic impacts of alien invasive fish; other aquatic vertebrates and purely conservational aspects, such as the extent to which aliens have endangered the existence of indigenous species, are summarized by Bruton and Van As (this volume).

In examining the economic impacts of alien fish we must assess whether the economic benefits, either tangible in terms of the generation of money, food or job opportunities, or intangible in the form of aesthetic values or the provision of recreation for city dwellers, have been greater than the disadvantages. In several cases the original expectations have been justified and the economic impacts of alien species have generally been beneficial. However, in cases such as those of *Perca fluviatilis* and *Tinca tinca*, the introduced species have only established negligible populations in a few localized areas. Most alien fish were introduced between 50 and 100 years ago and the original argument that no indigenous species served the original purpose as well as the introduced alien still holds good. An excellent example of this has been the development of an economically important fishery on lake Kariba after the introduction of the pelagic *Limnothrissa miodon* (Jackson 1966).

Very few indigenous species, such as *Oreochromis mossambicus*, *Barbus aeneus*, *Clarias gariepinus* and some *Barbus* species, are comparable economically with the aliens. It is significant that none of these, with their values in aquaculture, as food, or as sporting or mosquito-destroying fish, are threatened by introduced alien species. An exception to this generalization is the genetic integrity of indigenous species which might be at risk from hybridization with alien species (see Bruton and Van As this volume).

Northern Hemisphere species of *Salmo* were introduced in the expectation that they would occupy a habitat not inhabited by any indigenous fish of comparable size, feeding ecology and sporting or edible qualities; this belief has since proved correct (Jackson in press). A subsequent considerable economic bonus has been the use of *S gairdneri* in aquaculture. Over 400 tons of *S gairdneri* were produced by some 60 fish farmers in 1982 (Safriel and Bruton 1984). Based on an estimated ex-pond price of R4 kg^{-1} in 1985, this amounts to an economic benefit of R1,6 million per year. Further, unquantified benefits in the form of job opportunities in the production and marketing sectors must also be added.

Today, *Salmo* species occur in most clear, well-oxygenated waters where summer water temperatures rarely exceed 20° C and are thus confined mainly to colder upland areas. Apart from formal aquaculture, *S gairdneri* are extensively stocked into reservoirs of appropriate water quality where they succeed better than the alien black basses (*Micropterus* species) through their more catholic feeding habits (Jackson 1983a). For many years, *Salmo* species were distributed free of charge from Government hatcheries and introduced into any reservoir and river where they might succeed. Since the raising, transportation and stocking of fish was 'free', large sums of public money were expended uneconomically (Pott 1984). Introduced *Salmo* species have often reduced markedly the numbers of several small indigenous fish. However, no indigenous fish were actually exterminated because remnant populations survived in shallow or inaccessible areas. Nowadays, modern management methods protect indigenous fish from trout predation (Safriel and Bruton 1984).

The three black basses of the genus *Micropterus* were also imported primarily for their sporting and eating qualities. This has remained their main economic role as they are not suited to aquaculture in the same way as *Salmo* species. The most successful and invasive of the three species is the largemouth bass *M salmoides* which flourishes in less-turbid, warm-temperate reservoirs (Jackson 1983b). The economic benefits of *M salmoides* have not been quantified but are great in terms of the recreation provided by these excellent game fish. On the debit side, *Micropterus* species have done considerably more damage to indigenous fish than *Salmo* species.

Bruton and Merron (1985) list indigenous fish species which have been translocated to areas other than where they originally occurred. To date, there is no evidence that such invasions have caused any economic disadvantage. Indeed, there may be economic advantages in that a previously absent sport-angling benefit may have been conferred. This has happened with *Barbus aeneus* which is now well established in the Kei and Limpopo systems (Jubb 1967) and with *Labeo umbratus* which is now an important commercial species in the Laing Dam, eastern Cape.

The carp, *Cyprinus carpio*, was probably the earliest alien fish introduced into South Africa (Jackson in press). Since its introduction it has become established in virtually every river system of the region except where the water is too cold, for example mountain streams, or too hot as in most subtropical or tropical areas. Despite less than ideal attributes in its poorly-flavoured flesh and habit of muddying the bottom, the species has considerable economic benefits and its reputation of eliminating more desirable fish and fouling water is often exaggerated (eg Harrison 1959).

Cyprinus carpio seldom compete with *Salmo* species because warm-temperate *C carpio* cannot tolerate the cold water of the average trout stream. Similarly, there is little evidence that *C carpio* ousts other fish though the indigenous *Labeo umbratus* has ousted *Micropterus* species from a small Transvaal impoundment (Noble and Hemens 1978). On the contrary, *C carpio* probably invaded the Orange-Vaal system around the turn of the century and has coexisted with indigenous fish there without apparent ill effects ever since. A recent detailed study of the P K le Roux Dam on the Orange River (Allanson and Jackson 1983) has shown that *C carpio* forms a very small part of the total fish population. The reason

Table 4. Numbers and mass of fish caught by sport fishermen in the Transvaal during the 1977/78 season (from Cadieux 1980). Asterisk indicates totals subject to rounding.

Species	Total number ($\times 10^3$)	Percentage	Total mass (kg) × 1000	Percentage of total mass	Average weight (kg)
Tilapia (*Tilapia* and *Oreochromis* spp)	3 185	33,6	2 411	25,0	0,76
Carp (*Cyprinus carpio*)	2 738	28,9	2 412	25,0	0,88
Labeos (*Labeo* spp)	1 132	12,0	861	9,0	0,76
Sharptooth catfish (*Clarias gariepinus*)	881	9,0	2 358	24,0	2,68
Yellowfish (*Barbus* spp)	730	7,7	1 052	11,0	1,44
Bass (*Micropterus* spp)	376	4,0	305	3,0	0,81
Butter catfish (*Eutropius*)	204	2,0	56	0,6	0,27
Eels (*Anguilla mossambica*)	38	0,4	49	0,5	1,29
Others	213	2,2	144	1,5	0,68
All species*	9 498	99,8*	9 650*	99,6*	1,02

probably lies with *C carpio's* bottom-feeding habit (Allanson and Jackson 1983) and its specialized spawning behaviour (Hume and Pribble 1980). Indeed, suitable breeding conditions do not occur regularly in South Africa and then usually only in shallow reservoirs; the reproduction of *C carpio* is therefore restricted. Increasing eutrophication of southern African reservoirs has favoured *C carpio* and other detritivorous fish, confirming the observations made by Welch (1980).

In spite of its drawbacks, *C carpio* is the most popular angling fish in South Africa other than trout. Anglers at the P K le Roux Dam fished primarily for *C carpio* (Allanson and Jackson 1983). A detailed survey of angling (other than trout) in the Transvaal showed that *C carpio* is the fish most anglers (29,5%) preferred to catch; 26,1% preferred tilapia (*O mossambicus* and *Tilapia rendalli*), while 25,1% preferred *Barbus* species (Cadieux 1980). Of an estimated 9,5 million fish caught by anglers during the 1977-78 season, 2,7 million or 28,9% were *C carpio*, with a total mass of 2 412 tonnes (Table 4). Since the two tilapias *O mossambicus* and *T rendalli* were considered together, *C carpio* was probably the most abundant single species caught. Between them, the aliens *C carpio* and *Micropterus* species made up nearly one third of the total catch. The survey revealed further (Table 5) that anglers possessed equipment worth R130 million at the beginning of that season and spent R32 million (equivalent to R230 million and R56,6 million, respectively, at 1985 Rand values) during the course of it (Cadieux 1980). Based on 1985 Rand values therefore, anglers spent approximately R5,9 kg^{-1} of fish caught.

Table 5. Investment (1985 Rand values) in fishing equipment possessed by anglers at the beginning of the 1977/78 season (adapted from Cadieux 1980). Totals are subject to rounding.

Equipment	Average investment (R)	Total investment (R'000s)	Percentage invested
Rods, reels, hooks, lines	273	30 939	13,5
Lures	19	2 267	1,0
Tackle boxes, keep-nets, etc	66	7 580	3,3
Boats	273	31 028	13,5
Outboard motors	230	26 175	11,4
Special clothing	30	3 471	1,5
Caravans	854	96 874	42,1
Tents	124	14 115	6,1
All other camping equipment	154	17 444	7,6
Total	2 024	229 911	

Contrary to the majority of the experience with invasive alien organisms in southern African aquatic ecosystems, the impact of alien fish has often had considerable economic benefits. The future expansion of aquaculture in southern Africa will extend the use of suitable alien vertebrates and increase these benefits. However, any adverse effects of alien species on conservation values must be minimized in the future by careful management (see Bruton and Van As this volume).

Conclusions

Effective, cost-efficient management of southern Africa's scarce water resources is often hampered by the presence of invasive alien organisms. The situation can be further complicated by the number and variety of State, Provincial and Municipal authorities who control water supplies in their areas of jurisdiction. In the case of alien plants and invertebrates, for example, action to control a troublesome invasive species is often not taken at an early enough stage because of a lack of information or uncertainty over which organization is ultimately responsible. Such delays can dramatically increase the subsequent costs of control or eradication programmes (Ashton et al 1979). In sharp contrast to plants and invertebrates, however, considerable financial benefits have been derived from alien fish species. Indeed, the fledgling South African aquaculture industry is based almost entirely on introduced species. Alien fish are thus seldom the target of control or eradication measures. Nevertheless, stricter controls on the importation, propagation and distribution of all alien species are necessary if adverse ecological and economic impacts are to be avoided.

In this regard, the development and enforcement of effective legislation regarding the importation to and transport of alien species within each country of the subcontinent should be implemented immediately. Possible legal loopholes in, for instance, the South African legislation related to weeds (Acts 38 and 43 of 1983) must be eliminated. Far stricter controls must also be enforced with regard to local nursery, aquarium and aquaculture trades since the unsupervised importation, sale and distribution of alien species can only cause further problems. A case in point is the continued sale of *S molesta* plants to the public at certain Pretoria nurseries. In addition, the attention that is currently being given to the aquarium trade (Bruton and Merron 1985) should also be directed to alien organisms kept in laboratories either because they are pathogens of interest to research or because they are good hosts for such pathogens. The release or escape of aquatic or waterborne organisms could have drastic ecological and economic consequences.

A basic requirement for the effective management of the region's water resources is the need to conduct further research into those alien species about which little is known, for example the invertebrate disease vectors. In many instances, development and implementation of comprehensive, integrated control programmes is not feasible without adequate fundamental research on the ecology of the organism concerned. It is acknowledged that situations do occur in which some form of action must be taken as soon as possible to control or eradicate an undesirable invasive alien species. In the absence of suitable information, action may have to be based on an 'educated guess'. This situation is obviously undesirable and yet it may be the only option available to water management authorities. The sooner this lack of information is rectified, the greater is the likelihood that management will become more cost-effective and will not simply be a hurried response to a series of ecological crises.

Acknowledgements

Sincere thanks are due to several colleagues for providing data relating to water usage and summaries of the costs incurred during various control programmes.

Act No 38 of 1983. Republic of South Africa, Agricultural Pests Act, 1983. Government Gazette Vol 214, No 8661, 20 April 1983.

Act No 43 of 1983. Republic of South Africa, Conservation of Agricultural Resources Act, 1983. Government Gazette Vol 214, No 8673, 27 April 1983.

Allanson B R and Jackson P B N 1983. Limnology and fisheries potential of Lake le Roux. South African National Scientific Programmes Report No 77. CSIR, Pretoria.

Appleton C C 1977. The exotic fresh-water snail *Helisoma duryi* (Wetherby 1879) (Planorbidae) in southern Africa. Zoologische Mededelingen 52, 125-135.

Appleton C C 1983. Wetlands and public health. Journal of the Limnological Society of southern Africa 9, 117-122.

Ashton P J, Scott W E, Steyn D J and Wells R J 1979. The chemical control programme against the water hyacinth *Eichhornia crassipes* (Mart) Solms on Hartbeespoort Dam: historical and practical aspects. South African Journal of Science 75, 303-306.

Ashton P J and Walmsley R D 1976. The aquatic fern *Azolla* and its *Anabaena* symbiont. Endeavour 35, 39-43.

Ashton P J and Walmsley R D 1984. The taxonomy and distribution of *Azolla* species in southern Africa. Botanical Journal of the Linnean Society 89, 239-247.

Bond W J and Roberts M G 1978. The colonization of Cabora Bassa, Mozambique, a new man-made lake, by floating aquatic macrophytes. Hydrobiologia 60, 243-259.

Boray J C 1974. The influence of the ecology and susceptibility of the intermediate host snails of *Fasciola hepatica* on the distribution of fascioliasis in Australasia. Proceedings of the 3rd International Congress of Parasitology. Munich, West Germany. pp 506-507.

Brandt F de W, Van As J G, Schoonbee H J and Hamilton-Attwell V L 1981. The occurrence and treatment of bothriocephalosis in the common carp, *Cyprinus carpio*, in fish ponds with notes on its presence in the largemouth yellowfish *Barbus kimberleyensis* from the Vaal Dam, Transvaal. Water SA 7, 35-42.

Brown C J, Macdonald I A W and Brown S E 1985. Invasive alien organisms in South West Africa/Namibia. South African National Scientific Programmes Report No 119. CSIR, Pretoria.

Brown D S 1980. Freshwater snails of Africa and their medical importance. Taylor and Francis, London.

Brown D S and Rollinson D 1982. The southern distribution of the freshwater snail *Bulinus truncatus*. South African Journal of Science 78, 209-293.

Bruton M N and Jackson P B N 1983. Fish and fisheries of wetlands. Journal of the Limnological Society of Southern Africa 9, 123-133.

Bruton M N and Merron S V 1985. Alien and translocated aquatic animals in southern Africa: a general introduction, checklist and bibliography. South African National Scientific Programmes Report No 113. CSIR, Pretoria.

Cadieux J 1980. Freshwater angling in the Transvaal: its environmental and economic impact. South African Journal of Science 76, 492-493.

Coetzee D J 1985. Verslag oor die kunsmatige aanhouding van die marron, *Cherax tenuimanus*, by die Jonkershoek Natuurbewaringstasie, Stellenbosch. Report of the Cape Department of Nature and Environmental Conservation.

Du Toit L H 1954. The geology of South Africa. Oliver and Boyd, London.

Frandsen F 1979. Studies of the relationships between *Schistosoma* and their intermediate hosts. II. The genus *Bulinus* and *Schistosoma haematobium* from the Sudan, Zaire and Zambia. Journal of Helminthology 53, 205-212.

Fripp P J 1979. An introduction to human parasitology with reference to southern Africa. MacMillan, Johannesburg.

Hamilton-Attwell V L, De Kock K N and Van Eeden J A 1970. The occurrence and distribution of *Physa acuta* Drapernaud in the Republic of South Africa. Wetenskaplike Bydraes van die PU vir CHO, Reeks B: Natuurwetenskappe 26, 1-11.

Harrison A C 1959. Editorial: the menace of carp. Piscator 46, 46.

Hart R C 1983. Some considerations on the introduction of exotic species. Journal of the Limnological Society of Southern Africa 9, 49-51.

Hofmeyr H P 1976. Techno-economic survey of eutrophication of the Hartbeespoort Dam. CSIR Special Report WAT 53. CSIR, Pretoria.

Hume D J and Pribble H J 1980. The biology and behaviour of carp (*Cyprinus carpio*): a brief review. Fisheries and Wildlife Division, Ministry for Conservation, Victoria, Australia, Carp Program 5, 1-30.

Jackson P B N 1966. The establishment of fisheries in man-made lakes in the tropics. In: Lowe-McConnell R H (ed) Man-made lakes. Academic Press, London.

Jackson P B N 1983a. The fish as a farm animal Part 1. Ichthos 2, 5-7.

Jackson P B N 1983b. The fish as a farm animal Part 2. Ichthos 4, 6-8.

Jackson P B N (in press). The history of aquaculture in South Africa. In: Hecht T (ed) Aquaculture in South Africa. Haum, Johannesburg.

Jacot Guillarmod A F M G 1979. Water weeds in southern Africa. Aquatic Botany 6, 377-391.

Jubb R A 1967. Notes on the popular angling yellow fishes and distribution outside their native range. Piscator 68, 120-125.

Kruger I, Van As J G and Saayman J E 1983. Observations on the occurrence of the fish louse *Argulus japonicus* Thiele, 1900 in the western Transvaal. South African Journal of Zoology 18, 408-410.

Küstner H G V 1984. Analysis of cholera IV. Epidemiological Comments 11, 1-22.

Küstner H G V, Gibson I H N, Carmichael T R, Van Zyl L, Chouler C A, Hyde J P and Du Plessis J N 1981. The spread of cholera in South Africa. South African Medical Journal 60, 87-90.

Lo C T, Burch J B and Schutte C H J 1970. Infection of diploid *Bulinus s s* with *Schistosoma haematobium*. Malacological Review 3, 121-126.

Mandahl-Barth G, Frandsen F and Jelnes J E 1976. *Bulinus* sp (2n=36) from Salisbury, Rhodesia, a close relative of *Bulinus truncatus* (Audouin) being a potential intermediate host of *Schistosoma haematobium* in south-east Africa. Transactions of the Royal Society for Tropical Medicine and Hygiene 70, 88.

Matthews L J 1967. Seedling establishment of water hyacinth. Pest Articles and News Summaries (PANS), Section C13, 7-8.

Mayet F G H and Cooppan R M 1983. Schistosomiasis. South African Journal of Continuing Medical Education 1, 37-44.

Mitchell D S (ed) 1974. Aquatic vegetation and its use and control. UNESCO, Paris.

Mitchell D S and Tur N M 1975. The rate of growth of *Salvinia molesta (S auriculata* Auct.) in laboratory and natural conditions. Journal of Applied Ecology 12, 213-225.

Musil C F and Breen C M 1977. The influence of site and position in the plant community on the nutrient distribution in, and content of *Eichhornia crassipes* (Mart) Solms. Hydrobiologia 53, 67-72.

Noble R G and Hemens J 1978. Inland water ecosystems in South Africa: review of research needs. South African National Scientific Programmes Report No 34. CSIR, Pretoria.

Pott R McC 1984. Present constraints in the trout industry. In: Hecht T et al (ed) Aquaculture South Africa. Occasional Report No 1, Ecosystem Programmes, FRD, CSIR, Pretoria.

Safriel O and Bruton M N 1984. Aquaculture in South Africa: a cooperative research programme. South African National Scientific Programmes Report No 89. CSIR, Pretoria.

Schlettwein C H G 1985. Distribution and densitites of *Cyrtobagous singularis* Hustache (Coleoptera: Curculionidae) on *Salvinia molesta* Mitchell in the Eastern Caprivi Zipfel. Madoqua 14, 291-293.

Schulze R E and McGee O S 1978. Climatic indices and classifications in relation to the biogeography of southern Africa. In: Werger M J A and Van Bruggen A C (eds) Biogeography and ecology of southern Africa. W Junk, The Hague. pp 21-52.

Science Committee 1983. Demographic trends in South Africa. Report of the President's Council. Government Printer, Pretoria.

Seedat M A 1983. Endemic cholera in South Africa. South African Journal of Continuing Medical Education 1, 22-26.

Toerien D F, Hyman K L and Bruwer M J 1975. A preliminary trophic status classification of some South African impoundments. Water SA 1, 15-23.

Twinch A J and Ashton P J 1983. Nutrient cycling in wetlands. Journal of the Limnological Society of Southern Africa 9, 104-109.

Van As J G, Schoonbee H J and Brandt F de W 1981. Further records of the occurrence of *Bothriocephalus* (Cestoda: Pseudophyllidae) in the Transvaal. South African Journal of Science 77, 343.

Van Eeden J A and Brown D S 1966. Colonization of freshwaters in the Republic of South Africa by *Lymnaea columella* Say. Nature, London 210, 1172-1173.

Van Eeden J A, De Kock K N and Pretorius S J 1983. Introduction of a freshwater crayfish in South African waters. Journal of the Limnological Society of Southern Africa 9, 49.

Viljoen F C and Haynes R E 1985. Financial implication of eutrophication. Proceedings of a Symposium on the Impact of Phosphate on South African Waters, 22 November 1985. CSIR, Pretoria. pp 17. (cyclostyled).

Welch E B (ed) 1980. Ecological effects of waste water. Cambridge University Press, Cambridge.

Managing invasive alien plants

Most of the alien species that have been subject to some form of management in southern Africa have been plants. What little is known about managing faunal invasions has been covered in the second section of this volume. The chapters that follow cover what is known about the scientific management of alien plant invasions in southern Africa.

In general they include little on the mechanical and chemical control measures that have been used. These measures have infrequently been the subject of sound, investigative research; rather they have arisen through a process of trial and error. Some of the information that has been accumulated through local experience is summarized in the reports of the workshop meetings held as part of the southern African contribution to the SCOPE project. There is little in the way of general principles that can be synthesized from these results; each prescription tends to be specific to the particular case. The general principles that underlie the action and efficacy of herbicides have not been the subject of southern African research.

Similarly, there is no chapter synthesizing the results of legislative control measures. This should not be taken to mean that legislation has not been used as an option in alien plant control in southern Africa. There have simply been no scientific studies of the application or efficacy of the laws and regulations that are currently in existence. Mention is made of certain of this legislation in some of the chapters comparing invasions in the different biomes (Chapters 6 - 10).

What does emerge as the central theme of this section is the biological control of invasive alien plants. This emphasis has arisen as a result of the long and generally successful history of biocontrol in the subcontinent. Even within this theme of biological control, attention has been focused on the area where southern Africa's contribution is most pronounced; this is the biological control of woody and cactoid plant invaders, in particular, through the utilization of seed-attacking insects.

Moran and co-authors' chapter provides the background for the evaluation of insects as potential control agents for southern Africa's alien plants. They offer several useful hypotheses which may be used to guide approaches to biological control. The chapter of **Zimmermann and co-authors** then proceeds to show how a long-established biocontrol agent can become a major determinant of the distribution pattern of its target plant species. In doing so the authors emphasize the fact that the distribution of alien plants must reflect the role of their imported herbivores as well as that of the physical environment.

An example of the conflict of interest that can arise when an invasive alien plant is valued by at least one sector of the community is illustrated by **Geldenhuys**, using the timber species *Acacia melanoxylon*. As such conflicts are increasing in southern Africa, **Neser and Kluge** then detail how seed-attacking biocontrol agents can profitably be used to reduce the invasions of potentially useful alien woody plants.

Finally, **Kluge and co-authors** define the principles underlying the establishment of integrated control programmes for invasive alien species. In this exercise they draw heavily on the practical experience that has been gained in the formulation of such programmes for species in southern Africa, as well as the careful biological studies that have led to successful and cost-effective integrated programmes.

What emerges from the chapters in this section is the tremendous value that biological control measures can have in the management of alien plants. This is most relevant where the invasions affect vast tracts of extensively utilized rangelands or large conservation areas, where unit costs of management must be kept low. Under such circumstances the more costly chemical and mechanical control measures can only be applied where they form part of a carefully planned and scientifically sound integrated strategy. It is in the development of such strategies that the long-term solutions to the management of invasive alien plants will be found.

Chapter 21

The potential of insect herbivores for the biological control of invasive plants in South Africa

V C Moran*, S Neser† and J H Hoffmann*
*University of Cape Town and †Plant Protection Research Institute.

Introduction

Many plant species are weeds in South Africa (Wells et al 1986) and many more are incipient weeds. Most are aliens but some are native. Any prognosis on the potential of insect herbivores for the biological control of weeds in South Africa will depend on historical perspectives provided by previous biological control programmes in this country and elsewhere. We attempt to provide these perspectives and then discuss some generalities and principles, in a South African context, that we believe may improve predictions for the success or failure of future weed biological control programmes. The emphasis in our account is on the attributes of the plants and not on the biological characteristics or population dynamics of the insect agents themselves.

The earliest recorded case of the biological control of a weed followed the inadvertent introduction to India in 1795 of a cochineal insect, *Dactylopius ceylonicus* (which was mistakenly thought to be the innocuous, commercially important species, *Dactylopius coccus*) that killed the cactus *Opuntia vulgaris* (Tryon 1910; Rao et al 1971; Goeden 1978; Julien 1982). The first deliberate use of an insect herbivore in the biological control of a weed, however, followed in 1863 when cactus pads infested with *D ceylonicus* were distributed manually on *O vulgaris* in southern India and this led to the spectacular control of these cactus infestations (Tryon 1910). Since then, in excess of 100 species of weeds have been the targets of biological control attempts worldwide (Julien 1982). Plants from 33 families have been subjected to biological control, and weeds in 22 of these families have been controlled to a greater or lesser extent (Julien et al 1984).

Recent compilations for the "Silwood International Project on the biological control of weeds" (Moran 1986) indicate that there have been at least 630 attempted introductions of agents for the biological control of weeds in 75 countries around the world. The vast majority (98%) of these releases have involved the use of herbivorous insects (Julien et al 1984). A few additional programmes have involved the use of mites, nematodes, fish and fungi but these organisms will not be included in the discussion which follows. About 25% of the total number of initial releases of insect herbivores against weeds in different countries have been on *Lantana camara* (Verbenaceae), nearly 20% on 22 species of cactus weeds (Cactaceae), eight per cent on about 12 species of thistles and knapweeds (Compositae) and six per cent on St John's Wort (*Hypericum perforatum*, Clusiaceae). Overall, insect herbivores have achieved some level of success in about 50% of all attempted weed biological control programmes.

Biological control attempts against weeds have involved the use of about 180 species of insect herbivores of which about 94% were beetles, moths, flies and bugs (Julien et al 1984), and the remainder involved wasps, grasshoppers or thrips. About two-thirds of the insect species introduced became established. In a few cases native insect herbivores have been manipulated successfully to control alien weeds, but mostly alien insects have been used.

The problems involved in quantifying and evaluating the successes and benefits achieved by the biological control of weeds have been discussed by several authors including Harris (1981a) and Schroeder (1983). Even the most successful cases, for example the biological control of the cactus weeds *Opuntia inermis* and *Opuntia stricta* in Queensland and New South Wales, Australia are difficult to express in economic terms. Nevertheless a number of different evaluation schemes have been suggested (eg Harris 1979 for weeds in Canada; Marsden et al 1980 for skeleton weed in Australia; Moran and Zimmermann 1984, 1986 for cacti worldwide). There is no doubt that the biological control method is often

Address for correspondence: V C Moran, Dean of Science, University of Cape Town, Rondebosch, 7700.

highly cost effective (Batra 1981). Control has been achieved with negligible risks, and although there have been at least 35 cases of inadvertent introductions of insects, mites and fungi during attempts at weed biological control world-wide, these have not been a problem (Julien 1982).

Recently, with the increasing emphasis on safety and the desire to protect native plants that might be in danger, quarantine procedures and meticulous screening of all introduced organisms used for biological control are usual. Consequently the risks in biological control of weeds are minimal, except where possible conflicts of interest are ignored or not anticipated. Conflicts of interest arise when the weeds selected for biological control also have desirable or attractive attributes. This aspect of biological weed control in the South African context has been discussed by Neser and Moran (1986).

Conflicts of interest should not inhibit research on biocontrol because satisfactory compromises can sometimes be achieved by the judicious selection of the correct biological control agents, for example those that reduce the aggressiveness of the weeds (eg seed attacking insects) without diminishing the useful attributes of the plant.

Precedents in South Africa and elsewhere, and the safety of the method, suggest that there is still great potential for continued and successful biological control of weeds in South Africa. The purpose of this paper is to discuss some generalities and principles that may enable a rational choice of weed types which would be most susceptible to attack by insect herbivores, and suggest those weed types that seem to be more refractory to biological control.

Alien or native target weeds

The cornerstone of biological control of weeds using imported insect herbivores is the notion that these alien plants are released from their native natural enemies when they are introduced into a foreign land and thus become invasive weeds. In their home country these plants are subjected to attack by numerous insect herbivore species that bring about defoliation, bore in the stems and roots, and kill buds and seeds. Released of herbivore pressure they become superior competitors and often achieve weed status. Introduced insect herbivores (themselves released from parasitoids, predators and as far as possible from pathogens) have a greater chance of increasing in number and reducing populations of the target weed.

Native plants usually become weeds only in disturbed areas or under unusual circumstances, because otherwise they are kept in check by a suite of herbivorous insect species. Consequently, the biological control of indigenous weeds is most unusual. This approach was tried unsuccessfully in New Zealand, for example, with releases of a sawfly against the native weed piripiri, *Acaena anserinifolia* (= *Acacia sanguisorbae*) (Holloway 1964; Goeden 1978). Currently, however, there are attempts to control, biologically, the native plants mesquite (*Prosopis* species) (De Loach 1986) and silver leaf nightshade *Solanum elaeagnifolium* (Cuda 1986) in Texas, and the bracken fern, *Pteridium aquilinum*, in England (J H Lawton personal communication). It remains to be seen whether any success will be achieved in these cases.

Notable exceptions are the biological control successes achieved with *Cactoblastis cactorum* against indigenous cacti that had assumed weed status in their native lands (Moran and Zimmermann 1984), including *Opuntia dillenii*, *Opuntia triacantha* and an unidentified *Opuntia* species in the West Indies (Simmonds and Bennett 1966; Bennett 1971), and on *Opuntia littoralis* and *Opuntia oricola* on Santa Cruz island (Goeden et al 1967, 1968; Goeden and Ricker 1981). Furthermore, the accidental introduction of the scale insect *Eriococcus orariensis* controlled a native weed, *Leptospermum scoparium*, in New Zealand (Julien 1982).

We conclude that biocontrol of native weeds normally has diminished chances of success, but that this is worthy of consideration even if this option is finally rejected or fails.

Abundance of the weed species in its native land

Intuitively it may seem that plants which are uncommon or rare in their native lands, presumably partly because of insect herbivore pressure, should be more susceptible to biological control in their country of introduction than plants which are widespread and common in their native lands. This supposition is difficult to quantify because of the paucity of information on the abundance and distribution of many weed species in their native lands and the lack of information on the status of the insect species that feed on them. One of the objectives of the Silwood Project (Moran 1986) is to explore this aspect, but in the interim we can only point to a few examples that fit the generality and then mention some obvious exceptions.

Hakea sericea (Proteaceae) is relatively rare in Australia where it is native and subject to the attack of numerous insect species (Neser 1968). In South Africa the rapid spread of this plant has been curtailed, and its weed status diminished, through the action of a seed-feeding weevil *Erytenna consputa* (Kluge 1983). St John's Wort, *Hypericum perforatum*, is also relatively rare in its native Europe and has been successfully controlled biologically in South Africa (Gordon et al 1986) and elsewhere (Goeden 1978). Jointed cactus, *Opuntia aurantiaca*, is not common in Central South America where it is a native species, except in botanically disturbed areas where it can achieve weed status (Moran et al 1976). In South Africa the introduction of the cochineal insect, *Dactylopius austrinus*, has greatly reduced the density of *O aurantiaca* over large areas (Moran and Annecke 1979) but, as is the case in South America, the plant is particularly troublesome in overexploited pasture lands, which seems to suggest that management is the key to control of this species (Robertson 1985).

In contrast, some thistles in the genus *Carduus*, *Centaurea* and *Cirsium* are abundant and widespread in parts of their native Europe, and *Carduus nutans* has been successfully controlled biologically in Canada (Schroeder 1980; Kelleher and Hulme 1984). Similarly, *Lantana camara* is a widespread and common plant in South America where it is native, and yet biological control through imported insect herbivores, particularly hispine beetles and tingids, has been outstandingly successful in Hawaii (Goeden 1978), and partly successful in Australia (Harley 1974), South Africa (Cilliers 1982) and elsewhere (Julien 1982).

Perhaps the real issue is not the relative abundance and aggressiveness of the plant in its native land but rather the attributes of its associated fauna of insect herbivores or other characteristics of the plant itself.

Annual and biennial plants

Crop weeds, such as *Cyperus rotundus* (Cyperaceae), have seldom been the target of biological control attempts using insect natural enemies and have rarely been successfully controlled biologically except for puncture vine *Tribulus terrestris*, in Hawaii and the south-western United States of America (Goeden 1978). It may be that annuals and biennials are poor subjects for biological control by specialist insect herbivores because they are relatively ephemeral and do not provide a stable, permanent habitat for these insects. This is an unconvincing argument because specialist herbivores would be expected to have developed mechanisms to overcome these problems. Generalist herbivores may circumvent these difficulties because they can feed on several different host plants or alternate hosts seasonally, but they are most unlikely to be seriously considered as biological control agents for importation because they are not specific to the target weed. The implication is that perennial weeds and woody plants that are in sites that are not often disturbed should be better targets for biological control by insects than annuals or biennials.

There are, however, two further considerations. Even if an annual or biennial weed species is successfully controlled biologically, it may be replaced within a short time by other weed species. Secondly, a number of weeds in South Africa, such as the thistle, *Cirsium vulgare*, exploit overgrazed or otherwise disturbed vegetation and this emphasizes that the nub of the problem may not relate to attributes of the plant but rather to incorrect management and we stress this point in the discussion.

Taxonomic isolation of the target weed

Because the prerequisite for release of an introduced insect herbivore species is its specificity on the target weed, plants that are taxonomically unrelated to beneficial or native plants in their country of introduction require simpler screening procedures during the introduction of insect herbivores.

For example, there are no species of plants in the family Cactaceae (except perhaps for *Rhipsalis baccifera)* which are native to South Africa: all are native to the Americas. It is unlikely therefore that the highly specialized insect herbivore species on Cactaceae (Moran 1980) will feed on beneficial or native plants when introduced to other countries (Mann 1970; Moran and Zimmermann 1984).

In contrast, consider the potential problems in importing insect herbivores from South America for the control of the legume *Sesbania punicea* in South Africa. There are native plants in the genus *Sesbania* in South Africa and there are numerous economically important legumes that are at risk. Meticulous screening and host specificity tests are imperative in this case; any insects that develop on plants other than *S punicea* must be rejected and destroyed.

Similar problems and difficulties surround the recent importations of insects for the biological control of *Solanum elaeagnifolium* in South Africa. There are many native *Solanum* species in South Africa and many beneficial plants such as potatoes, egg plant, tomatoes and tobacco in the same family as the target weed, a fact that will protract screening procedures and militate against the release of these agents in this country.

On the other hand, there are numerous examples of herbivorous insects that attack target weed species without any damage to closely related native or useful plants that occur sympatrically. An example is the gall-wasp *Trichilogaster acaciaelongifoliae* that attacks the alien Australian plants *Acacia longifolia* and *A floribunda* in South Africa, but none of the many native or other introduced acacias in this country.

So, the relatedness of the target weed to indigenous or useful plants requires additional precautions (Turner 1986) but should not be used as an argument to preclude biological control.

The aggressive attributes of the weed

We suggest that if the reasons for the aggressiveness of a particular weed in the country of its introduction are understood then it should be possible to increase the chances of successful biological control by selecting insect herbivores that are aimed at counteracting the specific aggressive and noxious qualities of the plant.

At least three important weeds in South Africa, *Hakea sericea*, *Acacia longifolia* and *Sesbania punicea*, produce prolific quantities of seeds and, in the absence of any insect herbivores, virtually 100% of the seeds set by the plants are viable.

In the case of *H sericea,* prolific seeding, in this fire-adapted species, may be regarded as the main reason for its success as an invasive plant. Larvae of the weevil *Erytenna consputa* destroy immature hakea fruits and the insect is strikingly adapted to its host and to surviving fires. Following establishment in South Africa, *E consputa* dispersed effectively and is greatly reducing seed accumulation by *H sericea* and consequently the regeneration of populations of the weed following fires (Kluge and Siebert 1985). Similarly, after the introduction of the bud-galling wasp, *Trichilogaster acaciaelongifoliae*, from Australia, against *A longifolia*, Dennill (1985) reported that, apart from almost total prevention of seeding, vegetative growth and plant vigour was greatly reduced, even at low gall densities.

Also, the case of *Sesbania punicea* is relevant. This plant is a South American shrub that has invaded the riverine habitats and is expanding its range throughout South Africa. The larvae of the tiny weevil *Trichapion lativentre*, from South America, develop in the flower buds and prevent seed-set. Preliminary surveys (Harris and Hoffmann 1986) showed that *T lativentre* destroyed most of the flower-buds on the plants in the hinterland of Natal and no seeds were set. Along the coast, the effects of the beetle were almost as striking although some seeds were produced even where beetle densities were high. In all three cases the potential for spread and invasion by these plants has been drastically reduced by the introduced insect herbivores.

A final example reinforces the point. The silver leaf nightshade, *Solanum elaeagnifolium* (known as satansbos in South Africa) is a weed of increasing importance in irrigated crop lands and pastures. Although it produces many seeds, it seems to owe its success to its large, deep, storage tap-root which tides the plant over during dry periods and allows it to outcompete the planted crops. The leaf-feeding chrysomelid beetles, *Leptinotarsa juncta* and *Leptinotarsa defecta* and the leaf-galling nematode *Orrina*, are presently undergoing specificity tests for release against this weed. The beetles are extremely damaging to the plant and eat large quantities of foliage. The hope is that by depleting the above-ground resources of the plant they will inhibit the growth of the root stock and prevent the plant from surviving dry periods.

The more insect herbivores the better

We have suggested, above, that insect agents are likely to be more effective if directed against a 'critical' stage or part of the plant. If several species of insect herbivores apply stress to different growth stages of the plant (eg the seeds, the roots, the stems and the leaves) then the chances for biological control will probably be enhanced (Harris 1981b, 1984), that is to say the more herbivores the better. This hypothesis however, may not stand critical scrutiny. There are numerous examples in which spectacular control of cactus weeds (*Opuntia* species) has been achieved after the introduction of a single agent species, usually the cactus-feeding moth *Cactoblastis cactorum* or one of the cochineal insects, *Dactylopius* species (Moran and Zimmermann 1984). For example, in India and in South Africa *Opuntia vulgaris* was completely controlled by *Dactylopius ceylonicus* alone. In other cases, however, for example *O streptacantha* in Australia, satisfactory control was not achieved even after the introduction and establishment of several species of herbivores.

It seems that the opuntias are less tolerant of insect stress than are many other plants. *Lantana camara*, for example, has outstanding regenerative powers and although at least four insect species have become established on this weed in South Africa (Cilliers 1982), biological control is not equally successful in all areas or on all varieties of the plant. Work is in progress to introduce more species to stress the plant further and to supplement the action of the established biological control agents.

The dictum 'The more the better' is therefore not a satisfactory simplification in the context of the numbers of herbivores required to achieve biological control of alien plants, and the sequence of introduction may also be important (Myers 1986).

There is some evidence to suggest that a greater number of herbivorous natural enemies will lead to a more seasonally consistent, widespread and stable attack on some target weeds. For example, St John's Wort, *Hypericum perforatum*, in the south-western Cape was initially controlled by the leaf-feeding beetles, *Chrysolina quadrigemina*, which completely destroyed dense stands of the weed at several sites. Subsequently, the weed resurged at these sites because recolonization by the beetles was slow (Gordon 1980). Stable biological control of *H perforatum* was eventually achieved after the introduction of a gall midge, *Zeuxidiplosis giardi*, which greatly reduced the survival, growth and flowering of the target weed (Gordon et al 1986). A more stable biological control resulted because *Z giardi*, by stunting plant growth and reducing survival of the seedlings, markedly decreased the density of the infestation. Thus populations of *C quadrigemina* were prevented from reaching devastating population numbers.

A similar situation seems to be the case with the biological control of jointed cactus, *Opuntia aurantiaca*, in South Africa, where the insect herbivore *Dactylopius austrinus* is so effective in dense infestations of the weed that the host plant and the insect are locally almost eliminated. Subsequent dispersal by *D austrinus* is limited and resurgence of the weed often occurs (Zimmermann and Moran 1982). The search for a more mobile insect herbivore, such as a cactus-feeding moth, to supplement the action of *D austrinus* has been in progress for some time but with little hope of success (Moran and Annecke 1979). In this case, and in others, the insect agent is more effective in dry climates so that additional herbivore species may achieve a more widespread biological control of the weed under different climatic conditions (Zimmermann and Moran 1982).

Adaptations, long associations and coevolution

Coevolution between herbivorous insects and their host plants, that is a reciprocal evolutionary interaction in which each organism affects the fitness of the other, is the most contentious of the generalities that we discuss in this paper, and this has been the subject of a recent account by Strong et al (1984).

In the context of biological control of weeds, the assumptions are as follows: herbivorous insects that have had a long evolutionary association with a particular species of host plant may have profoundly influenced the course of evolution of the plant, and vice versa. During this association the insects develop morphological, physiological and behavioural adaptations or specializations which lead to host-specificity (monophagy) and they become 'finely tuned' to features and the phenology of the host plant. These insect herbivores (analogous to plant parasites), like some parasites of vertebrates, are not very damaging to their host because it does not make 'evolutionary sense' to destroy the host, and the plant itself may have evolved mechanisms to tolerate, minimize or overcome attacks by these herbivores. Burdon and Marshall (1981) suggest that asexually reproducing plants are more likely to be successfully controlled biologically than sexually reproducing plant species because the latter are presumably genetically more variable and adaptable.

The biological control of cacti seems to bear out some of these suppositions. Cacti reproduce mainly asexually (the dislodged pads may root) and also sexually through seed. A feature of the biological control of Cactaceae is that success has often been achieved by using agents that do not include the target weed species in their native host range. An example is the cactus feeding moth, *Cactoblastis cactorum* which controlled *Opuntia inermis* and *O stricta* in Australia although these weeds are of North American origin and *C cactorum* occurs in South America only and on different species of cactus. Moran and Zimmermann (1984) estimated that more than 50% of the total contribution to the biological control of Cactaceae worldwide has been due to agents that fall into this category.

Hokkanen and Pimentel (1984) have suggested that success in weed biological control may be increased, at least twofold, by using insect herbivores from closely related species of plants and not those from the target weed species itself "New exploiter-victim associations expand opportunities for the biological control of both introduced and native pests and should be used as the preferred method in selecting biological control agents".This hypothesis may be valid for the biological control of insect pests and it is mostly these cases that were the basis of the analysis by Hokkanen and Pimentel (1984). However, because only 12 cases of biological control of weeds were considered in the data base used by Hokkanen and Pimentel (1984) and because eight of the cited introductions involved the special and unusual case of *Opuntia* species and the single natural enemy *C cactorum*, this hypothesis does not apply to the biological control of weeds (Goeden and Kok 1986).

On the contrary, there are many examples where a close association between an insect species and its host plant has been the key to successful biological control. Thus, although the status of the species *Lantana camara* is in doubt, the most effective biological control agents used against this weed (that is to say the herbivorous insects that are most damaging to the plant) are those species which are adapted to specific "biotypes" of the host. Populations of the same species of insects collected from a different "biotype" or "strain" of the host plant are ineffective and often fail to establish.

Furthermore, the weevil *Trichapion lativentre* is remarkably 'finely tuned' to the features and phenology of its host plant, *Sesbania punicea*. The adult weevils overwinter while the host is devoid of leaves but start feeding in synchrony with the new flush in spring. Oogenesis (egg production) only commences once the females have fed on the flower buds and this ensures that the insects mature synchronously with buds that are suitable for oviposition and large enough to sustain the developing insect through to pupation. *Trichapion lativentre* is undoubtedly a most specialized monophagous herbivore that has probably had a long association with its host, yet when released in South Africa and removed from its predators and parasitoids, it built up rapidly in numbers and is now extremely damaging to the host. *Trichilogaster acaciaelongifoliae* is another case in point - this insect is clearly very closely attuned to its host plant and yet it is a spectacular biological control agent when freed of its natural parasitoids and predators (Dennill 1985; Neser 1986).

Cacti constitute an extraordinary case of weed biological control (Moran and Zimmermann 1984) and we conclude that specialist monophagous insects that are closely adapted to the host plant are usually the most effective weed biological control agents once they are freed of their natural enemies. This point of view disagrees with Harris (1973) but it supports Goeden (1983) and Goeden and Kok (1986).

Conclusions

We have discussed the validity of several generalities that have been chosen to provide a perspective on the potential and limitations of insect herbivores for the biological control of invasive plants in South Africa. We have used these generalities to emphasize the range and scope of weed biological control programmes in South Africa. We have not discussed the controversial issue of close 'climatic matching' between the country of origin of the insect biological control agents and their country of introduction (Wapshere 1981, 1985) because there are numerous examples showing that introductions of insect herbivores can be successful despite climatic differences (Kluge 1983; Harris 1984). We believe that no potential weed biological control agent should be rejected because the climate in the country of introduction does not seem to be suitable or similar.

We believe, further, that a consideration of these generalities may help in deciding priorities from the long list of weeds (Wells et al 1986) that are potential targets for biological control. In choosing biological control agents, the pragmatic preference should be for agents that have proved successful against the target weed in other parts of the world (Harris 1973) although this is refuted by Goeden (1983) and Wapshere (1985). We stress that these considerations should be coupled with an understanding of management principles, including pasture management, and the possibilities for integration of chemical and mechanical control with biological control.

Having emphasized the principles and generalities involved, we confidently await surprises in the future when 'no-hopers', contradicting most or all of the principles we have discussed, become classics in successful weed biological control, and when 'sitting-ducks' are never bagged but remain as puzzling and embarrassing contradictions in the history of weed biological control in South Africa.

Acknowledgements

We acknowledge the perceptive and useful comments of three anonymous referees and in particular we thank Professor R D Goeden.

Batra S W T 1981. Biological control of weeds: principles and prospects. In: Papavizas G C (ed) Beltsville Symposia in Agricultural Research (5) Biological Control in Crop Production. Allanheld, Osmun, Totowa. pp 45-59.

Bennett F D 1971. Some recent successes in the field of biological control in the West Indies. Revista Peruana de Entomologia 14, 369-373.

Burdon J J and Marshall D R 1981. Biological control and the reproductive mode of weeds. Journal of Applied Ecology 18, 649-658.

Cilliers C J 1982. An evaluation of the effects of imported insects on the weed *Lantana camara* L. in South Africa. PhD Thesis, Rhodes University, Grahamstown.

Cuda J P 1986. Biology and impact of *Trichobaris texana* (Coleoptera: Curculionidae) on silverleaf nightshade, *Solanum elaeagnifolium* Cav. in central Texas. Proceedings of the 6th International Symposium on Biological Control of Weeds. Vancouver, Canada. Agriculture Canada.

De Loach C J 1986. Conflicts of interest in the biological control of mesquite (*Prosopis*). Proceedings of the 6th International Symposium on Biological Control of Weeds. Vancouver, Canada. Agriculture Canada.

Dennill G B 1985. The effect of the gall wasp *Trichilogaster acaciaelongifoliae* (Hymenoptera: Pteromalidae) on reproductive potential and vegetative growth of the weed *Acacia longifolia*. Agriculture, Ecosystems and Environment 14, 53-61.

Goeden R D 1978. Biological control of weeds. In: Clausen C B (ed) Introduced parasites and predators of arthropod pests and weeds: a world review. United States Department of Agriculture Handbook 480, 357-414.

Goeden R D 1983. Critique and revision of Harris' scoring system for selection of insect agents in biological control of weeds. Protection Ecology 5, 287-301.

Goeden R D, Fleschner C A and Ricker D W 1967. Biological control of prickly pear cacti on Santa Cruz Island, California. Hilgardia 38, 579-606.

Goeden R D, Fleschner C A and Ricker D W 1968. Insects control prickly pear cactus. California Agriculture 22, 8-10.

Goeden R D and Kok L T 1986. Comments on a proposed "new" approach for selecting agents for the biological control of weeds. Canadian Entomologist 118, 51-58.

Goeden R D and Ricker D W 1981. Santa Cruz Island - revisited. Sequential photography records the causation, rates of progress, and lasting benefits of successful biological control. Proceedings of the 5th International Symposium on Biological Control of Weeds. Brisbane, 1980. pp 355-365.

Gordon A J 1980. St John's Word *Hypericum perforatum* L. In: Stirton C J (ed) Plant invaders-beautiful but dangerous. Department of Nature and Environmental Conservation of the Cape Provincial Administration. Cape Town.

Gordon A J, Kluge R L and Neser S 1986. The effect of the gall-midge, *Zeuxidiplosis giardi* (Diptera: Cecidomyiidae) on seedlings of *Hypericum perforatum* (Clusiaceae) in South Africa. Proceedings of the 6th International Symposium on Biological Control of Weeds. Vancouver, Canada. Agriculture Canada. pp 743-748.

Harley K L S 1974. Biological control of Lantana in Australia. Proceedings of the 3rd International Symposium on Biological Control of Weeds. Montpellier, 1973, 23-29.

Harris P 1973. The selection of effective agents for the biological control of weeds. Canadian Entomologist 105, 1495-1503.

Harris P 1979. Cost of biological control of weeds by insects in Canada. Weed Science 27, 242-250.

Harris P 1981a. Evaluating biocontrol of weeds projects. Proceedings of the 5th International Symposium on Biological Control of Weeds. Brisbane 1980, 345-353.

Harris P 1981b. Stress as a strategy in the biological control of weeds. In: Papavizas G C (ed) Beltsville Symposia in Agricultural Research (5) Biological Control in Crop Production. Allanheld, Osmun, Totowa. pp 333-340.

Harris P 1984. Current approaches to biological control of weeds. In: Kelleher J S and Hulme M A (eds) Biological control programmes against insects and weeds in Canada 1969-1980. Commonwealth Agricultural Bureaux, Farnham Royal. pp 95-103.

Harris M S and Hoffmann J H 1986. The weed *Sesbania punicea* (Leguminosae) in South Africa nipped in the bud by the weevil *Trichapion lativentre*. Proceedings of the 6th International Symposium on Biological Control of Weeds. Vancouver, Canada. Agriculture Canada. pp 757-760.

Hokkanen H and Pimentel D 1984. New approach for selecting biological control agents. Canadian Entomologist 116, 1109-1121.

Holloway J K 1964. Projects in biological control of weeds. In: DeBach P (ed) Biological control of insect pests and weeds. Chapman and Hall, London. pp 650-670.

Julien M H (ed) 1982. Biological control of weeds: a world catalogue of agents and their target weeds. Commonwealth Agricultural Bureaux, Farnham Royal. 108 pp.

Julien M H, Kerr J D and Chan R R 1984. Biological control of weeds: an evaluation. Protection Ecology 7, 3-25.

Kelleher J S and Hulme M A (eds) 1984. Biological control programmes against insects and weeds in Canada 1969-1980. Commonwealth Agricultural Bureaux, Farnham Royal. 410 pp.

Kluge R L 1983. The hakea fruit weevil, *Erytenna consputa* Pascoe (Coleoptera: Curculionidae), and the biological control of *Hakea sericea* Schrader in South Africa. PhD Thesis, Rhodes University, Grahamstown.

Kluge R L and Siebert M W 1985. *Erytenna consputa* Pascoe (Coleoptera: Curculionidae) as the main mortality factor of developing fruits of the weed, *Hakea sericea* Schrader, in South Africa. Journal of the Entomological Society of Southern Africa 48, 241-245.

Marsden J S, Martin G E, Parham D J, Ridsdill Smith T J and Johnston B G 1980. Returns on Australian agricultural research. The Joint Industries Assistance Commission - CSIRO benefit-cost study of the CSIRO Division of Entomology, Canberra. 107 pp.

Mann J 1970. Cacti naturalized in Australia and their control. Government Printer, Brisbane. 128 pp.

Moran V C 1980. Interactions between phytophagous insects and their *Opuntia* hosts. Ecological Entomology 5, 153-164.

Moran V C 1986. The Silwood International Project on the biological control of weeds. Proceedings of the 6th International Symposium on Biological Control of Weeds. Vancouver, Canada. Agriculture Canada. pp 65-68.

Moran V C and Annecke D P 1979. Critical reviews of biological pest control in South Africa, 3. The jointed cactus *Opuntia aurantiaca* Lindley. Journal of the Entomological Society of Southern Africa 42, 299-329.

Moran V C and Zimmermann H G 1984. The biological control of cactus weeds: achievements and prospects. Biocontrol News and Information, 5. Commonwealth Agricultural Bureaux, Farnham Royal. pp 297-320.

Moran V C and Zimmermann H G 1986. The biological control of Cactaceae: success ratings and the contribution of individual agent species. Proceedings of the 6th International Symposium on Biological Control of Weeds. Vancouver, Canada. Agriculture Canada. pp 69-75.

Moran V C, Zimmermann H G and Annecke D P 1976. The identity and distribution of *Opuntia aurantiaca* Lindley. Taxon 25, 281-287.

Myers J H 1986. How many insect species are necessary for successful biological control of weeds? Proceedings of the 6th International Symposium on Biological Control of Weeds. Vancouver, Canada. Agriculture Canada.

Neser S 1968. Studies on insect enemies of needle bushes. PhD Thesis, Australian National University, Canberra.

Neser S 1986. The bud-galling wasp *Trichilogaster acaciaelongifoliae* (Pteromalidae), an effective biological control agent against the woody shrub *Acacia longifolia* in South Africa. Proceedings of the 6th International Symposium on Biological Control of Weeds. Vancouver, Canada. Agriculture Canada. pp 797-803.

Neser S and Moran V C 1986. Tactics for evading conflicts in the biological control of South African weeds. Proceedings of the 6th International Symposium on Biological Control of Weeds. Vancouver, Canada. Agriculture Canada. pp 359-363.

Rao V P, Ghani M A, Sankaran T and Mathur K C 1971. A review of the biological control of insects and other pests in South-East Asia and the Pacific Region. Technical Communication No 6. Commonwealth Agricultural Bureaux, Farnham Royal. 149 pp.

Robertson H G 1985. The ecology of *Cactoblastis cactorum* (Berg) (Lepidoptera: Phycitidae) in relation to its effectiveness as a biological control agent of prickly pear and jointed cactus in South Africa. PhD Thesis, Rhodes University, Grahamstown.

Schroeder D 1980. The biological control of thistles. Biocontrol News and Information, 1. Commonwealth Agricultural Bureaux, Farnham Royal. pp 9-26.

Schroeder D 1983. Biological control of weeds. In: FletcherW W (ed) Recent advances in weed research. Commonwealth Agricultural Bureaux, Farnham Royal. pp 41-78.

Simmonds F J and Bennett F D 1966. Biological control of *Opuntia* spp. by *Cactoblastis cactorum* in the Leeward Islands (West Indies). Entomophaga II, 183-189.

Strong D R, Lawton J H and Southwood T R E 1984. Insects on plants. Community patterns and mechanisms. Blackwell Scientific Publications, Oxford. 313 pp.

Tryon H 1910. "The wild cochineal insect", with reference to its injurious action on prickly pear (*Opuntia* spp) in India, etc, and to its availability for the subjugation of this plant in Queensland and elsewhere. Queensland Agricultural Journal 25, 188-197.

Turner C E 1986. Biological control, conflicting interests, and the issue of risk to non-target native plants. Proceedings of the 6th International Symposium on Biological Control of Weeds. Vancouver, Canada. Agriculture Canada.

Wapshere A J 1981. Recent thoughts on exploration and discovery for biological control of weeds. Proceedings of the 5th International Symposium on Biological Control of Weeds. Brisbane, 1980. pp 75-79.

Wapshere A J 1985. Effectiveness of biological control agents for weeds: present quandries. Agriculture, Ecosystems and Environment 13, 261-280.

Wells M J, Balsinghas A A, Joffe H, Engelbrecht V M, Harding G and Stirton C H 1986. A catalogue of problem plants in southern Africa incorporating the national weed list of South Africa. Memoirs of the Botanical Survey of South Africa, 53. Government Printer, Pretoria.

Zimmermann H G and Moran V C 1982. Ecology and management of cactus weeds in South Africa. South African Journal of Science 78, 314-320.

Chapter 22

Insect herbivores as determinants of the present distribution and abundance of invasive cacti in South Africa

H G Zimmermann*, V C Moran† and J H Hoffmann†
*Plant Protection Research Institute and †University of Cape Town.

Introduction

All cacti (Cactaceae) in South Africa are aliens that came originally from the Americas, and all are invasive to a greater or lesser extent. Of the 50 or more cactus species that have become naturalized in South Africa, 13 are recognized as weeds (Table 1). Undoubtedly other established species will be added to the weeds list in the future. In the past one alien cactus species has been proclaimed as a weed approximately every three years. Cactophagous insects have been introduced and manipulated as biological control agents against some of these cacti in South Africa since 1910, and biological control by cactus-feeding insects has frequently been augmented by chemical and mechanical control methods (Table 1).

The purpose of this paper is to emphasize that the distribution and abundance of the major alien invasive cacti in South Africa has been fundamentally affected by the action of one or more species of imported insect herbivores. Apart from the gross influence of climate, our contention is that the present-day distribution of major alien invasive Cactaceae is principally governed by pressure from insect herbivores and not primarily, or directly, by other factors such as soil type or competition. We provide evidence for this contention by discussing the biological control of three species of cacti, the prickly pears *Opuntia vulgaris* and *Opuntia ficus-indica*, and the jointed cactus *Opuntia aurantiaca*. Details of the history, biology and control of these invasive species in South Africa are provided by Annecke and Moran (1978), Moran and Annecke (1979) and Zimmermann and Moran (1982).

We also consider the role of the third trophic level (insect predators) as important elements that modify the performance of the herbivores as biological control agents and thus affect the distribution and abundance of the host plant. Finally, we conclude that the demonstrably important role of herbivores as determinants of plant abundance and distribution is not a unique feature of alien Cactaceae, but almost certainly applies to other alien and native plant elements in South Africa.

Address for correspondence: H G Zimmermann, Plant Protection Research Institute, Weeds Laboratory, P O Box 330, Uitenhage, 6230.

Opuntia vulgaris

The present distribution of *O vulgaris* is restricted to isolated plants or small clumps along the eastern coastal strip of South Africa. Apparently this species was already present in South Africa by 1772. By the end of the 19th century *O vulgaris* was a major weed, probably more important than *O ficus-indica* is today, and occurred at high densities along the coastal strip from Cape Town to northern Natal. In 1913, the plant-sucking cochineal insect, *Dactylopius ceylonicus*, was deliberately introduced for biological control of this weed. In a few years the abundance and distribution of the plant had been spectacularly reduced (Neser and Annecke 1973). As a result of the impact of this single species of insect herbivore, an alien plant that was previously aggressively invasive has now become scarce. Clearly, herbivore pressure has been the primary factor restricting the present distribution and invasiveness of *O vulgaris*.

Opuntia ficus-indica

Although the case of *O vulgaris* is striking and relevant to the point we are making, the information is mainly descriptive and anecdotal. Our hypothesis is supported by more substantial evidence from the distribution and abundance of *O ficus-indica* before and after the introduction of insect herbivores to South Africa.

Opuntia ficus-indica is a native of Mexico. It is not known precisely when the species was first introduced into South Africa but it has been an invasive element in the Cape Province for at least 250 years (Annecke and Moran 1978; Wells et al this volume).

Table 1. Alien Cactaceae of American origin that have become weeds in South Africa with information on their abundance, the insects released for their biological control and the status of control. The information has been up-dated and modified from Moran and Zimmermann (1984).

Name of weed	Distribution and weed status	Biological control: Insect established	Biological control: Degree of control
Cereus peruvianus	Localized in Transvaal. Minor importance	None	Controlled chemically.
Cylindropuntia (*Opuntia*) *imbricata*	Localized infestations only, of intermediate importance	*Dactylopius tomentosus* (Homoptera: Dactylopiidae)	Substantial (supplemented by herbicides).
Cylindropuntia (*Opuntia*) *rosea*	Localized. Minor importance	*D tomentosus*	Ineffectual.
Eriocereus martinii	Localized. Minor importance	*Hypogeococcus festerianus* (Homoptera: Pseudococcidae)	Recently established. Promising.
Opuntia aurantiaca	About 830 000 ha infested. South Africa's most costly weed	(i) *Dactylopius austrinus*	Substantial (supplemented by herbicides).
		(ii) *Cactoblastis cactorum* (Lepidoptera: Phycitidae)	Partial.
		(iii) *Tucumania tapiacola* (Lepidoptera: Phycitidae)	Ineffectual. Establishment uncertain.
		(iv) *Mimorista puchellalis* (Lepidoptera: Pyraustidae)	Ineffectual, recently established.
		(v) *Nanaia* sp (Lepidoptera: Phycitidae)	Unlikely to establish.
Opuntia dellenii	Localized. Minor importance	Unknown because not surveyed	Unknown.
Opuntia exaltata	Localized. Minor importance	Unknown because not surveyed.	Unknown.
Opuntia ficus-indica	Formerly very widespread, about 900 000 ha densely infested. Following biological control it now infests less than 100 000 ha but is still an important weed	(i) *Dactylopius opuntiae*	Substantial (supplemented by limited herbicidal control and by hand-felling).
		(ii) *Cactoblastis cactorum*	Effective on small plants only
		(iii) *Archlagocheirus funestus* (Coleoptera: Cerambycidae)	Ineffectual.
		(iv) *Metamasius spinolae* (Coleoptera: Curculionidae)	Partial.
Opuntia lindheimeri	Localized. Minor importance.	*Dactylopius opuntiae*	Substantial.
Opuntia spinulifera	Extremely restricted distribution. Minor importance.	Unknown because not surveyed.	Unknown.
Opuntia stricta	Localized infestations only. Minor importance.	*Cactoblastis cactorum*	Recently established. Partial (supplemented by herbicides).
		Dactylopius opuntiae	Unknown.
Opuntia vulgaris	Coastal distribution but of no importance following biocontrol.	*Dactylopius ceylonicus*	Complete, remaining plants now colonised by *D opuntiae* not *D ceylonicus*.
Pereskia aculeata	Subtropical, localized. Important in forest areas.	None	Not yet controlled.

Prior to the introduction of insect herbivores for the biological control of this plant, and despite vigorous attempts at mechanical and herbicidal control dating from the 1890's, the plant multiplied and spread, and came to occupy about 900 000 ha, more or less densely (Figure 1A), the heaviest infestations being in the eastern Cape and Karoo (Du Toit 1942; Annecke and Moran 1978).

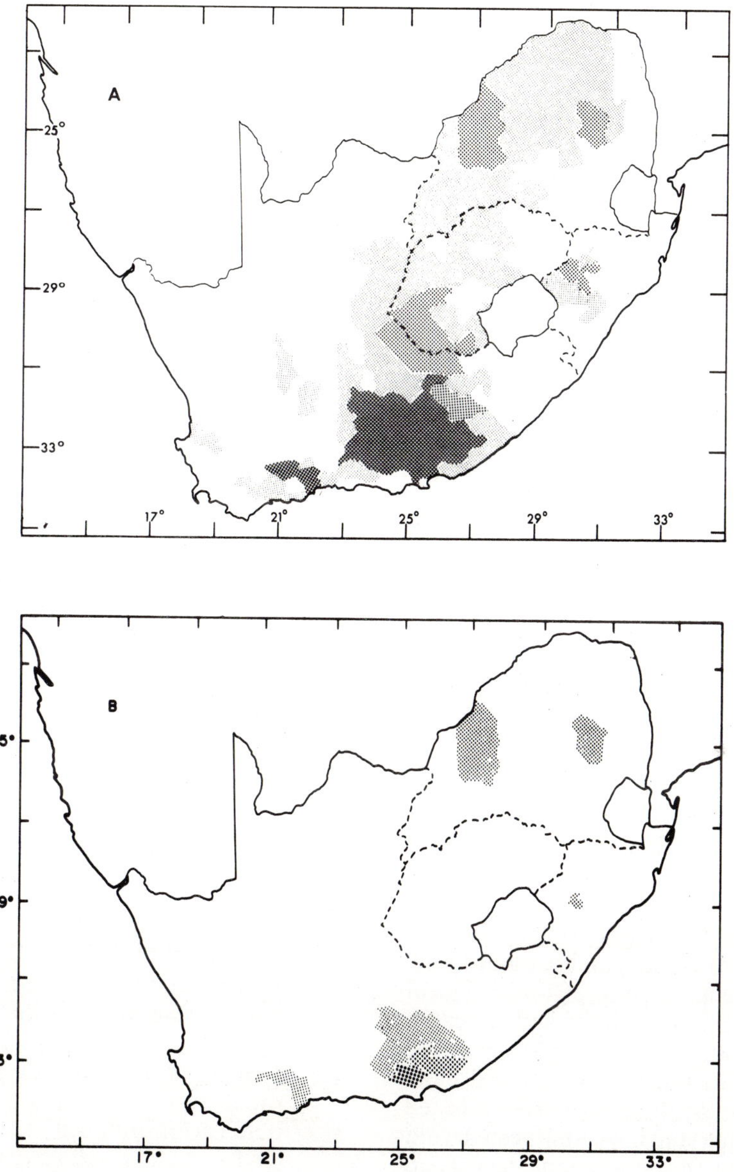

Figure 1. The distribution and abundance of *Opuntia ficus-indica* in South Africa, (A) before (about 1935) and (B) after (1946) release of the biological control agents *Dactylopius opuntiae* and *Cactoblastis cactorum*. Modified from Pettey (1948). Heavy stipple = dense infestations. Medium stipple = medium density infestations. Light stipple = sparse infestations.

A moth, *Cactoblastis cactorum*, whose larvae feed on and destroy prickly pear cladodes has proved successful in Australia for the biological control of *Opuntia stricta* and *O inermis* (Dodd 1940). This species was released in large numbers on *O ficus-indica* in South Africa from 1933-1941. Later, from 1938-1944, the cochineal insect, *Dactylopius opuntiae*, was distributed manually in areas occupied by *O ficus-indica* (Pettey 1948). Two species of beetle (Table 1) were also subsequently released on *O ficus-indica* (Pettey 1953) but their impact was insignificant.

The biology of *C cactorom* and of *D opuntiae*, the effects of these natural enemies on *O ficus-indica*, and the influence of climate, predators, disease and other factors on the performance of these two herbivore species have been extensively documented (Pettey 1948; Annecke and Moran 1978; Zimmermann and Moran 1982). The result of the introduction of these two insect species on *O ficus-indica* was to decrease the distribution of dense infestations of the plant by about 90%. The cochineal insects, aided by hand-felling of colonized plants, were the main contributors to the spectacular clearance of most of the areas invaded by *O ficus-indica* including many of the densest populations, whereas *C cactorum* killed a substantial proportion of the more isolated plants (Pettey 1948; Annecke and Moran 1978; Zimmermann and Malan 1981). Today dense populations of *O ficus-indica* occupy less than 100 000 ha and are confined mostly to the coastal area of the eastern Cape and to isolated pockets in the Karoo (Figure IB).

The present distribution of *O ficus-indica* in the Karoo is limited to the southern slopes and tops of mountains and protected valleys where the climate is relatively cool and humid. The observed distribution is mainly a reflection of the effects of climate on the insect herbivores, particularly *D opuntiae*, and not a direct influence of climate on the plant itself. In cool, wet areas, the abundance of the cochineal insects is limited by reduced fecundity and increased mortality (Pettey 1948). Also, in moist areas the plants are better able to resist insect-induced mortality (Zimmermann 1981). Previously infested regions in the drier hinterland have now been devoid of *O ficus-indica* plants for decades and reinvasion by *O ficus-indica* is prevented by the imported insect herbivores (Zimmermann and Malan 1981). We stress that it would be fallacious to explain the present observed distribution of *O ficus-indica* in South Africa by referring to the plant only, or to the influence of climate, soil-type or competition. The insects *D opuntiae* and, to a lesser extent, *C cactorum*, are mainly responsible for the observed patterns of distribution because the plant is now absent from many areas where it previously grew in abundance. The present status and distribution of *O aurantiaca* provides further evidence to support this theme.

Opuntia aurantiaca

In contrast to *O vulgaris* and *O ficus-indica*, *O aurantiaca* is an inconspicuous, low-growing plant comprising many elongated cylindrical cladodes, armed with long barbed spines, that are easily detached from the plant. It is South Africa's most costly invasive weed and consequently has attracted considerable research interest.

Opuntia aurantiaca was introduced to South Africa from Argentina, probably via England, in 1843 (Moran et al 1976) and has been the subject of vigorous mechanical and chemical control attempts from about 1892 (Moran and Annecke 1979). The present-day distribution of *O aurantiaca* is over about 830 000 ha, mostly in the eastern half of the country, and concentrated in the eastern Cape and Karoo. Prior to the release in 1935 of the cochineal insect, *Dactylopius austrinus*, for the biological control of *O aurantiaca*, dense populations of the plant were abundant throughout this distribution range. Following biological control, dense populations of this weed are now scarce and are confined to riverine bush, where the insects are less effective (Zimmermann 1981), and to a few isolated patches mainly in Natal and in the Transvaal, where *D austrinus* has not yet been introduced.

The current observed distribution and density of *O aurantiaca* in South Africa is the result of complicated interactions between the climate, growth and population dynamics of the plant, and the biological characteristics of *D austrinus* (Zimmermann 1981). There are several contributory factors:

(1) growth of *O aurantiaca* is more vigorous at coastal sites than in the hotter, drier inland areas;
(2) growth of *D austrinus* populations is more rapid on *O aurantiaca* on inland sites;
(3) desiccation-induced mortality of isolated cladodes and small plants of *O aurantiaca* is a key mortality factor operating independent of density at all sites, and is greatest in the dry interior;
(4) plant mortality as a result of desiccation is exacerbated by *O austrinus* attack, primarily in the interior (Hosking and Deighton 1981);

(5) proportionately fewer plants are killed by *D austrinus* in moist areas or during periods of high rainfall; and
(6) *D austrinus* is killed or is washed from the host plants by rainfall which is more frequent in the coastal areas of South Africa.

The overall effect of these interactions is that *O aurantiaca* populations are reduced by desiccation and by *D austrinus* attack; mortality of the weed is greatest in the warmer inland areas. Throughout the range of distribution of *O aurantiaca* in South Africa, populations of the plant increase during wet years which favour the plant but not *D austrinus*. During dry seasons, *D austrinus* populations build up rapidly and reduce the density of the plant so that its abundance in South Africa is markedly cyclical and dynamic (Zimmermann and Moran 1982).

There is a further complication. Dry seasons favour *D austrinus* and lead to increased mortality of the plants. Reduced *O aurantiaca* densities affect the dispersal characteristics of the insects themselves. The dispersive stages of *D austrinus* are windborne, and dispersal distances from the low-growing *O aurantiaca* host plants are short: this in turn determines the lowest plant densities achieved by the combination of insect attack and desiccation (Moran et al 1982). Clearly, the pressure of one insect herbivore species is of major significance in the abundance and distribution of the invasive weed *O aurantiaca* in this country.

Finally, our synthesis of the interaction between *O aurantiaca* and *D austrinus* ignores the minor, but complicating, impact of *C cactorum* on this plant and makes no mention of the effects of the herbicidal control programme on the population dynamics of *O aurantiaca*. A wider perspective is provided by Moran and Annecke (1979), Zimmermann (1979), Zimmermann and Moran (1982), and Robertson (1986).

Insect predators

The outcome of any interaction between an invasive plant species and one or more introduced natural enemies will depend on the intervention and influence of native predators, parasitoids and diseases on the population numbers of the introduced insect herbivores. These interactions influence the distribution and invasive potential of the host plant. To illustrate the point, we record examples of how predatory beetles and ants have influenced populations of cactus-feeding insects in South Africa, enhancing the outcome of biological control and ultimately the distribution of the host plant.

The first example involves predation by coccinellid beetles (Coccinellidae) on *Dactylopius opuntiae* resident on *O ficus-indica*. We quote the account by Zimmermann and Moran (1982): 'Annecke et al (1969) ... tested the hypothesis that (beetle) predation limited the effectiveness of the cochineal insects *D opuntiae* as biological control agents. In essence, they showed that when the cochineal insects were protected from predation by the application of a dilute solution of DDT to the prickly pear plants, that killed the predators but not the cochineal insects, biological control was greatly enhanced." "... mortality of prickly pear cladodes was far higher when the cochineal insects were protected from predation by DDT application. These impressive results led to a trial in which 5 942 large prickly pear plants in an area of approximately 50 ha were treated with a total of about 9,5 kg of DDT over a period of 18 months. The cochineal populations increased so greatly that all the plants shed their cladodes ...". This example illustrates the important modifying role of the third trophic level on insect herbivores and thus eventually on the abundance and distribution of an invasive plant.

Secondly, we have mentioned that *Cactoblastis cactorum* is of minor importance in determining population levels of *O aurantiaca* and that it is of secondary importance as a biological control agent of *O ficus-indica* in South Africa. Robertson (1986) has shown that at one site in the eastern Cape, predation by at least eight species of ants accounted for between 55 and 78% of mortality in *C cactorum* eggs on *O ficus-indica* and *O aurantiaca* plants. This is certainly a major factor determining the outcome of the interaction between the herbivore *C cacatorum* and its *Opuntia* hosts and may explain the lesser impact of *C cactorum* on *O ficus-indica* and *O aurantiaca* in South Africa compared to its spectacular success in other parts of the world (Moran and Zimmermann 1984) particularly Australia (Dodd 1940; Robertson 1986). Perhaps even more striking is the fact that ant predation has probably prevented the establishment of two recently introduced herbivores, namely *Tucumania tapiacola* (Hoffmann and Moran 1977; Hoffmann 1981) and *Nanaia* species which were released against *O aurantiaca* in South Africa.

Conclusions

Our conclusion, that insect herbivores and their acquired predators are a major determinant of the present abundance, distribution and invasiveness of alien Cactaceae in South Africa, certainly applies to the three species of cacti that we have discussed in this paper, and probably also applies to other interactions between herbivores and alien plants here and elsewhere. These principles may also be extrapolated to native plant communities. Generally, plant ecologists may have been guilty of underestimating the role of phytophagous insects in the population dynamics and distribution of alien invasive plants. Harper (1977) has made the same point using as an example the biological control of St John's Wort, *Hypericum perforatum*, in California. In this case the herbivorous beetles, *Chrysolina quadrigemina*, are demonstrably destructive to plants growing in the sun, but are far less damaging in shady habitats because the beetles seldom lay eggs on plants in the shade. Harper (1977) concludes as follows: "It is indeed quite likely that the ecologist would seek to explain the 'shade-loving' character of *Hypericum* by growing it in growth chambers at a range of light intensities under various climatic regimes and sooner or later would discover something that the plant did better at low light intensities. He can then be imagined shouting 'eureka' as he rushes to publish his false causal explanation."

Annecke D P, Karny M and Burger W A 1969. Improved biological control of the prickly pear, *Opuntia megacantha* Salm-Dyck, in South Africa through the use of an insecticide. Phytophylactica 1, 9-13.

Annecke D P and Moran V C 1978. Critical reviews of biological pest control in South Africa. 2. The prickly pear, *Opuntia ficus-indica* (L.) Miller. Journal of the Entomological Society of Southern Africa 41, 161-188.

Dodd A P 1940. The biological campaign against prickly pear. Commonwealth Prickly-pear Board Bulletin, Brisbane.

Du Toit R 1942. The spread of prickly pear in the Union. Farming in South Africa 17(194), 300-304.

Harper J L 1977. Population Biology of Plants. Academic Press, London.

Hoffmann J H 1981. Release of *Tucumania tapiacola* (Lepidoptera: Pyralidae) in South Africa against *Opuntia aurantiaca*: the value of detailed monitoring. Proceedings of the 5th International Symposium on Biological Control of Weeds, Brisbane, 1980. pp 367-373.

Hoffmann J H and Moran V C 1977. Pre-release studies on *Tucumania tapiacola* Dyar (Lepidoptera: Pyralidae), a potential biocontrol agent against jointed cactus. Journal of the Entomological Society of Southern Africa 40, 205-209.

Hosking J R and Deighton P J 1981. Biological control of moisture stressed *Opuntia aurantiaca* using *Dactylopius austrinus*. Proceedings of the 5th International Symposium on Biological Control of Weeds, Brisbane, 1980. pp 483-487.

Moran V C and Annecke D P 1979. Critical reviews of biological pest control in South Africa. 3. The jointed cactus *Opuntia aurantiaca* Lindley. Journal of the Entomological Society of Southern Africa 42, 299-329.

Moran V C, Gunn B H and Walter G H 1982. Wind dispersal and settling of first instar crawlers of the cochineal insect *Dactylopius austrinus* (Homoptera: Coccoidea: Dactylopiidae). Ecological Entomology 7, 409-419.

Moran V C and Zimmermann H G 1984. The biological control of cactus weeds: achievements and prospects. Commonwealth Agricultural Bureaux, Biocontrol News and Information 5, 297-320.

Moran V C, Zimmermann H G and Annecke D P 1976. The identity and distribution of *Opuntia aurantiaca* Lindley. Taxon 25, 281-287.

Neser S and Annecke D P 1973. Biological control of weeds in South Africa. Department of Agricultural Technical Services, Entomology Memoir 28, 27 pp.

Pettey F W 1948. The biological control of prickly pears in South Africa. Union of South Africa, Department of Agriculture, Science Bulletin 271, (Entomology Series No 22). 163 pp.

Pettey F W 1953. The boring beetles of prickly pear in South Africa and their importance in the control of *Opuntia megacantha*. Union of South Africa, Department of Agriculture, Science Bulletin 340, (Entomology Series No 38). 35 pp.

Robertson H G 1986. Egg predation by ants as a partial explanation of the difference in the performance of *Cactoblastis cactorum* on cactus weeds in South Africa and Australia. Proceedings of the 6th International Symposium on the Biological Control of Weeds. Vancouver, 1984.

Zimmermann H G 1979. Herbicidal control in relation to distribution of *Opuntia aurantiaca* Lindley and effects on cochineal populations. Weed Research 19, 89-93.

Zimmermann H G 1981. The ecology and control of *Opuntia aurantiaca* in South Africa in relation to the cochineal insect *Dactylopius austrinus*. PhD Thesis, Rhodes University, Grahamstown. 154 pp.

Zimmermann H G and Malan D E 1981. The role of imported natural enemies in suppressing re-growth of prickly pear, *Opuntia ficus-indica*, in South Africa. Proceedings of the 5th International Symposium on Biological Control of Weeds, Brisbane, 1980. pp 375-381.

Zimmermann H G and Moran V C 1982. Ecology and management of cactus weeds in South Africa. South African Journal of Science 78, 314-320.

Chapter 23

Costs and benefits of the Australian Blackwood *Acacia melanoxylon* in South African forestry

C J Geldenhuys
South African Forest Research Institute.

Introduction

Many plants introduced into the South African environment have served a useful purpose, and will continue to do so. However, several species have also become undesirable in many situations and are causing conflicts of interest. Poynton (1984) has tabulated the characteristics and uses of 223 indigenous and 385 introduced trees and shrubs cultivated in South Africa. He identified the climatic regions to which each was suited for cultivation by reason of its moisture requirements and susceptibility to drought and frost, and indicated which species have become invasive in natural and modified ecosystems.

The uses of the introduced species and their classification as being invasive or not, are summarized in Table 1. Most of these useful aliens have not been found to spread, or have not yet begun to spread. On the other hand, the proportion of invasive species rated as excellent for a given use is greater than in the case of those which do not invade. The risk of invasion by a species introduced for a given use must be a complex function of several factors, one of which may be the degree to which a species is favoured for a given purpose.

Because the introduction and husbandry of useful plant species must continue it is necessary to develop adequate policies upon which to base decisions about prospective introductions and the management of species already introduced. Policies for the control of the introduction and management of alien species need to accommodate the following considerations: (a) availability of suitable species among the indigenous or noninvasive alien flora; (b) whether land-use and management practices have contributed to the spread of the alien and how alternative practices could improve the situation; (c) whether manipulation rather than eradication of the alien invader would contribute to establishment of more suitable vegetation and result in more successful and less expensive control of the alien. The history of *Acacia melanoxylon* in South Africa illustrates the importance of these considerations and exemplifies the frequent conflict of interest in such cases. Analysis of

Table 1. Analysis of the uses of 385 alien trees and shrubs according to the degree of invasion in South Africa, based on data from Poynton (1984). Numbers in brackets indicate the species which are rated as excellent for the particular use. There is often more than one use for a given species.

	Number of species		
		Plants spreading	
Uses	Plants not spreading	In modified ecosystems	In undisturbed natural ecosystems
Processed timber	182 (18)	38 (12)	8 (3)
Posts and poles	139 (10)	34 (5)	6 (–)
Firewood	148 (21)	38 (6)	12 (4)
Food and fodder	69 (21)	12 (4)	3 (2)
Honey	180 (36)	47 (13)	11 (1)
Windbreaks and shelterbelts	261 (42)	57 (17)	13 (6)
Clipped hedges	53 (10)	16 (4)	4 (1)
Reclamation of gullies	15 (7)	7 (–)	4 (–)
Reclamation of driftsands	14 (1)	7 (–)	8 (2)
Ornament	306 (54)	62 (8)	13 (2)
Total number of species	309 (169)	62 (46)	14 (13)
Average uses per species	4,4 (1,3)	5,1 (1,5)	5,8 (1,6)

Address for correspondence: C J Geldenhuys, Saasveld Forestry Research Centre, Private Bag X6515, George, 6530.

the case serves to show how conflicts of this kind could be resolved.

The conflicting interests

Acacia melanoxylon (Australian blackwood) was introduced to South Africa from Australia in 1848. It was planted in the southern Cape forests from 1856 and later also in other parts of South Africa. Concern about the Australian bug, *Icerya purchasi*, a serious pest, prevented its large-scale introduction until about 1909. Blackwood became popular once the bug was controlled by the introduction of a ladybird *Vedalia cardinalis* (Phillips 1928, 1931; I A W Macdonald personal communication). The tree has since naturalized and is now widespread in the humid regions of South Africa (Poynton 1984; Geldenhuys et al this volume). Blackwood has caused a threefold management conflict. Firstly, it has naturalized in open and frequently disturbed ecosystems outside forests and this has entailed expensive control measures. Secondly, by meeting a strong demand it relieves pressure for the harvesting of indigenous timbers arising from the requirements of the furniture industry. Thirdly, it can be useful as a nurse tree for the rehabilitation of disturbed natural forest.

The current extent of invasion

Acacia melanoxylon invades modified forest and other ecosystems in humid regions lacking severe frost, but not in subhumid regions (Poynton 1984). It establishes extensively under mature stands of *Pinus* and *Eucalyptus* species, on edges of indigenous forests, on farmlands and wasteland, and along streams and roadsides. *Acacia melanoxylon* together with *A mearnsii* has invaded the following areas of *Pinus* plantations on state land in South Africa: Cape 7,2% of 118 500 ha, Natal 0,9% of 130 600 ha, and Transvaal 0,8% of 89 100 ha (le Roux 1982).

Thus in the Transvaal and Natal *A melanoxylon* has not yet spread far, but has the potential to do so throughout the cooler and moister parts of these regions. Small plants have been seen far from any planting, particularly along roadsides and in clearings in plantations and forests (Henderson and Musil 1984). There is little evidence of spread into closed indigenous forest. Although the species was planted in Ngoye forest in Zululand during 1912-1922, it has spread little into the forest (Huntley 1965; Ward 1985).

In lowland shrublands of the Cape Province the tree grows mainly in riverine habitats and the potential for further invasion is regarded as low. The potential for invasion of riverine habitats and forested valleys in mountain shrublands has been judged to be high, though present levels are moderate (Macdonald and Jarman 1984). In mountain catchments on state land in the western Cape the species occurs on 0,08% of the area. In the southern Cape it has been recorded on about one per cent of the area of the inland mountains near Ladismith and on Longmore State Forest (Macdonald et al 1985). On Table Mountain and other reserves the distribution pattern of blackwood is related to the history of deliberate plantings (Campbell and Moll 1977; Macdonald et al 1985). The main areas of infestation occur in the southern Cape between the Gouritz River and Port Elizabeth (Le Roux 1982; Macdonald and Jarman 1984).

The tree in its native habitat

The species has a wide range in eastern Australia, occurring at altitudes above 500 m on the Atherton Tableland at 17° S but between sea level and 1 350 m in Tasmania at 43° S (Farrell and Ashton 1978). This range includes regions with winter, equinoctial, and summer maxima in annual rainfall, with means of 450-1 800 mm. The tree grows to 35 m tall and 1,3 m diameter at breast height (DBH) in good sites, but reaches only three metres in height in the harshest environments. Deep humic soils, a high rainfall, and a plentiful supply of aerated groundwater enhances vigorous growth and good stem form both in Australia and South Africa (Farrell and Ashton 1978; De Zwaan 1983; Grey 1983). Blackwood is a pioneer species in the rainforest succession in Australia. It is an important component of regrowth forest, but senesces and dies in older regrowth stands and is absent from mature rainforest (Hopkins et al 1977; Farrell and Ashton 1978; McDonald and Whiteman 1979). In the swamps in north-western Tasmania even-aged stands, associated with tall *Leptospermum lanigerum* and *Melaleuca ericifolia*, probably resulted from fires, exceptional flooding or logging (Farrell and Ashton 1978; Geldenhuys 1980; De Zwaan 1983).

Aspects of the biology of *Acacia melanoxylon*

The wide range of conditions tolerated by *A melanoxylon* in its native continent and several other characteristics of the plant indicate considerable potential for spread in South Africa. Seed production is regular and prolific and germination capacity is high (90-100%). Van den Berg (1982a,b,c) listed several insects attacking blackwood in Australia. De Zwaan (1983) reported heavy predation of seed collected in Australia but found no seed predation in South Africa. The enlarged red funicle surrounding the black seed promotes extensive dispersal of the seed by birds, eg Knysna loerie *Tauraco corythaix* and rameron pigeon *Columba arquatrix*, and enhances flotation and spread of the seed by streams (Phillips 1928; C J Geldenhuys personal observation). The effects of this form of dispersal are seen in local patterns of distribution of seed and seedlings. Soil-stored seed reserves of *A melanoxylon* in the Groenkop forest averaged 320 ($n=13$, $s_x=347$, range=20-1120) and 583 ($n=5$, $s_x=656$, range = 0-1406) seeds m^{-2} under the canopy of two dominant trees. At least 20 seeds m^{-2} were present at five-metre intervals up to 30 m from the parent trees. In the 25-ha study site in the Groenkop forest near George a survey indicated 27 saplings and small trees up to 450 m from the nearest parent tree. These had grown from soil-stored seed after logging (C J Geldenhuys unpublished).

Sound hard-coated seed remain dormant for long periods and germinate after fire, logging, or other soil disturbance, to form dense pioneer stands. Abundant seedlings of *A melanoxylon* germinated from soil-stored seed after fire in Australian rainforest (Farrell and Ashton 1978; Hill 1982). Seed germinated only at the interface between burnt and unburnt humus. High mortality of the seedlings was evidently related to heavy litter accumulation from leaves of trees killed by the fire (Hill 1982). Scattered etiolated seedlings under closed forest canopies in the southern Cape (C J Geldenhuys personal observation) indicate that light or high temperature are not always required for germination. Scarification of the hard-coated seed during passage through the gut of frugivores probably induced germination. The species establishes in abundance in mature stands of *Pinus* and *Eucalyptus* species from bird-dispersed seed (Le Roux 1982; Geldenhuys et al this volume).

In Australia, experiments with exclosures have shown that the young, protein-rich foliage of unprotected plants is severely browsed by many animals including native marsupials and introduced rats (De Zwaan 1983; C J Geldenhuys personal observation). *Acacia melanoxylon* plants are able to survive outside exclosures only if protected by dense clumps of shrubs and bushes. In South Africa elephant *Loxodonta africana*, bushbuck *Tragelaphus scriptus*, other smaller antelope, bush-pig *Potamochoerus porcus*, and cattle destroy or damage a large proportion of seedlings and of sprouts and root-suckers from cut stumps or windblown parent trees (Phillips 1928; De Zwaan 1977; Odendaal 1977; Koen 1984). Most *A melanoxylon* regeneration in the forest is apparently destroyed, mainly by bushbuck, and few reach maturity. Many of the blackwood plants in the Groenkop forest study site were heavily browsed (C J Geldenhuys unpublished). The browsers are, however, confined mostly to forested habitats. Their densities are relatively low, and they tend to concentrate on sites with dense regeneration. Scattered single plants or small groups in disturbed forest and many plants outside forest therefore escape browsing and grow to maturity.

Blackwood suckers profusely after disturbance of the roots. Coppice shoots develop on the stem when young plants are browsed. Felled trees regenerate by coppicing from the stump. Some of this coppice may escape browsing to grow into trees (Phillips 1928; C J Geldenhuys personal observation).

Economic value as furniture timber

Blackwood is highly rated as a sawtimber, but also yields useful building poles, serviceable fence posts, and good firewood. It is very effective as a windbreak, since open-grown trees have well-developed root systems, is moderately suitable for gully reclamation, probably due to its very dense surface root system, and is moderately useful as an ornamental tree (Poynton 1984). The local furniture industry depends on this species as a supplement to the limited supply of traditional indigenous timbers such as stinkwood *Ocotea bullata* and the two yellowwoods *Podocarpus* species. During 1975-1985 1 500-2 700 m^3 of timber harvested from the southern Cape indigenous forests were auctioned annually. Blackwood comprised 40-65% of this timber. In 1985 blackwood contributed 51% to the 1 730 m^3 of timber and 35% to the R0,75 million income (Geldenhuys 1982, 1983; unpublished departmental records). This timber is used mainly in the furniture trade, and its economic value must be considered relative to the costs of its invasion of modified and natural habitats.

Establishment and growth in indigenous forests

During 1909-1930 the tree was extensively planted in gaps in logged forest to suppress ruderals and to assist the regeneration of indigenous tree species (Phillips 1928; Stapleton 1962; Stirton 1978). Most of those trees are currently being harvested for timber. Maintenance of a productive stock requires suitable establishment techniques and rapid early growth. Experimental plantings in the Bloukrans, Lottering and Storms River forests during 1968-1973 indicated that establishment and growth was successful if canopy gaps of 0,06 ha and greater were selected and if the plants were protected against browsing by bushbuck (C J Geldenhuys unpublished). Ten stands of size 0,02-0,2 ha were measured 13 years after establishment. Stand densities of 475-1 100 stems ha^{-1} and canopy heights of 21-24 m were comparable to figures for the surrounding forest. Average annual diameter growth of the stands was 20 mm with a maximum of 35 mm. This growth rate is high relative to the 7-10 mm yr^{-1} for this species in Goudveld forest and in Tasmania or the 2,5 mm yr^{-1} for local indigenous trees (Van Daalen 1982; De Zwaan 1983; Geldenhuys et al 1986). Blackwood trees on the edge of gaps were significantly smaller (50-150 mm DBH) than trees towards the centre of the gap and trees that were not overtopped by indigenous trees (200-450 mm DBH). Suppression of this kind was more marked in smaller gaps where the majority of trees were edge trees. However, the large number of small, often closely spaced gaps involved in a production system of this kind was strongly conducive to the further spread of blackwood. The system was abandoned in 1973 because the small stands did not contribute materially to increasing the blackwood resource, affected the forest adversely, and would in any event require intensive management as in the case of conventional plantations.

Growing stock in indigenous forest

The present productive stock of *A melanoxylon* is confined mostly to the southern Cape and to the indigenous forests in particular. A few scattered stands, both inside and outside forest, exist in other parts of the country. Most of the blackwood in the indigenous forests occur as single trees. The growing stock is concentrated in a few of the forests (Table 2; Geldenhuys et al 1986), with the largest concentrations in the Storms River and Goudveld forests (16,5% each). The relatively large growing stock in the Goudveld forests near Knysna has resulted from planting at Jubilee Creek and other areas in about 1913 in forest heavily exploited during goldmining towards the end of the 19th century.

Most trees (92%) had diameters at breast height exceeding 300 mm, indicating little recruitment to the population subsequent to planting. The young trees, and a further 11% considered of no use because of poor form or senescence were to be killed standing. The balance of utilizable trees were to be harvested and eliminated over a period adequate to allow the establishment of plantations as alternative sources of timber. Of these, 45% would be harvested by helicopter or other means dictated by the sensitivity of the site (Geldenhuys 1983; Luyt 1983). Thus, provision has been made for the rational use and elimination of blackwood from indigenous forests.

Table 2. Summary results of a survey of the growing stock of *Acacia melanoxylon* within the indigenous forests on state forest land of the southern Cape (ie Southern Cape and Tsitsikamma Forestry Regions). The survey was conducted by the Forestry Branch, Department of Environment Affairs, in 1983 and included all trees of 100 mm DBH and larger. The survey covered 42 000 ha of indigenous forest. Total growing stock of blackwood was 42 500 m^3.

	Per cent of total contributed by stock on given group of forests	
Region and forest complex	Area	Timber volume
Southern Cape Forestry Region		
Buffelsnek, Diepwalle, Gouna	19,3	10,2
Kaffirkop, Harkerville	18,5	14,3
Goudveld, Karatara	16,0	25,0
Bergplaas, Witfontein, Jonkershoek	12,8	3,2
Total	66,6	52,7
Tsitsikamma Forestry Region		
Kromrivier, Witelsbos	4,0	5,0
Blueliliesbush, Storms River	8,9	26,5
Lottering, Bloukrans, Keurboomsrivier	20,5	15,8
Total	33,4	47,3

Establishment of plantations outside indigenous forests

Acacia melanoxylon has been planted in pure stands outside the forest to expand the growing stock. By March 1981, 227 ha of pure stands of this species existed outside the forest in the southern Cape, of which 145 ha were established prior to 1950. However, these stands were often subjected to windfalls, and alternative better sheltered sites were needed. The southern Cape was singled out as the region for the future establishment of blackwood and the target area has been set at 1 000 ha. However, it has proved difficult to find sites with deep soils which are protected against strong wind.

Ecological conflicts in the indigenous forest

Ecologically, *A melanoxylon* has both useful and detrimental effects on the indigenous forest. Phillips (1928), Stirton (1978), Geldenhuys (1983) and others have argued that the planting and subsequent spread of this species jeopardizes the long-term existence of the indigenous forests. Degradation was seen to be effected through the rapid growth of blackwood, through its depletion of local soil-moisture reserves, through its suppression of the canopy of the indigenous forest as well as of any regeneration, and through excessively large and frequent canopy gaps caused because of its proneness to windfall and its unusually large crown. Recent data and experience allow a partial test of these assertions.

Potential for spread in forests

The size-class distribution of *A melanoxylon* populations in different sites in Goudveld forest reflects recruitment and mortality patterns and this indicates the potential for spread into the forest (Geldenhuys et al 1986). Although the statistics indicated an expanding population, the relatively low frequencies in the smallest size classes indicated slow expansion. In the closed forest the population distribution was bell-shaped, a curve typical of a pioneer population in maturing vegetation. Average DBH ranged between 400 and 500 mm but trees of 1,2 m DBH were present. Pioneer stand density declined due to low recruitment and high mortality. Small trees were most abundant in windfall gaps. By contrast, the population structure along rivers and forest margins was negatively exponential ie included large numbers of small trees and small numbers of large trees. This indicated rapidly expanding populations in these habitats. The latter curves are similar to those of two common indigenous canopy trees of the plateau forests, *Olea capensis* subspecies *macrocarpa* and *Podocarpus latifolius* (Geldenhuys and Van Laar 1980). Both indigenous species regenerate continuously under the closed forest canopy, but *A melanoxylon* can only do so in frequently disturbed forest. Even so, the chances for spread of blackwood in small logging gaps are minimal. For example, thirteen years after logging in the Groenkop forest, 74% of the 27 blackwood plants which had established in small logging gaps had a DBH of less than 100 mm, with the smaller plants less than one metre high (Geldenhuys unpublished).

Potential to alter prevailing disturbance regime in forests

It is possible that the presence of scattered blackwood trees could alter the disturbance regime prevailing in indigenous forests, to the detriment of the latter. The trees are often subject to windfalls, within as well as outside the forest. During heavy rains and strong winds in 1981 several small stands of this species on the edge of the forest were completely uprooted. Blackwood trees are taller and larger than dominant native species such as *Olea capensis* subspecies *macrocarpa*. This species is the most common indigenous forest canopy tree in the southern Cape, has a large crown, and is prone to windfalls (Geldenhuys and Maliepaard 1983). Comparison of 21 paired samples showed that mature trees of *A melanoxylon* emerge above the forest canopy, and have a longer crown and greater crown volume than canopy trees of *O capensis* subspecies *macrocarpa*, (Table 3; C J Geldenhuys unpublished). General observations suggest that *A melanoxylon* is more prone to windfalls than *O capensis* subspecies *macrocarpa*, but even if both were equally prone to windfalls the larger crown of *A melanoxylon* could cause a larger gap and damage more of the surrounding trees than would happen under natural disturbances (Geldenhuys and Maliepaard 1983). Such damage could best be reduced by felling the trees when they are smaller.

Table 3. Statistics on the dimensions of 21 paired samples of dominant trees of the alien *Acacia melanoxylon* and canopy trees of the common indigenous tree *Olea capensis* subspecies *macrocarpa* in indigenous forests.

	Acacia melanoxylon		*Olea capensis* subsp *macrocarpa*	
	Average	Range	Average	Range
Total height m	29,4	23 – 36	22,7	17 – 30
Crown depth m	13,7	10 – 20	12,0	7 – 18
Bole length m	15,7	11 – 21	10,7	5 – 18
Crown diameter m	12,6	7 – 18	10,4	5 – 17
Crown volume m^3	1249,5	238 – 2969	761,5	73 – 1994
Diameter breast high mm	757	460 – 1000	491	220 – 840

Potential to facilitate establishment of indigenous trees

Abundant recruitment of indigenous tree species is often observed beneath canopies of *A melanoxylon*. Seagrief (1965) recorded the establishment of large numbers of indigenous seedlings under a monospecific, 40-year old stand of blackwood near Hogsback in the Amatole mountains of the eastern

Cape. Stand density was 605 stems ha^{-1} for trees exceeding 100 mm diameter at 0,3 m height (60% possessed diameters exceeding 200 mm). The understorey layer two to nine metres high included 1 770 stems ha^{-1} of which blackwood contributed 82% followed by *Podocarpus latifolius* (10%), *Kiggelaria africana*, *Pittosporum viridiflorum*, *Pterocelastrus tricuspidatus* and *Rapanea melanophloeos*, all local indigenous forest trees. In the lower understorey blackwood was almost absent, but *P latifolius* contributed 82% of the 9 280 plants ha^{-1} in the one to two metre height class and the 23 900 plants ha^{-1} in the 0,3-1,0 m height class.

Several other indigenous tree and shrub species (24) were present. All the species except *Cunonia capensis*, represented by a single individual, had seed dispersed by birds. Indigenous forest on the lower margin of this stand provided a seed source. Wilson (1982) noted the destruction of this regeneration during subsequent clear felling of the stand. She also noted the establishment of *Podocarpus* species under *Acacia mearnsii*, *A melanoxylon*, *Pinus pinaster*, *P radiata* and *Rubus (cuneifolius?)* in the same area, even on relatively dry sites. The dynamics of recruitment in these stands is similar to that observed in the indigenous pioneer stands of species such as *Virgilia oroboides*, *Acacia karroo*, and *Trema orientalis*. Establishment of forest species is delayed until the stands mature and the canopies open (Phillips 1926, 1928; Seagrief 1965; Scheepers et al 1968; Weisser and Marques 1979; Wilson 1982). Blackwood seed is more readily dispersed by birds, establishes easily from seed in disturbed sites and along forest margins, and grows fast. As such blackwood complements the less mobile, less available, indigenous pioneer tree species in restoring disturbed forest. Active manipulation of such blackwood stands, rather than eradication, would contribute to regeneration of indigenous trees and to successful and less expensive alien control (Geldenhuys et al this volume).

Resolving the conflicts

Acacia melanoxylon invades disturbed forest only. Once established the plant seems to survive even if regularly browsed. However, established plants cannot compete successfully in the closed forest because the tree requires a large gap for adequate growth. Nursery-grown seedlings of *A melanoxylon* grow slower but are more shade-tolerant relative to three other alien *Acacia* species (Milton 1982). The seedlings require a minimum of 20% daylight for normal growth but can survive in eight per cent daylight though they become severely stunted. Shading results in etiolation and slower development from pinnate leaves to phyllodes though the plants have a greater total photosynthetic area. The shoot/root ratio is consistently larger in shaded plants implying that shoot development proceeds at the expense of root development when light is limited. This might explain the slow growth and etiolation of blackwood plants growing in closed forest, the instability in strong wind of trees growing in stands, and survival in strong wind of open-grown trees. Blackwood may therefore be controlled in the forest if the number and size of canopy gaps can be minimized through suitable logging techniques and if the 10-year felling cycle with regular follow-up clearing operations is followed. Increment studies suggest that a plantation system provides the best means of producing good quality timber within a reasonable period. Blackwood growing scattered within the forest may be systematically removed by a combination of control methods and be progressively replaced as a timber source by plantations.

Acacia melanoxylon, like other alien invader species associated with the forest (Geldenhuys et al this volume), is a pioneer plant and will continue to invade disturbed and modified forest ecosystems. Potentially, all sites receiving more than 600 mm rain and which do not experience severe frost might be invaded. Once established, a store of hard-coated, long-lived seed will accumulate and will respond to future disturbance of those sites. A balance is required between useful silviculture of blackwood where it is desired, and its elimination where it interferes with other land-use objectives.

Integrated ecosystem management

An informal study group has been formed in the Southern Cape Forestry Region in 1982 to resolve the options and conflicts of blackwood timber utilization and environmental conservation. The group included researchers, managers and planners in conservation and production forestry. They developed a detailed plan to establish blackwood plantations on suitable sites outside the forest, to sustain the production of timber from trees in the forest until the established stands would supply the demand, and to eventually remove blackwood from the forest ecosystem. The survey of *A melanoxylon* timber resources in the southern Cape forests (Geldenhuys et al 1986) formed the basis of this programme.

Control methods

Control in indigenous forests

Control of the unwanted spread of *A melanoxylon*

in forest has begun only recently. The current practice is to harvest only the oldest trees from the forest while eradicating trees of less than 100 mm DBH and larger trees of inferior quality. The practice of removing only the oldest trees leaves smaller, often suppressed trees in the stand. Such trees provide a continuous seed source and are possibly more prone to windfall due to turbulance in the gap caused by felling. Future logging in the gap could hamper the regeneration of forest trees by damaging saplings, as well as providing conditions for further recruitment of blackwood.

There are several options for harvesting blackwood in the forest.

(1) Indigenous trees are harvested by means which minimize damage to the surrounding forest, eg by delimbing before felling and the use of careful extraction methods.
(2) Priority is given both to harvesting in areas with a high density of mature trees and in areas where individuals are scattered. The high-density areas form the basis for yield control, while felling in the sparse areas supplements the annual cut.
(3) Each area is systematically cleared of all blackwood plants after extraction of all usable timber. All inaccessible trees are killed in situ unless alternative, appropriate extraction methods such as helicopter logging are available.
(4) Each area allocated to harvesting is cleared completely of usable and non-usable trees during a single operation, to eliminate the seed source and to allow for effective regeneration of the forest. The regeneration of indigenous trees is stimulated while that of alien species is discouraged by removal.

Control of unwanted *A melanoxylon* at the forest edge will only be achieved by replacing it with some useful crop or a permanent vegetative cover which will inhibit its establishment. McDowell and Moll (1981) suggested that fast-growing *Virgilia oroboides* or other suitable indigenous species should be used as a subsidiary control measure against invasive aliens of the genera *Albizia* and *Acacia*. The alternative is to manipulate the stand of aliens to encourage the establishment of indigenous seedlings before final removal of the alien species (Geldenhuys et al this volume). Trees could be killed by ring-barking larger trees, bark-peeling smaller trees to prevent coppice, and hand-pulling saplings and seedlings.

Control in other natural ecosystems

The control methods used in any particular area will depend on management objectives. The long-term objectives for control in all natural ecosystems should be aided by selecting and releasing suitable seed predators that will not affect the growth of the tree (Macdonald et al 1985). Suitable candidates have been found for this purpose (Neser et al this volume). Many agencies attempt to control *A melanoxylon* and other woody pioneers with long-lived, soil-stored seed by expensive eradication programmes which involve mechanical and chemical methods (Macdonald et al 1985). Control will be achieved only if the seed store can be eliminated or if a permanent vegetation cover can be established on the site or both.

Research

Development of a plantation management system for blackwood requires knowledge of the genetic variation, the effects of growth factors on timber quality, and the ecology of *A melanoxylon*. A research programme, initiated during 1973, comprised tree improvement, timber quality and silvicultural techniques (De Zwaan 1982). Tree improvement includes provenance and progeny trials, breeding strategies, hybridization and an attempt to produce sterile clones (De Zwaan 1980b). Timber quality was investigated in relation to growth rate and site conditions (Harrison 1975; De Zwaan 1982). Silvicultural techniques cover vegetative and reproductive propagation (De Zwaan 1978); establishment by planting (De Zwaan 1977) or natural regeneration; formation of mixed stands with *Pinus* and *Eucalyptus* species to overcome windfall problems; spacement, thinning and pruning trials (De Zwaan 1980a); and studies of site requirements in relation to growth rate and windfall susceptibility.

Integrated utilization of resources: useful alien invaders

Large parts of South Africa receive low rainfall and carry few indigenous woody plants that are useful. For example there are presently few suitable substitutes for *Atriplex nummularia* and *Prosopis* species in the arid regions (Poynton 1984). In the humid zones several *Pinus* and other tree species indicated as naturalized aliens form the backbone of the timber industry in South Africa. Examples of the other uses could be listed. The approach described for *Acacia melanoxylon* is equally applicable to the other species. It requires careful, integrated management of limited resources. It also requires knowledge of the ecology of the alien organism, in relation to its spread in natural or modified ecosystems.

Acknowledgements

This report was funded through the Conservation Forestry Research Programme of the Department of Environment Affairs (Forestry Branch). The useful comments offered by colleagues are acknowledged with thanks.

Campbell B M and Moll E J 1977. The forest communities of Table Mountain, South Africa. Vegetatio 34, 105-115.

De Zwaan J G 1977. Protection against buck damage in blackwood (*Acacia melanoxylon*). South African Forestry Journal 102, 81-82.

De Zwaan J G 1978. The effects of hot-water-treatment and stratification on the germination of blackwood (*Acacia melanoxylon*) seed. South African Forestry Journal 105, 40-42.

De Zwaan J G 1980a. The correlation between crown shape and height growth in young blackwood trees. South African Forestry Journal 112, 20-22.

De Zwaan J G 1980b. Is a seedless blackwood tree a possibility? South African Foretry Journal 113, 59-61.

De Zwaan J G 1982. The silviculture of blackwood (*Acacia melanoxylon*). South African Forestry Journal 121, 38-43.

De Zwaan J G 1983. Report on blackwood study tour: Tasmania, Victoria, New South Wales and Queensland: February-March 1983. Unpublished report, Saasveld Forestry Research Centre, George. 38 pp.

Farrell T P and Ashton D H 1978. Population studies on *Acacia melanoxylon* R.Br. I. Variation in seed and vegetative characteristics. Australian Journal of Botany 26, 365-379.

Geldenhuys C J 1980. Report on a study tour of rainforests in Australia, New Zealand and Chile: 13 August 1979 to 22 January 1980. Unpublished report, Saasveld Forestry Research Centre, George. 45 pp.

Geldenhuys C J 1982. The management of the southern Cape forests. South African Forestry Journal 121, 1-7.

Geldenhuys C J 1983. A criticial evaluation of indigenous forest management. In: Proceedings of the Jubilee Symposia. Faculty of Forestry, University of Stellenbosch, Stellenbosch. pp 497-524.

Geldenhuys C J and Maliepaard W 1983. The causes and sizes of canopy gaps in the southern Cape forests. South African Forestry Journal 124, 50-55.

Geldenhuys C J, Reynell D F and Furgeson M A 1986. Survey and analysis of the growing stock of *Acacia melanoxylon* in the southern Cape Forests. Saasveld Forestry Research Centre, George.

Geldenhuys C J and Van Laar A 1980. Interrelationships between vegetation and site characteristics in the Gouna forest. South African Forestry Journal 112, 3-9.

Grey D C 1983. Forest soils and site requirements of exotic tree species. In: Odendaal P B, Owen D L, Bosman D L, Schönau A P G, von Gadow K and van der Zel D W (eds) South African Forestry Handbook. South African Institute of Forestry, Pretoria. pp 55-62.

Harrison C M 1975. The relative influence of genetics and environment upon certain timber quality characters of *Acacia melanoxylon* in South Africa. Forestry in South Africa 17, 23-27.

Henderson L and Musil K J 1984. Exotic woody plant invaders of the Transvaal. Bothalia 15, 297-313.

Hill P S 1982. Rainforest fire in western Tasmania. Australian Journal of Botany 30, 583-589.

Hopkins M S, Kikkawa J, Graham A W, Tracey J G and Webb L J 1977. An ecological basis for the management of rainforest. In: The Border Ranges: a land use conflict in regional perspective. Brisbane. Royal Society of Queensland. pp 57-66.

Huntley B J 1965. A preliminary account of the Ngoye Forest Reserve, Zululand. Journal of South African Botany 31, 177-205.

Koen J H 1984. A study of the distribution, population composition, movements and feeding of the Knysna elephants *Loxodonta africana africana*. Unpublished Report No S.84/6, Saasveld Forestry Research Centre, George. 38 pp.

Le Roux P J 1982. Plant invader species in pine plantations in South Africa. In: Van de Venter H A and Mason M (eds) Proceedings of the 4th National Weeds Conference of South Africa. Balkema, Cape Town. pp 155-168.

Luyt V J 1983. Exploitation of indigenous timber with helicopter. Unpublished report. Southern Cape Forestry Region, Knysna.

Macdonald I A W and Jarman M L (eds) 1984. Invasive alien organisms in the terrestrial ecosystems of the fynbos biome, South Africa. South African National Scientific Programmes Report No 85. CSIR, Pretoria.

Macdonald I A W, Jarman M L and Beeston P (eds) 1985. Management of invasive alien plants in the fynbos biome. National Scientific Programmes Report No 111. CSIR, Pretoria.

McDonald W J F and Whiteman W G 1979. Moreton region vegetation map series: Explanatory booklet for Murwillumbah sheet. Botany Branch, Department of Primary Industries, Brisbane.

McDowell C R and Moll E J 1981. Studies of seed germination and seedling competition in *Virgilia oroboides* (Berg.) Salter, *Albizia lophantha* (Willd.) Benth. and *Acacia longifolia* (Andr.) Willd. Journal of South African Botany 47, 653-685.

Milton S J 1982. Effects of shading on nursery grown *Acacia* seedlings. Journal of South African Botany 48, 245-272.

Odendaal P B 1977. Some aspects of the ecology of bushbuck (*Tragelaphus scriptus* Pallas 1776) in the southern Cape. Unpublished MSc Thesis, University of Stellenbosch, Stellenbosch. 200 pp.

Phillips J F V 1926. *Virgilia capensis* Lamk. ('Keurboom'): A contribution to its ecology and sylviculture. South African Journal of Science 23, 435-454.

Phillips J F V 1928. The behaviour of *Acacia melanoxylon* R.Br. ('Tasmanian Blackwood') in the Knysna forests. Transactions of the Royal Society of South Africa 16, 31-43.

Phillips J F V 1931. Forest ecology and succession in the Knysna region. Botanical Survey Memoir of South Africa 14. Government Printer, Pretoria.

Poynton R J 1984. Characteristics and uses of selected trees and shrubs cultivated in South Africa. Bulletin No 39, Department of Environment Affairs, Pretoria. 202 pp.

Scheepers J C, Van der Schijff H P and Keet J D M 1968. An ecological account of the *Trema* plantations of Westfalia Estate.

Tydskrif vir Natuurwetenskappe 8, 105-120.

Seagrief S C 1965. Establishment of *Podocarpus latifolius* in Blackwood plantations at the Hogsback. South African Journal of Science 61, 433-437.

Stapleton C C 1962. Letter. Journal of the South African Forestry Association 41, 31.

Stirton C H 1978. Plant invaders beautiful, but dangerous. The Cape Department of Nature and Environmental Conservation, Cape Town.

Van Daalen J C 1982. Growth rate of indigenous tree species - a review. Unpublished report, Saasveld Forestry Research Centre, George. 17 pp.

Van den Berg M A 1982a. Lepidoptera attacking *Acacia dealbata* Link., *A decurrens* Villd., *A longifolia* (Andr.) Villd., *A mearnsii* De Wild. and *A melanoxylon* R.Br. in eastern Australia. Phytophylactica 14, 43-46.

Van den Berg M A 1982b. Hemiptera attacking *Acacia dealbata* Link., *A decurrens* Villd., *A longifolia* (Andr.) Villd., *A mearnsii* DeWild. and *A melanoxylon* R.Br. in Australia. Phytophylactica 14, 47-50.

Van den Berg M A 1982c. Coleoptera attacking *Acacia dealbata* Link., *A decurrens* Villd., *A longifolia* (Andr.) Villd., *A mearnsii* DeWild. and *A melanoxylon* R.Br. in Australia. Phytophylactica 14, 51-55.

Ward M C 1985. The current extent of alien infestations: KwaZulu areas. In: Macdonald I A W and Jarman M L (eds) Invasive alien plants in the terrestrial ecosystems of Natal, South Africa. South African National Scientific Programmes Report No 118. CSIR, Pretoria.

Weisser P J and Marques F 1979. Gross vegetation changes in the dune area between Richards Bay and the Mfolozi River, 1937-1974. Bothalia 12, 711-721.

Wilson M 1982. The re-establishment of yellowwood forest. Veld and Flora 68, 111-112.

Chapter 24

The importance of seed-attacking agents in the biological control of invasive alien plants

S Neser and R L Kluge
Plant Protection Research Institute.

Introduction

Many invasive plants reproduce exclusively by seeds. If their effective seed production can be reduced below a critical level, their populations should eventually decline. This is especially true if the species has a limited life-span or succumbs readily during periodic "catastrophes" such as fires. Attempts at control using this approach are indirect and will be slow if the plants are long-lived, or if there is a large seed bank.

Holloway (1964) suggested that almost total seed destruction would have to be achieved before any effects on the herbaceous *Xanthium strumarium* in Australia could be expected. In the past it was usually considered that even the destruction of large proportions of seeds of perennial plants would have little value in biological control, as illustrated in reviews by Wilson (1964) and Huffaker (1964). This view persisted as is shown by the low score awarded to seed-attacking insects by Harris (1973) in his rating of biocontrol agents. According to this scoring system, modified by Goeden (1983), agents that inflict direct damage to the vital vascular or mechanical support system of the plant are more likely to be effective control agents; traits such as seed destruction of long-lived perennials, specialization towards strict host-specificity, obligate univoltinism, a relatively small number of offspring per generation, solitary feeding and a small size generally afford low marks to an agent. Seed-attackers would rate poorly on these grounds. Despite this we propose to show firstly that seed-attacking agents may have been underestimated as a factor in the invasive success of plants in a new environment and secondly why their use is an important option in the control of alien plants.

For brevity and convenience, "seed" will be used in a broad sense to include developing and mature fruits, and "seed attack" will refer to any stage of seed production from flower-bud initiation onwards.

Many biocontrol agents reduce seed production and viability indirectly by reducing plant vigour (eg stemborers and defoliators), by acting as nutrient sinks (various gall-formers in vegetative parts), or by causing increased abortion of flowers or young fruits (eg sap-sucking species). Although their effects on seed production may sometimes be as important as that on growth and competitive ability, such agents will not be considered here as primarily affecting seed production. Polyphagous insects including certain ants, and larger seed-eating organisms such as rodents and birds may also be important seed mortality factors but are not likely to be used in biological control programmes.

Unchecked, prolific seed production as a factor in invasiveness

Intrinsic vigour and growth rate, superior rates of photosynthesis or nutrient utilization, or better adaptation to drought, fire or other adverse conditions encountered in their new environment are sometimes cited as likely reasons why some introduced plants became successful and "aggressive" invaders (Roux and Middlemiss 1963; Milton 1980a). The reasons for the success of different invader plants in their new environment are complex and governed by so many local and other factors, that generalizations must be made cautiously. Although it is agreed that preadaptive features enable plants to become invaders, we believe that the importance of the absence of specialized seed-attacking organisms is still underestimated. Various types of natural enemies may directly affect the production and survival of seeds. Many of these organisms are very poorly known and are easily overlooked because of their small size and often cryptic effects in their natural habitat. However, they can have large effects: for example *Hakea sericea* was found to have a seed store

Address for correspondence: S Neser, Plant Protection Research Institute, Private Bag X134, Pretoria, 0001.

of up to 75 000 seeds m^{-2} in a typical dense stand of 15-year old plants in the south-western Cape Province (M W Siebert personal communication) while this species in its native range in Australia is comparatively very unsuccessful in accumulating viable seeds (Neser 1968, 1984b; Gill and Neser 1984). Milton (1980b) reported average seed densities of 1 200-7 400 seeds m^{-2} in the leaf-litter under stands of *Acacia cyclops*, *A melanoxylon*, *A saligna* and *A longifolia* in the south-western Cape. In contrast, the seed bank of the south African *Chrysanthemoides monilifera* was about 50 times greater than that of the indigenous *Acacia longifolia* in Australia, while in South Africa the reverse is true (Weiss and Milton 1984). Predation by mainly insects is implicated in both instances (Neser and Morris 1985; Neser unpublished).

Experience has shown that specialized organisms affecting seed production are likely to be found for most species in their natural ranges (Neser 1968; Zimmermann 1974; Van den Berg 1977, 1980b, 1982a,b,c; Erb 1980; Neser and Morris 1985; Zwölfer in press). A few examples of those directly affecting seed production and survival which were encountered in searches for biological control agents for Australian acacias illustrate the diversity (summarized by Neser 1984c): the rust fungus *Uromycladium tepperianum* (M J Morris personal communication) and microscopic eriophyid mites (*Aceria* species) (M K P Meyer Smith personal communication) affecting the developing inflorescences; pteromalid wasps of the genus *Trichilogaster* (Van den Berg 1980a;) forming galls in buds; various unidentified midges, the larvae of which develop specifically in inflorescence axes, flower buds, unopened individual florets, newly-set podlets, and, in an apparent association with a fungus, in young seeds (Neser 1984c and unpublished); various species of *Melanterius* snout beetles and eurytomid wasps (below) developing in immature seeds, as well as members of different hemipterous families which minutely puncture (and thus destroy) seeds on the plant and in the leaf-litter (Van den Berg 1977, 1980a,b, 1982a,b,c). Of many of the insects the adults are seldom reared and practically unknown because of their small size, ephemeral nature, cryptic damage and extremely heavy parasitism. Similar ranges of insects, but including other orders and families were encountered on *Hakea* species (Neser 1968, 1984b), and preliminary searches on *Leptospermum laevigatum* revealed comparable seed-affecting organisms, including a specialized nematode and fly (Sphaerulariidae, Fergusoninidae) association in galled buds.

Studies such as those of the acacias and *Hakea sericea* within their natural range in Australia and in South Africa, have illustrated the importance of insects in their seed production. This led to the hypothesis that certain seed-attacking species introduced by themselves may be sufficient to reduce the plants' invasiveness in the new environment and eventually accomplish acceptable levels of control (Neser 1968, 1984a).

Prolific seed production may give an invasive species an advantage over the indigenous species and, in particular, over ecological analogues, whose seed production is most likely to be constrained by various natural enemies. Not only is a large seed crop important in overcoming post-dispersal mortality, but it also gives the plant an advantage in reaching, and competing in as many as possible germination sites. Furthermore the initial competitive advantage afforded to the invader by prolific seeding is compounded by the increased vigour of the alien plant in the absence of its natural enemies. This may further reduce seeding by the indigenous plants, as illustrated by Weiss (1985) for the locally indigenous *Acacia longifolia* when growing intermingled with the introduced *Chrysanthemoides monilifera* on coastal dunes in Australia. Once a foothold is gained by the invader even in a relatively undisturbed situation, the invasion is likely to be expanded during any subsequent disturbance. Few other species are likely to survive the competition where seedling densities hundreds of times the number that will eventually mature occur, as has been observed in the typically dense regeneration of hakea, acacias and other invasive plants.

The importance of seed numbers in the regeneration of plants is also emphasized by Harper (1977) who distinguishes two important effects that seed reduction can have on a plant population. Firstly, seed loss proportionately reduces the distance at which a given density of seeds occur from the seed source, ie the rate of dispersal. Secondly, if seed production is reduced below a critical threshold level this can result in a decrease in the local density of the plant. Theoretically, therefore, any level of seed attack reduces the rate of spread of a plant, whereas a reduction in plant density is only possible in cases where seed attack reaches a specific level.

The effects that seed-attacking organisms have on plant populations could vary depending on characteristics of both the host-plant and the organism. The influence of seed-attackers is most likely to be severe when (Kluge 1983):

(1) there is no seed dormancy ie no large, long-lived seed banks;
(2) there is no vegetative reproduction;
(3) the insect achieves a high level of attack;
(4) the level of attack is consistently maintained throughout the generation time of the plant, since the size of the accumulated seed store determines the regeneration success;
(5) the host-plant does not have effective defense mechanisms against seed-attackers such as "mast fruiting";
(6) populations of the plant are homogeneous, ensuring consistency of attack; and
(7) the habitat is stable to ensure continuity of the insect populations in time.

Many plants have evolved in the presence of seed-attackers or organisms which cause consistently high mortality in the seed store and yet are dependent on seed for regeneration. Seed losses of up to 90% are encountered within the native range of many plants (Janzen 1971). The ability to compensate for heavy seed predation appears to have developed for instance in a number of *Hakea* species in Australia where seed destruction caused by the specialized acanthosomatid bug *Sangarius paradoxus*, which is only one of a complex of seed-attackers, can be as high as 70-99% (Neser 1968, 1977; Gill and Neser unpublished). The survival of such plants in their native environments would be severely threatened if they had not acquired mechanisms to offset seed destruction. These may include:

(1) a capacity to produce an excess of seeds to allow for the losses;
(2) having effective mechanisms for seed dispersal allowing at least temporary establishment and further seeding before being located by the predators;
(3) a variety of "anti-predator" traits such as irregular or unseasonal fruiting, thus thwarting seed-attackers which have regular, annual life-cycles;
(4) a capacity in some populations to resprout from ground level after fires; and
(5) various modifications to the fruit which may help to reduce damage by seed-attackers (Neser 1968).

A plant may replace early losses in propagules; for example *H sericea* may produce a second flowering flush if the first fruits are lost at an early stage (Neser 1968; Fugler 1979). There is, however, no way in which it may actively compensate for post-dispersal losses unless it is through an evolved compensatory or "excessive" seeding capacity. In a new habitat such plants are likely to be released from the pressure of pre- and post-dispersal seed losses induced by specific predators, yet inherently they will continue with the now totally excessive seed production and the resultant increased invasive success.

Feasibility of the use of seed-attackers in biological control

Since most species produce many more seeds than are required to replace the parent plants, it would seem that very high levels of seed destruction would be needed in the new environment to prevent regeneration. However, mortalities in seeds through physical, density-independent factors may often be extremely high. Such losses have been evaluated for various plants in invaded and natural habitats and may often approach 100% (Milton 1980b; Weiss 1983; Kluge 1983; Van Zyl and Kluge 1983; Pieterse 1984; Richardson and Van Wilgen 1984). Reasons may be short-lived seeds, exacting requirements for germination and establishment, predation by generalists and dispersal to unfavourable situations. These same losses will be superimposed on those caused by specific biological factors such as deliberately introduced seed-attackers. If the post-dispersal losses are not density dependent, seed survival will be directly proportional to the number of seeds produced. Further, if this number has already been reduced by a factor of 10-20, as may easily be achieved - see for example the analysis of Harley (in press) of 11 control programmes in which seed-attacking insects were used – then the number of seeds reaching establishment sites may be reduced accordingly.

If prolific seeding is a crucial factor in making a plant invasive, then curbing this seeding, perhaps to levels similar to those of its indigenous analogues, would enable indigenes to compete more successfully with the invaders.

It may be argued that, in view of the longevity of acacia seeds, reducing the production of these seeds is a futile biocontrol strategy. It has, however, been shown that the seed banks under *A longifolia* stands in South Africa may not be the threat they were previously believed to be. Pieterse and Cairns (1986) found that about 90% of the seeds near the soil surface were destroyed during a fire and speculated that many seedlings from scarified seeds in deeper layers would succumb because of the depth of burial, leaving relatively few seeds to survive. Theoretically this number will be proportionate to the number of seeds escaping introduced seed-attackers, suggesting that there is indeed a good case to be made for their use.

Hardly any insect species in its natural habitat escapes attack by predators and parasitoids and high levels of mortality are usual, so that they also are rarely abundant in their natural environments. However, in the absence of competing species and of specific natural enemies, the specialized seed-attacking insects may rapidly become abundant and particularly destructive, as has been found in various biological control projects throughout the world (Julien 1982). The fact that different introduced seed-attackers have undergone ''population explosions'' illustrates the potential of these agents (Williams 1960; Greathead 1971; Kluge 1983; Dennill 1985; Neser and Kluge in press).

Present projects involving seed-attackers in South Africa

South African researchers have been working on seed-attacking natural enemies of several invasive plants over the last two decades. Conflicts of interest (discussed below) were conducive to this approach in many, but not in all instances. The first results that have been achieved are most encouraging, as are the prospects for several projects which are still in the early stages.

Hakea sericea

The snout beetle *Erytenna consputa*, the larvae of which destroy developing fruits of *H sericea*, was distributed to all the mountain ranges in the southern Cape Province where this woody shrub, regenerating only from seeds released after fires, occurred in dense populations (Kluge 1983). *Erytenna consputa* has become a major cause of loss of developing hakea fruits at many sites (Kluge and Siebert 1985). The effects and value of this insect were clearly illustrated at a study site where only 20 beetles had been released in 1975, and where steadily increasing fruit losses resulted: from 39% in 1979 to 81% in 1981 (Figure 1). The level of attack appeared to be higher still (M W Siebert personal communication) when a fire swept through the site in 1982. Two years later the density of the new generation of *H sericea* was only 10% that of the parent population (Kluge unpublished), whereas characteristically in the Cape mountains hakea seeds released after fires gave rise to populations denser than that of the parent generation. This was a well-quantified and clear demonstration of a reduction in the density of an invader plant affected by an introduced seed-attacking insect, and it is believed that this pattern will also be evident at other release sites. This beetle is now making an important contribution to the integrated control programme being implemented for *H sericea* (Kluge et al this volume).

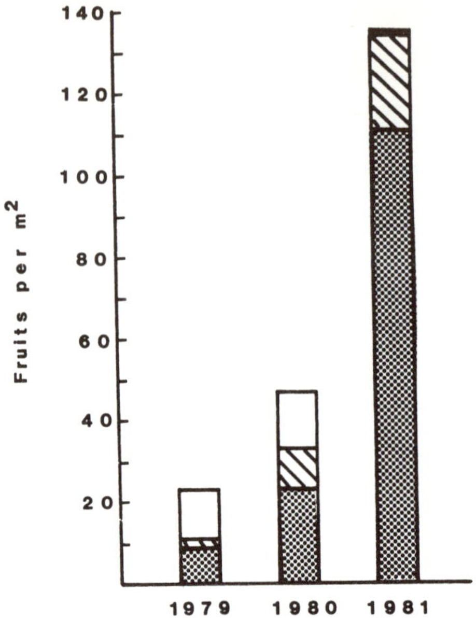

Figure 1. Incidence of seed-attack by *Erytenna consputa* (stippled) in seed crop of young plants of *Hakea sevicea* at Goudini in three successive years. Hatching indicates maturing fruits attacked, but seeds not destroyed.

The moth *Carposina autologa*, the larvae of which destroy the seeds in mature hakea follicles was introduced to augment the seed attack of the snout beetle. The larvae develop in fruits which may accumulate on the plant and destroy seeds which may have escaped beetle attack (Neser 1968; Neser and Annecke 1973). The moth became established (Neser and Kluge in press) but has apparently not multiplied sufficiently to make it a major factor in seed destruction. This was mainly because the pathogenic fungus *Colletotrichum gloeosporioides* (Morris 1982, 1983; Richardson and Manders 1985) killed the hakea plants at the release sites.

Acacia longifolia

The bud-galling wasp *Trichilogaster acaciae-longifoliae* was the first insect to be considered for introduction into South Africa against *A longifolia* (Van den Berg 1980a). Following its successful establishment (Neser in press), seeding on infested plants was reduced in proportion to the percentage of flower branches with galls, with more than 95% reduction in the seeding when more than 75% of the branches contained galls. When as few as 50% of the branches on a tree had galls, seeding was reduced by an average of 89% (Figure 2); this was accompanied

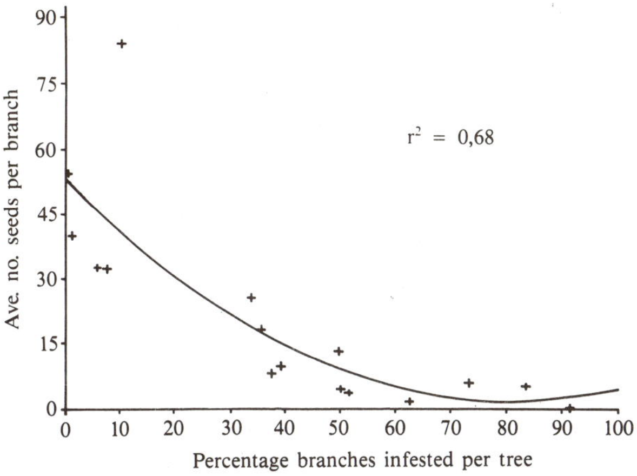

Figure 2. Reduction of seeding in *Acacia longifolia* induced by different levels of galling by the wasp *Trichilogaster acaciae-longifoliae* (after Denill 1985).

by marked suppression of growth (Dennill 1985).

This wasp is now widely established at more than 200 sites from the western Cape through the southern and south-eastern Cape to Natal. The increase in the wasp populations has been spectacular. It is, however, still too early to predict whether the insect will be consistently successful every year throughout the range of the host plant. Present results are promising, especially in the light of the findings (Pieterse and Cairns 1986) that the seed store may not be as inexhaustible as it was believed to be.

A second seed-destroying insect, the snout beetle *Melanterius ventralis* has also been released recently to augment the damage of the gall-wasp (D Donnelly personal communication). A seed-attacking eurytomid wasp *Bruchophagus* species is also being investigated for possible future use.

Sesbania punicea

The first agent to become widely established was the tiny snout beetle *Trichapion lativentre*. The adults feed on leaves, flowers and young pods and the larvae develop in young flower buds (Harris and Hoffmann in press). This beetle spread rapidly and entirely unaided throughout Natal and parts of the eastern Cape. It was also recently released and successfully established at many other sites in the south-western Cape and the Transvaal. In the hinterland of Natal it prevented seeding in various stands of the weed during 1984 and 1985 (Harris and Hoffmann in press). However, pods were produced at coastal sites where the initial effect of the insect appeared to be limited (J H Hoffmann personal communication). Another Argentinian snout beetle, *Rhyssomatus marginatus*, which develops in immature sesbania pods was also recently successfully released in the eastern Cape to augment *T lativentre* (J H Hoffmann personal communication).

Graaff and Van Staden (1983) found that the vast majority of seeds released germinate within a few months and that few are likely to survive in the soil for more than two years. If it is true under all conditions where seeds accumulate, control of this weed by seed destruction becomes feasible, since severe limitation of seed production for more than two years will result in low levels of regeneration. The sparse plants on which the beetles are likely to continue flourishing, may then be easily controlled, or may be left to die in veld fires or of old age, the shrub being short-lived and sensitive to fire.

Other invasive plants

Studies on seed-attackers of various other invasive plants are in progress in the Plant Protection Research Institute. Specificity studies have been completed on *Melanterius acaciae*, a snout beetle developing in immature pods of *Acacia melanoxylon* in Australia (D Donnelly personal communication). The use of this beetle may reduce the spread of *A melanoxylon* from areas where it is grown commercially (see Geldenhuys this volume).

A similar beetle, *M servulus*, from *Paraserianthes* (formerly *Albizia*) *lophantha* has also been subjected to specificity studies with a view to release (D Donnelly personal communication) and may be valuable in curbing this widespread, but still localized, invader which may assume more serious proportions in South Africa if left unchecked.

Bruchid beetles, such as *Algarobius prosopis* which develop in, and destroy the seeds of *Prosopis velutina*, without affecting the fodder value of the pods, are also under investigation (H G Zimmermann personal communication) and may eventually allow the continued use of *Prosopis* species without the present problems with spread and the development of stands too dense for utilization (Neser and Moran in press).

Seed-attackers of herbacious weeds like the thistles *Cirsium arvense* and *Silybum marianum* and of *Solanum elaeagnifolium* have also been released recently (H G Zimmerman personal communication) in an attempt to limit their spread.

Seed reduction as a useful compromise in biological control

Various situations have arisen in South Africa

where potentially damaging agents that attack the vegetative parts of invader plants may be available, but where some of their effects would be undesirable (Neser and Moran in press). In some instances seed-attacking agents may be preferable to others attacking vegetative parts, even if results may not be obtained as rapidly as with other agents or methods.

(1) In situations where conflicts of interest exist, seed-attackers may provide an acceptable alternative to agents which act more directly and destructively on the plant.

  (a) The weed itself may be valuable or useful in certain situations eg *Acacia melanoxylon* (blackwood), *A mearnsii* (blackwattle) and *A cyclops* (rooikrans) for timber, bark extracts, firewood and windbreaks. An alien species fulfilling a useful function in a certain situation may have become unacceptably dense there and/or invasive elsewhere, eg *A saligna* (Port Jackson) on the sandy Cape Flats. Although *Prosopis* species growing as single trees may be desirable for the provision of shade and stock fodder in the Karoo, their uncontrolled spread and occurrence in dense stands are undesirable.

  (b) An invader may be prickly or unpalatable and thus resistant to grazing. Plants like some thistles may be the only vegetation left preventing uncontrolled erosion in badly degraded areas and may act as nurse plants in these situations where no recovery of the vegetation would otherwise be possible.

  (c) Some of the more destructive agents may also pose a threat to nontarget plants, such as related crops or commercially grown alien plants. Examples are certain stemborers of *Solanum mauritianum*, which would attack solanaceous crops (Neser unpublished); a phyllode- attacking bug *Rayieria* species capable of killing *Acacia longifolia* which might attack commercially valuable alien *Acacia* species (Donnelly and Neser 1983; Donnelly in press), and a destructive defoliator of *Leptospermum laevigatum* (Neser and Annecke 1973) which may attack cultivated *Eucalyptus* species (Neser unpublished).

(2) Gradual removal of the target plants through the slow, indirect action of seed-attackers may be advantageous.

  (a) Virtual monocultures of invaders such as *Pinus pinaster*, *Acacia* species and *Leptospermum laevigatum* may occur on sites where accelerated erosion or sand dune formation could follow if dense stands of the alien are removed rapidly without allowing gradual recovery of the original vegetation.

  (b) Apart from the dominant invader, there may be other, more undesirable, invasive species in the same vegetation complex, and relatively rapid removal of the dominant species, eg by "non-biological" means may result in an aggravated problem for which there may be no quick solution. An example is the case where the relatively easily controlled *H sericea* was rapidly replaced by *A longifolia* after manual clearing of the former in the Banhoek Valley near Stellenbosch.

(3) Seed-attacking agents may be used as an interim measure to contain the spread of alien plants which are at an early stage of invasion.

  (a) Seed-attacking insects are generally very specialized and host-specific (monophagous) or at most stenophagous (attacking only a few closely related hosts) (Janzen 1980). In contrast, the more destructive natural enemies which attack vegetative parts of the plants are often less host-specific (Moran et al this volume) and are unfortunately sometimes deemed unacceptable for release in the new country only after lengthy studies. As a result, obtaining specific seed-attackers, and confirming their safety for introduction, is likely to be a quicker and less costly option, allowing attention to be given to many more problems.

  (b) Although little financial support is usually forthcoming for research programmes on plants which have not yet become serious problems, such work may be a most rewarding investment for the future. Whereas seed-attacking agents may not give rapid results against long-lived woody invaders which are already widespread and occur in dense stands, it is probable that, had the seed-attackers been introduced at an early stage, the plants may not have developed into a problem, or at least not at an uncontrolled rate.

The introduction of seed-attacking agents for those species showing signs of becoming invasive, should therefore be given greater priority. Many possible candidates occur in South Africa, eg some less well-known Australian acacias, *Ageratina adenophora*, *Agonis flexuosa*, *Ailanthus altissima*, *Alhagi camelorum*, *Cassia didymobotrya*, *Casuarina equisetifolia*, *Cestrum laevigatum*, certain *Eucalyptus* species, *Jacaranda mimosifolia*, *Leptospermum laevigatum*, *Leucaena leucocephala*, *Litsea sebifera*, *Melia azedarach*, *Pereskia aculeata*, *Pinus pinaster*, *Psidium cattleianum*, *Pueraria* species, *Schinus terebinthifolius*, *Spartium junceum*, *Ulex europaeus* and undoubtedly many others. For some of these specific seed-attacking agents may already be known in their countries of origin.

Conclusions

We have discussed the different ways in which natural enemies may curb seed production and the accumulation of seed stores. We infer that the role of seed-attacking organisms on their host-plants in their native habitats, and the importance of their absence in new situations, may have been underestimated and hence also their potential value in control programmes.

We attempted to show how seed-attacking organisms, even though they may act indirectly and slowly, can have an effect on populations of invasive plants by reducing their rate of spread and/or density. There is merit in, and considerable scope, for their use in biological control programmes, especially where prolific seeding of the target plant is suspected to be an important factor in its success as an invader. Results achieved so far with seed-attacking insects against *Hakea sericea*, *Acacia longifolia* and *Sesbania punicea* have confirmed the potential of seed-attacking insects.

Because seed-attacking organisms are often specialized and host-specific, many may be studied and introduced at relatively low costs and as a most valuable investment.

Certain situations have been identified in which the use of seed-attacking agents is to be preferred to other organisms which cause direct damage to the plant. These include situations where there are conflicts of interest around a particular invader, where gradual removal of the target is advisable, and where plants are still in an early stage of invasion.

Acknowledgements

We thank our collegues and referees who made useful comments on drafts of this paper.

Dennill G B 1985. The effect of the gall wasp *Trichilogaster acaciaelongifoliae* (Hymenoptera: Pteromalidae) on the reproductive potential and vegetative growth of the weed *Acacia longifolia*. Agriculture, Ecosystems and Environment 14, 53-61.

Donnelly D (in press). *Rayieria* sp (Heteroptera: Miridae) - Host specificity, conflicting interests, and rejection as a biological control agent against the weed *Acacia longifolia* (Andr.) Willd. in South Africa. Journal of the Entomological Society of Southern Africa 47(2).

Donnelly D and Neser S 1983. Preliminary host specificity studies of *Rayieria* sp (Heteroptera: Miridae), a potential biocontrol agent against the weed *Acacia longifolia* in South Africa. In: Van de Venter H A and Mason M (eds) Proceedings of the 4th National Weeds Conference of South Africa. Balkema, Cape Town, 188-199.

Erb H E 1980. The natural enemies and distribution of *Sesbania punicea* (Cav.) Benth. in Argentina. In: Neser S and Cairns A L P (eds) Proceedings of the 3rd National Weeds Conference of South Africa. Balkema, Cape Town, 205-210.

Fugler S R 1979. Some aspects of the autecology of three *Hakea* species in the Cape Province, South Africa. MSc Thesis, University of Cape Town, Cape Town. 295 pp.

Gill A M and Neser S 1984. *Acacia cyclops* and *Hakea sericea* at home and abroad. In: Dell B (ed) MEDECOS IV: Proceedings of the 4th International Conference on Mediterranean Ecosystems, Perth. pp 56-57.

Goeden R D 1983. Critique and review of Harris' scoring system for selection of insect agents in biological control of weeds. Protection Ecology 5, 287-301.

Graaff J L and Van Staden J 1983. Germination of *Sesbania punicea*. In: Van de Venter H A and Mason M (eds) Proceedings of the 4th National Weeds Conference of South Africa. Balkema, Cape Town, 35-44.

Greathead D J 1971. A review of biological control in the Ethiopian Region. Technical Communication 5, CIBC, Commonwealth Agricultural Bureaux, Farnham Royal. 162 pp.

Harley K L S (in press). Suppression of reproduction of woody weeds using insects which destroy flowers or seeds. In: Delfosse E S (ed) Proceedings of the 6th International Symposium on Biological Control of Weeds. Vancouver, Canada. Agriculture Canada.

Harper J C 1977. Population Biology of Plants. Third edition. Academic Press, London.

Harris M S and Hoffmann J H (in press). The weed *Sesbania punicea* (Leguminosae) in South Africa nipped in the bud by the weevil *Trichapion lativentre*. In: Delfosse E S (ed) Proceedings of the 6th International Symposium on Biological Control of Weeds. Vancouver, Canada. Agriculture Canada. pp 757-760.

Harris P 1973. The selection of effective agents for the biological control of weeds. The Canadian Entomologist 105, 1495-1503.

Holloway J K 1964. Prospects in biological control of weeds. In: DeBach P (ed) Biological Control of Insect Pests and Weeds. First edition. Chapman and Hall, London. pp 650-670.

Huffaker C B 1964. Fundamentals of biological weed control. In: DeBach P (ed) Biological Control of Insect Pests and Weeds. First edition. Chapman and Hall, London. pp 631-649.

Janzen D H 1971. Seed predation by animals. Annual review of Ecology and Systematics 2, 465-492.

Janzen D H 1980. Specificity of seed-attacking beetles in a Costa Rican deciduous forest. Journal of Ecology 68, 929-952.

Julien M H (ed) 1982. Biological Control of weeds: a world catalogue of agents and their target weeds. Commonwealth Agricultural Bureaux, Farnham Royal. 108 pp.

Kluge R L 1983. The hakea fruit weevil, *Erytenna consputa* and the biological control of *Hakea sericea* in South Africa. PhD Thesis, Rhodes University, Grahamstown. 132 pp.

Kluge R L and Siebert M W 1985. *Erytenna consputa* (Coleoptera: Curculionidae) as the main mortality factor of developing fruits of the weed, *Hakea sericea* Schrader, in South Africa. Journal of the Entomological Society of Southern Africa 48, 241-245.

Milton S J 1980a. Australian acacias in the S W Cape: pre-adaptation, predation and success. In: Neser S and Cairns A L P (eds) Proceedings of the 3rd National Weeds Conference of South Africa. Balkema, Cape Town, 69-78.

Milton S J 1980b. Studies on Australian acacias in the south western Cape, South Af.ica. MSc Thesis, University of Cape Town, Cape Town. 283 pp.

Morris M J 1982. Gummosis and die-back of *Hakea sericea* in South Africa. In: Van de Venter H A and Mason M (eds) Proceedings of the 4th National Weeds Conference of South Africa. Balkema, Cape Town, 51-54.

Morris M J 1983. Evaluation of field trials with *Colletotrichum gloeosporioides* for the biological control of *Hakea sericea*. Phytophylactica 15, 13-16.

Neser S 1968. Studies on some potentially useful insect enemies of the needle-bush, (*Hakea* spp Proteaceae). PhD Thesis, Australian National University, Canberra. 194 pp.

Neser S 1977. *Sangarius paradoxus* Stal (Hemiptera: Acanthosomatidae): a paradoxical bug on *Hakea* species in Australia. In: (Summaries) Proceedings of the 2nd Entomological Congress of the Entomological Society of Southern Africa, Pretoria 1977, 51-52.

Neser S 1984a. Theories on why only some introduced plants become "aggressive" weeds. In: (Summaries) Proceedings of the 6th National Weeds Conference, Southern African Weeds Science Society, Nelspruit, 1984. p 2.

Neser S 1984b. Insect enemies of *Hakea sericea* (Proteaceae) in Australia. In: Dell B (ed) MEDECOS IV: Proceedings of the 4th International Conference on Mediterranean Ecosystems, Perth, 126-127.

Neser S 1984c. Natural enemies of *Acacia* species in Australia. In: Dell B (ed) MEDECOS IV: Proceedings of the 4th International Conference on Mediterranean Ecosystems, Perth, 124-125.

Neser S (in press). A most promising bud-galling wasp, *Trichilogaster acaciaelongifoliae* (Pteromalidae) established against *Acacia longifolia* in South Africa. In: Delfosse E S (ed) Proceedings of the 6th International Symposium on Biological Control of Weeds. Vancouver, Canada. Agriculture Canada. pp 797-803.

Neser S and Annecke D P 1973. Biological control of weeds in South Africa. Department of Agricultural Technical Services, Republic of South Africa, Entomology Memoir No 28. 27 pp.

Neser S and Kluge R L (in press). A seed-feeding insect showing promise in the control of a woody invasive plant: the weevil *Erytenna consputa* Pascoe on *Hakea sericea* (Proteaceae) in South Africa. In: Delfosse E S (ed) Proceedings of the 6th International Symposium on Biological Control of Weeds. Vancouver, Canada. Agriculture Canada. pp 805-809.

Neser S and Moran V C (in press). Tactics for evading conflicts in the biological control of South African weeds. In: Delfosse E S (ed) Proceedings of the 6th International Symposium on Biological Control of Weeds. Vancouver, Canada. Agriculture Canada. pp 359-363.

Neser S and Morris M J 1985. Preliminary observations on natural enemies of *Chrysanthemoides monilifera*. In: Love A and Dyason R (eds) Proceedings of a National Conference on *Chrysanthemoides monilifera*, Port Macquarie 1984, 105-109.

Pieterse P J 1984. Aspects of the seed production of *Acacia longifolia* in the western Cape. In: (Summaries) Proceedings of the 6th National Weeds Conference, Southern African Weeds Science Society, Nelspruit, 1984. p 4.

Pieterse P J and Cairns A L P 1986. The effect of fire on an *Acacia longifolia* seedbank in the south-western Cape. South African Journal of Botany 52, 233-236.

Richardson D M and Manders P 1985. Predicting pathogen induced mortality in *Hakea sericea*, an aggressive alien plant invading South Africa. Annals of Applied Biology 106, 143-254.

Richardson D M and Van Wilgen B W 1984. Factors affecting the regeneration success of *Hakea sericea*. South African Forestry Journal 131, 63-68.

Roux E R and Middlemiss E 1963. The occurrence and distribution of *A cyanophylla* and *A cyclops* in the Cape Province. South African Journal of Science 59, 286-293.

Van den Berg M A 1977. Natural enemies of certain acacias in Australia. In: Annecke D P (ed) Proceedings of the 2nd National Weeds Conference of South Africa. Balkema, Cape Town, 75-82.

Van den Berg M A 1980a. *Trichilogaster acaciaelongifoliae* (Froggatt) (Hymenoptera: Pteromalidae): a potential agent for the biological control of *Acacia longifolia* in South Africa. In: Neser S and Cairns A L P (eds) Proceedings of the 3rd National Weeds Conference of South Africa. Balkema, Cape Town, 61-64.

Van den Berg M A 1980b. Natural enemies of *Acacia cyclops* A. Cunn. ex G Don and *Acacia saligna* (Labill.) Wendl. in Western Australia. 1. Lepidoptera. Phytophylactica 12, 165-167.

Van den Berg M A 1982a. Lepidoptera attacking *Acacia dealbata* Link., *Acacia decurrens* Willd., *Acacia longifolia* (Andr.) Willd., *Acacia mearnsii* de Wild. and *Acacia melanoxylon* R. Br. in Eastern Australia. Phytophylactica 14, 43-46.

Van den Berg M A 1982b. Hemiptera attacking *Acacia dealbata* Link., *Acacia decurrens* Willd., *Acacia longifolia* (Andr.) Willd., *Acacia mearnsii* de Wild. and *Acacia melanoxylon* R.Br. in Australia. Phytophylactica 14, 47-50.

Van den Berg M A 1982c. Coleoptera attacking *Acacia dealbata*

Link., *Acacia decurrens* Willd., *Acacia longifolia* (Andr.) Willd., *Acacia mearnsii* de Wild. and *Acacia melanoxylon* R.Br. in Australia. Phytophylactica 14, 51-55.

Van Zyl W A and Kluge R L 1983. Aspects of the seed production of St John's wort (*Hypericum perforatum* L.) in the south-western Cape (Abstract). In: Neser S and Cairns A L P (eds) Proceedings of the 5th National Weeds Conference of South Africa, Stellenbosch 1983. p 13.

Weiss P W 1983. Invasion of coastal *Acacia* communities by *Chrysanthemoides*. PhD Thesis, Australian National University, Canberra.

Weiss P W 1985. The invasiveness of *Chrysanthemoides monilifera*. In: Love A and Dyason R (eds) Proceedings of a National Conference on Chrysanthemoides monilifera. Port Macquarie 1984, 83-87.

Weiss P W and Milton S J 1984. *Chrysanthemoides monilifera* and *Acacia longifolia* in Australia and South Africa. In: Dell B (ed) MEDECOS IV: Proceedings of the 4th International Conference on Mediterranean Ecosystems, Perth, 159-160.

Williams J R 1960. The control of black sage (*Cordia macrostachya*) in Mauritius: the introduction, biology and bionomics of a species of *Eurytoma* (Hymenoptera, Chalcidoidea). Bulletin of entomological Research 51, 123-132.

Wilson F 1964. The biological control of weeds. Annual Review of Entomology 9, 225-244.

Zimmermann H G 1974. Preliminary work on the biological control of satansbos *Solanum elaeagnifolium* Cav. In: Papers presented at the 1st National Weeds Conference of South Africa, Pretoria, 151-168.

Zwolfer H (in press). Insects and thistle heads: resource utilization and guild structure. In: Delfosse E S (ed) Proceedings of the 6th International Symposium on Biological Control of Weeds, Vancouver 1984.

Chapter 25

Integrated control for invasive alien weeds

R L Kluge, H G Zimmermann, C J Cilliers and G B Harding
Plant Protection Research Institute.

Introduction

Integrated control may vary in complexity. At its simplest level integrated control is understood to refer to procedures where the implementation of two or three methods of control is rationalized to increase the cost-benefit ratio of the control programme as a whole. Here, however, it is applied in its broadest sense, to mean the integration of any number of control options in any possible way, to achieve the most favourable outcome in terms of the predictable economic, sociological and ecological consequences.

Procedures for integrated weed control (IWC) programmes are shown in Figure 1. Such a programme consists of three main components: the target plant; the control options; and the implications of control. Also involved are four main procedures, namely basic research, planning, implementation and management of the programme (Allan and Bath 1980). In planning IWC programmes it is important that equal consideration be given to all three components, starting always with the target plant. These components must also not be considered in isolation. The outcome of any integrated control programme will only be as good as the combined understanding of all three main components. Finally, it is important to distinguish clearly between the four procedures, and to realize that they follow in sequence.

Although integrated pest management is widely practiced in agriculture, especially for the control of insects and pathogens (Huffaker 1980), this approach is not yet widely used against invasive alien plants in rangeland and natural vegetation. Examples of such targets for IWC are *Mimosa pigra*, which has recently become the subject of an IWC programme in Thailand (Harley et al 1985), and *Senecio jacobeae* in the United States of America (Isaacson 1976).

Address for correspondence: R L Kluge, Plant Protection Research Institute, Sub-division Weeds and Pesticides Research, Private Bag X5017, Stellenbosch, 7600.

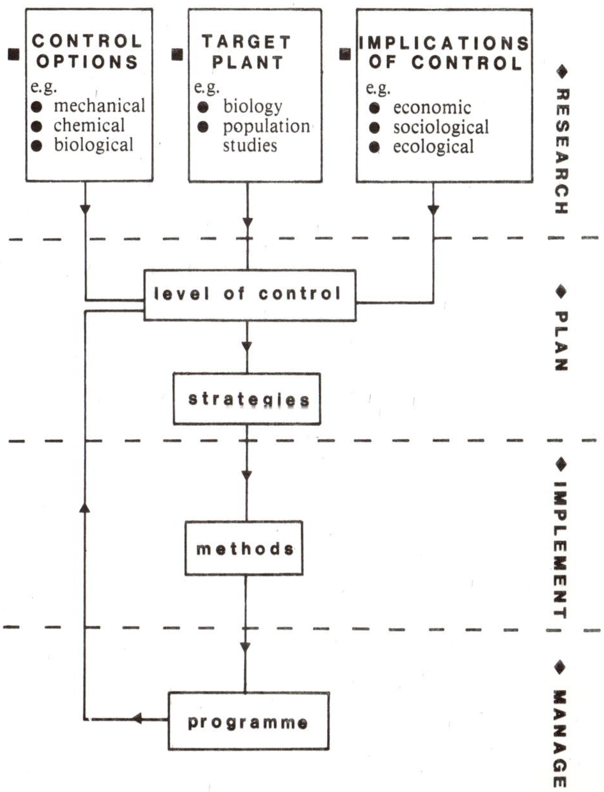

Figure 1. The components (■) and procedures (◆) of an integrated weed control programme.

South Africa has a history of alien weed control dating back to 1890 (Moran and Annecke 1979). Instances of integrated control were first recorded in the 1940's by Petty (1943), and continued to occur intermittently thereafter. Over the last decade, however, there has been a strong trend towards integrated control of weeds such as jointed cactus *Opuntia aurantiaca*, prickly pear *Opuntia ficus-indica*, silky hakea *Hakea sericea*, lantana *Lantana camara*, nassella tussock *Stipa trichotoma* and mesquite *Prosopis* species (Table 1). This trend has involved more control options, a greater degree of integration of the different options, and a greater regard for all the possible implications of control. Some general and specific points that have emerged from the history of the control of these plants, and which can profitably be considered in the planning of future IWC programmes against alien plants in South Africa, are highlighted in this paper.

General characteristics of past control programmes

In the past, there has been a preoccupation with the "control" option (Figure 1), probably because of the naive notion that invasive alien plants can be eradicated. With the trend towards IWC, more consideration is being given to those aspects of the biology of the weed and the weed populations which are relevant to control. More recently the broader implications of control, eg political and ecological, have

Table 1. A summary of all the main options presently available for the integrated control of six invasive alien plants in South Africa. Descriptions and other background information of the plants are given in Stirton (1978).

	Control options				
Weed	Chemical	Mechanical	Biological	Cultural	Other
Jointed cactus, *Opuntia aurantiaca*	Subsidized MSMA herbicide for spot spraying all plant parts. Compulsory control vigorously enforced	Hand collecting of small plants and loose cladodes	Cochineal, *Dactylopius austrinus*, for plants and cladodes mainly in hot and dry areas	Sound grazing management to conserve natural ground-cover as a preventative measure	None
Prickly pear, *Opuntia ficus-indica*	Subsidized MSMA stem-injection for large plants	None	Large & small plants adequately controlled by cochineal, *Dactylopius opuntiae* and the moth, *Cactoblastis cactorum*	None	Vast potential as a vegetable or for fruit, fodder and carminic acid production in areas where biological control has not been successful
Silky hakea, *Hakea sericea*	Tebuthiuron registered but rarely used	Felling combined with burning for clearing of mature infestations	Fruit-attacking weevil, *Erytenna consputa*, to limit spread. Fungal pathogen, *Colletotrichum gloeosporioides*, kills seedlings and large plants	Management of fire to reduce rate of invasion	Limited use in dried flower industry
Lantana, *Lantana camara*	Various herbicides for foliar and soil application	Slashing and grubbing of mature plants and hand weeding of seedlings	One seed-attacking and five defoliating insects to limit aggressiveness of mature plants	Management of natural vegetation or oversowing with pastures	None
Nasella tussock, *Stipa trichotoma*	Subsidized tetrapion herbicide for all infestations. Compulsory control being vigorously enforced	Grubbing of isolated plants	None	Sound grazing management and oversowing with suitable pasture plants	None
Mesquite, *Prosopis* spp	Tebuthiuron registered but rarely used	Felling together with stump treatment for large plants	Seed-attacking beetle being considered	None	Great potential in dry areas for pods as fodder and wood for fuel

begun to receive attention as will be shown below. Because of the urgency to control the target plant, research and programme planning have been neglected in the past so that all the available control options were not effectively used. Nor has there been much proper management in IWC.

Control progammes for all the invasive alien plants can be divided into short-and long-term phases. Short-term control is aimed at removing existing stands of the plant, usually by means of mechanical and chemical methods. Long-term control, usually involving biological agents or some form of vegetation management or both, is then aimed at maintaining the alien plant population at an acceptable level.

Understanding the target plant

Ideally an IWC programme should be based on the exploitation of as many vulnerable aspects of the biology of the target plant as possible. The features exploited in the control of *Hakea sericea*, for example, are its complete dependency on seed for reproduction (Neser 1968), the absence of a soil-stored seed bank, and the vulnerability of the thin-coated seed. Hakea seeds are normally protected from fire and seed predators in woody follicles which accumulate on the tree through its life. On the death of the plant, usually after fire, the follicles dehisce and the seeds are dropped en masse to germinate in the ash bed. According to Fugler (1979) mature three-metre tall trees may each drop as many as 5 000 seeds. With the fell-and-burn method that is used (Fenn 1980), the plants are first felled so that the seeds are released from the follicles. Six to eight months later the area is burnt, destroying any seedlings and seed that have not been consumed by predators (Richardson and Van Wilgen 1984). Recruitment can be markedly reduced in this way. The characteristics of hakea mentioned above similarly influenced the initial choice of seed-attacking insects for the biological control of *H sericea* (Neser 1968).

Although it is important initially to understand the biological characteristics of the individual plant, control is ultimately aimed at plant populations. Population parameters to be considered because of their likely significant influence on the outcome of the chosen control programme, are, for example, the distribution pattern of the plants, the population growth rate and the age and size structure of the population. Although age composition, size distribution and density of the alien plant population have been considered in the past, population parameters such as spatial distribution patterns and growth rate have been neglected. This is reflected in the history of *Opuntia aurantiaca* control as reviewed by Moran and Annecke (1979). The most significant improvements in the control of the weed occurred after 80 years of unsatisfactory control involving alternatively mechanical, chemical and biological methods, and resulted from population studies of jointed cactus, which strongly influenced the move towards integrated control. These studies showed that *O aurantiaca* populations were highly aggregated (Zimmermann 1977, 1979), and that this affected the efficiency of the control programme then current. Firstly, the spray teams readily located larger plants but often overlooked loose, unrooted cladodes and small rooted plants, which were responsible for perpetuating the problem. Furthermore, the occurrence of the biological control agent cochineal *Dactylopius austrinus* was positively correlated with aggregations of the plant, with the inadvertent consequence that the spray teams tended to eliminate cochineal, thereby reducing its contribution to the control of surviving plants in chemically treated areas.

Factors affecting the growth of the population were also investigated for the first time. It was found that the desiccation of small plants and isolated cladodes, together with mortality caused by *D austrinus* were the key mortality factors (Zimmermann 1981; Zimmermann and Moran 1982), and that in hot, dry areas these were sufficient to maintain satisfactory control without chemical control (Zimmermann et al this volume). Using this information it was possible to modify conventional control programmes by reserving certain areas for biological control only.

Past attempts at control were frustrated because of the regeneration from residual populations of the weed. For example, an attempt to control *O aurantiaca* in the 1950's over an area of 300 000 ha by the mechanical removal of the plants failed ultimately because of regeneration from the remaining joints and underground tubers (Moran and Annecke 1979). In the case of *Stipa trichotoma* (Joubert 1983) and *Prosopis* species (Harding 1978) the longevity of the seed in the seed banks is seen as a major obstacle to long-term control. In the absence of really effective methods to reduce the seed bank, the programme for *S trichotoma* has until now been aimed merely at preventing successful regeneration from the seed banks by establishing and maintaining a dense grass cover (Wells 1978). Recently, however, it has been found that burning nassella tussock reduces its seed bank by 20% (Joubert 1983). Another problem with *Prosopis* species is that the plants can sucker from buds occurring up to 50 mm below the ground (Hass et al 1973).

Variation among different populations of the same invasive species can have important implications for IWC. Although the geographical distribution of these weeds has always been well documented, the influence of climate and other ecological factors on the characteristics of the local weed population was not appreciated. Because of this, the early control programmes against *O aurantiaca,* for example, tended to be uniformly designed for application to the weed over its entire range. Only recently has it been recognized that jointed cactus occurs in eleven different vegetation types, in climatically distinct areas and that control programmes needed to be adjusted to the population processes operating in the different environments (see above; Zimmermann 1981; Zimmermann and Moran 1982). To utilize all the available control options optimally, different subpopulations of the target plant should therefore be specified with regard to population variables such as the growth rate, age- and size-class distributions and density, and control prescriptions should then be modified appropriately.

The use of different control options could also be influenced by the genotypic variation in weed populations. *Lantana camara* in South Africa, for example, comprises 50 genetically distinct taxa (Spies and Stirton 1982). Because of the specificity of the introduced insects to the different taxa, biological control is only effective in the control of the pink-flowered taxon. Similarly *H sericea* is accepted as being largely homogeneous in South Africa, except for an isolated population at Grahamstown. Although the plants in this population are attacked by the hakea fruit weevil *Erytenna consputa* (Kluge 1983) they are more resistant to attack by the fungal pathogen *Colletotrichum gloeosporioides* (S Neser personal communication) which is an important control agent in other areas (Morris 1982). The confusion surrounding the taxonomy of the introduced *Prosopis* species in South Africa also has important implications in deciding on control methods; two species (or varieties) tentatively identified as *Prosopis glandulosa* and *Prosopis velutina*, and what are possibly their hybrids, are invasive whereas populations referred to as *Prosopis pubescens* are not (Harding 1977).

Optimizing the use of control options

Before energy and money is expended on implementing various control methods, consideration should be given to any possible role played by local environmental factors as in the case of *O aurantiaca.* Local plant communities may also be important in the control of alien plants since it has been recognized that a reduction in plant diversity or cover through intensive or selective grazing predisposes veld to invasion by weeds such as *Opuntia aurantiaca* (Serfontein 1964) and *Stipa trichotoma* (Wells 1978). *S trichotoma* can, for example, be successfully controlled in areas that have been lightly or recently infested, by firstly removing the existing tussocks chemically or mechanically, and by then applying judicious grazing management aimed at maintaining a dense, vigorous stand of native, perennial grass species. The success of *Hakea sericea* regeneration from seed might also be considerably reduced by felling the plants at the start of the arid summers, thereby increasing the chances of predation by rodents before most of the seeds germinate in winter (Richardson and Van Wilgen 1984). Past evidence and predictive models (Richardson and Manders 1985) are also making it feasible to confidently leave certain *H sericea* infestations for control by the presumably native fungal pathogen, *Colletotrichum gloeosporioides* (Morris 1982).

The effectiveness of control can be improved through the combined use of existing control methods. Petty (1943) found that the destructive effect of cochineal *Dactylopius opuntiae* on *Opuntia ficus-indica* was greatly increased when cochineal-infested plants were first felled. This method was successfully employed in treating 450 000 ha of veld. Similarly the complex of introduced insects collectively contribute to the control of the pink-flowered form of lantana by suppressing the growth and vigour of the plant (Cilliers 1982), thereby increasing the feasibility of control using the existing mechanical and chemical methods. Dense infestations of *O aurantiaca* in certain areas are best controlled by firstly using *D austrinus to* reduce the density of the population to a level where chemical control becomes more economical (Zimmermann 1981).

IWC has also been improved by gaining a better insight into the different control options. Ever since 1946, after the initial impact it had on *O aurantiaca*, *D austrinus* was considered to be ineffective. However, in addition to the adverse effect which the spray teams had on the *D austrinus* populations (see above), the apparent ineffectiveness of cochineal was also found to be largely attributable to the limited dispersal of the wind-borne crawler stage (Moran et al 1982). Hand dispersal of infested cladodes was encouraged which greatly increased the effectiveness of the insect. Annecke et al (1969) and Zimmermann and Moran (1982) also showed that *D opuntiae* increased dramatically on *O ficus-indica* when it was protected from predators by the application of a dilute solution of DDT, which killed the predators but not the cochineal insects. Heavy cladode mortality resulted and the effectiveness of biological control was greatly enhanced (see also Zimmermann et al this volume).

By carefully analysing herbicide spraying methods for *O aurantiaca,* Zimmermann and Malan (1980) were able to suggest a procedure which reduced application costs by 47%. They found that with the existing methods small *O aurantiaca* plants were being heavily overdosed. Not only did this increase the herbicide spraying costs, but the productivity of the spray operators was limited by the time wasted in searching for small plants hidden in the vegetation. To improve the cost-benefit ratio of chemical control it was recommended that the sprayers double their walking pace and that they ignore the small plants and loose cladodes which could later be collected by hand and treated collectively in a dump. Similarly it has been recommended that small *O ficus-indica* plants (less than 14 cladodes) should rather be left to control by the established biological control agents than be treated chemically (Zimmermann and Malan 1980).

Compatibility of different control methods is important in enhancing the overall effectiveness of an integrated control program. The biological control of *Hakea sericea* has been accommodated in the mechanical clearing program of hakea by maintaining breeding colonies of *Erytenna consputa* in specially protected 12-15 ha hakea reserves (Kluge and Richardson 1983). These reserves are located on the perimeter of the 800-1 500 ha blocks which have been cleared mechanically and then burnt. The recolonization of regenerating *H sericea* in these burnt blocks by the fruit weevil has also been accelerated by making supplementary releases of the weevil.

One of the important features of the integrated control of *Stipa trichotoma* is the selectivity of the herbicide tetrapion, which makes compatibility between chemical and cultural control possible. When applied at a rate of 880 g of active ingredient per hectare, tetrapion not only kills the existing tussocks, but continues to kill seedlings for as long as 12 months after application. Furthermore valuable non-target pasture grasses such as *Themeda triandra* and *Andropogon appendiculatus* are less severely affected thus enhancing the recovery of the native grassland (Viljoen and Harding 1985).

Apart from the selective elimination of *Dactylopius austrinus* colonies by spray teams in *Opuntia aurantiaca* clumps (see above), it was also found that *D austrinus* were killed by the paraffin carrier in the Tordon herbicide. The switch to the water-based monosodium methylarsenate (MSMA), which has no insecticidal action, has helped to retain the beneficial effects of biological control (Zimmermann and Moran).

Existing control methods can sometimes be substituted by less costly alternative forms of control. Consideration is being given to replacing the expensive hand weeding of *H sericea*, which normally follows the initial fell-and-burn procedure, with the large-scale introductions of *E consputa* and the fungal pathogen *C gloeosporioides*. This could reduce labour costs in hakea control by up to 30%.

A controversial control option for alien weed control is that of changing the status of the plant as a weed by capitalizing on any useful properties which it may have. This could lead to more favourable perception of the plant or to a reduction of its invasion through exploitation, or both. Utilization is seen as a potentially important solution for the 'control' of the remaining *O ficus-indica* infestations in those areas where biological control has not been successful. Although the sale of fruits already provides an income for many poor people (Annecke and Moran 1978), the fruit crop is underexploited because current legislation prohibits the transport and marketing of the fruit. South Africa also has the potential to become a world leader in the production of carminic acid which presently has an international market worth R20 million annually. The mass-rearing of another species of cochineal *Dactylopius coccus* on prickly pear is being investigated for this purpose. Young *O ficus-indica* cladodes may also be used as a vegetable. In Mexico, where this is known as "nopalitos", 20 000 ha of prickly pear are under cultivation for this purpose. 'Nopalitos" could be an abundant, inexpensive and nourishing food source in many areas of South Africa. Prickly pear is also used by many farmers in South Africa as emergency fodder for stock during drought.

Considering the implications of control measures

An important consideration for good IWC is that the weed and its control should be seen in the context of the environment. For example, even though the fell-and-burn method has been crucial to the success of the short-term control of *Hakea sericea*, increasing concern is being expressed about the possible detrimental effect that these intense fires may have on the regeneration of the indigenous plant species. It has been suggested that *H sericea* should be burnt standing, or that fire prediction models should be used to select the most favourable conditions for burning when stands are slashed (Richardson and Van Wilgen 1985).

Prosopis species occur as weeds in arid regions with a mean annual rainfall of 150 mm. Although the open country where prosopis grows makes it ideally suited for the aerial application of herbicides, the associated mortality of slow-growing, natural plants would result in large-scale denudation of the veld which would be revegetated very slowly. Other means of control such as biological control and the utilization of the weed are therefore being investigated.

After the successful chemical control of lantana the plants are often replaced by the even more invasive triffid weed *Chromolaena odorata*. Control programmes against lantana should therefore utilize methods that prevent this unwanted succession, such as defoliating biological control agents which allow increased light penetration and thereby encourage the establishment of grasses and other useful plants. The use of slow-acting selective herbicides is also being considered.

Determining the required level of control

A definition of the required level of control is important in an integrated control programme. Firstly it determines the way in which the problem is perceived and therefore the whole approach to control, and secondly it sets an objective for the control programme. For many years it was thought that *O aurantiaca* could and should be eradicated so that light infestations were regarded as being serious. This notion was based on the potential threat of the weed rather than on realistic considerations. Now that it has been accepted that eradication is unattainable, and that the crisis infestations that were once common prior to the release of *Dactylopius austrinus* and the moth *Cactoblastis cactorum* cannot recur, there is a more rational approach to the control of this weed. In the light of this experience it would therefore seem unlikely that the integrated control of nassella tussock which is now being attempted can be successfully implemented until the present objective for its control is changed from eradication to a feasible target density level.

Because of the difficulty in determining an acceptable threshold level for the control of invasives, most of the programmes in Table 1 have poorly defined objectives. For *O aurantiaca* this is seen as a serious shortcoming in the formulation of an integrated control programme. Because most of the land affected by the invasives has little or no agricultural value and its other values cannot be expressed satisfactorily in commercial terms, thresholds based on the profitability of control, such as the "economic threshold" and "optimal level threshold" (Mumford and Norton 1984), are of little use in setting objectives. Furthermore, the large-scale subsidization of many of these programmes makes nonsense of economic thresholds. The costs of herbicides applied to many farms in the Karoo, for example, are in excess of their market value. The objective should therefore be based on a threshold density of the plant, either at a level where it is tolerable, or where it ceases to be a threat. Predictions of the latter will have to be based on a knowledge of the growth rate, growth regulatory factors and the nature of the equilibrium (ie the size and pattern of fluctuations around the average abundance) of the weed population. The value of such a threshold is not so much to know when to apply control but to know when to stop. Eradication is rarely a feasible target.

Deciding on a level of control can sometimes be difficult where there is a conflict of interest regarding the status of the weed, as in the case of *Prosopis* species. Apart from being invasive these species are important in many areas of the northern Cape Province for their nutritious pods which are used as fodder. Mesquite is also a source of fuel and supplies shade for animals in an area where few other trees will grow. For all its drawbacks some farmers are reluctant to lose the benefits of mesquite by having it drastically controlled. As a possible solution to this dilemma it has been proposed that host-specific, seed-attacking bruchid beetles should be introduced. It is intended that these will suppress the reproductive potential of the plant without interfering with the utilization of the nutritious pods and the other beneficial attributes of the plant.

Planning the integrated control programme

Planning should follow once objectives have been decided on, and information about the weed, the control options and all the possible implications of control has been collected. Many of the shortcomings of the past could be avoided here if the planning were to be done as objectively as possible by, for example, using a systems analysis approach as shown in Figure 2 in which all the possible options for each situation are considered. Techniques such as the decision theory model of Mumford and Norton (1984) should also be considered, since they are designed to arrive at the best possible decision, despite a large degree of uncertainty, by considering the probabilities of the various outcomes.

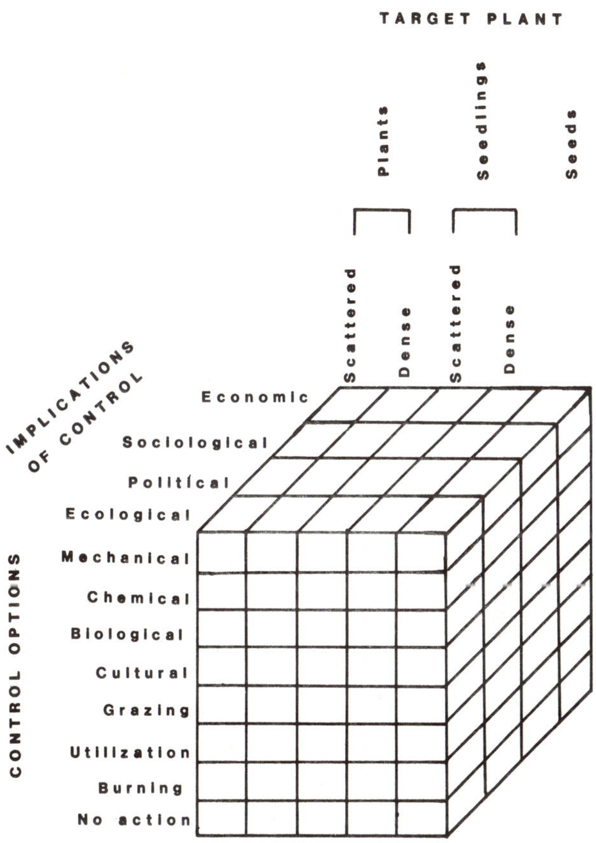

Figure 2. An example of the type of conceptual matrix that could be used to increase objectivity in the planning of integrated weed control programmes.

Implementing the integrated control programme

A potential weakness of integrated control is that, because of the many facets involved, the end result depends to a large extent on the way in which the programme is implemented. Inadequate extension, ineffective application of legislation, and a lack of motivation, often induced by well-intentioned State assistance, are limiting the implementation of integrated control of *Opuntia aurantiaca* (Moran and Annecke 1979). For example, the 80% subsidy on the purchase price of herbicide discourages landowners from considering alternative control methods. A substantial decrease in this subsidy would possibly make integrated control an economically attractive proposition, and encourage landowners to invest in better control methods.

Because of the degree of sophistication involved in integrating various control methods for *O aurantiaca*, it is imperative that landowners be given more insight into the factors that regulate jointed cactus and *Dactylopius austrinus* populations. This will enable them to make decisions about the type of control to be implemented. Alternatively, the number of weed inspectors will have to be increased, and their training upgraded so that more decisions can be left to them. The successful integration of the biological control of *Hakea sericea* into the existing management plans for the mountain catchment areas (Kluge and Richardson 1983) can largely be attributed to the fact that it was mostly trained foresters that were involved in implementing the control programme.

Managing the integrated control programme

An integrated control programme is an ongoing process and constant management of the constituent parts is a prerequisite for success. Regular monitoring is required to evaluate the outcome of the various control decisions, and to assess the efficacy of the different control methods. The value of such work is clear from the benefits derived from the careful monitoring of *D austrinus* populations and the herbicide spraying on *O aurantiaca* (Zimmermann 1979) (see above).

The future of integrated control of invasive alien plants

Despite the many instances of integration of control methods that have been cited above, it is questionable whether integrated control is yet being rationally practised in South Africa. A possible exception is the *Hakea sericea* control programme of the Directorate of Forestry, where both the fell-and-burn method and biological control are being incorporated into the overall conservation plan for the mountain catchments (Fenn 1980). Although comprehensive plans for the integrated control of *Opuntia aurantiaca* have been drawn up (Zimmermann 1982), these have not yet been implemented, for various of the reasons mentioned above, eg the lack of motivation on the part of the land-owners. Sufficient information is available to be able to make a start in designing integrated control programmes for *Stipa trichotoma*, *O ficus-indica* and *Prosopis* species. Prospects for the integrated control of *Lantana camara* remain poor, largely because of the lack of knowledge on, for example, its population biology.

It is clear from all these case histories that because of the many variables involved, an integrated control strategy is almost certain to evolve in any programme aimed at combating invasive plants. This should be borne in mind from the outset to ensure a methodical and disciplined approach in the planning of IWC programmes along the lines suggested by Figures 1 and 2. The future progress of integrated control would also benefit greatly if such objectively designed programmes could then be documented and published. Not only will this enhance their implementation, but the outcome of past control decisions could be evaluated as the programme progresses.

Summary

In summary it is worth emphasizing that successful integration of control methods requires an understanding of the biology of the weed at both an individual and population level. The IWC programme should be modified to accommodate the variation between different populations of the target plant especially as affected by variations in the environment. Any programme should have a well-defined objective ie a level of control. Where necessary this should be redefined for different areas. Optimal integration of the different control options depends on a good insight into the mode of action of each option. The outcome of an IWC programme should be seen in the context of the environment, and with due regard to all the possible consequences that the control measures may have. Implementation and management are crucially important in the ultimate success of an IWC programme.

An integrated approach is almost certain to evolve in any programme aimed at controlling alien invasives. Planning should therefore proceed with this in mind.

Allan G E and Bath J E 1980. The conceptual and institutional aspects of integrated pest management. Bioscience 30, 658-664.

Annecke D P, Karny M and Burger W A 1969. Improved biological control of the prickly pear, *Opuntia megacantha* Salm-Dyck, in South Africa through the use of an insecticide. Phytophylactica 1, 9-13.

Annecke D P and Moran V C 1978. Critical reviews of biological pest control in South Africa. 2. The prickly pear, *Opuntia ficus-indica* (L) Miller. Journal of the Entomological Society of Southern Africa 41, 161-188.

Cilliers C J 1982. An evaluation of the effects of imported insects on the weed *Lantana camara* L. in South Africa. PhD Thesis, Rhodes University, Grahamstown. 107 pp.

Fenn J A 1980. Control of hakea in the Western Cape. Proceedings of the 3rd National Weeds Conference of South Africa, Pretoria 1979. pp 167-173.

Fugler S R 1979. Some aspects of the autecology of three *Hakea* species in the Cape Province, South Africa. MSc Thesis, University of Cape Town, Cape Town.

Harding G B 1977. Survey of *Prosopis*. Unpublished Department of Agriculture Report, Plant Protection Research Institute, Pretoria.

Harding G B 1978. Mesquite. In: Stirton C H (ed) Plant invaders - beautiful but dangerous. The Department of Nature and Environmental Conservation of the Cape Provincial Administration, Cape Town. pp 128-131.

Harley K L S, Miller I L, Napompeth B and Thamasara S 1985. An integrated approach to the management of *Mimosa pigra* L in Australia and Thailand. Proceedings of the 10th Conference of the Asian Pacific Weed Science Society, Chiang Mai, Thailand 1985.

Hass R H, Meyer R E, Scifris C J and Brock J H 1973. Growth and development of mesquite. Research Monograph No 1, Texas A and M University. pp 10-19.

Huffaker C B 1980. New technology of pest control. John Wiley and Sons, New York.

Isaacson D L 1976. The role of biological agents in the integrated control of tansy ragwort. In: Freeman T E (ed) Proceedings of the 4th International Symposium on the Biological Control of Weeds. Gainesville, Florida. pp 189-192.

Joubert D C 1983. The soil seed bank under nasella tussock infestations at Boschberg. South African Journal of Plant and Soil 1, 1-3.

Kluge R L 1983. The hakea fruit weevil, *Erytenna consputa* Pascoe (Coleoptera: Curculionidae), and the biological control of *Hakea sericea* Schrader in South Africa. PhD Thesis, Rhodes University, Grahamstown.

Kluge R L and Richardson D M 1983. Progress in the fight against hakea. Veld and Flora 69, 136-138.

Moran V C and Annecke D P 1979. Critical reviews of biological pest control in South Africa. 3. The jointed cactus, *Opuntia aurantiaca* Lindley in South Africa. Journal of the Entomological Society of Southern Africa 42, 229-329.

Moran V C, Gunn B H and Walter G H 1982. Wind dispersal and settling of first-instar crawlers of the cochineal insect *Dactylopius austrinus* (Homoptera: Coccoidea: Dactylopiidae). Ecological Entomology 7, 409-419.

Morris M J 1982. Gummosis and die-back of *Hakea sericea* in South Africa. Proceedings of the 4th National Weeds Conference of South Africa, Pretoria, 1981. pp 51-54.

Mumford J D and Norton G A 1984. Economics of decision making in pest management. Annual Review of Entomology 29, 157-174.

Neser S 1968. Studies on some potentially useful insect enemies of the needle-bush, (*Hakea* spp Proteaceae). PhD Thesis, Australian National University, Canberra.

Petty F W 1943. Prickly-pear eradication by insects and felling of plants. Farming in South Africa 18, 743-746.

Richardson D M and Manders P T 1985. Predicting pathogen-induced mortality in *Hakea sericea* (Proteaceae), an aggressive alien plant invader in South Africa. Annals of Applied Biology 106, 243-254.

Richardson D M and Van Wilgen B 1984. Factors affecting the regeneration success of *Hakea sericea*. South African Journal of Forestry 131, 63-68.

Richardson D M and Van Wilgen B (in press). The effects of fire in felled *Hakea sericea* and natural fynbos and implications for weed control in mountain catchments. South African Forestry Journal.

Serfontein J 1964. A prickly problem: control jointed cactus with correct veld management. Farming in South Africa 40, 23-24.

Spies J J and Stirton C H 1982. Meiotic studies of some South African cultivars of *Lantana camara* (Verbenaceae). Bothalia 14, 101-111.

Stirton C H (ed) 1978. Plant invaders beautiful but dangerous. The Department of Nature and Environmental Conservation of the Cape Provincial Administration, Cape Town.

Viljoen B and Harding G B 1985. Control of nassella. Unpublished Department of Agriculture Report, Plant Protection Research Institute, Pretoria.

Wells M J 1978. Nassella tussock. In: Stirton C H (ed) Plant invaders - beautiful but dangerous. The Department of Nature and Environmental Conservation of the Cape Provincial Administration, Cape Town. pp 140-143.

Zimmermann H G 1977. A sampling system for field infestations of jointed cactus, *Opuntia aurantiaca* Lindley. Proceedings of the 2nd National Weeds Conference of South Africa, Stellenbosch. pp 203-220.

Zimmermann H G 1979. Herbicidal control in relation to distribution of *Opuntia aurantiaca* Lindley and its effects on cochineal populations. Weed Research 19, 89-94.

Zimmermann H G 1981. The ecology and control of *Opuntia aurantiaca* in South Africa in relation to the cochineal insect *Dactylopius austrinus*. PhD Thesis, Rhodes University, Grahamstown.

Zimmermann H G 1982. Control of prickly pear. Farming in South Africa. Weeds series. BII/1982. Department of Agriculture, Pretoria.

Zimmermann H G and Malan D E 1980. The role of imported natural enemies in suppressing regrowth of prickly pear, *Opuntia ficus-indica*, in South Africa. Proceedings of the 5th International Symposium on the Biological Control of Weeds, Brisbane. pp 375-381.

Zimmermann H G and Moran V C 1982. The ecology and management of cactus weeds in South Africa. South African Journal of Science 78, 314-320.

Authors

Abbott S K, South African Forestry Research Institute, P O Box 727, Pretoria, 0001

Appleton C C, Department of Zoology, University of Natal P O Box 375, Pietermaritzburg, 3200

Ashton P J, National Institute for Water Research, CSIR, P O Box 395, Pretoria, 0001

Balsinhas A A, Botanical Research Institute, Private Bag X101, Pretoria, 0001

Bate G C, University of Port Elizabeth, P O Box 1600, Port Elizabeth, 6000

Breytenbach G J, Saasveld Forestry Research Centre, Private Bag X6515, George, 6530

Brooke R K, Percy FitzPatrick Institute of African Ornithology, University of Cape Town, Rondebosch, 7700

Brown C J, Department of Agriculture and Nature Conservation, Private Bag 13306, Windhoek, 9000 South West Africa/Namibia

Bruton M N, JLB Smith Institute of Ichthyology, Private Bag 1015, Grahamstown, 6140

Cilliers C J, Plant Protection Research Institute, Private Bag X134, Pretoria, 0001

Cooper J, Percy FitzPatrick Institute of African Ornithology, University of Cape Town, Rondebosch, 7700

Cooper K H, Wildlife Society of Southern Africa, 100 Brand Road, Durban, 4001

de Villiers A L, Jonkershoek Nature Conservation Station, Private Bag X5014, Stellenbosch, 7600

Deacon J, Department of Archaeology, University of Stellenbosch, Stellenbosch, 7600

Dean S J, Department of Botany, University of Cape Town, Rondebosch, 7700

Erasmus D J, Plant Protection Research Institute, Private Bag X9059, Pietermaritzburg, 3200

Ferrar A A, CSIR, Foundation for Research Development, P O Box 395, Pretoria, 0001

Geldenhuys C J, Saasveld Forestry Research Centre, Private Bag X6515, George, 6530

Gubb A A, McGregor Museum, P O Box 316, Kimberley, 8300

Harding G B, Plant Protection Research Institute, P O Box 330, Uitenhage, 6230

Henderson L, Botanical Research Institute, Private Bag X101, Pretoria, 0001

Hoffmann J H, Department of Zoology, University of Cape Town, Rondebosch, 7700

Holmes P M, Department of Botany, University of Cape Town, Rondebosch, 7700

Jackson P B N, JLB Smith Institute of Ichthyology, Private Bag 1015, Grahamstown, 6140

Joffe H, Botanical Research Institute, Private Bag X101, Pretoria, 0001

Kluge R L, Plant Protection Research Institute, Private Bag X5017, Stellenbosch, 7600

Knight R S, Percy FitzPatrick Institute of African Ornithology, University of Cape Town, Rondebosch, 7700

Kruger F J, South African Forestry Research Institute, P O Box 727, Pretoria, 0001

Le Roux R J, Directorate of Forestry, Department of Environment Affairs, Private Bag X447, Pretoria, 0001

Lloyd P H, Jonkershoek Nature Conservation Station, Private Bag X5014, Stellenbosch, 7600

Macdonald I A W, Percy FitzPatrick Institute of African Ornithology, University of Cape Town, Rondebosch, 7700

Moran V C, Faculty of Science, University of Cape Town, Rondebosch, 7700

Musil C F, Botanical Research Institute, Private Bag X16, Rondebosch, 7700

Musil K J, Botanical Research Institute, Private Bag X101, Pretoria, 0001

Neser S, Plant Protection Research Institute, Private Bag X134, Pretoria, 0001

Powrie F J, Percy FitzPatrick Institute of African Ornithology, University of Cape Town, Rondebosch, 7700

Poynton R J, South African Forestry Research Institute, P O Box 727, Pretoria, 0001

Pressinger F M, Botanical Research Institute, Private Bag X16, Rondebosch, 7700

Richardson D M, Jonkershoek Forestry Research Centre, Private Bag X5011, Stellenbosch, 7600

Rutherford M C, Botanical Research Institute, Private Bag X16, Rondebosch, 7700

Shaughnessy G, Australian National Parks and Wildlife Service, GPO Box 636, Canberra ACT 2602, Australia

Siegfried W R, Percy FitzPatrick Institute of African Ornithology, University of Cape Town, Rondebosch, 7700

Van As J, Department of Zoology, Rand Afrikaans University, P O Box 524, Johannesburg, 2000

Van Hoepen E, Botanical Research Institute, Private Bag X101, Pretoria, 0001

Van Wilgen B W, Jonkershoek Forestry Research Centre, Private Bag X5011, Stellenbosch 7600

Versfeld D B, Jonkershoek Forestry Research Centre, Private Bag X5011, Stellenbosch, 7600

Weiss P W, Woden College of Technical and Further Education, P O Box 666, Woden ACT 2606, Australia

Wells M J, Botanical Research Institute, Private Bag X101, Pretoria, 0001

Whiting B H, Department of Botany, University of Port Elizabeth, P O Box 1600, Port Elizabeth, 6000

Zimmermann H G, Plant Protection Research Institute, P O Box 330, Uitenhage, 6230

Referees

Dr A H Arthington, Griffith University, Nathan, Australia
Prof H Aschmann, University of California, Riverside, USA
Dr D S Ashton, University of Melbourne, Parkville, Australia
Dr P Bridgewater, Bureau of Flora and Fauna, Canberra, Australia
Dr P Christensen, Department of Conservation and Land Management, Perth, Australia
Dr B A Colvin, Bowling Green State University, Ohio, USA
Prof W R Courtenay, Florida Atlantic University, Boca Raton, USA
Dr R M Cowling, University of Cape Town, South Africa
Dr J M Cullen, CSIRO, Montpellier, France
Dr E S Delfosse, CSIRO, Canberra, Australia
Dr C Field, Carnegie Institution of Washington, Stanford, USA
B J Fox, University of New South Wales, Kensington, Australia
Dr G Fryer, Freshwater Biological Association, Ambleside, England
Professor E R Fuentes, Pontifical Catholic University of Chile, Santiago, Chile
Prof R D Goeden, University of California, Riverside, USA
Dr D J Greathead, Imperial College, Ascot, England
Prof J P Grime, University of Sheffield, England
Dr R H Groves, CSIRO, Canberra, Australia
Prof D R Harris, University of London, England
Dr P Harris, Agriculture Canada, Regina, Canada
Prof J D Hewlett, University of Georgia, Athens, USA
Dr M W Holdgate, Department of the Environment, Cambridge, England
Dr F M Jaksic, Pontifical Catholic University of Chile, Santiago, Chile
Dr G W Johnstone, Department of Science and Technology, Kingston, Tasmania
Prof R N Mack, Washington State University, Pullman, USA
Dr D Mitchell, CSIRO, Griffith, Australia
Dr A M Mortimer, University of Liverpool, England
Prof P B Moyle, University of California, Davis, USA
Dr I R Noble, Australian National University, Canberra, Australia
Dr R O'Connor, British Trust for Ornithology, Tring, England
Prof G Orshan, Hebrew University of Jerusalem, Israel
Prof D Pimentel, Cornell University, Ithaca, USA
Dr R J Poynton, SA Forestry Research Institute, Pretoria, South Africa
Dr H A Roberts, National Vegetable Research Station, Wellesbourne, England
Dr G Shaughnessy, Australian National Parks and Wildlife Service, Canberra, Australia
Dr R Snaydon, University of Reading, England
Prof R L Specht, University of Queensland, St Lucia, Australia
Prof N M Tainton, University of Natal, Pietermaritzburg, South Africa
Dr F von Breitenbach, SA Forestry Research Institute, Pretoria, South Africa
Mr C J Ward, University of Durban-Westville, Durban, South Africa
Dr P W Weiss, Woden College, Woden, Australia
Prof M H Williamson, University of York, England
Prof P H Zedler, San Diego State University, USA
Prof Dr H Zwölfer, Universitat Bayreuth, West Germany

Plus two referees who prefer to remain anonymous

Biological index

This index is set out as a guide for the reader, not only to locate the mention of a species in the text but also to assist in its correct indentification. It does not assume to be a completely authoritative and accurate taxonomic list of all the species referred to, especially in respect of those taxa mentioned in a peripheral sense to the main topics discussed. Where taxa are not identified to species within the text they are listed here by the generic term only.

Abies 37
Acacia 30, 32, 75, 95, 111, 115–116, 122, 126, 127, 135, 148, 150–152, 153, 165, 190, 194, 239, 242, 280, 281
Acacia americana [= *Prosopis* sp?] 29
Acacia baileyana F. Muell. 34, 167
Acacia cultriformis A. Cunn. 147, 167
Acacia cyclops A. Cunn. ex G. Don 34, 38, 39–42, 79, 81, 83, 85–86, 124, 146, 148–152, 157, 161, 164, 172–176, 178, 190–195, 227, 230, 233, 239–241, 286, 290
Acacia dealbata Link 31, 34, 110, 112–115, 121, 124, 146, 157, 161, 165, 191, 195
Acacia decurrens Willd. 34, 112, 124, 161
Acacia erioloba E. Mey 95
Acacia floribunda Sieber 264
Acacia hereroensis Engl. 95
Acacia implexa Benth. 164
Acacia karroo Hayne 7, 95, 129, 280
Acacia longifolia (Andr.) Willd. 34, 38–39, 42, 81–82, 123, 124, 126, 148, 160, 164, 190–191, 193–194, 196, 241, 264, 286–291
Acacia mangium Willd. 32
Acacia mearnsii De Wild 15, 30, 34, 38, 40–42, 75, 81–82, 110, 112–115, 121–124, 126, 129, 146, 149, 151, 161, 165, 195–196, 239, 242, 244, 276, 280, 290
Acacia melanoxylon R. Br. 31–32, 34, 39–42, 75, 112, 121–124, 126–129, 146, 161, 164–165, 178, 193–194, 196, 241, 259, 275–281, 286, 289–290
Acacia mellifera (Vahl) Benth. subsp *detinens* (Burch.) Brenan 95
Acacia nilotica (L.) Willd. ex Del. 164
Acacia podalyriifolia A. Cunn. ex G. Don 32, 34, 147, 167
Acacia pycnantha Benth. 34, 38, 40–42, 124, 194
Acacia saligna (Labill.) Wendl. 34, 38–42, 79, 81–83, 85–86, 124–125, 146, 148–149, 160, 164–165, 190–195, 239–242, 286, 290
Acacia schweinfurthii Brenan & Exell 129
Acacia tortilis (Forssk.) Hayne subsp *heteracantha* (Burch.) Brenan 95
Acacia victoriae Benth. 164
Acaena anserinifolia (J R & G Forster) [= *Acaena sanguisorbae*] 262
Acaena magellanica (Lamarck) 140
Acaena sanguisorbae Vahl [see *Acaena anserinifolia*]
Acanthospermum 23, 160
Acanthospermum australe (Loefl.) Kuntze 31, 105
Acanthospermum glabratum (DC.) Wild 31
Acanthospermum hispidum DC. 31, 105
Achyranthes aspera L. 23
Achyranthes sicula (L.) All. 30
Acomys subspinosus (Waterhouse) 233
Aceria 286
Acridotheres cristatellus (L.) 65
Acridotheres tristis (L.) 65, 108, 140
Actinidia 178
Addax nasomaculatus (Blainville) 70, 108
Aepyceros melampus (Lichtenstein) 70, 108
Aeschynomene indica L. 30
Aesculus 178
Agapornis cana (Gmelin) 66
Agapornis nigrigenis Sclater 67
Agapornis roseicollis (Vieillot) 66
Agathis australis (D. Don.) Salisb. 121, 124
Agave 105, 124
Agave americana L. 99
Ageratina adenophora (Spreng.) King & Robinson [see *Eupatorium adenophorum*]
Agonis flexuosa (K. Spreng.) Schauer 291
Agrimonia odorata Mill. 29
Agriotherium africanum 3
Agrostis avenacea Gmelin 31
Agrostis magellanica Lam. 181, 184
Agrostis stolonifera L. 140–141, 181, 184, 186
Ailanthus altissima (Mill.) Swingle 30, 291
Aira caryophyllea L. 30
Albizia [sensu *A lophantha* see *Paraserianthes lophantha*] 37, 75, 281
Albizia lophantha (Willd.) Benth. [= *Paraserianthes lophantha*] 30, 38–42, 81, 86, 124–125, 161, 165, 289
Alcelaphus lichtensteinii (Peters) 70
Alectoris chukar (Gray) 67, 140
Alestes lateralis Boulenger 48
Aleurites fordii Hemsl. 32
Algarobius prosopis (LeConte) 289
Alhagi camelorum Fisch 31, 105, 161, 291
Alopochen aegyptiacus (L.) 56
Alternanthera pungens H.B.K. 23, 30, 105, 160
Alternanthera sessilis (L.) DC. 105
Amandava amandava (L.) 67
Amaranthus blitoides S. Wats. 105
Amaranthus cruentus L. 105
Amaranthus deflexus L. 30
Amaranthus hybridus L. 105
Amaranthus viridis L. 31, 105
Ambrosia artemisiifolia L. 31
Amelanchier 178
Amitermes 236
Ammotragus lervia (Pallas) 70, 108
Ampelopsis 178
Anacardium occidentale L. 124, 178
Anagallis arvensis L. 105
Anas platyrhynchos L. 48, 56
Anas platyrhynchos x *Anas undulata* 56

Anas rubripes Brewster 48
Anas undulata Dubois 56
Anchusa azurea Mill. 31
Andromeda 178
Andropadus importunus (Vieillot) 119, 172
Andropogon appendiculatus Nees 299
Anguilla mossambica Peters 254
Anoplolepis custodiens (Smith) 231
Anthephora pubescens Nees 95
Anthoxanthum odoratum L. 31
Antidorcas bondi 4
Antidorcas marsupialis (Zimmermann) 70
Antilope cervicapra (L.) 70
Apiosoma pisciola Blanchard 53
Apium graveolens L. 105
Apium leptophyllum Pers. [= *Ciclospermum leptophyllum* (Pers.) Sprague] 23, 105
Arachis hypogaea L. 22
Aratinga jandaya (Gmelin) 67
Aratinga pertinax (L.) 67
Aratinga weddellii (Deville) 67
Araucaria cunninghamii D. Don 124–125
Araucaria excelsa (Lamb.) R. Br. 121
Araujia sericifera Brot. 32
Arbutus 178
Archlagocheirus funestus (Thomson) 270
Arctocephalus pusillus (Schreber) 135
Arctostaphylos 178
Ardisia 178
Argemone mexicana L. 23, 160
Argemone subfusiformis G.B. Ownbey 99–100, 101, 105
Argulus japonicus Thiele 48, 53, 54, 252
Aristida 115
Aristida junciformis Trin. & Rupr. var *junciformis* 23
Aristolochia elegans Mast. 32
Aristotelia 178
Aronia 178
Arundo donax L. 35, 107, 110, 112, 115, 124
Asimina 126
Aster subulatus [misidentification, = *A squamatus* (Spreng.) Hieron] 105
Atriplex muelleri Benth. 31
Atriplex nummularia Lindl. 105, 281
Atriplex semibaccata R. Br. 31, 106
Aucuba 178
Austroglanis gilli (Barnard) 56
Austroglanis sclateri (Boulenger) 48
Avena byzantina C. Koch 31
Avena sativa L. 107
Axis porcinus [see *Cervus porcinus*]
Aythya fuligula (L.) 48
Azedirachta indica A. Juss. 166
Azara 178
Azolla filiculoides Lam. 32, 34, 160, 167, 190, 192, 194, 196, 248–249
Azolla nilotica Decne. ex Mett. 248

Bambusa balcooa Roxb. 124
Barbus 55, 254–255
Barbus aeneus (Burchell) 48, 253–254
Barbus anoplus M. Weber 48
Barbus barnardi Jubb 48
Barbus bifrenatus Fowler 48
Barbus burchelli A. Smith 56
Barbus burgi Boulenger [= *B bergi*, invalid] 55–56
Barbus calidus Barnard 56
Barbus eutaenia Boulenger 48
Barbus gurneyi Günther 48, 50, 51
Barbus kimberleyensis Gilchrist & Thompson 54
Barbus marequensis A. Smith 55
Barbus mattozi Guimaraes 48, 55
Barbus natalensis Castelnau 48
Barbus phlegethon Barnard 56
Barbus tangandensis Jubb 48
Barbus treurensis Groenewald 56
Barbus trevelyani Günther 56
Bauhinia variegata L. 124
Bedeva paivae (Crosse) 48
Berberis 178
Bidens 23, 99
Bidens bipinnata L. 30, 160, 163
Bidens biternata (Lour.) Merr. & Sherff 29–30, 100, 106
Bidens formosa (Bonato) Sch. Bip. 31
Bidens pilosa L. 29, 106, 124, 160, 163
Biomphalaria 48, 252
Blackiella inflata (F. Muell.) Aell. 30, 106
Blighia 119
Boerhavia diffusa L. 30, 106
Boerhavia pterocarpa Watts [= *B cordobensis* Kuntze] 106
Borassus aethiopum Mart. 23
Bos 12, 108
Bos taurus L. 4, 9, 11
Boselaphus tragocamelus (Pallas) 70, 108
Bostrychia hagedash (Latham) 85, 233
Bothriocephalus acheilognathi Yamaguthi 48, 53–54, 252
Brachiaria eruciformis (J. E. Sm.) Griseb. 23
Brachiaria nigropedata (Fical. & Hiern) Stapf 95
Brachystegia spiciformis Harms 191
Brassica juncea (L.) Czern. 31
Brayulinea densa (Willd.) Small 106
Briza maxima L. 107
Bromus diandrus Roth 31, 107
Bromus japonicus Thunb. 107, 147
Bromus unioloides H.B.K. 107, 183
Bruchophagus 289
Bubalus arnee [= *B bubalis*] 108
Bubalus bubalis (L.) 70, 108
Bulinus truncatus Audouin 252
Buxus macowanii Oliv. 190

Cacatua sulphurea (Gmelin) 67
Cactoblastis cactorum (Berg) 101, 115, 262, 264–265, 270–273, 296, 300
Caesalpinia 111
Caesalpinia decapetala (Roth) Alston 31, 112, 124, 127–129, 161

Caesalpinia gilliesii (Wall. ex Hook.) Benth. 105
Cairina moschata L. 48
Cajanus cajan (L.) Millsp. 8, 13
Callicarpa 178
Callitris 39
Camelus dromedarius L. 70, 108
Canis familiaris L. 4, 9, 11, 68, 108
Canis mesomelas Schreber 108, 218
Canna 165
Canna indica L. 29
Cannabis sativa L. 22–23, 105
Capra hircus L. 4, 9, 11–12, 102, 108
Capra ibex L. 108
Capsella bursa-pastoris (L.) Medik. 106
Carassius auratus (L.) 48, 50, 51, 55, 205
Carassius carassius L. 48, 50
Carcinus maenas Lamark 48
Cardiospermum grandiflorum Schwartz 106
Carduelis carduelis (L.) 67
Carduus 263
Carduus marianus L. [see *Silybum marianum*]
Carduus nutans L. 263
Carica benghalensis 124
Carpobrotus edulis (L.) L. Bol. 39
Carposina autologa Meyr. 288
Cassia 161
Cassia bicapsularis L. 32, 112, 124
Cassia coluteoides Collad. 124
Cassia corymbosa Lam. 32
Cassia didymobotrya Fresen. 112, 124, 291
Cassia floribunda Cav. 124
Cassia hirsuta L. 30
Cassia tomentosa L.f. 30
Casuarina 122, 124
Casuarina equisetifolia G. Forst. 105, 122, 291
Casuarina stricta Ait. 37
Catharanthus roseus (L.) G. Don 23
Ceanothus 178
Cedrela 122
Cedrela odorata L. 121
Cedrela toona Roxb. ex Rottl. & Willd. [see *Toona ciliata*]
Cedrus 122
Celastrus 178
Celtis 119, 127, 178
Celtis africana Burm. f. 127
Cenchrus brownii Roem. & Schult. 124
Cenchrus ciliaris L. 13, 23
Cenchrus incertus M.A. Curtis 31, 107
Centaurea 263
Centaurea cyanus L. 26
Centaurea repens L. 31
Centella asiatica (L.) Urb. 23
Cephalophus monticola (Thunberg) 70, 108, 127
Cephalophus natalensis A. Smith 70
Ceratotherium simum Burchell 108
Cercopithecus albogularis (Sykes) 119
Cercopithecus pygerythrus (F. Cuvier) 119
Cereus 30
Cereus peruvianus (L.) Miller 34, 105, 112, 114, 150–153, 161, 166, 270
Cervus dama L. [= *Dama dama*] 69, 108, 141, 220
Cervus elaphus L. 70, 108
Cervus nippon Temminck 70
Cervus porcinus (Zimmermann) 70, 108
Cervus unicolor (Kerr) 70
Cestrum 126, 166, 178
Cestrum aurantiacum Lindl. 30
Cestrum laevigatum Schlechtd. 34, 112, 124, 127, 129, 161, 167, 178, 291
Chaenomelas 178
Chamaecyparis lawsoniana (A. Murr.) Parl. 124–125
Chara globularis Thuill. 191
Chenopodium album L. 106
Chenopodium ambrosioides L. 106
Chenopodium carinatum R. Br. 31, 106
Chenopodium glaucum L. 106
Chenopodium multifidum L. 106
Chenopodium murale L. 106
Chenopodium pumilio R. Br. 31
Cherax tenuimanus Smith. 253
Chersine angulata (Schweigger) 136
Chilodonella cyprini Moroff 53
Chilodonella hexasticha Kiernik 53–54
Chionanthus 178
Chionis minor Hartlaub 139
Chloris gayana Kunth 30
Chlorocichla flaviventris (A. Smith) 119
Chromolaena 30, 163
Chromolaena odorata (L.) King & Robinson [= *Eupatorium odoratum*] 33, 34, 75, 112–114, 116, 124, 126–128, 152–153, 160, 163, 242, 300
Chrysanthellum americanum (L.) Vatke 106
Chrysanthemoides monilifera (L.) T. Norl. 172, 175, 286
Chrysemys scripta (Schoepff) 48, 54–55
Chrysolina quadrigemina (Suffrian) 265, 274
Chukrasia tabularis A. Juss. 122
Cichorium intybus L. 106
Cinnamomum 121–122, 124, 126–127, 129
Cinnamomum camphora (L.) J. Presl 178
Cirsium 263
Cirsium arvense (L.) Scop. 31, 289
Cirsium vulgare (Savi) Ten. 116, 124, 263
Cisticola subruficapilla (A. Smith) 85
Citharexylum 178
Citrullus 8,12
Citrus 124
Citrus limon (L.) Burm. f. [= *C limonia* & *C limonum*] 23, 178
Clarias gariepinus (Burchell) 48, 51, 55, 108, 204, 206, 253–254
Clerodendrum 178
Clutia pulchella L. 126
Cnicus benedictus L. 29
Cocos nucifera L. 22–23
Coix lachryma-jobi L. 124
Cola 119
Colius colius (L.) 173
Colius striatus Gmelin 173

Colletotrichum gloeosporiodes (Penz.) Sacc. 227, 288, 296, 298, 299
Colocasia antiquorum Schott [= *C a esculenta*] 22
Colophospermum mopane (Kirk ex Benth.) Kirk ex J. Leonard 95
Colpoon compressum Berg. 172, 175
Columba arquatrix Temminck 119, 127, 152, 173, 277
Columba guinea L. 64, 102
Columba livia L. 64, 102, 108, 140
Combretum apiculatum Sond. 95
Commelina benghalensis L. 23, 124
Commiphora 95
Connochaetes 71
Connochaetes gnou (Zimmermann) 70, 108
Connochaetes taurinus (Burchell) 70, 108
Conyza bonariensis (L.) Cronq. 30, 106
Conyza canadensis (L.) Cronq. 29, 106
Conyza floribunda Kunth [= *C sumatrensis* (Retz.) E.H. Walker] 106
Coprosma 178
Coptostomabarbus wittei David & Poll 48
Cordyline 178
Cordylus jonesii (Boulenger) 64
Cornus 178
Corokia 178
Coronopus didymus (L.) Sm. 29
Coronopus integrifolius (D.C.) Spreng. 106
Cortaderia 167
Cortaderia jubata (Lem.) Stapf 160
Cortaderia selloana (Schult.) Aschers. & Graebn. 32
Corvus frugilegus L. 71
Corvus splendens Vieillot 67
Cotoneaster 166, 178
Cotoneaster franchetii Boiss. 178
Craspedacusta sowerbyi Lankester 48, 54
Crataegus 178
Crocodylus niloticus L. 48
Crotalaria agatiflora Schweinf. 124
Crotalaria juncea L. 105
Crotalaria laburnifolia L. 22
Croton 119
Cryptotermes brevis Walker 79
Ctenopharyngodon idellus Valenciennes 50, 53, 252
Cunonia capensis L. 280
Cupressus 124
Cupressus lusitanica Mill. 122–123, 128
Cupressus macrocarpa Hartw. 149
Curtisia dentata (Burm. f.) C.A. Sm. 126
Cuscuta 29
Cuscuta campestris Yunck. 31
Cygnus atratus (Latham) 48
Cygnus olor (Gmelin) 48
Cylindropuntia rosea (D.C.) Backeberg [see *Opuntia rosea*]
Cymbopogon validus (Stapf) Stapf ex Burtt Davy 183
Cynodon aethiopicus Clayton & Harlan 32
Cynodon nlemfuensis Vanderryst 32
Cynodon plectostachyus (K. Schum.) Pilg. 32
Cynosurus echinatus L. 31
Cyperus rotundus L. 23, 263
Cyphomandra 178
Cyphostemma 95
Cyprinus carpio L. 48–52, 54–55, 102, 108, 204–205, 252, 254–255
Cyrtobagous salviniae (Caulder & Sands) 48, 251
Cyrtobagous singularis (Hustactie) 48
Cysticercus bovis [invalid, see *Taenia saginata*]
Cysticercus cellulosae [invalid, see *Taenia solium*]

Dactylis glomerata L. 31
Dactylopius 264
Dactylopius austrinus De Lotto 263, 265, 270, 272–273, 296–301
Dactylopius ceylonicus (Green) 261, 264, 270
Dactylopius coccus Costa 261, 299
Dactylopius opuntiae (Cockerell) 101, 115, 270–273, 296, 298–299
Dactylopius tomentosus (Lamarck) 270
Dama dama [see *Cervus dama*]
Damalinia 108
Damaliscus dorcas subspp 71, 108
Damaliscus dorcas dorcas Pallas 70
Damaliscus dorcas phillipsi Harper 71
Daphne 178
Datura 99–101
Datura ferox L. 101–106
Datura innoxia Mill. 101, 106, 160
Datura stramonium L. 29, 101, 106
Dichroa 178
Dichrostachys cinerea (L.) Wight & Arn. 95
Digitaria didactyla Willd. 32
Digitaria sanguinalis (L.) Scop. 107
Dioscorea 22
Diospyros 178
Diospyros dichrophylla (Gand.) De Winter 8
Diphyllobothrium latum (L.) 252
Diplodia pinea [see *Sphaeropsis sapinea*]
Dodonaea viscosa Jacq. var *angustifolia* Benth. [= *D angustifolia* L.f.] 106, 157
Dombeya burgessiae Gerr. ex Harv. 127
Dracaena 178
Drimys 178
Duchesnea indica (Andr.) Focke 32
Duranta repens L. 23, 178

Easculus 178
Echinochloa stagnina (Retz.) Beauv. 23
Echium vulgare L. 31
Eclipta prostrata (L.) L. 106
Ehrharta calycina J.E. Sm. 229
Eichhornia crassipes (Mart.) Solms-Laub. 5, 32, 34, 162, 167, 190, 192–196, 248–251
Elaeagnus 178
Elaeocarpus 178
Elaeocarpus kirtonii F. Muell. 121
Elaphurus davidianus (Milne-Edwards) 70
Elephas 6
Eleusine 4

Eleusine coracana (L.) Gaertn. 8, 12, 23
Eleusine indica (L.) Gaertn. 22–23
Eleusine tristachya (Lam.) Lam. 32
Elodea canadensis Michx. 162–167
Ensete 124
Entada spicata (E. Mey.) Druce 129
Epigaea 178
Epomophorus 119
Equus 3
Equus asinus L. 4, 68, 102, 108
Equus burchelli (Gray) 108, 230
Equus caballus L. 4, 102, 108
Equus capensis 4
Equus quagga Gmelin 5, 10
Equus zebra hartmannae Matschie 71, 102, 108
Equus zebra subspp 71, 108
Equus zebra zebra L. 69, 71, 230
Eragrostis 183
Eragrostis pilosa (L.) Beauv. 107
Eragrostis tef (Zucc.) Trotter 30, 107
Eragrostis tef x *E* spp 107
Eragrostis tenuifolia (A. Rich.) Steud. 32
Eragrostis virescens Presl 107
Eriobotrya 178
Eriocephalus 95
Eriocereus martinii (Labouret) Riccobono [see *Harrisia martinii*]
Eriococcus orariensis Hoy 262
Erodium moschatum (L.) L'Herit. ex Ait. 29
Erytenna consputa Pascoe 263–264, 288, 296, 298–299
Erythropygia coryphaeus (Lesson) 85
Eucalyptus 30–31, 79, 100, 105, 112, 115, 121, 122, 124–125, 127–129, 146, 148, 150–151, 153, 159, 176, 190, 239, 240, 276, 281, 290–291
Eucalyptus camaldulensis Dehn. 105, 121, 149
Eucalyptus cladocalyx F. Muell. 34, 160
Eucalyptus cornuta Labill. 148
Eucalyptus diversicolor F. Muell. 126, 149
Eucalyptus globulus Labill. 121, 126, 149
Eucalyptus grandis W. Hill ex Maid. 34, 122, 160, 190, 192–196, 244–245
Eucalyptus gummifera (Soland. ex Gaertn.) Hochr. 148
Eucalyptus lehmannii (Schav.) Benth. 34, 160
Eucalyptus microtheca F. Muell. 105
Eucalyptus nitens (Seane & Maid.) Maid. 153, 159
Eucalyptus regnans F. Muell. 191
Eucalyptus robusta Sm. 149
Eucalyptus saligna Sm. 122–123, 128
Eucalyptus sideroxylon A. Cunn. ex Woolls. 105
Eucalyptus viminalis Labill. 147
Euclea racemosa Murray 172, 175
Euclea undulata Thunb. 95
Eugenia 178
Euonymus 178
Eupatorium adenophorum Spreng. [see *Ageratina adenophora*] 112, 116, 185, 291
Eupatorium macrocephalum Less. 32
Eupatorium odoratum L. [see *Chromolaena odorata*]
Euphorbia 126
Euphorbia chamaesyce L. 23, 106
Euphorbia guerichiana Pax 95
Euphorbia peplus L. 29, 106
Euphorbia tirucalli L. 23, 35
Eupodotis afra (L.) 148
Eutropius 254
Eutropius depressirostris (Peters) 48

Fagopyrum esculentum Moench 106
Fannia canicularis (L.) 139
Fasciola 252
Fasciola gigantica Cobbold 54, 251
Fasciola hepatica L. 54, 251
Felis catus L. 4, 68, 79, 102, 108, 136, 219, 230
Felis lybica Forster 69, 102, 230
Ficus 119, 178
Flaveria bidentis (L.) Kuntze 31, 106
Flindersia australis R. Br. 121
Flindersia brayleyana F. Muell. 121
Flindersia oxleyana F. Muell. 121
Foeniculum vulgare A.W. Hill 29
Forpus passerinus (L.) 66–67
Fraxinus 122
Fraxinus angustifolia Vahl 121
Fringilla coelebs L. 66, 79, 82
Fuchsia 178
Fumaria officinalis L. [= *F muralis* (Sond.) ex Koch] 106

Galinsoga parviflora Cav. 106
Gallus domesticus L. 4, 10–12, 67
Gallus gallus (L.) [see *G domesticus*]
Gambusia affinis (Baird & Girard) 48, 50–51, 55, 205–206
Gardenia 178
Gasterophilus intestinalis (De Geer) 108
Gaultheria 178
Gaura sinuata Nutt. ex Ser. 31
Geochelone pardalis (Bell) 64
Giraffa camelopardalis (L.) 70
Gleditsia triacanthos L. 30, 105
Glycyrrhiza glabra L. 106
Gmelina 122
Gnaphalium luteo-album L. [see *Pseudognaphalium luteo-album* (L.) Hilliard & Burt]
Gnaphalium pensylvanicum Willd. 29
Gomphrena celosioides Mart. 106
Gossypium 4
Gossypium arboreum L. 14
Gossypium herbaceum L. 23
Gossypium herbaceum var *africanum* (Watt) Hutch. & Ghose 14
Grevillea robusta A. Cunn. 122–124
Grewia 178
Griselinia 178

Haemmatopinus 108
Hakea 5, 30, 37, 40, 75, 151–152, 159, 167, 190, 233–235, 239–240
Hakea gibbosa (Sm.) Cav. 34, 38–39, 42, 152, 159–160

Hakea salicifolia (Vent.) B.L. Burtt [= *H saligna* Knight] 151, 159, 167
Hakea sericea Schrader 34, 38, 42, 79, 80, 81, 84–86, 124–125, 149, 150–153, 159, 160, 190–191, 196, 227, 230–231, 233–235, 239–242, 262, 264, 285–286, 288, 290–291, 294–300
Hakea suaveolens R. Br. 31, 34, 38–39, 42, 149, 152–153, 159–160
Harpephyllum 119
Harrisia martinii (Labouret) Britton [= *Eriocereus martinii*] 32, 116, 178, 270
Hedera 178
Heliotropium curassavicum L. 29
Helisoma duryi (Weatherby) 48, 252
Helix adspersa (Müller) 82
Hemidactylus mabouia (Moreau de Jonnes) 63
Hemigrammocharax machadoi Poll 48
Hemitragus jemlahicus (H. Smith) 5, 17, 69, 79, 82, 85–86, 108, 220–221, 230
Heteromeles 178
Hibiscus sabdariffa L. 23
Hibiscus trionum L. 106
Hipparion 6
Hippophae 178
Hippotragus leucophaeus (Pallas) 4, 10
Hippotragus niger (Harris) 70
Homalanthus populifolius R.A. Grah. 32, 124
Homo 3, 6
Homo sapiens L. 6
Homoroselaps lacteus (L.) 64
Hordeum murinum L. 107
Hordeum vulgare L. 22, 29
Hydrochaeris hydrochaeris (L.) 70
Hydrocynus vittatus Castelnau 48
Hymenanthera 178
Hypericum 163
Hypericum perforatum L. 31, 86, 146, 160, 163, 261, 263, 265, 274
Hypogeococcus festerianus (Lizer & Trelles) 270
Hypophthalmichthys molitrix Cuvier & Valenciennes 50

Ibicella lutea (Lindl.) V. Eseltine 31
Icerya purchasi Maskell 276
Ichthyobodo necator Henneguy 53
Ichthyophthirius multifiliis Fouquet 48, 53–54
Idesia 178
Ilex 119, 178
Imbrasia cytherea (F.) 70, 123
Inula graveolens Desf. 31
Iochroma 178
Ipomoea batatas (L.) Lam. 22
Ipomoea purpurea (L.) Roth 106
Iridomyrmex humilis (Mayr) 5, 17, 45, 79, 82, 85–86, 227, 230–231
Ixora 178

Jacaranda mimosifolia D. Don. 35, 112, 123–124, 159, 161, 166, 291
Jasminum 178
Jatropha 178
Jatropha curcas L. 22–23
Juncus bufonius L. 23
Juniperus 122, 178

Kadsura 178
Khaya nyasica Stapf ex Bak. f. 122
Kiggelaria africana L. 125–126, 190, 280
Kneria auriculata (Pellegrin) 48, 55–56
Kobus ellipsiprymnus (Ogilby) 70
Kobus leche (Gray) 70, 108

Labeo 254
Labeo capensis A. Smith 48
Labeo cylindricus Peters 48
Labeo umbratus (A. Smith) 48, 55, 254
Lablab purpureus (L.) Sweet 23
Lactuca serrioloa L. 106
Lama glama (L.) 70
Lamium amplexicaule L. 106
Lamprothamnium papulosum (Wallr.) J. Gr. 191
Lanius collaris L. 85
Lantana 166, 178
Lantana camara L. 5, 23, 30, 35, 75, 105, 111–116, 124, 126–128, 143, 161, 167, 172, 175, 178, 261, 263–264, 266, 295–296, 298, 301
Lantana trifolia L. 178
Lappula echinata Gilib. 106
Lates niloticus (L.) 206
Laurus 178
Ledum 178
Leea 178
Lepidium bonariense L. 31, 106
Lepomis cyanellus Rafinesque 48, 50
Lepomis macrochirus Rafinesque 48, 50, 55, 205
Leptinotarsa defecta Stal 264
Leptinotarsa juncta Schaeffer 264
Leptospermum 37, 75, 159
Leptospermum laevigatum F. Muell. 30, 34, 38–39, 41–42, 124, 149, 160, 194, 244, 286, 290–291
Leptospermum lanigerum (Ait.) Sm. 276
Leptospermum scoparium Forster & Forster f. 262
Leucadendron salignum Berg. 233–235
Leucaena leucocephala (Lam.) De Wit 30, 112, 124, 291
Leycesteria 178
Ligustrum 178
Ligustrum japonicum Thunb. 124
Ligustrum lucidum Ait. 178
Lilium formosanum (Bak.) Wallace 32
Limnophyes pusillus Eaton 138
Limnothrissa miodon Boulenger 205, 253
Linaria vulgaris Mill. 30
Linognathus spp 108
Lithospermum arvense L. 30, 106
Litsea glutinosa (Lour.) C.B. Robinson var *brideliifolia* (Hayata) Merrill 178
Litsea sebifera Pers. 124, 291
Lolium multiflorum Lam. 107

Lolium temulentum L. 29, 107
Lonicera 178
Loxodonta africana (Blumenbach) 151, 277
Lybius leucomelas (Boddaert) 85, 233
Lycium 95
Lycopersicon pimpinellifolium (L.) Mill. 178
Lygodactylus capensis (A. Smith) 64
Lymnaea columella Say 48, 54, 251–252
Lymnaea natalensis Krauss 54, 251–252
Lymnaea truncatula (Müller) 251
Macadamia 178
Macleania 178
Maclura 178
Mahonia 178
Malus 178
Malva 163
Malva parviflora L. 29, 106
Malva pusilla Sm. 30
Mangifera indica L. 112, 124, 178
Manihot 22, 29
Marcusenius macrolepidotus (Peters) 48
Maytenus acuminata (L.f.) Loes. 172, 175
Maytenus oleoides (Lam.) Loes. 126
Medicago 146
Medicago aschersoniana Urb. [= *M lanciniata* (L.) Mill.] 106
Medicago denticulata Meisn. 106
Medicago nigra Willd. 106
Medicago polymorpha L. 4, 8–9
Medicago sativa L. 106
Megalotragus priscus 4
Melaleuca ericifolia Sm. 276
Melanterius 286
Melanterius acaciae Lea 289
Melanterius servulus Pascoe 289
Melanterius ventralis Lea 289
Melia 166, 178
Melia azedarach L. 35, 75, 99–100, 105, 110–114, 116, 124, 126–127, 161, 166, 178, 291
Melilotus alba Medik. 106
Melilotus indica (L.) All. 106
Meliosma 178
Melopsittacus undulatus (Shaw) 67, 108
Menispermum 178
Meranoplus peringueyi Emery 231
Merxmuellera 182
Merxmuellera stricta (Schrad.) Conert 183
Mesobola brevianalis Boulenger 48
Mespilus 178
Metamasius spinolae (Gyllenhal) 270
Micropterus 254
Micropterus dolomieu Lacépède 48, 50, 55–56, 205–206
Micropterus punctulatus Lacépède 48, 50, 55–56
Micropterus salmoides (Lacépède) 48, 50, 51, 55–56, 102, 108, 205, 254
Mimetes cucullatus (L.) R. Br. 85
Mimorista pulchellalis (Dyar) 270
Mimosa pigra L. 295
Mirabilis jalapa L. 23
Modiola caroliniana (L.) G. Don 106
Mollugo nudicaulis Lam. 29
Monomorium schultzei Forel 231
Montanoa bipinnatifolia (Kunth) C. Koch 124
Montanoa hibiscifolia Benth. 32, 124
Morus 178
Morus alba L. 112, 124, 127, 178
Morus nigra L. 173
Motacilla capensis L. 66
Mugil cephalus L. 48
Murraya 178
Mus musculus L. 4, 68, 82, 102, 108, 135, 210–211, 216, 218, 219–220
Musa paradisiaca L. 22
Musa sapienta L. 22
Mustela vison Schreber 69
Myocastor coypus (Molina) 49, 69
Myoporum 178
Myoporum serratum R. Br. 30, 172–173, 175–176, 178
Myrica 178
Myrica cordifolia L. 172, 175
Myriophyllum aquaticum (Vell.) Verdc. 32, 35, 162, 167, 248–251
Myro kerguelenensis Cambridge 138
Myrsine africana L. 172, 175
Myrtus 178
Mytilus galloprovincialis Latreille 48, 52
Myxus capensis Valenciennes 48

Nanaia 270, 273
Nandayus nenday (Vieillot) 67
Nandina 178
Nectarinia chalybea (L.) 85
Neochetina eichhorniae Warner 48
Neotragus moschatus (von Dueben) 70
Nerium oleander L. 5, 160, 163, 167
Nicandra physaloides Gaertn. 29, 106
Nicotiana glauca R.A. Grah. 5, 30, 35, 100, 105, 124, 145, 146, 160, 163
Nicotiana tabacum L. 4, 14, 22
Nothobranchius orthonotus (Peters) 48
Nothoscordium inodorum (Ait.) Nicholson 106
Numida meleagris (L.) 68, 79, 220
Nymphicus hollandicus (Kerr) 67
Nyssa 178

Ocotea 127, 173
Ocotea bullata (Burch.) Baill. 125, 277
Oenothera 163
Oenothera biennis L. 31
Oenothera indecora Cambess. 106
Oenothera parodiana Munz 31
Oenothera rosea L'Harit. ex Ait. 106
Oenothera tetraptera Cav. 30
Oestrus ovis L. 102, 108
Olea 119, 127, 173
Olea capensis L. subsp *capensis* 125
Olea capensis L. subsp *macrocarpa* (C.H. Wr.) Verdoorn 279
Olea europaea L. subsp *africana* (Mill.) P.S. Green 95, 126

Onychognathus morio (L.) 65, 148, 173
Onycocephalus capensis A. Smith 63
Opuntia 1, 30, 75, 99–100, 101, 105, 143, 151, 166, 181, 190, 194, 231, 264
Opuntia aurantiaca Lindl. 5, 30, 35, 105, 110, 112–114, 116, 124, 146, 162, 166–167, 181–182, 184, 185, 263, 265, 269–270, 272–273, 295–301
Opuntia dillenii (Ker-Gawl). Haw. 32, 262, 270
Opuntia engelmannii [see *O ficus-indica*] 101, 105
Opuntia exaltata A. Berger 270
Opuntia ficus-indica (L.) Mill. 4, 29–30, 35, 101, 105, 110–116, 124, 151, 161, 167, 178, 269–271, 272–273, 295–296, 297–298, 301
Opuntia imbricata (Haw.) DC. 32, 105, 112, 116, 184, 185, 270
Opuntia inermis DC. 261, 265, 272
Opuntia lindheimeri Engelm. 116, 270
Opuntia littoralis (Engelm.) Cockerell 262
Opuntia oricola Philbrick 262
Opuntia rosea DC. 105, 112, 116, 184, 185, 270
Opuntia spinulifera Salm-Dyck 116, 270
Opuntia streptacantha Lem. 264
Opuntia stricta Haw. 105, 261, 265, 270, 272
Opuntia triacantha (Willd.) Sweet 262
Opuntia vulgaris Mill. 32, 110, 112, 116, 261, 264, 269–270
Oreochromis 56, 254
Oreochromis andersonii (Castelnau) 48, 55
Oreochromis aureus (Steindachner) 56
Oreochromis macrochir (Boulenger) 48, 55
Oreochromis mossambicus (Peters) 48, 50–51, 54–56, 102, 108, 203, 205–206, 253, 255
Oreochromis niloticus (L.) 56
Oreochromis niloticus x *O aureus* 50
Oreodaimon 55
Oreodaimon quathlambae (Barnard) 55–56
Oriolus larvatus Lichtenstein 173
Orobanche minor Sutton 31
Orobanche ramosa L. 106
Orrina 264
Oryctolagus cuniculus (L.) 41, 71, 108, 135
Oryx dammah (Cretzschmar) 70
Oryx gazella (L.) 70
Oryza sativa L. 22
Osmanthus 178
Osmaronia 178
Osteomelas 178
Otomys 233
Otomys irroratus (Brants) 233
Ovis 108
Ovis aries L. 4, 9, 11, 68
Ovis musimon Pallas 70
Oxalis corniculata L. 106
Oxalis latifolia H.B.K. 106
Oxytenanthera abyssinica (A. Rich) Munro 23

Pachydactylus bibronii (A. Smith) 64, 79
Paliurus 178
Panax 178
Panicum maximum Jacq. 31
Panthera 3
Papaver hybridum L. 107
Papaver rhoeas L. 107
Papio ursinus (Kerr) 119, 148
Paraserianthes lophantha (Willd.) Nielsen [see *Albizia lophantha*]
Parkinsonia aculeata L. 30, 105
Paroaria 67
Paroaria coronata (Miller) 67
Paroaria dominicana (L.) 67
Parthenium hysterophorus L. 30, 75
Parthenocissus 178
Paspalum dilatatum Poir. 107
Passer domesticus (L.) 65–66, 82, 102, 108, 135, 220
Passer melanurus (Statius Mueller) 66, 85, 173
Passiflora 178
Passiflora caerulea L. 178
Passiflora edulis Sims 124, 178
Paulinia acuminata De Geer 48
Pavo cristatus L. 67, 140
Pelecanoides urinatrix (Gmelin) 140
Pelorovis antiquus 4
Pennisetum 4, 10
Pennisetum clandestinum Chiov. 32
Pennisetum glaucum (L.) R. Br. 8, 11, 12
Pennisetum purpureum Schumach. 32
Pennisetum setaceum (Forssk.) Chiov. 32, 107
Pennisetum villosum R. Br. ex Fresen. 32
Perca fluviatilis L. 48, 50, 55, 206, 253
Pereskia aculeata Mill. 112, 124, 127–128, 161, 178, 270, 291
Pernettya 178
Persea 178
Persoonia 178
Petrocephalus catostoma (Günther) 48
Phacochoerus aethiopicus (Pallas) 70
Phalaris arundinacea L. 107
Phalaris caerulescens 31
Phalaris minor Retz. 107
Phalaris tuberosa L. 31
Pharyngochromis darlingi (Boulenger) 48
Phasianus colchicus L. 67
Pheidole capensis Mayr 231
Phellodendron 178
Phillyrea 178
Phoenix 178
Photinia 178
Phragmites 250–251
Phragmites australis (Cav.) Steud. 192
Phrynomerus bifasciatus A. Smith 48
Physa acuta Draparnaud 48, 251
Physalis angulata L. 23, 107
Physalis minima L. 30
Physalis peruviana L. 29, 105, 107, 161, 178
Physalis viscosa L. 31
Phytolacca 124, 128
Phytolacca dioica L. 178
Phytolacca octandra L. 173
Phytophthora cinnamomi Rands. 82

Pinus 31, 38, 82, 86, 115, 121, 124, 127–129, 146, 148, 150–151, 159, 190, 240, 276, 281
Pinus canariensis Sweet ex K. Spreng. 79, 121, 126
Pinus caribaea Mor. 146
Pinus elliottii Engelm. 192–193
Pinus halepensis Mill. 37–39, 41–42, 121, 160
Pinus oocarpa Schiede 148
Pinus palustris Mill. 122
Pinus patula Schlechtd. & Cham. 31, 35, 112, 122–124, 128, 152, 190, 192–193, 239–242, 244–245
Pinus pinaster Ait. 4, 35, 37–42, 69, 79, 81–82, 84, 86, 121, 123, 125, 147, 150, 153, 158, 160, 190, 192–196, 229, 233, 240–241, 280, 290–291
Pinus pinea L. 35, 37, 69, 75, 122, 126, 150, 191
Pinus pseudostrobus Lindl. 122
Pinus radiata D. Don 30, 35, 39, 79, 83, 121, 123, 126, 147–151, 158, 160, 190–193, 195, 229, 231–232, 239–241, 244, 280
Pinus sylvestris L. 37
Pinus taeda L. 121
Pistacia 178
Pistia stratiotes L. 248
Pittosporum 119, 116, 178
Pittosporum undulatum Vent. 30, 84, 122, 124, 126–127, 129, 147–148, 151, 157, 161, 166, 172–173, 175–176, 178, 244
Pittosporum viridiflorum Sims 280
Pityrogramma calomelanos (L.) Link 30
Plantago lanceolata L. 107
Plantago major L. 107
Plectranthus barbatus Andr. 32
Ploceus 67
Ploceus capensis (L.) 173
Poa annua L. 107, 138–139
Podocarpus 119, 123, 125, 178, 277
Podocarpus henkelii Stapf ex Dallim. & Jacks. 190
Podocarpus latifolius (Thunb.) R. Br. ex Mirb. 279–280
Poecilia reticulata Peters 48, 50, 55, 205
Pogoniulus 119
Polygonum aviculare L. 29, 107
Polygonum convolvulus L. 107
Polypogon monspeliensis L. Desf. 107
Populus 121–122, 124, 167
Populus alba L. 35
Populus alba x *canescens* 162, 167
Populus canescens (Ait.) J.E. Sm. 35, 112–115, 123–124, 240
Populus deltoides Baart. ex Marsh 105
Portulaca oleracea L. 29, 107, 163
Potamonautes perlatus (Milne Edwards) 253
Potamochoerus porcus (L.) 70, 119, 148, 277
Potamogeton 250
Potamogeton crispus L. 194
Potamogeton pectinatus L. 191
Praomys verreauxii (A. Smith) 232
Prinia maculosa (Boddaert) 233
Prionium serratum (L.f.) Drege ex E. Mey. 82
Procambarus clarkii (Girard) 253
Promerops cafer (L.) 233
Prosopis 5, 29, 71, 75, 99, 100, 103, 105, 111, 113, 146, 149, 164–165, 167, 262, 281
Prosopis chilensis (Mol.) Stunz
Prosopis glandulosa Torrey 35, 161, 194, 297
Prosopis pubescens Benth. 298
Prosopis velutina Wooton 35, 289, 298
Protea 153, 240
Protea eximia (Salisb. ex Knight) Fourc. 233
Protea lorifolia (Salisb. ex Knight) Fourc. 233–235
Protea neriifolia R. Br. 240, 244
Protea punctata Meisn. 233
Protea repens (L.) L. 190, 233, 240
Protopterus annectens brieni Poll 48
Protorhus 119
Prunus 127, 173, 178
Prunus africana (Hook. f.) Kalkm. 127
Prunus persica (L.) Batsch. 30, 124, 178
Pseudocrenilabrus philander (Weber) 48, 55
Pseudognaphalium luteo-album (L.) Hilliard & Burt 29, 106
Psidium 166
Psidium cattleianum Sabine 124, 178, 291
Psidium guajava L. 22–23, 29, 35, 75, 112–113, 124, 126–127, 129, 161, 178
Psittacula cyanocephala (L.) 66
Psittacula krameri (Scopoli) 66
Psychotria 178
Pteridium aquilinum (L.) Kuhn 262
Pterocelastrus tricuspidatus (Lam.) Sond. 125, 280
Pterocles bicinctus Temminck 101
Pueraria 291
Punica granatum L. 178
Pycnonotus barbatus (Desfontaines) 173
Pycnonotus capensis (L.) 173
Pyracantha 161, 166, 178
Pyracantha angustifolia Schneid. 35, 124, 178
Pyracantha coccinea M.J. Roem. 178
Pyrus 178

Quercus 122
Quercus canariensis Willd. 121
Quercus robur L. 69, 82, 121–122, 124, 149
Quercus suber L. 121

Ramphotyphlops braminus (Daudin) 63, 79
Rapanea 119
Rapanea melanophloeos (L.) Mez 125–126, 172, 175, 280
Raphiolepis 178
Rattus norvegicus (Berkenhout) 68, 82, 108, 220
Rattus rattus (L.) 4, 10–12, 14, 68, 79, 82, 103, 108, 135, 210–211, 215, 218–220, 230
Rauvolfia caffra Sond. 127
Rayiera 290
Redunca arundinum (Boddaert) 108
Rhabdalestes maunensis Fowler 48
Rhabdomys pumilio (Sparrmann) 85
Rhamnus 178
Rhipsalis baccifera (J. Mill.) Stearn 23, 263
Rhodotypos 178
Rhus 95, 176, 178
Rhus angustifolia L. 126

Rhus crenata Thunb. 178
Rhus glauca Thunb. 176
Rhus laevigata L. 176
Rhus lucida L. 176
Rhus tomentosa L. 176
Rhus undulata Jacq. 95
Rhynchelytrum repens (Willd.) C.E. Hubbard 22, 31
Rhyssomatus marginatus Fahraeus 289
Ribes 178
Richardia brasiliensis Gomes 30
Ricinus communis L. 4, 8–9, 13
Rivina humilis L. 124
Robinia pseudo-acacia L. 30
Rorippa nasturtium-aquaticum (L.) Hayek 29
Rosa 124, 166, 178
Rosa eglanteria L. 30, 178
Rosa odorata Sweet 178
Rosa rubiginosa L. [see *R eglanteria*] 161, 167
Rousettus 119
Rubus 32, 115, 128, 166, 167, 172, 175, 176, 178
Rubus affinus Wight & Arn. 178
Rubus cuneifolius Pursh. 32, 35, 112–113, 116, 123–124, 126, 127, 161, 178, 185, 280
Rubus fruticosus L. 172, 178
Rubus hybrid 178
Rubus idaeus L. 178
Rubus niveus Thunb. 124, 178
Rubus phoenicolasius Maxim. 178
Rubus rosifolius J.E. Sm. 178
Rumex 163
Rumex angiocarpus Murb. 29, 30, 107
Ruppia spiralis L. ex Dum. [= *R cirrhosa* (Petagna) Grande] 191
Ruscus 178

Saccharum officinarum L. 22
Sagina apetala Ard. 140–141
Salix 122
Salix babylonica L. 30, 35, 105, 113–114, 124, 162, 167
Salmo 253–254
Salmo gairdneri Richardson 48, 50, 52, 54–56, 204–205, 253
Salmo salar L. 48, 50
Salmo trutta L. 48, 50, 55–56, 141, 205
Salsola kali L. 99–100, 107
Salvelinus fontinalis (Mitchell)
Salvinia molesta D.S. Mitchell 32, 162, 167, 192–196, 248–249, 250–251, 255
Sambucus 178
Sandelia bainsii Castelnau 55, 57, 204
Sandelia capensis (Couvier & Valenciennes) 48, 56
Sangarius paradoxus Stal 287
Sarcococca 178
Sassafras 178
Schinus molle L. 105, 161, 178
Schinus terebinthifolius Raddi 31, 124, 161, 166, 178, 291
Schistosoma haematobium (Bilharz) 252
Schizandra 178
Schkuhria pinnata (Lam.) Cabr. 107
Schmidtia kalihariensis Stent 95
Scirpus littoralis Schrad. 192
Sciurus carolinensis Gmelin 5, 17, 69, 79, 81–82, 220
Senecio jacobaea L. 295
Sequoia 122
Sequoia sempervirens (D. Don) Endl. 121, 124–125
Serinus 67
Serinus gularis (A. Smith) 173
Serinus sulphuratus (L.) 173
Serissa 178
Serranochromis 50
Serranochromis codringtoni (Boulenger) 48
Serranochromis giardi (Pellegrin) 48
Serranochromis macrocephalus (Boulenger) 48
Serranochromis robustus jallae (Boulenger) 48, 50
Serranochromis robustus robustus (Günther) 48, 50
Serranochromis cf *steindachneri* (Boulenger) 48
Serranochromis thumbergi (Castelnau) 48
Sesamum indicum L. 23
Sesbania 111
Sesbania punicea (Cav.) Benth. 32, 35, 79, 81, 82, 105, 112–113, 116, 124, 146, 161, 164, 165, 263, 264, 266, 289, 291
Setaria verticillata (L.) Beauv. 23
Sideroxylon inerme L. 172, 175
Silene gallica L. 107
Silybum marianum (L.) Gaertn. [= *Carduus marianus*] 106, 289
Sivatherium 3
Smodingium 178
Solanum 126, 166, 178, 263
Solanum elaeagnifolium Cav. 116, 178, 262, 263, 264, 289
Solanum giganteum Jacq. 129
Solanum hispidum Pers. 178
Solanum jasminoides Paxt. 178
Solanum mauritianum Scop. 26, 35, 107, 110, 112–113, 116, 124, **126–129**, 148, 151–152, 161, 166, 172–173, 175, 178, 243, 290
Solanum nigrum L. 107, 178
Solanum rostratum Dun. 178
Solanum seaforthianum Andr. 32, 124, 161, 178
Solanum sisymbrifolium Lam. 31, 178
Sonchus asper (L.) Hill 107
Sonchus maritimus L. 107
Sonchus oleraceus L. 30, 107
Sorbus 178
Sorghum 4, 8, 10
Sorghum almum Parodi [see *S halepense*]
Sorghum halepense (L.) Pers. 30, 107
Sorghum verticilliflorum 23
Spartina 23
Spartium junceum L. 30, 161, 291
Spergula arvensis L. 29
Spergularia media (L.) Presl 107
Spergularia rubra (L.) J. & C. Presl 107
Sphaeropsis sapinea (Fr.) Dyko & Sutton [= *Diplodea pinea* (Desm.) Kickx] 151
Spheniscus demersus (L.) 140
Sphenoeacus afer (Gmelin) 85
Sporobolus africanus (Poir.) Robyns & Tournay 183

Spreo bicolor (Gmelin) 65
Stachyurus 178
Stellaria media (L.) Vill. 107, 139
Stipa clandestina Hack. 31
Stipa neesiana Trin. & Rupr. 31
Stipa tenuissima Trin. 26, 31, 35
Stipa trichotoma Nees 26, 31, 35, 75, 86, 112–116, 146, 147, 160, 163, 182–183, 191, 193, 296–299, 300
Stipa variabilis Hughes 31
Stipagrostis 95
Stipagrostis amabilis (Schweick.) De Winter 95
Stipagrostis sabulicola (Pilg.) De Winter 95
Stipagrostis uniplumis (Licht.) De Winter 95
Stranyaesia 178
Streptopelia capicola (Sundevall) 85, 101, 173
Streptopelia semitorquata (Rueppell) 68
Streptopelia senegalensis (L.) 85, 101, 173
Striga asiatica (L.) Kuntze 23
Struthio camelus L. 68, 101, 108, 141
Strychnos 119
Sturnus vulgaris L. 64, 79, 80, 81–82, 102, 108, 135, 173, 220–221, 230
Sus scrofa L. 70, 108, 123, 220–221
Sylvicapra grimmia L. 152
Symphoricarpus 178
Syncarpia glomulifera (Smith) Niedz. 124–125
Syncerus caffer (Sparrman) 108
Synodontis leopardinus Pellegrin 48
Synodontis macrostigma Boulenger 48
Synodontis woosnami Boulenger 48
Syzygium cuminii Skeels 178

Tadorna tadorna (L.) 48
Taenia saginata Goeze 108
Taenia solium L. 108
Tagetes minuta L. 31, 99, 107, 160, 163
Tamarindus indica L. 23
Tamarix gallica L. 122
Tamarix ramosissima Ledeb. 105
Tamarix usneoides E. Mey. ex Bunge 95
Taraxacum dens-leonis Desf. [see *T officinale*]
Taraxacum officinale Weber 29, 107
Tarchonanthus camphoratus L. 95
Tauraco 119
Tauraco corythaix (Wagler) 277
Tauraco porphyreolophus (Vigors) 173
Taxodium 122
Taxus 178
Tecoma stans (L.) H.B.K. 124
Terminalia sericea Burch. ex DC. 95
Ternstroemia 178
Thais haemostoma L. 48
Theba pisana (Müller) 45, 79, 82, 86
Themeda 182, 183, 245
Themeda triandra Forssk. 152, 183, 299
Thuja 122
Tilapia 48, 254
Tilapia rendalli Boulenger 55, 255
Tilapia rendalli rendalli Boulenger 48, 50
Tilapia rendalli swierstrae Gilchrist & Thompson 48, 55
Tilapia sparrmanii A. Smith 48, 55, 108
Tilia 178
Tinca tinca L. 48, 50, 55, 206, 253
Tithonia diversifolia (Hemsl.) A. Gray 32
Tithonia rotundifolia (Mill.) Blake 32
Toona ciliata [= *Cedrela toona*] 121–124, 161, 166
Torreya 178
Tragelaphus angasii Gray 70, 108, 123, 127, 152
Tragelaphus scriptus (Pallas) 70, 108, 127, 277
Tragopogon porrifolius L. 107
Tragus berteronianus Schult. 23
Trema orientalis (L.) Blume 127, 129, 190, 280
Trianthema hereroensis Schinz 95
Tribulus terrestris L. 263
Trichapion lativentre (Benguin-Billecocq) 264, 266, 289
Trichilogaster 286
Trichilogaster acaciaelongifoliae (Froggatt) 164, 264, 266, 288
Trichodesma zeylanicum (Burm. f.) R. Br. 30
Trichodina acuta Lom 48, 53, 54
Trichodina nigra Lom 53
Trichodinella epizootica (Raabe) 53
Triticum aestivum L. 22, 29
Triumfetta pilosa Roth. var *tomentosa* Szyszyl. ex Sprague & Hutch. 23
Tucumania tapiacola Dyar 270, 273
Turdus merula L. 67
Turdus olivaceus L. 67
Turdus philomelos Brehm 67
Typha latifolia L. [= *T capensis* (Rohrb.) N.E. Br.] 192
Tyto alba (Scopoli) 211–215, 219

Ulex europaeus L. 30, 291
Umbellularia 178
Uromycladium tepperianum (Sacc.) McAlp. 286
Urtica urens L. 107
Utricularia 54

Vedalia cardinalis Mulsant [= *Rodolia cardinalis* (Mulsant)] 276
Verbena bonariensis L. 29
Verbena officinalis L. 107
Verbesina encelioides (Cav.) Benth. & Hook. 107
Vespula germanica (F.) 79, 86
Vestia 178
Vibrio cholerae Pacini 252
Viburnum 178
Vicia atropurpurea Desf. [= *V benghalensis* L.] 107
Vicia sativa L. 107
Vigna unguiculata (L.) Walp. var *sinensis* [= *V u sesquipedalis* (L.) Verdc.] 8, 13
Virgilia oroboides (Berg.) Salter 129, 157, 280–281
Vitex 178
Vitis 124, 178
Voandzeia subterranea (L.) Thouars [= *Vigna subterranea* (L.) Verdc.] 8, 13

Xanthisma texanum DC. 107
Xanthium spinosum L. 29–30, 107, 115
Xanthium strumarium L. 107, 285
Xenopus laevis Daudin 48, 52
Xiphophorus helleri Heckel 48, 50, 55, 205
Xymalos 119

Yersinia pestis 218

Yucca 178
Yucca aloifolia L. 178

Zanthoxylum 178
Zea mays L. 4, 14, 22, 29
Zeuxidiplosis giardi Kieffer 265
Zinnia peruviana (L.) L. 107
Zosterops pallidus Swainson 119, 173

Subject index

Page numbers in italic face indicate figures or tables.

Addo National Park 64, 181, 209
Afromantaine forest 119
agriculture
  early, impact *4, 5,* 9–17, 22–23, 25–26, 29–31
  commercial 70–71
Alexandria 83
Angola *xiii, 94*
Amatole Mountains 121
amphibians, alien 63, 71
amphibians, indigenous *48,* 52, 78, *136, 137*
aquaculture industry 32, 56
aquatic ecosystems 47–59, 209, 227, 247–256
Arabs, role in invasions 12, 13, 14, 22
Assegaaibos National Reserve 210
Auob River 94

Baviaanskloof Mountains 230, 231, 232
Bedford *182*
Beira 68
Berg River 51, *55,* 81
Big Elephant Shelter 8
Bird Island, Algoa Bay 133, *134, 135,* 136
Bird Island, Lambert's Bay *134, 135*
biological control
  agents
    invertebrates 140, 261–266, 269–274, 285–290
    vertebrates *69,* 70
  targets
    invertebrates 252, 276
    plants 85–87, 115–116, 233–234, 261–262, 269–274, 285–290, 296–298
    vertebrates 51–53
  success 101, 115, 261, 263–264, 269–274, 279–280, 297
birds, alien
  feral populations 64, 66, 67, 68, 100, 102
  impact 54, 64–65, 67–68, 69, 100, 230, 285
  introduction 136–137, 138
    intentional 56, 65–68, 71, 79, *80,* 100, 102, 141, 220, 253
    translocated 66, 67–68, 135, 220
    unintentional 65–68, 71, 102
  invasive success 64–67, 79, 102, 140, 214, 221
  management 141, 261
  origins 64–68, 79, 220, 230
  status 65, 66
birds, indigenous
  impact of alien species on, 7, 85, 230, 231, 233
  invasion susceptibility 67
  translocation *48,* 68, 71
Blinkklipkop *12*
Bloemfontein *xiii,* 210
Bokkeveld Series 77
Bolus Herbarium 27
Boomplaas Cave 9, *11*
Border Cave 6, 8
Boschberg *183*
Botswana *xiii,* 9, 13, 14, 24, 66, *94,* 204
Bowie, J 37, 38
Bredasdorp 67
Breede River *xiii*
Breutel, Bishop 30
Broederstroom 11, *11,* 12, *12*
Brown, J C 40

Cape Agulhas 5
Cape Department of Nature and Environmental Conservation 96
Cape Flats 39–40, 231
Cape Fold Mountains 9, 77
Cape of Good Hope 15, 17, 22, 29, 37–39
  Nature Reserve 64, 210
Cape Peninsula 15, 37–41, 63, 70, 81, 82, 121, 125, 176
Cape Province
  Eastern 77–84, 109–117
  Northern 93–103
  Southern 77–87
  South-Western 37–42, 77–87, 229–233
Cape St Francis 9, *12*
Cape Supergroup 77
Cape Town *xiii,* 9, 15–18, 33, 37–42, 63, 64, 68, 69, 79, 81, 82, 122, 210
Cape Town Botanic Gardens 30, 37, 39
Cape Town, Municipality of 41
Cathcart 122
Cederberg 82
Ceres 67
chemical control programmes
  plants, alien 78, 86, 100, 115–116, 140–141, 250, 251, 269, *296,* 297, 299–301
  risks 116, 250
climate
  amelioration of 41
  palaeoclimatic record 3–6
  influence
    fire regime 77, 242
    invertebrates, alien 47
    man 5–6, 16, 17
    plants, alien 5–6, 98–99, 146, 193–195, 221, 222, 298
    vertebrates, alien 5–6, 65
Coastal Fynbos 80–81, 149, 210
Concordia Forestry Nursery *149*
Constantiaberg 40
Citrusdal control programmes *296*
Cunene River *xiii,* 52, 93, 94, 102
Cunene-Zambezi catchment 48, *49*
Cuvelai River 52, 102

Damaraland 95
Damara sequence 94
Dassen Island 134, *134, 135,* 136, 139, 140
De Aar *94*
De Hangen 8
De Hoop Nature Reserve 82, 64
Department of Agriculture and Forestry 28
Department of Environment Affairs (Forestry Branch) 86, 128
desert biome *xiv,* 93–103, *105–108*
    description 93–95
    plants, alien *105–107, 214,* 219
        introductions 96–97
        invasion process 98–101
    invasion susceptibility 97–98, 100, 101, *214*–219
    vertebrates, alien 101–103, *108, 214,* 216, *217,* 218
Devil's Peak 40, 41
Die Kelders 9, *11,* 12
Diepwalle Forest 125
Downton, Nicholas 33
Drakensberg 14, 119
Dudley Pringle Dam 56
Durban *xiii,* 17, 33, 63, 65, 66, 67, 210
Dwarf Shrub Zone 93
Dyer Island *135*

East London *xiii,* 13, 66, 210
Eerste River 51, 126
Elands Bay 8
Elgin 67, 68
Enkwazini, Natal 11
Escarpment, Great 11
Escarpment Zone, semi-arid 93
Etosha Pan National Park 94, 95, 102
European Colonization 14–17, 29–32, 97, 121
    British 1, 15, 17, 30, 37, 51, 97
    Dutch 9, 15, 16, 17, 37, 63
European Inland Fisheries Advisory Commission 57
extinction 3, *4,* 6, 7, 10, 84, 232
    local 55, *55, 56,* 66, 67, 84

Falls Rock Shelter 9
False Bay 81
fire xv, 1, 6, 7, 11
    hazard 127, 241, 242
    impact 81, 151, 158, 236, *236,* 242–244, 297, 300
    regime 77, 78–79, 235
    resistance to 244
fish, alien
    impact
        colonization 48, 51, 57, 247, 248
        economic 47, 57, 247–251, 253–255
        habitat 55, 57
        parasite introduction 52–53, *54*
        species diversity 55–56, 57
        role of man 48
    introductions
        history of 51–52
        intentional 48, 51–52, *53, 55,* 100, *139,* 248, 253, 254
        translocation 52, 55, 102, 206
        unintentional 49, 52, 53, 102
        invasion process 55–56, 204–206
        invasive success 49–51, *54, 55*
    management
        appraisal of further introductions 57–59, 141, 207, 225, 256, 261
    origins 51–54, 204–206
    status 55
Fish Hoek 66
fish, indigenous
    economic importance 253
    parasites, alien, impact on 52–53, *54*
    translocation *48, 49,* 51–53, 55
fish parasites 52–53, *54*
fish, reproductive guilds 201–207
Fish River, Great *xiii,* 16
Fish River *xiii,* 93
Fisheries Development Corporation 56
forest biome *xiv,* 119–130
    description 119, *120*
    deforestation 11, 14, 120–121, 297
    invasion susceptibility 123, *214,* 219
    plants, alien
        impact 123–127, 277
        introductions 121–123, *214,* 219
        invasion process 121, 123–127
        status *124, 125*
    vertebrates, alien 123, 214
Forestry Research Institute 27
forestry zones *120,* 146, 150
Fort Beaufort *182*
Fort Grey 122
fossil record 3–14
Francistown, Botswana 210, 218
Franschhoek 79, 81
freshwater ecosystems 47–59, 209, 227, 247–256
frugivory 127, 129, 148, 150, 151–152, 164–166, 171–178
fynbos biome
    description 77–78
    fire, role in 77, 78, 83, 159, 236,
    invasion susceptibility 79, 87, 151, 209, 210, *214,* 216, *217,* 219–222
    invertebrates, alien 79, 82, 85, 210, 230–231
    plants, alien
        impact 79–81, 82–85, 87, *149,* 196, 230, 232, 239–245
        invasion process 81, 148–153, 171, 173, *175,* 176, 190–191
    vertebrates, alien *69,* 79, 82, 210, 230–233, 235

Gaborone *xiii*
Gamtoos River *xiii*
Genadendal 30
George 79, 81, 121
George Forestry Nursery *149*
Geyser Island *134, 135*
Ghaap Plateau 94
Gordon's Bay 63

Gouritz River *xiii*, 9
Grahamstown 15, *182*
grassland biome *xiv*, 109–116
 animals, alien 210, *214*
 description 109
 invasion susceptibility 111–112, *214*, 216, *217*, 219
 management 114, 115–116
 plants, alien 110, *214*
  impact 112–116, 233, *239*, 240, *242*, 244, 245
  introduction 110
Gravelotte 67
Grey, G 39
Griqualand West 17
Griquatown *94*, 97
Groote Schuur Estate 66
Groote Schuur 200, 69
Grootvaderbosch 121

Halifax Island *135*
Happy Rest 11, *11*, *12*
Harare *xiii*, 66
Hardap Dam 93, 95
Harrison, Captain C 121
Hartebeesport Dam 51, 55, *250*
Harts River 94
Hawston 9, *11*,
Heidelberg 67
Hermanus 67
herbivory 78, 151, 232, 234–235, 262–266, 269, *270*, 277
Highland Savanna 94–98
Hluhluwe-Umfolozi Game Reserve 67, 127, 151, 209
Hoanib River 94, 101
Holocene Age 5, 6
Hollamsbird Island 133, *134*, *135*
Hooker, Sir Joseph 33
Hooker, William 37
Hottentot's Holland 15
Hout Bay 40
Howick 67
Huab River 94
human population density 17, 31, 103, 247
human population distribution *10*, 16
human population prior to 1652 xv, *4–5*, 6–17, 22–23
 Bantu 1, 22
 Iron Age 10–14
 Khoi 1, 9–10, 15, 29
 San 79
 Stone Age 6–10
 Xhosa 16
Hutchins, D E 41
hybridization 55–56, 64, 68–69, 70, 71, 100, 102, *108*, 115, 167, 168, 230, 253, 298

Ichaboe Island *135*
ichthyofaunal zones 47
impoundments 47, 51, 53, 54, 102, 103, 247, *249*, 250–251, 254–255
industrial activity 13, 31–32, 247
intercatchment (water) connections xv, 47, 52, 53, 55, 57, 247
 Cunene-Cuvelai = Ombalanta
 Eastern National Water Carrier 52, 102
 Okavango System 52
 Ombalanta Canal 52, 102
 Orange-Fish 52, 55, 66, 206
 Vaal-Tugela 52
International Council of Scientific Unions (ICSU) v
International Council for the Exploitation of the Sea 57
invertebrates, alien *136–137*, 138, 139, *139*
 introductions
  intentional *48*, 251–253, 261–266, 269
  translocation 48, 52
  unintentional 79, 210, 253
 origins 79, 230–231, 251–253, 264, 269–272, 288–289
 See also parasites, alien
introduced species
 definition xiv, 145
 history of 8–17, 21–33, 51–52, 63–72, 79, 110, 119–122
islands, continental 133–141
 description 133, *134*, *135*
 introductions 135, 136–137, 139
 management, alien species 140–141
 plants, alien 136–138, 139, 140
 species diversity 136–138
 vertebrates 136–138, 140
islands, oceanic
 description *134*, *135*, 136
 introductions 138, 139
 management of alien species 140–141
 plants, alien 138, *139*, 184
 vertebrates, alien 138, *139*

Johannesburg *xiii*, 28, 65, 210
Jutten Island *134*, *135*
Jessievale *147*

Kalahari Savanna 93–98
Kalahari-Gemsbok National Park 95, 210
Kaokoland 95
Karoo 65, 271, 272
Karoo home *xiv*, 93–103, *105–107*
 invasion susceptibility 97–98, 101, *214*, 216, *217*, 218, 219
 plants, alien 97, 99–100, 101, 103, *105–107*, 149, 214
 vertebrates, alien 102–103, *108*, 214, 216–218
Karoo, Little 68, 218
Karoo, Upper 93–103
Kasteelberg 9, *11*, *12*
Katberg 122
Keetmanshoop *94*
Kei River *xiii*
Kimberley 15, 64, 65, 66, *94*, 95, 97
Kirstenbosch National Botanical Gardens 42, 79
Klasies River Mouth 6
Kluitjieskraal Forestry Nursery *149*
Knysna 67
Kommetjie 64
Kosi 51

Kouga Mountains 230, 231, 232
Kromdraai 3
Kruger National Park 64, 69, 218
Kuiseb River 93, 101
Kuruman *94*
Kuruman – Asbestos Mountains 94
Kuruman River 94

Lake Chilwa 207
Lake Kariba *xiii*, 57, 205, 253
Lake Midmar 51, 54
Lake Ngami 207
Lake St Lucia 127
Lake Sibaya 51, 203
Lake Verwoerd 51
Lake Victoria 206
Langebaanweg 3
Langeberg 94
Legislation 45, 245, 255, 259
  Crown Lands Act, 1860 16
  ineffective application of 116, 301
  need for 57, 103, 255
  Weeds Act of 1937 86
  Weeds Acts 38 and 43 of 1983 255
Lesotho xiii, 7, 14, 24, 204
Limpopo River xiii, 12, 13, 14, 24
Lister, J S 41
Loerie 182
Luderitz *94*, 102
Luderitz Bay 218

Mafikeng *94*
Magogo, Natal 11, *11*, 12, *12*
Makapansgat 3
Malgas Island *134*, *135*
Malmesbury Shales 77
Mammals, alien
  feral populations 68–69, 71, 100, 102, 136, 230
  impact
    fauna 127, 140
    flora 85, 140
    species diversity 68–69, 70, 100, 230
    soil erosion 85
  introductions
    intentional 9–13, 16–17, 70–71, 123, 135, 136
    translocation 70–71, 136, 220
    unintentional *12*, 17, 35, 49, 68, 69, 79, 82, 100, 102, 136 210, 220
  invasion process 71, 82, 216–219, 220–221
  invasive success 71, *108*, *136*, *137*, *139*, 140, 210, *214*, *215*, *216*, 230
  management 49, *69*, 70, 86, 140–141
  origins 17, 68–71, 79, *108*, 230–231, 235
mammals, indigenous
  impact of alien plants 230, 232–233
  translocation 70–71, 123
Mapungubwe Hill 14
Maputo *xiii*, 28, 33, 66, 210
Maputoland 51
Marcus Island 133, *134*, *135*
Margate 116
Marion Island 133, 135, 138, 139–140, 141, 145
Matopos National Park 218
Mauritius 65
McGregor Museum Herbarium 95
mechanical control programmes 115–116, 250–251, 269–272, *296*, 299–301, 234
Meeuw Island 134, *135*
Mercury Island *135*
Middleburg 66
military activity
  Anglo-Boer War, 1899–1902 16, 30–31
  Zulu Wars 14
mining 11, 13, 14–16, 30, 95
Mirabib *11*
Molapo Oog 55
Molopo River 94
Mopane Savanna 94–98
Mountain Fynbos 81, 149
Mountain Sourveld 151
Mozambique *xiii*
Mzonjani, Natal 11

Namaqualand 9, 16, 17, 64
Nama System 77
Namib desert 6, 9, 93–103, 210, 218
Namib-Nauklift Park 95, 100
Natal Herbarium 27
Natal Parks Board 56, 123
National Herbarium 27, 109, 110, 113
National Institute for Water Research 51, 55
National Zoological Gardens 69
Nature reserves, invasions in 209, 221
Ndondondwane 11, *11*, *12*
Nelson Bay Cave 9, *11*, *12*
Northern Cape Savanna 94–98
Nossob River 94
Ntendeka Wilderness Area 126

Okavango River and System *xiii*, 51, 94, 102
Olifants River, Cape *xiii*, *49*, 51, *56*
Olifants River, Transvaal, *xiii*
Omaruru River, 101
Omatako River 51, 94, 101
Ombalanta Canal 52
Orange River 51, 66, 93, 94, 95, 100
Orange-Fish Tunnel 52, 55
Orange Free State *xiii*, 15, 16, *94*
Oranjemund 64, 102
Origins of alien species
  Africa, elsewhere in
    plants 23, *24–26*, *32*, *34*, 79, 96, *97*, 110
    vertebrates 51, 52, 66, 68, 206
  Americas
    invertebrates 79, 230, 231, 251–253, 264, 269, 288–289
    plants 29–35, 75, 96, *97*, 100, 103, *105–107*, 110, 159–

163, 185, 264, 269, 272
vertebrates 51, 52, 54, 67, 69, 204
Asia
invertebrates 252, 264, 272, 289, 291
plants 17, 25–28, 30–32, 37, 75, 96, *97, 103, 105, 106, 110, 151, 160–162,* 194–196, 262, 263, 275–283, 286
vertebrates 204, 253, 264, 272, 289, 291
Australia
invertebrates 253, 264, 272, 289, 291
plants 17, 25, *25, 26,* 27, 28, 30, 31, 32, 37, 38, 39, 40, 75, 96, *97,* 103, *105, 106,* 110, 151, 160–162, 164, 194–196, 262, 263, 275, 276–283, 286
vertebrates 204, 253, 264, 272, 289, 291
Europe
invertebrates 252
plants 17, *24–26,* 28–32, 34, 37, 38, 79, 96, *97, 105, 107,* 110, 262, 263, 272
vertebrates 51–53, 68–71, 79, 253
Otjiwarongo *94*
Oudsthoorn 218
owl pellet/alien rodent data 211–213

Paardevlei 56
parasites, alien 14, 49, 53–54, 101–102, *105,* 251, 252
parasites, indigenous 102, 163
pathogens xv, 53, 57, 71, 82, 140, 163, 218, 227, 299
Pearston 182
Penguin Island *134, 135*
Percy Fitzpatrick Institute of African Ornithology 215
Phalaborwa 13
Phoenician explorers 22
Pietermaritzburg 65, 66
Pinelands 39
P K le Roux Dam 254, 255
plants, alien
aquatic 167, 190, 192, 193, 194, *195,* 196, 248–251
impact
carbon cycle 179–181, 183, 229–232, 239–241, 248, 277
economic 249–251, 277
energy balance 243
fauna 230–232
fire regime 77, 78, 83, 127, 159, 236, 239, 241, *241*
soil erosion 82, 236, 244
invasion process
biotic factors 147, 151–152, 153, 232, 234–235
comparison: Australia 147, 152, 164, 195, 277
dispersal 81, 163–167, 171, 173, 176, 231–232, 248, 277
growth 189–196, 278
habitat factors 81, 146–148, 150–151, 276
humans, role of 42, 121, 125, 149, 164, 165, 175–176
physiology 179–185, 280
reproduction 157–166, 171–176, 233, 235, 264, 276–277, 280, 285–286, 297
invasive success 42, 96–99, 100, 139, 167, 248–249, *275*
management 236–237, 259, 275, 280–281, 294–299
biological control 115, 164, 251, 254, 261–266, 269–274, 285, *296,* 297
chemical control 115, 128, 235, 250–251, *296,* 297
cultural control 128, *296*
integrated weed control (IWC) 295–301
mechanical control 128, 233–234, 235, 250, 281, 296, 297
Pleistocene Age 3, 5
Plumpudding Island *134, 135*
Pomona Island *134, 135*
Pongola River *xiii,* 194, 203
Pont Drift *12,* 14
Port Alfred *182*
Port Elizabeth *xiii,* 15, 17, 33, 82, 83, *182,* 210
Possession Island *134, 135*
predation 71, 140, 218, 219, 299
seed 151, 158–159, 164, 232–233
Pretoria 66, 69
Prieska 96
Prince Edward Island 133, *134,* 135, *135,* 138, 139, 141
public education, need for 57, 58

Randburg 67
Redcliff, Zimbabwe 6
Rehoboth *94,* 97, 100, 101
Renbaai 8
Renosterveld 17, 149, 210
reptiles, alien *48,* 136, *136, 137, 214*
introduction
intentional 79, 253
unintentional 63
reptiles indigenous
invasion susceptibility 67
translocation *48,* 64
Rhodes, Cecil John 64, 66, 67, 68, 69
Riet River 94, 95
Riviersonderend 37
Robben Island 29, 71, 79, 133, 135, 136–137, 139, 140, 141
Rooiport Estate 67
Rosherville 67

St Croix Island *134, 135,* 136, 140
St Helena 9, 82
Savanna biome *xiv,* 93–103, *105–107,* 109–116
description 93–95, 109–110
invasion susceptibility 97–98, 100, 101, 110–111, 216, *217*
invertebrates, alien 101–102
plants, alien
impact 115, 99–100, 103, *105–107*
introductions 97, 100, 101, 110
invasion process *158,* 159, *160–162*
invasive success 96–100, 101, 103, *110, 112, 113, 214,* 219
plants, indigenous
invasion by 114
vertebrates, alien 101–103, *214,* 216
Sandvis Harbour 103, 210, 217, 218
Save River *xiii*
Scientific Committee on Problems of the Environment (SCOPE) v, xv

Schaapen Island 140, 141
Schoda 12, 13, 14
Scott's Cave 9, *12*
Seal Island, False Bay 133, *134, 135*
Seal Island, Lüderitz *134, 135*
Seymour *182*
Shire River 48
Shongweni South Cave 8, 12
Silver Leaves, Transvaal 11
Simonstown 63
Sinclair Island *134, 135*
Skeleton Coast Park 95, 100, 101
Smith, Dr Sir Andrew 63
Smith, Sir Harry 39
Sodwana Bay 54, 66
Sofala 22
Somerset East *182*
South African Division of Seed Control 31
Southern Africa, map *xiii*
South West Africa/Namibia *xiii,* 93–103, 157–163
Soutpansberg 119
Stellenbosch 67
State Forest 78, 129
  authorities 40–41, 87
Sterkfontein 3
Strandveld 80–81, 149, 210
Sundays River *xiii*
Swakopmund *94,* 95, 102
Swakop River 94, 100
Swakop schist 94
Swartkrans 3
Swaziland *xiii,* 32, 204
Swellendam 121

Table Mountain 38, 40–41, 66, 69, 71, 82, 125, 231
terrestrial vertebrates, alien 63–72
  see also amphibians, birds, fish, mammals, reptiles
Thornbush savanna 94–98
Tokai Arboretum 40, 112, *149*
translocation
  definition 47
transport systems
  role in introductions 15, 16, 31,-32, 67, 210
Transvaal Bontanist 31
Tugela River *xiii*
Tyume River 55

Ugab River 100
Uitenhage *182*
Uitenhage Weed Laboratory 101
Uitvlugt Plantation 40
Umbogintwini River 67
Umgeni River 56
Uniab River 94
Upington *94,* 97

Vaal Barrage 251, *251*
Vaal River *xiii,* 52, 53, 94 95
Van der Plank, J 30
Van der Stel, Governor W A 29
Van Riebeeck, J 29, 37, 121
Vertebrates xv
  see also amphibians, birds, fish, mammals, reptiles
Victoria West 102
Vondeling Island *134, 135*
Von Ludwig, Baron C 30, 38
Von Mueller, Baron F J H 39

Walvis Bay *xiii, 94,* 218
weeds
  classification 146–147
  integrated control 295–303
  see also plants, alien
White sands project 39–40
Windhoek *xiii,* 52, 56, *94,* 97, 100, 102
Windhoek State Herbarium 95
Witfontein State Forest 126, *126*
Witteberg Group 77
Witwatersrand 15, 109, 111
Woodbush State Forest 122
World Conservation Strategy 57

Zambezi River *xiii,* 51, 205
Zambia *xiii,* 66
Zimbabwe *xiii,* 9, 10, 13, 14, 24, 28, 55, 66